STUDIES IN INTERFACE SCIENCE

Electrorheological Fluids

The Non-aqueous Suspensions

STUDIES IN INTERFACE SCIENCE

SERIES EDITORS
D. Möbius and R. Miller

Vol. 1 Dynamics of Adsorption at Liquid Interfaces. *Theory, Experiment, Application.* By S.S. Dukhin, G. Kretzschmar and R. Miller
Vol. 2 An Introduction to Dynamics of Colloids. By J.K.G. Dhont
Vol. 3 Interfacial Tensiometry. By A.I. Rusanov and V.A. Prokhorov
Vol. 4 New Developments in Construction and Functions of Organic Thin Films. Edited by T. Kajiyama and M. Aizawa
Vol. 5 Foam and Foam Films. By D. Exerowa and P.M. Kruglyakov
Vol. 6 Drops and Bubbles in Interfacial Research. Edited by D. Möbius and R. Miller
Vol. 7 Proteins at Liquid Interfaces. Edited by D. Möbius and R. Miller
Vol. 8 Dynamic Surface Tensiometry in Medicine. By V.M. Kazakov, O.V. Sinyachenko, V.B. Fainerman, U. Pison and R. Miller
Vol. 9 Hydrophile-Lipophile Balance of Surfactants and Solid Particles. *Physicochemical Aspects and Applications.* By P.M. Kruglyakov
Vol. 10 Particles at Fluid Interfaces and Membranes. *Attachment of Colloid Particles and Proteins to Interfaces and Formation of Two-Dimensional Arrays.* By P.A. Kralchevsky and K. Nagayama
Vol. 11 Novel Methods to Study Interfacial Layers. By D. Möbius and R. Miller
Vol. 12 Colloid and Surface Chemistry. By E.D. Shchukin, A.V. Pertsov, E.A. Amelina and A.S. Zelenev
Vol. 13 Surfactants: Chemistry, Interfacial Properties, Applications. Edited by V.B. Fainerman, D. Möbius and R. Miller
Vol. 14 Complex Wave Dynamics on Thin Films. By H.-C. Chang and E.A. Demekhin
Vol. 15 Ultrasound for Characterizing Colloids. *Particle Sizing, Zeta Potential, Rheology.* By A.S. Dukhin and P.J. Goetz
Vol. 16 Organized Monolayers and Assemblies: Structure, Processes and Function. Edited by D. Möbius and R. Miller
Vol. 17 Introduction to Molecular-Microsimulation of Colloidal Dispersions. By A. Satoh
Vol. 18 Transport Mediated by Electrified Interfaces: Studies in the linear, non-linear and far from equilibrium regimes. By R.C. Srivastava and R.P. Rastogi
Vol. 19 Stable Gas-in-Liquid Emulsions: Production in Natural Waters and Artificial Media Second Edition By J.S. D'Arrigo
Vol. 20 Interfacial Separation of Particles. By S. Lu, R.J. Pugh and E. Forssberg
Vol. 21 Surface Activity in Drug Action. By R.C. Srivastava and A.N. Nagappa

Electrorheological Fluids
The Non-aqueous Suspensions

Tian Hao

Cambridge, Massachusetts, USA

ELSEVIER

Amsterdam – Boston – Heidelberg – London – New York – Oxford
Paris – San Diego – San Francisco – Singapore – Sydney – Tokyo

ELSEVIER B.V.
Radarweg 29
P.O. Box 211, 1000 AE Amsterdam
The Netherlands

ELSEVIER Inc.
525 B Street, Suite 1900
San Diego, CA 92101-4495
USA

ELSEVIER Ltd
The Boulevard, Langford Lane
Kidlington, Oxford OX5 1GB
UK

ELSEVIER Ltd
84 Theobalds Road
London WC1X 8RR
UK

First edition 2005

Library of Congress Cataloging in Publication Data
A catalog record is available from the Library of Congress.

British Library Cataloguing in Publication Data
A catalogue record is available from the British Library.

ISBN-13: 978-0-444-52180-4
ISBN-10: 0-444-52180-1
ISSN: 1383-7303

♾ The paper used in this publication meets the requirements of ANSI/NISO Z39.48-1992 (Permanence of Paper).
Printed in The Netherlands.

There are three philosophical principles that have had a great impact on me from my childhood:

The first is the theory of knowledge attributed to Albert Einstein: The knowledge that we have can be analogous to a circle. Inside the circle is what we know and is called knowledge; outside the circle is what we don't know and need to explore. As our circle of knowledge expands, so does the circumference of darkness surrounding it. So the more we know, the more we feel that we don't know.

The second is the relationship between content and the format of its expression. Any content has to be expressed in a proper format. A beautiful format is helpful to readers' eyes but doesn't enhance what the content is. The content is always more important than the format and non-compensable by the beauty of the format.

The third is also attributed to Albert Einstein: An intelligent fool can make things bigger and more complex. An intelligent wisdom can make complicated matters simple and beautiful.

If anybody feels while reading this book that I am uncertain regarding many firmly established tendencies or facts, careless of the proper expression format, and over-simplify many complicated and important issues, I apologize. These shortcomings are direct consequences of the principles mentioned above. However, the blame should bear on me where I have made improper utilization of the three principles.

Tian Hao

Cambridge, Massachusetts
July 15, 2005

Preface

An electrorheological (ER) suspension is made from an insulating liquid medium embodying either a semi-conductive particulate material or a semi-conductive liquid material (usually a liquid crystal material). The rheological properties (viscosity, yield stress, shear modulus, etc.) of an ER suspension could reversibly change several orders of magnitude under an external electric field with the strength of several kilovolts per millimeter. Since its mechanical properties can be easily controlled within a wide range (almost from a pure liquid to a solid), the ER fluid could be used as an electric and mechanical interface in various industrial areas. For example, it could be used in the automotive industrial for clutch, brake and damping systems. It also could be used in robotic arm joints and hands. In addition, the ER technique can be used to fabricate advanced functional materials such as photonic crystals, smart inks, and heterogeneous polymer composites. The potential applications have stimulated a great deal of interest both in academic and industrial areas since the ER effect was first described by Winslow in 1949. There is now a large body of literature on the mechanism of the ER effect and the design of industrially applicable ER devices.

The ER fluids are simply non-aqueous suspensions operating under an external electric field. Non-aqueous suspensions are widely used in daily life, such as liquid printing inks and various paints and coatings. For better understanding the ER fluids, non-aqueous suspensions are described in parallel with the ER fluids. The physical mechanisms behind the ER phenomena do not solely belong to the ER suspensions, instead they definitely can deepen our general understanding of non-aqueous suspensions. So this book is not written specifically for people who are working in the ER fluids area or related application fields, it has the more generic purpose of informing people who are interested in non-aqueous systems including polymer and polymer composites. Mathematical derivation is unavoidable and the best effort is to avoid any "hopping" in the derivation and to start from the earliest stage.

The major objective of this book is to present a comprehensive survey on the ER suspensions in term of screening high performance ER materials, physical mechanisms of ER effect, and the applications of ER technology. To achieve those goals, a large body of literature has been explored, and particular attention is given to the achievements made within recent decades.

However, many interesting publications may not even be mentioned in this book, which doesn't mean that they are unimportant. The choice is purely based on the author's own preference for the integrity and consistency of the whole book.

The outline of this book is shown below. A survey on the viscosity of pure liquids and colloidal suspensions with and without an external electric field is presented first. ER related effects including positive, negative, photo-ER effects, and the electromagnetorheological (EMR) effect, are introduced thereafter The materials that are already used for making ER fluids are then described., before the critical physical parameters that play a key role in controlling the ER response are presented. The physical processes occurring in ER fluids are addressed to open a way to introduce the ER mechanisms. Much attention is paid to the dielectric properties of ER suspensions, after the dielectric property of non-aqueous suspension in general is discussed in detail. Comparisons between models or theories are emphasized for presenting a clear roadmap of how ER theories are evolved. The potential applications of ER fluids in various industrial fields and the affiliated technological problems are summarized at the end.

Sincere gratitude is due to related colleagues, classmates, teachers, seniors, relatives and friends for their inspiration and help in keeping me in good psychological and physical health, in activating my intrinsic potential and wisdom, encouraging and enhancing my spirit, and cultivating my confidence, persistence and willpower, etc.. Most parts of this book were written after I moved from Piscataway, New Jersey to Cambridge, Massachusetts at the beginning of 2004. Without their continuous encouragement and stimulation, this book would not have been finished within such a short time frame.

I would like to thank Dr. Mikio Nakamura and Dr. Fumikazu Ikazaki for providing me with the opportunity to continuously work on electrorheological fluids and related subjects in Japan. Dr. Yuanze Xu and Professor Kunquan Lu are gratefully appreciated for introducing me to work in this exciting field. Their instruction and guidance have been invaluable to my gaining a better understanding of this subject. Dr. Richard E. Riman is greatly appreciated for providing me with a chance to work in the USA on various generic issues related to non-aqueous colloidal suspensions that has deepened and broadened my understanding of electrorheological systems in particular and non-aqueous systems in general.

I also would like to thank many friends for collecting the literature used in this book. Special thanks are due to Dr. Chunling Hu, Dr. Ting Hao, Dr. Yucheng Lan, Dr. Liwei Huang, Dr. Chunwei Chen, and many others whose names are not listed here. Their time and effort spent generously for helping me in a timely manner are greatly appreciated.

My editors for this book, Derek Coleman, and Louise Morris, and my colleagues, Dr. Shamus Patry, Dr. Lan Cao, and Dr. Bin Wu, are gratefully thanked for reading through the first version of this book. Their corrections and suggestions are invaluable to its success.

Finally I would like to thank my family members for their understanding and support during the writing period. Writing of the book took all of my spare time after my daily work, and would have been impossible to finish without the strong support of my family members.

Due to the complexity of the electrorheological effect and the large body of literature on this subject, mistakes may not be avoidable in the attempt to arrange the abundant achievements in a logical and simple manner. Any suggestions and comments are warmly welcomed and appreciated.

Tian Hao

Cambridge, Massachusetts
July 15, 2005

Table of Contents

Chapter 1

Colloidal suspensions and electrorheological fluids

1. COLLOIDAL SUSPENSIONS

A suspension is a liquid-like soft material in which solid particles are dispersed in a liquid [1]. If the dispersed particles are in the size range below 1 μm and above 10^{-3} μm, such a suspension is usually called a colloidal suspension [2]. The term dispersion is also frequently used for describing a system in which one non-continuous phase (solid, liquid, gas) is dispersed into a continuous phase (liquid, gas), forming a homogeneous and stable soft material. For the purpose of facilitating the dispersal of non-continuous phase into a continuous liquid phase, a surfactant or "surface active agent" is usually added into the dispersion to lower the surface tension between the two phases. The word "dispersant" is frequently used, instead of surfactant, to represent the small amount of additive which can facilitate the breakup of particle aggregates and stabilize the distribution of particles. Wetting agent is another term that is used more frequently in industry, referring to a substance that can reduce the surface tension of solid particles in a solid liquid type suspension. Clearly, a dispersant is a subclass of the surfactant, while the suspension is the subclass of the dispersion. Wetting agent is almost identical to dispersant.

Another interchangeable term for suspension is slurry, which is a concentrated solid-liquid mixture having a broad solid particle size distribution ranging from submicrometers to some hundreds of micrometers [3, 4]. The term "concentrated" is semiquantitative, vaguely referring to where the average separation distance between two similar particles is close to or less than the particle size [3]

If the dispersed non-continuous phase is a liquid and the dispersing continuous phase is also a liquid, this dispersion is called an emulsion for dispersed liquid sizes between 0.1 μm and 1 μm, a microemulsion for dispersed liquid sizes between 0.01 and 0.1 μm, and a micelle for dispersed sizes between 0.001 and 0.01 μm, respectively[5]. A micelle is the formation of surfactant molecular aggregates that remain in solution with properties distinct from those of the monomeric solution [6].

If the dispersing liquid is water, the suspension is sometimes called an aqueous suspension or system, parallel to the non-aqueous suspension which contains the particles dispersed into an organic medium rather than water. The classification of two-phase system is shown in Figure 1. Water is a very common solvent and dispersal medium, Extensive fundamental studies in aqueous system have been carried out, and modern colloidal chemistry is established mainly on the basis of the understanding of aqueous systems. Detailed description of important concepts of colloidal chemistry, such as particle surface charge, electric double layer, and Zeta potential, etc, would be exhaustive and beyond the scope of this book. Only a brief overview on the electric property related terms with relevance to the electrorheological suspensions will be given.

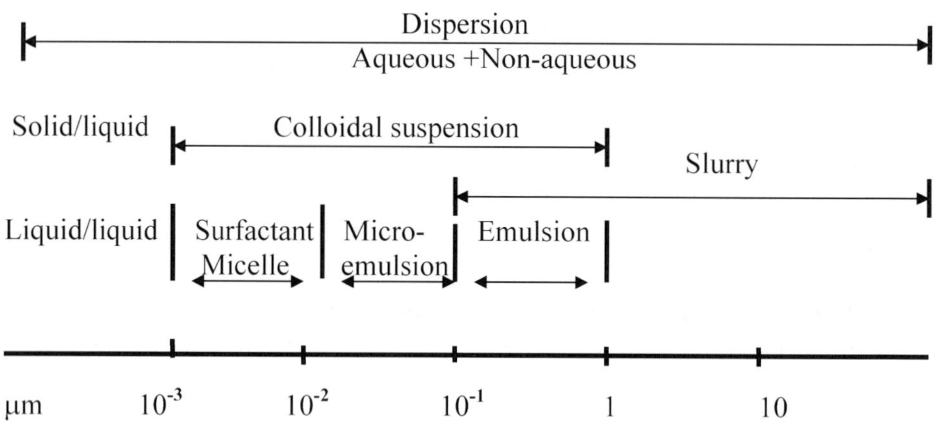

Figure 1 Classification of a two-phase system

1.1 Particle surface charge in aqueous system

Most solid particles will attain electric charge once they are dispersed into water or an aqueous solution. The sources of interfacial or surface charges, also called the mechanism of surface charge development, can be classified into the following four categories [6, 7]. 1) Different preference for the ions of two phases; 2) direct ionization of surface groups; 3) specific ion adsorption, and 4) defects of specific crystal structures. The former three mechanisms are associated with van der Waals and/or dispersion forces, and the last one is related to pure charge-charge electrostatic force. Van der Waals or dispersion forces cover the interactions between permanent and induced dipoles, as well as rapidly fluctuating dipoles interaction resulting from the movement of the electrons of atoms or molecules. It determines the ion solubility, ionization and ion substitution, and ion adsorption capability of a particle surface. Depending on the crystal face exposed, the anisotropic crystal surface may be charged positively at one site and negatively at another. Charged sites may physically entrap non-mobile charges, which in turn attract oppositely-charged species for balancing. In this case, zero surface charge cannot be attained by simply adjusting the pH. Thanks to the above four charge development mechanisms, particle surfaces are most likely charged in an aqueous system, creating electric potential near to the surface. Depending on the magnitude of the electrostatic interaction, the ions or other charge species in solution may be distributed adjacent to the particle surface, following Boltzmann's distribution law. Closely associated (bound) ions and loosely connected ions form a diffuse double layer, commonly referred to as the electric double layer (EDL). Zeta potential is usually used to represent the electric potential of charged surfaces, strictly speaking the electric potential at the interface between the bound ion layer and the diffuse layer. Detailed information on Zeta potential and EDL can be readily found in standard colloidal chemistry textbooks such as ref. [2].

1.2 Particle surface charge in non-aqueous systems

In contrast, the mechanisms of particle surface charge in non-aqueous systems should be different from the four mechanisms described above, as there is no obvious dissolvability issue usually encountered in aqueous systems. According to Kosmulsi [8], the mechanism of surface charging in non-aqueous suspension without a specific charging agent is the preferential adsorption of ions from the organic medium. Sources of the ions are, 1) the trace amount of water; 2) other inorganic or organic impurities originally staying in the pure organic medium; 3) the dissociation of surface groups of the dispersed particles, especially the proton dissociation, involving the so-called acid-base interaction introduced by Fowkes and co-workers [9, 10]; 4) ionic surfactants or stabilizers presented on purpose; 5) ions generated under an electric field due to the Debye-Falkenhagen effect [11]. When ions are absolutely absent, electron transfer between dispersing medium molecules and the particle surface is a possible mechanism, which is similar to the acid-base interaction concept if the Lewis acid-base concept is used (the Lewis acid is the electron pair acceptor and the base is the electron pair donor). However, this scenario cannot explain one phenomenon that commonly exists in non-aqueous systems: The small amounts of ions in an organic medium tend to coalesce together due to the Coulombic interaction between oppositely-charged ions [12]. The Coulombic interaction energy can be expressed as:

$$E_{Coul} = \frac{-q_1 q_2}{4\pi\varepsilon_m \varepsilon_0 d} \tag{1}$$

where q_1 and q_2 are the charge of two ions, ε_m is the dielectric constant of the medium, ε_0 is the permittivity of vacuum, 8.85×10^{-12} C/Vm, and d is the distance of two ions from center-to-center. Since the dielectric constant of organic media is usually around 2.5 and water is 80, the Coulombic interaction is almost 30 times stronger in organic media than that in aqueous systems. So in non-aqueous systems, the ions can only be dissociated if the ions are larger enough, or if they form some large structures, like inverse micelles, or complex macro-ions.

The inverse micelle, as the name indicates, has the inversed form of the micelle structure frequently observed in the aqueous system. The hydrophilic heads form the core structure, while the hydrophobic chains penetrate into the nonaqueous oil phase. Inside the core structure there is a water or polar impurity pool enveloped by the hydrophilic head groups. Figure 2 shows the inverse micelle structure of zirconyl 2-ethylhexanote formed in decane and determined via the small angle neutron and X-ray scatterings. The molecular structure of zirconyl 2-ethylhexanote is shown in Figure 3. The core radius is 6.3 Å and the shell thickness is 5.3 Å, which gives the core diameter 12.6 Å and an outer micelle radius 11.6 Å. From the experimental results it is calculated that the micelles have a mean aggregation number of 33 zirconyl 2-ethylhexanote molecules per micelle. The small angle neutron scattering measurements also show that the size and shape of the micelles are invariant to temperature over the temperature range 20 to 80 ^{o}C [46]. Formation of inverse micelles is believed to be quite important in stabilizing charge separation, and therefore exchanging the charges between micelle and particle surface [9, 13, 14]. Figure 4 shows a possible scenario of how a particle is charged in the presence of an amphoretic (zwitterionic) charging agent. The charging agent molecules will form inverse micelles without the presence of the particle. Once the particle is added in, there should be abundant polar sites on the particle surface. Those polar sites are either inherited from the particulate material or created with a polymeric coating material. The inverse micelles, no matter whether they are charged or uncharged, may take the particle as a big polar pool and rapidly stick to the particle surface. If the particle surface has negative polar sites or it is slightly negatively charged, the positive group of the charging agent molecules may directly adhere to the particle surface, leaving the negative groups outside and inducing the particle more negatively charged. In contrast, if the particle surface is slightly positive in origin, then the charging agent will make the particle become more positively charged. Choosing a right charging agent can definitely charge particle either negatively or positively. Particle surface charging thus can be fully controlled using the charging agent.

The micelle structure should be considered as a dynamic one with ions or molecules leaving and joining at a rapid rate [15]. Zeta potential of inorganic particle in non-aqueous medium is not as small as researchers have usually thought, above ±30mV [16]. Commonly used charging agents in non-aqueous systems for creating charges and detailed charging mechanism of inverse micelles in non-aqueous medium will be discussed in a future chapter.

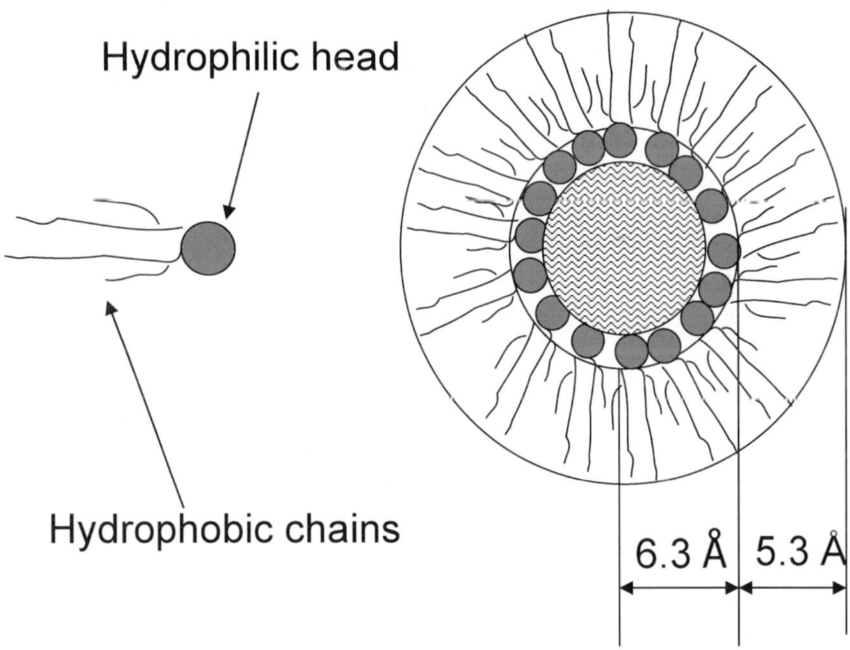

Figure 2 Proposed micelle structure of Zirconyl 2-ethylhexanote/decane. Redrawn from R.I. Keir, and J.N. Watson, Langmuir, 16(2000)7182.

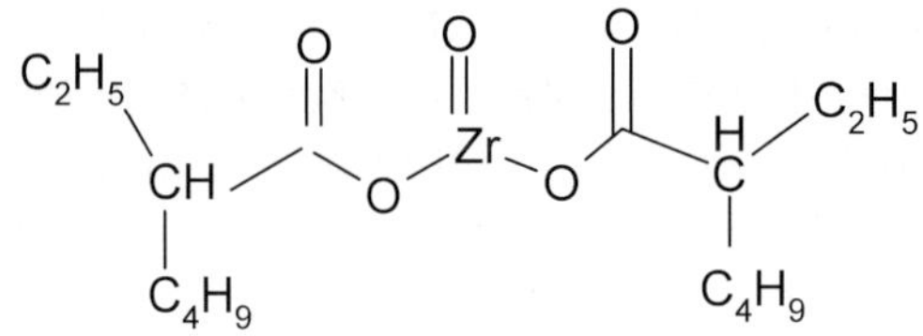

Figure 3 Zirconyl 2-ethyl hexanoate.

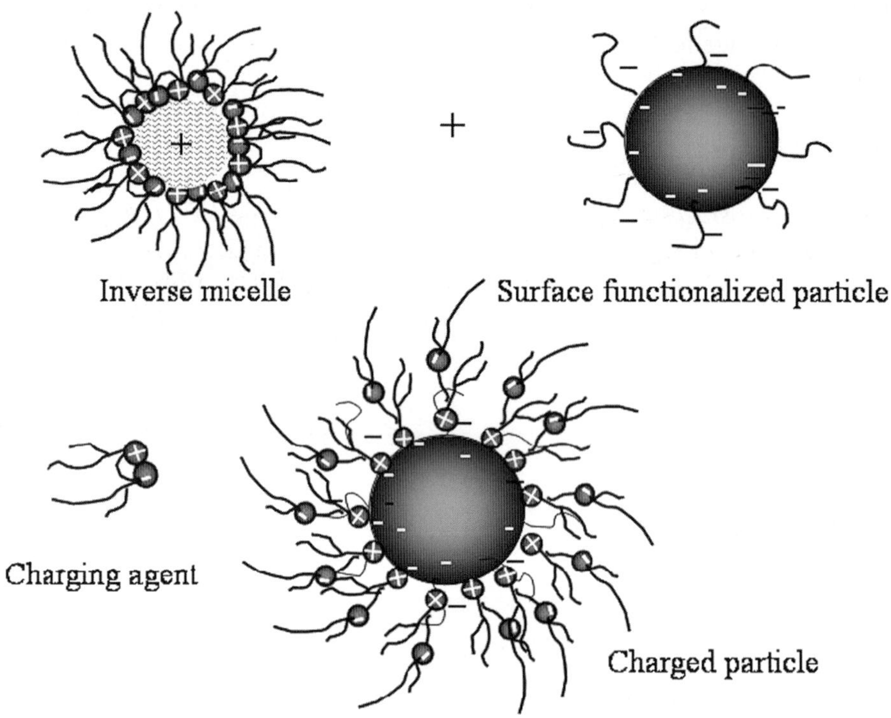

Figure 4 Schematic illustration of how a particle is charged in the presence of an amphoteric (and zwitterionic) charging agent.

Trace amounts of electrolytes, including water, plays an extremely important role in controlling surface charge in non-aqueous systems, when the pH value is unable to be measured and the Zeta potential can only be correlated to the concentration of electrolytes. For example, at sufficiently high CsCl concentration, the sign of the Zeta potentials of titania [17] and silica [16] was found to be reversed. Added electrolytes change: 1) the preferential adsorption of dissociated electrolytes; 2) the adsorption or structure of charged micelles; and 3) desorption of surface anions or cations due to the adsorption of electrolytes [9]. Although water cannot charge a particle surface directly in a non-aqueous system, trace concentrations of water have a great impact on the physical properties of the whole dispersion for at least two reasons. The first is that the water can enhance the formation of micelles in organic media substantially. A single water molecule was found to be sufficient to generate a micelle [18]. The second is that water can change the acid-base character of a particle surface, enhance autoprotolysis of organic media, and hence change the surface-electrolyte interaction [16, 19].

1.3 Relationship between surface charge density and Zeta potential

Surface charge density, d_q, is termed the charge quantity, q, of a spherical particle with radius, r, divided by the particle surface area

$$d_q = \frac{q}{4\pi r^2} \tag{2}$$

Based on the Stokes law, the particle surface charge and particle mobility, μ, under a dc electric field, E, are correlated as:

$$\mu = \frac{v}{E} = \frac{q}{6\pi\eta r} \tag{3}$$

Where v is the velocity of particle under an electric field, η is the viscosity of the medium. Zeta potential, ζ, is a parameter for characterizing particle surface charge and calculated on the basis of mobility measurement:

$$\zeta = \eta\mu / \varepsilon_m \varepsilon_0 \qquad \text{(Smoluchowski)} \tag{4}$$

$$= 3\eta\mu / 2\varepsilon_m \varepsilon_0 \qquad \text{(Hückel)} \tag{5}$$

Eq. (4) and (5) can only be valid if the reciprocal of Debye length $(\kappa r)^{-1}$ is very high >100 or very low <<1. If the zeta potential is not so high, e.g., smaller than 50 mV, Henry's equation should be used:

$$\zeta = 3\eta\mu / \left[2\varepsilon_m \varepsilon_0 f(\kappa r)\right] \qquad \text{(Henry)} \qquad (6)$$

Where f(κr) is a function given by Oshima [20]

$$f(\kappa r) = 1 + \frac{1}{2\left[1 + \left(\frac{2.5}{\kappa r\left[1 + 2\exp(-\kappa r)\right]}\right)\right]^3} \qquad \text{(Oshima)} \qquad (7)$$

Note that f(κr) approaches 1 for small κr (Hückel equation) and 3/2 for large κr (Smoluchowski equation). For an aqueous suspension, the double layer (DL) is considered as a "thin DL", usually defined as κr >10, thus the Smoluchowski equation holds. In a non-aqueous suspension, the double layer is considered as a "thick DL", usually defined as κr<0.1, thus the Hückel equation holds. For the region 10>κr>0.1, the situation becomes more complicated [21, 22]. A suspension with particles less than 100 nm, no matter which dispersal medium is used (aqueous or non-aqueous), belongs to this range [23]. Use of either the Hückel equation or Smoluchowski equation for this region will lead to significant error (>5%), especially for large values of ζ [7]. Solutions for calculating the mobility of an isolated, spherical solid non-conducting particle of arbitrary ζ for arbitrary κr have been worked out by Wiersema [24] and O'Brien [25], with incorporation of the relaxation effect-the distortion of the electric field induced by the particle movement. Detailed description and comparison is complicated and out of the scope of this book. Since a non-aqueous system is focused, the Hückel equation is used for deriving the relationship between the particle surface charge density and zeta potential. From Eq.(2), (3) and (5), one can easily obtain:

$$\zeta = \frac{q}{4\pi\varepsilon_m \varepsilon_0 r} = \frac{rd_q}{\varepsilon_m \varepsilon_0} \qquad (8)$$

or

$$d_q = \frac{\zeta \varepsilon_m \varepsilon_0}{r} \tag{9}$$

Eq.(8) and (9) indicate that Zeta potential has a linear relationship with the particle surface charge density. Note that Eq. (8) and (9) are only valid in a charge-free medium, where both the electrophoretic retardation effect and the relaxation effect are unimportant. The electrophoretic retardation effect results from the double layer ions that will move to the opposite direction of the particle and reduce the velocity of the migrating particle. The electrophoretic relaxation effect refers to the phenomenon that the double layer may be no longer spherically symmetrical during the electrophoresis at high zeta potentials. Eq. (8) and (9) are simply applied to the system in which there is no background electrolyte. As one knows, the Zeta potential is the potential between the shear plane and the bulk suspension, so strictly speaking, the d_q should be called the electrokinetic charge density, since it is determined in the electrokinetic experiment, where the Zeta potential is measured. The r in Eq.(8) and (9) should thus be the particle radius plus the fraction of the Debye-Hückel length $1/\kappa$, as only at the radius $[r + 1/(s\kappa)]$ does the surface charge correspond to the Zeta potential. Eq.(9) can be rewritten as:

$$d_q = \frac{\zeta \varepsilon_m \varepsilon_0}{[r + 1/(s\kappa)]} \tag{10}$$

Where s is an integral number. For a very thin double layer, i.e., $1/\kappa \ll r$, Eq.(10) returns to either Eq.(8) or (9), indicating that those two equations are only valid for a very thin double layer case, most likely happened in aqueous systems. For a very thick double layer, even $1/(s\kappa) \gg r$, Eq.(10) becomes:

$$d_q = \zeta \varepsilon_m \varepsilon_0 s\kappa \tag{11}$$

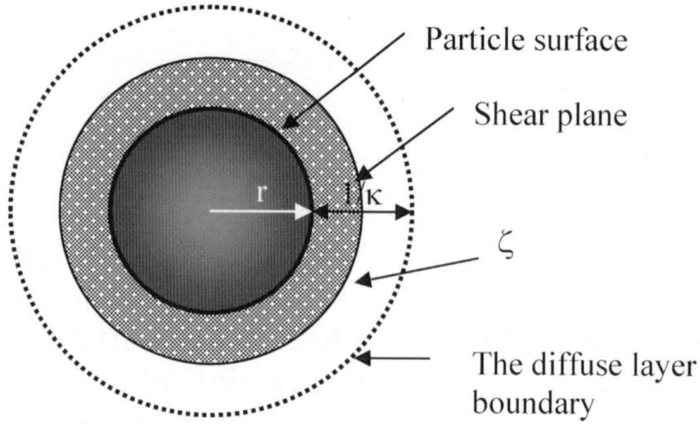

Figure 5 Illustration of Zeta potential, particle radius and Debye-Hückel length.

In this case, the system behaves as a parallel plate capacitor with a distance $1/(s\kappa)$ between the two plates, similar to the Helmholtz model [7]. If $1/(s\kappa)$ is comparable to r, Eq.(10) must keep its original form. Where the shear plane is defined in the system will determine how the surface charge density correlates with the Zeta potential. The constant s should be a function of particle concentration, ionic strength, and particle surface potential. Simply using either Eq.(9) or Eq.(11) to correlate the surface charge density and Zeta potential will induce significant error. This problem becomes very severe in non-aqueous systems. Suppose that the thickness of the electric double layer will not change while a particle is moving in an electric field, i.e., there is a dynamic equilibrium between the diffuse layer and the bulk suspension, the shear plane thus can be assumed to be quite close to the diffuse layer boundary. The constant $s \approx 1$, and Eq.(10) can be rewritten as

$$d_q = \frac{\zeta \varepsilon_m \varepsilon_0 \kappa}{(\kappa r + 1)} \tag{12}$$

The surface charge q

$$q = 4\pi r^2 d_q = \frac{4\pi\zeta\varepsilon_m\varepsilon_0\kappa r^2}{(\kappa r+1)} = 4\pi\zeta\varepsilon_m\varepsilon_0 r(1+\kappa r)\frac{\kappa r}{(1+\kappa r)^2}$$
$$= 4\pi\zeta\varepsilon_m\varepsilon_0 r(1+\kappa r)F(\kappa r) \qquad (13)$$

Where $F(\kappa r) = \frac{\kappa r}{(1+\kappa r)^2}$. Eq.(13) has a similar form to the accurate equation proposed and developed by Loeb [26] and Stigter [27]. In addition, Eq.(10) can be approximately simplified as:

$$d_q = \frac{\zeta\varepsilon_m\varepsilon_0\kappa}{(\kappa r + 1/s)} \approx \frac{\zeta\varepsilon_m\varepsilon_0}{2r} \qquad (14)$$

under the assumption that $1/(s\kappa) \approx r$. Eq.(14) clearly indicates that a half amount of surface charge density of a non-aqueous system will generate the same Zeta potential in comparison with an aqueous system. In other words, although there would be a smaller amount of ions existing in a non-aqueous system due to poor ionization, the Zeta potential could still be high.

Keep in mind that Eqs. (8-14) are only valid for small κr, when the electrophoresis retardation (electric-field-induced movement of ions in the electric double layer, which is opposite to the direction of particle movement) is unimportant [41]. This limitation is inherent to the Hückel equation. Practically, a colloidal suspension always contains charged particles dispersed in a medium with surfactants (or electrolytes) of both polarities. In this case the Poisson's equation must be used for deriving the surface charge density and Zeta potential relationship. Under the Debye-Hückel approximation, i.e., the small value of potential, $ze\psi << k_B T$, where ψ is the potential and z is the valency of ion, a simple relationship between the surface charge density and Zeta potential can be easily obtained [7]. The Poisson's equation simply says that the potential flux per unit volume of a potential field is equal to the charge density in that area divided by the dielectric constant of the medium. It can be mathematically expressed as:

$$\Delta\psi = -\frac{d_q}{\varepsilon_m\varepsilon_0} \qquad (15)$$

where Δ is divergence operator, $\Delta = \frac{\partial^2}{\partial x^2} + \frac{\partial^2}{\partial y^2} + \frac{\partial^2}{\partial z^2} = \nabla^2$. ∇^2 is the Laplace operator. If the number of ions per unit volume at the potential ψ is n , then according to the Boltzmann equation, n can be expressed as:

$$n = n_0 \exp(-\frac{ze\psi}{k_B T}) \tag{16}$$

and the ion density (or charge density) is:

$$d_q = nze \tag{17}$$

Substituting Eq. (17) into Eq. (15) and using Eq. (16) to replace n leads to:

$$\Delta\psi = -\frac{n_0 ze}{\varepsilon_m} \exp(-\frac{ze\psi}{k_B T}) \tag{18}$$

Eq. (18) is called the Poisson- Boltzmann equation. If $ze\psi << k_B T$, Eq. (18) can be expanded using the relation $e^{-x} \approx 1 - x$ for small x:

$$\Delta\psi = -\frac{n_0 ze}{\varepsilon_m} + \frac{n_0 z^2 e^2 \psi}{\varepsilon_m k_B T} \tag{19}$$

The first term in the right side of Eq. (19) must be zero for preserving the electroneutrality, so Eq. (19) can be rewritten as:

$$\Delta\psi = \kappa^2 \psi \tag{20}$$

Suppose that the potential ψ only varies one dimensionally with x, then Eq. (20) can be simplified as:

$$\frac{d^2\psi}{dx^2} = \kappa^2\psi \tag{21}$$

Solving Eq.(21) leads to:

$$\psi = \psi_0 r \frac{e^{-\kappa(x-r)}}{x} \tag{22}$$

where ψ_0 is the particle surface potential at the particle radius r. The total amount of charge on the particle surface Q is:

$$Q = -\int_r^\infty 4\pi r^2 d_q dr \tag{23}$$

From Eq.(15) and Eq.(20), one may obtain:

$$d_q = -\varepsilon_m \varepsilon_0 \kappa^2 \psi \tag{24}$$

Substituting Eq.(24) into Eq. (23) and integrating Eq.(23) leads to:

$$Q = 4\pi r \varepsilon_m \varepsilon_0 (1 + \kappa r)\psi_0 \tag{25}$$

since $d_q = Q/\left(4\pi r^2\right)$, thus Eq. (23) leads to:

$$d_q = \frac{\varepsilon_m \varepsilon_0 (1 + \kappa r)}{r} \psi_0 \tag{26}$$

For the electrokinetic charge on the particle, $\zeta = \psi_0$, so Eq.(24) becomes

$$\frac{d_q}{\zeta} = \frac{\varepsilon_m \varepsilon_0 (1 + \kappa r)}{r} \tag{27}$$

and Eq. (25) becomes:

$$Q = 4\pi\varepsilon_m \varepsilon_0 r(1 + \kappa r)\zeta \tag{28}$$

Note that Eq. (28) has a similar form as Eq. (13). However, those two equations cannot be equalized, as they are derived in different conditions. A more complicated relationship between Zeta potential and surface charge density in a non-aqueous system has been developed by Chen [28] for a suspension containing both counterions and coions and by Oshima [29] for a suspension containing only counterions (a salt-free medium). In a concentrated suspension, the Zeta potential/surface charge density relationship is also a function of the particle volume fraction [42-45]. The actual Zeta potential should be smaller than that expected from the linear relationship.

2. **ELECTRORHEOLOGICAL (ER) SUSPENSIONS**

An ER suspension is a kind of non-aqueous colloidal system, which is composed of a non-continuous phase (solid or liquid) dispersed into a non-aqueous organic medium, and has the capability of responding to an external electric field stimulation. Typically, the rheological properties of an ER suspension, including viscosity, shear stress, and modulus, either increase or decrease with the applied electric field strength, depending on the physical properties of both the dispersed and dispersing phases. Such a response usually takes place on a millisecond scale, with unlimited reversibility. The magnitude of rheological property change is quite remarkable, over several orders of magnitude. Such a phenomenon is called the ER effect. Since an ER suspension can be used as an electric-mechanical interface, it has received a great deal of attention since it was discovered by Winslow [30] in 1947. There is an extensive literature on the preparation of high performance ER suspensions, the mechanism of ER effect, and the design of industrially applicable ER devices. Many comprehensive reviews on the ER fluids and the ER mechanism have been published [31-39], summarizing the ER research achievements obtained at different periods. A systematic review, covering both previous achievements and recent findings, would be very helpful for people who want to get into this interesting field and for those who already are involved and want to gain a systematic understanding of ER technology.

The purpose of this book is to give an overview of this emergent research field, from materials preparation to the understanding of physical mechanism, and to the potential practical applications. The viscosities of pure liquids and colloidal suspensions with and without an external electric field are presented in Chapter 2. New approaches are used to describe the viscosities of pure liquids, colloidal suspensions, and polymer systems via a single free volume concept. The ER effect originates from the so-called electroviscous effect observed in a pure organic solvent [40]. A much stronger electroviscous effect was thereafter found in electrolytes and suspensions. Chapter 3 will introduce three major ER-related effects observed in various suspensions, including the positive ER effect, negative ER effect, and the photo-ER effect. The synergic electro-magneto-rheological (EMR) effect is briefly described. Chapter 4 will summarize the ER-active materials discovered so far. The dispersing continuous liquids, the dispersed particulates, and the additives for enhancing either the ER effect or suspension stability, or both, are focused upon. Critical physical parameters that control ER effect are summarized in Chapter 5. Those parameters include the electric field strength, frequency of the electric field, particle size, particle conductivity, particle dielectric property, particle volume fraction, temperature, water content, and liquid medium. Physical phenomena or processes occurring in ER suspensions are discussed in Chapter 6, in which forces relevant to the ER effect, phase transition, percolation transition, rheological properties, the conductivity mechanism, and the polarization process will be focused upon. Dielectric properties of non-aqueous heterogeneous systems are described in more detailed way in chapter 7. The Maxwell-Wagner model for diluted suspensions and the Hanai model for concentrated suspensions, as well as the relaxation times of various polarizations, receive a particular attention. The dielectric properties of ER suspensions, the response time of ER suspensions, and the dielectric properties under a high electric field are summarized in Chapter 8. After those physical aspects become clear, the mechanisms of the ER effect are reviewed in Chapter 9. The physical models, which were proposed for explaining observed ER phenomena, are introduced chronologically Those models include the fibrillation model, the electric double layer (EDL) model, the water/surfactant bridge mechanism, the polarization model, the conduction model, and the dielectric loss model. Finally, the potential applications of electrorheological fluids are presented in Chapter 10. ER suspensions can be used in the mechanical damping system and in car clutches for improving control of mechanical vibration or efficient transmission of the mechanical force. In materials science and engineering, the ER technology can be used to fabricate smart composite materials, ER inks and pigments, photonic crystals, ER sensors, and ER tactile displays. In the semiconductor industry, ER fluids can be used as intelligent mechanical polishing slurries. It is obvious that the ER technology will revolutionize many industrial areas. A prospective summary is given at the end of Chapter 10.

REFERENCES

[1] V.A. Hackley and C.F. Ferrais, The use of nomenclature in dispersion science and technology, National Institute of Standards and Technology, special publication 960-3, August, 2001

[2] P.C. Hiemenz and R. Rajagopalan, Principles of Colloid and Surface Chemistry, Marcel Dekker, New York, 1997

[3] M.Z. Sengun and R.F. Probstein, Rheologica Acta, 28(1989)382

[4] V.A. Hackley and C.F. Ferrais, The use of nomenclature in dispersion science and technology, National Institute of Standards and Technology, special publication 960-3, August, 2001

[5] M.Takeo, Disperse Systems, Wiley-VCH, 1998

[6] Drew Myers, Surfaces, Interfaces, and Colloids, VCH Publishers, 1991

[7] R.J. Hunter, Zeta Potential in Colloidal Science, Academic Press, 1981

[8] M. Kosmulski, Colloid Surf. A, 159, 277(1999); Electrical interfacial layer in nonaqueous solvents, in N. Kallay, Ed., Interfacial Dynamics, Marcel Dekker, 2000, pages 273-313

[9] F.M.Fowkes, Discuss. Faraday Soc., 46(1966)246

[10] R.J. Pugh, T. Matsunaga, and F.M.Fowkes, Colloids Surf., 7(1983)183

[11] P. Debye and H. Falkenhagen, Phys. Z, 29:121; 401(1928)

[12] I.D. Morrison, Colloids Surf. A, 71(1993)1

[13] A. Kitahara, Adv.Colloid Interface Sci., 12(1980)109

[14] F.M. Fowkes, The interaction of polar molecules, micelles, and polymers in non-aqueous media, in K.Shinoda, Ed., Solvent Properties of Surfactant Solutions, Dekker, New York, 1967, pp.65-115;

[15] D.F. Evans, S. Mukherjee, D.J. Mitchell, and B.W. Ninham, J.Colloid Interface Sci.,93(1983)184

[16] M.Kosmulski, Electrical interfacial layer in nonaquesous solvents, in N. Kallay, Ed., "Interfacial Dynamics", Marcel Dekker, New York, 2000, page 273-313

[17] M.Kosmulski, in E.Pelizzetti, Ed., Fine Particles Science and Technology, Kluwer, Dordrecht, 1996, p185

[18] K.A. Cogan, F.A. Leermakers, and A.P. Gast, Langmuir, 8(1992) 429
[19] J.Lyklema, Adv. Colloid Interface Sci., 2(1968)65
[20] H. Oshima, J. Colloid Interface Sci., 168(1994)269
[21] S.S. Dukhin, and N.M. Semenikhin, "Theory of double layer polarization and its effect on electrokinetic and electrooptical phenomena and the dielectric constant of dispersed systems". Kolloid. Zh., 32 (1970)360.
[22] R.W. O'Brien, and L.R. White, J. Chem. Soc. Faraday Trans. II, 74(1978)1607
[23] V.N. Shilov, Yu.B. Borkovskaja, and A.S. Dukhin, "Electroacoustic theory for concentrated colloids with arbitrary κa. nano-colloids, non-aqueous colloids", private communication.
[24] P.H. Wiersema, A.L. Loeb, and J.Th.G. Overbeek, J. Colloid Interface Sci., 22(1966)78
[25] R.W. O'Brien, and R.L. White, J. Chem.,Soc.,Faraday II, 74(1978) 1607
[26] A.L. Loeb, P.H. Wiesema, and J.Th.G. Overbeek, The Electrical Double Layer Around a Spherical Colloidal Particle, MIT press, Cambridge, MA, 1961.
[27] D. Stigter, J. Electroanal. Chem., 37(1972)61
[28] I. Chen, Langmuir, 12(1996)3437
[29] H. Ohshima, J.Colloid Interface Sci., 247(2002)18
[30] W.M. Winslow, U.S. Patent No. 2417850(1947)
[31] H. Block, J.P. Kelley, J. Phys. D.: Apply. Phys., 21(1988)1661
[32] A. P. Gast, C. F. Zukoski, Adv. Colloid Interface Sci. 30(1989)153
[33] T.C. Jordan, M.T. Shaw, IEEE Trans. Electr. Insul., 24(1989)849
[34] T. C. Halsey, Science, 23(1992)761
[35] C. F. Zukoski, Ann. Rev. Mater. Sci., 23(1993)45
[36] M. Parthasarathy, D. J. Klingenberg, Mat. Sci. Eng., R17 (1996)57
[37] F.E.Filisko, D.R. Gamota, ASME, 153(1992) 75
[38] T. Hao, Adv. Mat. 13(2001)1847
[39] T. Hao, Adv. Colloid Interface Sci. 97(2002)1
[40] A. W.Duff, Phys. Rev. 4(1896)23
[41] S.S. Dukhin, and B.V. Derjaguin, in E. Matijevic, Ed., "Electrokinetic Phenomena", in "Surface and Colloid Science",. Wiley, New York, Vol., 7, 1974
[42] H. Ohshima, J. Colloid Interface Sci., 195(1997)137;
[43] H. Ohshima, J. Colloid Interface Sci. 188(1997)481;
[44] H. Ohshima, J. Colloid Interface Sci. 212(1999)443;
[45] F. Carrique, F. J. Arroyo, and A. V. Delgado, J. Colloid Interface Sci. 243(2001)351
[46] R.I. Keir, and J.N. Watson, Langmuir, 16(2000)7182

Chapter 2

Viscosities of liquids and colloidal suspensions with and without an external electric field

The ER effect originates from the studies of how viscosity of a liquid (or solution) changes with an external electric field. A comprehensive review on this subject was made by Conway [1], where the term visco-electric effect was used to express the viscosity enhancement observed in pure liquids or homogeneous emulsions under an external electric field. Another concept, the electroviscous (EV) effect, was used to term the viscosity enhancement observed in various particulate-type suspensions or polyelectrolyte solutions. In the literature published in the early 1900's, the electroviscous effect referred to the viscosity enhancement due to the presence of an electric double layer, and there was no external electric field actually applied to the system. The EV effect was categorized into three different classes: The primary EV effect, the secondary EV effect, and the tertiary EV effect [1-4]. The primary EV effect refers to the effect arising from the distortion of electric double layer (EDL), under the assumption that there is no EDL overlapping between particles. In other words, the primary EV effect occurs in a very diluted suspension. The secondary EV effect is caused by the overlapping or interaction of EDL, which usually takes place in a relatively concentrated suspension. The tertiary EV effect describes the behaviors of polyelectrolytes or polymeric particulate suspensions under an electric field. The viscosity increase of those effects is within two times higher than the original value, which is not comparable with the ER effect discovered by Winslow in 1947 [5], where an increase by several orders of magnitude was obtained. Unfortunately, the EV effect and ER effect are interchangeably used in the literature for referring to the viscosity change arising from an external electric field. For avoiding further confusion, any viscosity (or other rheological parameters) change induced by an external electric field is termed the ER effect in this book. Sometimes the EV effect may also be used only for historical reasons. Also, the different types of EV effects mentioned above are no longer differentiated again. Creating many similar jargons will only, in the author's opinion, create more barriers for researchers to gain a better understanding in any scientific field. Instead, the difference between various systems from pure liquids, colloidal suspensions, and polymeric systems, is focused upon, and thus the ER effect is sectioned into three parts, the ER effects of pure liquids, polymeric electrolytes, and colloidal suspensions, respectively. Before the ER effect of each system is introduced, a brief review of the viscosity of such a system without the presence of an electric field will be presented for comparison reasons.

1. PURE LIQUIDS

1.1 Viscosity of pure liquids

A pure liquid means that this material is one single component and at room temperature it is in liquid form. Polymer solutions are obviously excluded from this section and will be discussed in section 3 of this chapter. Viscosity of liquid was treated theoretically by Andrade [6] and Glasstone [7]. Consider two layers of molecules in a liquid having a distance d apart and relative velocity between these two layers is v, under a shear force f, the viscosity, η, according to the definition, can be expressed as

$$\eta = \frac{f}{v/d} = \frac{fd}{v} \tag{1}$$

Since the flow of a liquid is a rate process, the rate process theory [7, 8] can be reasonably applied to the problem of viscosity. A molecule moving from one equilibrium position to another needs to overcome the energy barrier, the activation energy, E_0, as shown in Figure 1. The applied shearing force will reduce the height of the energy barrier in the flowing direction by ΔE, while it will raise the height of the energy barrier in the opposite direction by the same amount. The number of times a molecule passes over the barrier per second is given by [7]

$$k = \frac{k_B T}{h} \bullet \frac{F_a}{F_i} e^{-E_0 / k_B T} \tag{2}$$

Where T is temperature, k_B is Boltzmann constant, h is Planck's constant, F_a and F_i are the partition functions for unit volume of the molecule in activated and initial states. Since the energy barrier of the flowing direction is lowered, so the specific rate in the flowing direction, k_f

$$k_f = \frac{k_B T}{h} \bullet \frac{F_a}{F_i} e^{-(E_0 - \Delta E)/k_B T} = k e^{\Delta E / k_B T} \tag{3}$$

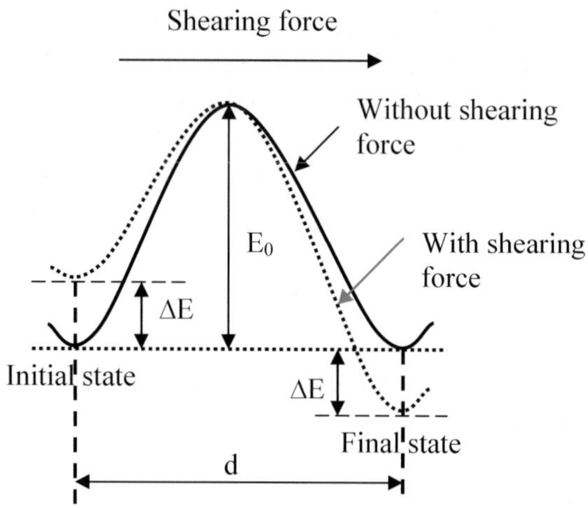

Figure 1 Potential energy barrier for the viscous flow

The specific rate in the backward direction k_b

$$k_b = ke^{-\Delta E / k_B T} \tag{4}$$

The net rate in the flowing direction (k_f - k_b) times by the distance between two equilibrium positions d_e should be the velocity of flowing

$$\begin{aligned} v &= d_e(k_f - k_b) = d_e k(e^{\Delta E / k_B T} - e^{-\Delta E / k_B T}) \\ &= 2d_e k \sinh \Delta E / k_B T \end{aligned} \tag{5}$$

Combining Eq.(1) with Eq.(5)

$$\eta = \frac{fd}{2d_e k \sinh \Delta E / k_B T} \tag{6}$$

Both d and d_e have the molecular dimension, and $\Delta E/k_B T \ll 1$ in a liquid, Eq.(6) can be simplified as ($d \approx d_e$, $\sinh x \approx x$ for small x)

$$\eta = \frac{fk_BT}{2k\Delta E} = \frac{hf}{2\Delta E} \bullet \frac{F_i}{F_a} e^{E_0 / k_BT} \tag{7}$$

Suppose that the molecule moving from one equilibrium position to another is a kind of rate controlled reaction, the equilibrium constant k_e

$$k_e = \frac{F_a}{F_i} e^{-E_0 / k_BT} \tag{8}$$

and the thermodynamic relationship

$$k_e = e^{-\Delta G / RT} \tag{9}$$

ΔG is the standard free energy, and R is the gas constant. The Eq.(7) can be therefore rewritten as:

$$\eta = \frac{hf}{2\Delta E} e^{\Delta G / RT} = (\frac{hf}{2\Delta E} e^{-\Delta S / R}) e^{\Delta H / RT} \tag{10}$$

ΔH is the enthalpy and ΔS is entropy. Since ΔE is dependent on f, f/ΔE thus can be taken as a constant. ΔS can also be taken as a constant under the assumption that the molar volume of liquid does not change greatly with temperature [7]. Eq.(9) can be written as:

$$\eta = Ae^{E / RT} \tag{11}$$

Eq.(11) is the commonly used Arrhenius equation.

It is reasonable to assume that there are a lot of holes in liquid and the free volume V_f of a liquid molecule is related to the molar volume of liquid V, the molecules packing constant c, and the energy of vaporization per molecule per mole, ΔE_{vap}. It can be expressed as [7]

$$V_f = \left(\frac{cRT}{\Delta E_{vap}} \right)^3 \frac{V}{N_A} \tag{12}$$

N_A is the Avogadro number. The partition function of a molecule of liquid may be expressed as [8]:

$$F = \frac{(2\pi mk_B T)^{3/2}}{h^3} V_f b_l e^{-\Delta E_{vap}/RT} \tag{13}$$

where b_l is the combined vibrational and rotational contribution, and m is the molar mass of liquid. Since at the activated state a molecule has one degree of translational freedom less than that at the initial state [7],

$$\frac{F_i}{F_a} = \frac{(2\pi mk_B T)^{1/2}}{h} V_f^{1/3} \tag{14}$$

Substituting Eq.(14) into Eq.(7) gives

$$\begin{aligned} \eta &= \frac{hf}{2\Delta E} \bullet \frac{(2\pi mk_B T)^{1/2}}{h} V_f^{1/3} e^{E_0/k_B T} \\ &= \frac{fcRT}{2\Delta E_{avp}\Delta E} (\frac{V}{N_A})^{1/3} (2\pi mk_B T)^{1/2} e^{E_0/k_B T} \end{aligned} \tag{15}$$

The shear force multiplied by the molar volume V and then divided by Avogadro's number N_A should be the energy applied to each molecular, ΔE

$$\begin{aligned} &fV / N_A = \Delta E, \\ &\frac{f}{\Delta E} = \frac{N_A}{V} \end{aligned} \tag{16}$$

Substituting Eq.(16) into Eq.(15)

$$\eta = \frac{cR}{2\Delta E_{avp}} (\frac{N_A}{V})^{2/3} (2\pi mk_B)^{1/2} T^{3/2} e^{E_0/k_B T} \tag{17}$$

Eq.(17) shows how the viscosity of a liquid changes with the temperature and other physical parameters. It also indicates that the constant A in Eq.(10) is dependent on the temperature. The free volume concept is important and will be used again to treat the colloidal suspension systems later.

1.2 The ER effect of pure liquids

At the end of the 19th century, the apparent viscosity of some pure insulating liquids was found to increase with the application of an electric field [9]. At that time this phenomenon was named the viscoelectric effect. Comprehensive investigations on the electroviscous effect were carried out by Andrade [10-13]. According to Andrade and Dodd, there is no measurable effect found in non-polar liquids, but with polar liquids there is a small effect that can be expressed as:

$$\frac{\Delta\eta}{\eta} = fE^2 \tag{18}$$

where $\Delta\eta$ is the viscosity increase due to the applied electric field E, i.e., the apparent viscosity difference of the liquid under an electric field and without an electric field, f is a factor named as viscoelectric constant and has the value about $2\text{x}10^{-16}$ $V^{-2}m^2$. A 10% viscosity increase may require the electric field of strength $20\text{x}10^3$ V/mm, which is close to the breakdown strength of many liquids, about 10^5 V/mm. However, for some strong polar liquids an exceptionally high viscosity increment induced by an electric field may be obtained. The viscosity increment $\Delta\eta/\eta$ against the applied electric field for acetonitrile is shown in Figure 2. The viscosity increases 115% observed at the electric field $1.8\text{x}10^3$ V/mm. It levels off with the further increase of the electric field. The viscosity increase was found to depend on the conductivity of the liquid. The maximum or called limiting value $\Delta\eta/\eta$ is plotted against the conductivity of acetonitrile in Figure 3. As acetonitrile is purified further, the conductivity of this liquid goes down, so does the maximum limiting value of $\Delta\eta/\eta$. If a liquid is so pure that it does not conduct, there would be a very small or no increase of the apparent viscosity. This trend was also observed in acetone, which is shown in Figure 4 and plotted as the viscosity increment vs. the electric field strength. The acetone of high conductivities has a much higher viscosity increment under an electric field.

Not all polar liquids show the ER effect under an electric field. For example, toluene, metaxylene, and anisole(methyl phenyl ether) do not show any detectable ER effect even under an electric field as high as $4\text{x}10^3$ V/mm [11]. Certain polar liquids, such as monochlorbenzene, show no viscosity increase when dry. After saturated with water they show a large viscosity increase with the electric field. Generally speaking, the liquids can be classified into three categories. 1) Non-polar liquids. No matter that they are dry or wet, they do not show any ER effect under an electric field; 2) Polar liquids that don't conduct or have a extremely low conductivity. They don't show any ER effect; 3) Polar liquids that have a reasonable conductivity either from impurity or from moisture. Those polar liquids do show the ER effect. The viscosity of liquid generally increases about 1.2 times and levels off at high electric fields. Successive purification may lead to the diminishing of the ER effect.

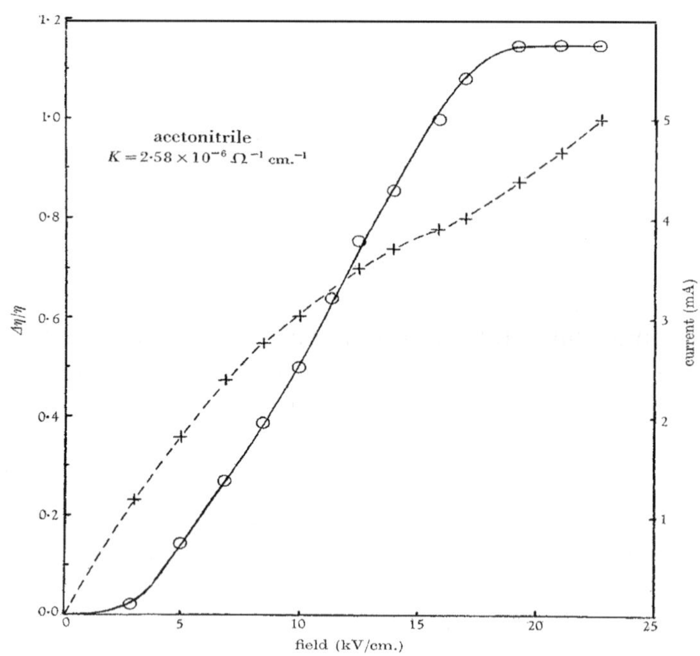

Figure 2 The viscosity increment, $\Delta\eta/\eta$, of acetonitrile having the conductivity $2.58x10^{-6}$ S/cm vs. the electric field strength. The broken line is electric current. Reproduced with permission from E. N. Andrade, C. Dodd, Roy. Soc. London Proc. A 187(1946)296

The frequency of the applied electric field was found to have a deep impact on the ER effect of liquids. The normalized $\Delta\eta/\eta$ of chloroform is plotted in Figure 5. There is a critical frequency, above which there is no detectable ER effect. This critical frequency is around 600 Hz for chloroform. There is a steep decrease of $\Delta\eta/\eta$ starting from 300 Hz and stopping at 600 Hz. When the frequency is less than 300 Hz, $\Delta\eta/\eta$ doesn't vary with frequency, while when the frequency is larger than 600 Hz, $\Delta\eta/\eta$ continues to slowly decrease with the frequency increase.

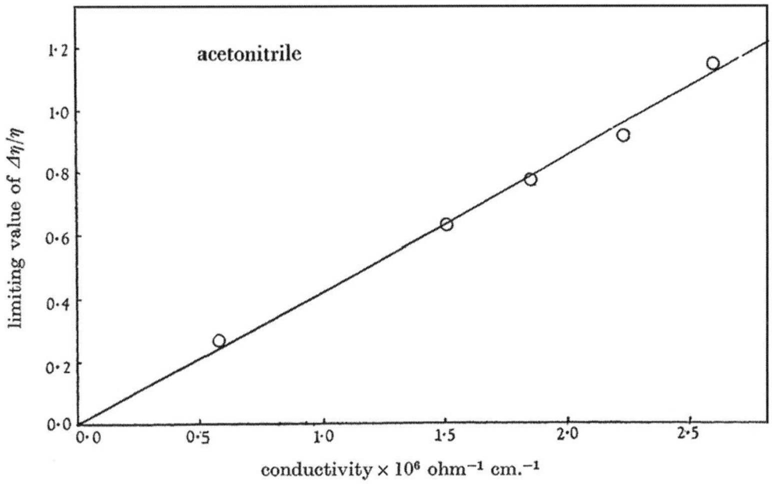

Figure 3 The maximum (or limiting) value of $\Delta\eta/\eta$ against the conductivity of acetonitrile. Reproduced with permission from ref. E. N. Andrade, C. Dodd, Roy. Soc. London Proc. A 187(1946)296.

There are two reasons proposed to explain the viscosity increase in liquids. Re-orientation of polarized molecules along the direction of the applied electric field and ion aggregation near the electrode surfaces were believed to be responsible for the viscosity increase. Local ion concentration increase nearby the electrode surface may provide possibility of forming large ion cluster, which leads to the viscosity increase. Free ions are definitely necessary, as the ER effect is associated with the reasonable conductivity of liquids. Polar molecules are necessary as only polar molecules can form large clusters and orient along the direction of the external electric field. Essentially the ER effect in a liquid should relate to the ions or polar molecules movement, which in turn depends on the frequency of the applied electric field. Ions or polar molecules cannot move fast enough in high frequency fields, thus the ER effect vanishes with the increase of frequency. According to this theory, non-polar liquids should not have the ER effect, as there are no ions or other charged entities in the systems.

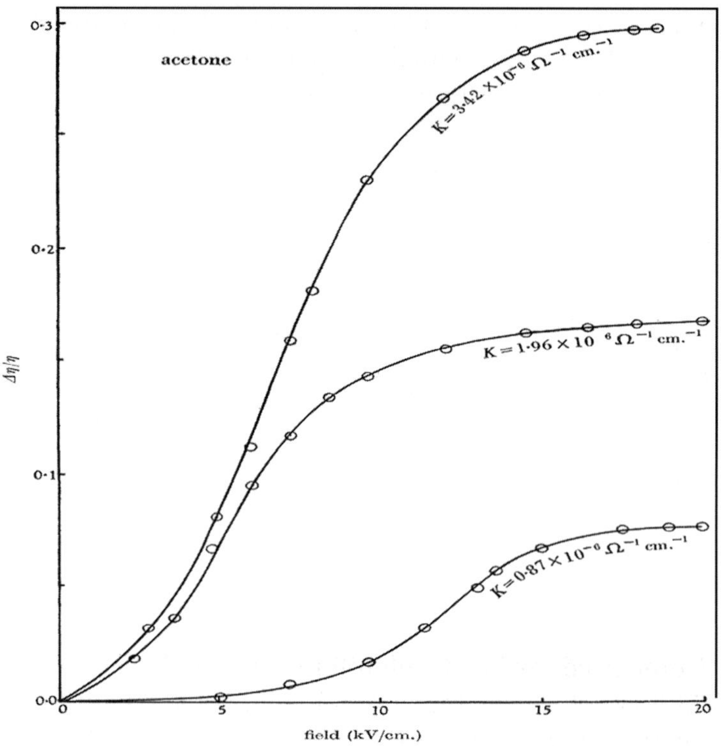

Figure 4 The viscosity increment, $\Delta\eta/\eta$, of acetone having different conductivity vs. the electric field strength. Reproduced with permission from E. N. Andrade, C. Dodd, Roy. Soc. London Proc. A 187(1946)296

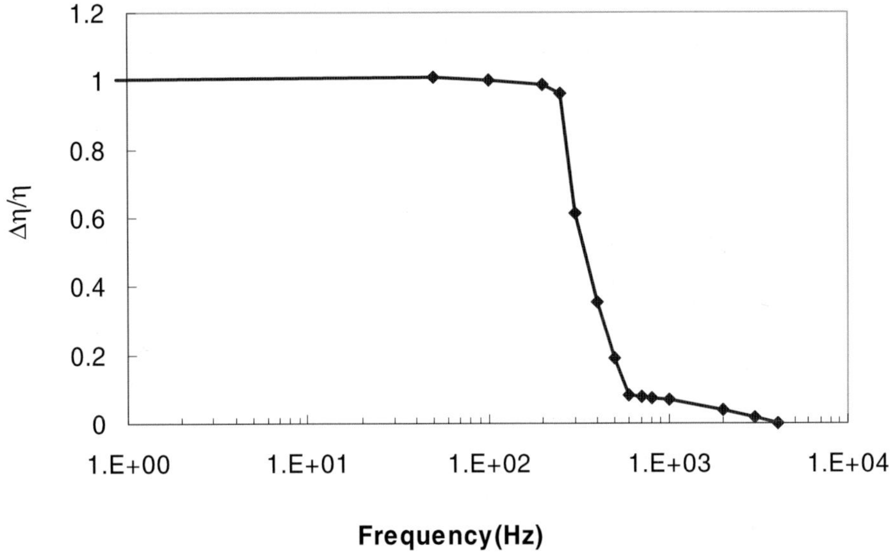

Figure 5 Normalized $\Delta\eta/\eta$ of chloroform vs. the frequency of applied electric field. Replotted from the data in E. N. Andrade and C. Dodd, Roy. Soc. London Proc. A 187(1946)296

2. COLLOIDAL SUSPENSIONS

Before the ER effect of suspensions is addressed, the viscosity of suspensions without the application of an external electric field is introduced for comparison. There is a tremendous amount of literature on the viscosity of colloidal suspensions [14,15]. An exhaustive literature survey on this subject is beyond the scope of this book. Instead, the free volume concept is a very important and intriguing idea, and is thus focused on in the following sections for deriving the viscosity equation.

2.1 The viscosity of colloidal suspensions

2.1.1. Derived from Eyring's rate theory

The viscosity of a suspension, η_s, should be the viscosity of the dispersing medium, η_m, plus that arising from the crowding of dispersed particles, η_ϕ, which obviously is proportional to the particle volume fraction, ϕ. If the particles are charged, there should be another viscosity contribution, η_p, due to the electrostatic interaction between particles.

$$\eta_s = \eta_m + \eta_\Phi + \eta_p \tag{19}$$

η_p will be only considered when the inter-particle spacing (IPS) is very small comparable to the particle radius. For the purpose of deriving viscosity equation, the IPS should be first derived. Hao [16] used Kuwabara's cell model [17], which was extended by many other researchers [18-20] for calculating the electrophoretic and electroacoustic mobility of particles, to estimate the IPS. The cell model assumes that each particle is surrounded by a virtual cell (see Figure 6) and the particle/liquid volume ratio in a unit cell is equal to the particle volume fraction throughout the entire system. Giving that the particle is spherical and monodisperse, the IPS should be zero when particles reach the maximum particle packing fraction, ϕ_m, as particles intimately contact each another at the maximum packing fraction. When the particle volume fraction, ϕ, is less than ϕ_m, there is a free volume unoccupied by particles. If the volume of a suspension is V_s, then the free volume of the particle should have in this suspension can expressed as:

$$V_s\phi_m - V_s\phi = V_s(\phi_m - \phi) \tag{20}$$

The free volume per particle should be:

$$V_s(\phi_m - \phi)/(V_s\phi/V_{ip}) = (\phi_m - \phi)V_{ip}/\phi \tag{21}$$

Where V_{ip} is the volume of individual particle, and is equal to $(4\pi r^3)/3$. r is the particle radius. The total volume that each particle occupies in the suspension is the volume of each individual particle plus the free volume per particle:

$$V_{ip} + (\phi_m - \phi)V_{ip}/\phi = \phi_m V_{ip}/\phi \tag{22}$$

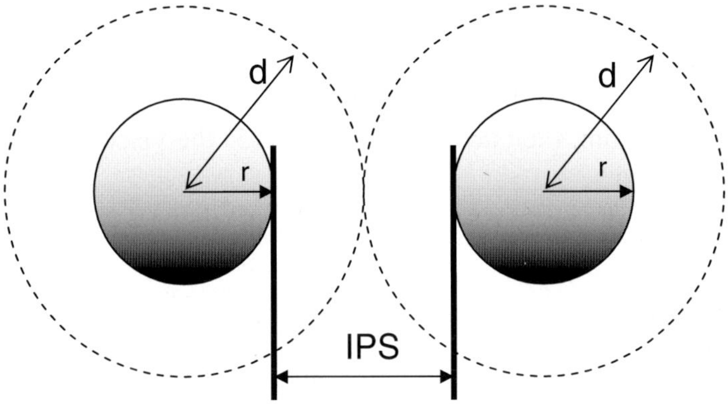

Figure 6. Illustration of Kuwabara's cell model used for calculating IPS in a particulate system.

If the radius of particle plus the virtual cell is d, then the IPS defined in Figure 6 can be expressed as:

$$IPS = 2(d- r) \tag{23}$$

Since d can be calculated from Eq.(22) using the following equation:

$$\frac{4\pi}{3}d^3 = \frac{\phi_m V_{ip}}{\phi} \tag{24}$$

Eq.(23) thus can be rewritten as:

$$IPS = 2\,(\sqrt[3]{\phi_m / \phi} - 1)r \qquad (\ \phi \leq \phi_m\ ) \tag{25}$$

Eq.(25) indicates that IPS is zero when the particle volume fraction reaches the maximum packing fraction, which is consistent with our assumption at the beginning. The parameter, ϕ, should be always less than ϕ_m. Once the maximum packing fraction and particle size of a powder system is known, the IPS can be easily estimated using Eq.(25).

For simplicity, one may assume that η_p is negligible. In other words, the interparticle spacing is so large that the particle interaction is unimportant and it doesn't contribute to the viscosity of whole system. In this case, $2(\sqrt[3]{\phi_m / \phi} - 1) \gg 1$. Eq.(19) thus becomes:

$$\eta_s = \eta_m + \eta_\Phi \tag{26}$$

Suppose that the presence of the dispersed particle doesn't change the free volume of the liquid medium, then according to Eq. (15) the viscosity of a molecule can be expressed

$$\eta = \frac{f(2\pi m k_B T)^{1/2}}{2\Delta E} \bullet V_f^{1/3} e^{E_0 / k_B T} \tag{27}$$

Since the viscosity is the energy spent per unit volume, the viscosity of the liquid η_m should be

$$\eta_m = \eta \times n_m \tag{28}$$

Where n_m is the number of the molecules per unit volume, which can be approximated as: [21]

$$n_m \approx 1/V_f \tag{29}$$

In other words, the number of molecules per unit volume is the unit volume divided by the free volume of each molecule. Substituting Eq. (29) into Eq. (28) and combining with Eq. (27) leads to

$$\eta_m = \frac{f(2\pi m_m k_B T)^{1/2}}{2\Delta E} \bullet V_f^{-2/3} e^{E_0 / k_B T} \tag{30}$$

Eq. (30) indicates that the viscosity of the liquid medium is inversely proportional to the free volume, i.e., the large free volume will result in a small viscosity, as interfrictional force between liquid molecules is small.

For calculating the viscosity contributed from the dispersed particle, it is necessary to estimate the free volume of the dispersed particle, V_{fp}. Suppose that the particle can freely move either to the left side or right side, as shown in Figure 7, with a distance of IPS until it touches the nearest neighbors, then the free volume of this particle should be

$$V_{fp} = (2IPS)^3 = 64(\sqrt[3]{\phi_m / \phi} - 1)^3 r^3 \tag{31}$$

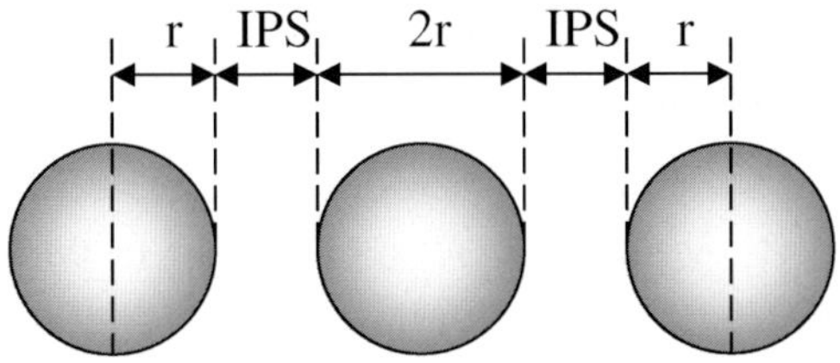

Figure 7. Determination of the free volume of the dispersed particle

as the particle can move three dimensionally. If the total number of the particle in a suspension is n_{tp}, then

$$\phi V_s = \frac{4\pi r^3}{3} \times n_{tp}$$

$$r^3 = \frac{3\phi V_s}{4\pi n_{tp}} \tag{32}$$

Suppose that the molecules of the liquid medium form a cubic packing structure, the free volume of such a packing structure is

$$V_f \times n_{tm} = (1-0.52)(1-\phi)V_s = 0.48(1-\phi)V_s \tag{33}$$

where n_{tm} is the total number of molecules of the liquid medium in the suspension. 0.52 is the maximum packing volume fraction of the cubic configuration [22]. Substituting Eq. (33) into Eq. (32) for removing V_s,

$$r^3 = \frac{0.5\phi}{1-\phi} \times \frac{n_{tm}}{n_{tp}} \times V_f \tag{34}$$

So Eq. (31) can be rewritten as

$$V_{fp} = 31.85(\sqrt[3]{\phi_m/\phi}-1)^3 \times \frac{\phi}{1-\phi} \times \frac{n_{tm}}{n_{tp}} \times V_f \tag{35}$$

Since there is not any inter-particle force between the particles, ΔE and E_0 will not be substantially altered after the particles are introduced into the liquid medium. The viscosity resulting from the particle volume fraction,

η_ϕ, can be obtained by substituting Eq. (35) into Eq. (30) with the change of m to m_p, the molar mass of the particle,

$$\eta_\phi = \frac{f(2\pi m_p k_B T)^{1/2}}{2\Delta E} \bullet \left[\frac{31.85\phi(\sqrt[3]{\phi_m/\phi}-1)^3 n_{tm}}{(1-\phi)n_{tp}}\right]^{-2/3} V_f^{-2/3} e^{E_0/k_B T} \qquad (36)$$

Assuming that the following relationship holds for the liquid medium and the dispersed particle used for making the suspension of any particle concentration:

$$m_p^{1/2} \times n_{tp}^{2/3} \approx m_m^{1/2} \times n_{tm}^{2/3} \qquad (37)$$

Note that m_p is the molar mass of the particle rather than the molar mass of the particulate material. In other words, the individual particle is considered as a huge molecule, behaving similarly to the liquid molecule and sharing the free volume created by each other. Eq. (36) can be rewritten as

$$\eta_\phi = \left[\frac{31.85\phi(\sqrt[3]{\phi_m/\phi}-1)^3}{(1-\phi)}\right]^{-2/3} \times \eta_m \qquad (38)$$

So the viscosity of suspension is:

$$\eta_s = \eta_m + \eta_\Phi = \eta_m \left\{1 + \frac{0.1(1-\phi)^{2/3}}{(\phi_m^{1/3} - \phi^{1/3})^2}\right\} \qquad (39)$$

Eq. (39) indicates that viscosity of a suspension is dependent on how the particle pack in a liquid medium and the particle volume fraction. When the particle volume fraction approaches to zero, the viscosity of the suspension almost reduces to the viscosity of the liquid medium. Suppose that particles will form a dense random packing in the suspension, thus ϕ_m=0.63 [22], the calculated η_s/η_m against the particle volume fraction ϕ is depicted in Figure 8. For the comparison reason, two widely used viscosity equations, the Kreiger-Dougherty equation [23] expressed in Eq. (40), and Frankel-Acrivos equation [24] expressed in Eq. (41), are also plotted in the same graph.

$$\eta_s = \eta_m(1-\phi/\phi_m)^{-2.5\phi_m} \tag{40}$$

$$\eta_s = \frac{9}{8}\eta_m\left[\frac{(\phi/\phi_m)^{1/3}}{1-(\phi/\phi_m)^{1/3}}\right] \tag{41}$$

Eqs. (39-41) predict a similar viscosity behavior, though Eq. (39) gives a relatively larger viscosity in comparison with the Kreiger-Doughty and Frankel-Acrivos equations. Note that both Eq.(39) and (41) cannot be reduced to Einstein's viscosity equation when the particle volume fraction approaches zero.

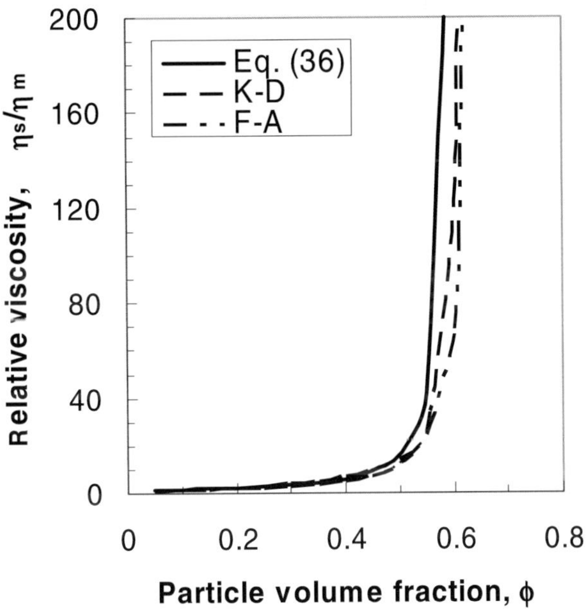

Figure 8. The calculated relative viscosity is plotted against particle volume fraction using Eq. (39). A dense random packing structure is assumed and ϕ_m=0.63. K-D represents Kreiger-Dougherty, and F-A the Frankel-Acrivos equation.

2.1.2. Derived from Einstein's equation

In 1906, Einstein [25] showed the viscosity of a suspension can be expressed as:

$$\eta_s = \eta_m(1+2.5\phi) \tag{42}$$

Eq. (42) is only valid when $\phi<1\%$. Ball [26] assumed that the effect of the particle in a concentrated suspension is the sum of the effect of the particle added sequentially. The free volume unoccupied by the particle relative to the maximum packing fraction is:

$$\frac{\phi_m-\phi}{\phi_m} = 1 - \phi/\phi_m \tag{43}$$

When the particle volume fraction is increased by an amount of $d\phi$, the particles already in the suspension suffer a crowding effect, which causes the viscosity increase by $(1-\phi/\phi_m)^{-1}$. A small free volume decrease results in a large viscosity increase. Since Eq. (42) can be written in the form:

$$d\eta_s = \eta_m(2.5)d\phi \tag{44}$$

Ball finally obtained

$$d\eta_s = 2.5\eta_s(1-\phi/\phi_m)^{-1}d\phi \tag{45}$$

Eq. (45) leads to the Kreiger-Dougherty equation [23], as shown in Eq. (40). Eq. (45) can be written in a more general form [27]:

$$d\eta_s = 2.5\eta_s(1-\phi/\phi_m)^{-n}d\phi \tag{46}$$

When n=0, Eq. (46) becomes Eq. (44), the Arrhenius equation [S.A. Arrhenius, Biochem. J., 11(1917) 112], shown in Eq. (11); when n=1, the Kreiger-Dougherty equation; when n=2, the Mooney equation [28].

In the previous section, the free volume concept is used to treat the viscosity of a pure liquid. This important concept will be used again to treat the viscosity of a suspension. The main idea is that the viscosity of a suspension should be inversely proportional to the free volume of the suspension. In other words, more free volume is available for the particle to move freely, a much less viscous force or viscosity will be generated. The term $(1-\phi/\phi_m)$ to some extent represents the normalized free volume in the equations shown above. However, this term may not accurately describe the actual free volume in a suspension. It may be better to use Eq. (31) to replace this term. Eq. (31) shows the free volume of one particle. The total free volume in a suspension, V_{tfp}

$$V_{tfp} = 64(\sqrt[3]{\phi_m / \phi} - 1)^3 r^3 \times \frac{\phi V_s}{\frac{4\pi r^3}{3}} = 15.29\phi(\sqrt[3]{\phi_m / \phi} - 1)^3 V_s \tag{47}$$

Suppose that the viscosity is related to the free volume per unit volume, i.e., V_{tfp}/V_s, Eq. (46) can be rewritten as

$$d\eta_s = 2.5\eta_s [15.29\phi(\sqrt[3]{\phi_m / \phi} - 1)^3]^{-n} d\phi \tag{48}$$

or

$$\frac{d\eta_s}{\eta_s} = \frac{2.5}{15.29^n} (\sqrt[3]{\phi_m} - \sqrt[3]{\phi})^{-3n} d\phi \tag{49}$$

A generalized viscosity equation then can be obtained by integrating Eq. (49)

$$\int_{\eta_m}^{\eta} \frac{d\eta_s}{\eta_s} = \frac{2.5}{15.29^n} \int_0^{\phi} (\sqrt[3]{\phi_m} - \sqrt[3]{\phi})^{-3n} d\phi \tag{50}$$

When n=1, Eq. (50) becomes

$$\ln \frac{\eta_s}{\eta_m} = 0.49 \left\{ \frac{\sqrt[3]{\phi_m}(4\phi^{1/3} - 3\phi_m^{1/3})}{(\sqrt[3]{\phi_m} - \sqrt[3]{\phi})^2} - \ln[1 - (\frac{\phi}{\phi_m})^{1/3}] + 1.5 \right\} \tag{51}$$

When n=1/3, Eq. (50) becomes

$$\ln \frac{\eta_s}{\eta_m} = -3.02 \left[\phi_m^{2/3} \ln[1 - (\phi / \phi_m)^{1/3}] + (\phi_m \phi)^{1/3} + 0.5\phi^{2/3} \right] \tag{52}$$

When n=2/3, Eq. (50) becomes

$$\ln \frac{\eta_s}{\eta_m} = -1.22 \left[2\phi_m^{1/3} \ln \frac{1}{1 - (\phi / \phi_m)^{1/3}} + \frac{\phi^{2/3} - 2\phi^{1/3}\phi_m^{1/3}}{\phi_m^{1/3} - \phi^{1/3}} \right] \tag{53}$$

When n≠1, n≠1/3, and n≠2/3, Eq. (50)

$$\ln\frac{\eta_s}{\eta_m}=\frac{7.5}{15.29^n}\left[-\frac{(\phi_m^{1/3}-\phi^{1/3})^{3-3n}}{3-3n}+\frac{2\phi_m^{1/3}(\phi_m^{1/3}-\phi^{1/3})^{2-3n}}{2-3n}-\frac{\phi_m^{2/3}(\phi_m^{1/3}-\phi^{1/3})^{1-3n}}{1-3n}+\frac{2\phi_m^{1-n}}{(1-3n)(2-3n)(3-3n)}\right] \quad (54)$$

The relative viscosity predicted with Eq. (51-54) is plotted against the particle volume fraction and depicted in Figure 9. The Kreiger-Dougherty and Frankel-Acrivos equations, shown in Eq. (40) and (41), respectively, are also depicted in Figure 9 for comparison. The dense random particle packing is assumed for all calculations and 0.63 is taken as the maximum packing fraction. When n=0.26 or 0.33, the calculated relative viscosity from Eq. (52) and (54) is similar to that predicted with the Kreiger-Dougherty equation at low particle volume fractions, and similar to that predicted with the Frankel-Acrivos equation at high particle volume fractions. Since the Eq. (37), Eq. (38), Eq.(52) at n=0.33, and (54) at n=0.26, present a similar viscosity behavior, the relative viscosity predicted with those equations are re-plotted against the particle volume fraction in Figure 10 for magnifying the difference between them.

All those equations except Eq. (41) reduce to 1 when the particle volume fraction approaches to zero. At low particle volume fractions, less than 0.4, the Kreiger-Dougherty equation gives a little bit higher relative viscosity than that predicted with either Eq. (52) or (54). The relative viscosities predicted with the Frankel-Acrivos equation and Eq. (54) at n=0.26 almost overlap at very high particle volume fractions, larger than 0.5. Figure 9 shows with the increase of the parameter n from 0.26 to 1, the relative viscosity increases substantially. The dramatic viscosity increase even happens in a low particle volume fraction, around 0.4 for n=1. The parameter n thus can be used to scale the interparticle force, which is the only reason causing the remarkable viscosity increase. Since the Kreiger-Dougherty equation is located in between n=0.26 and n=0.33 in Figure 10, it may be possible to find a suitable value of n, which results in more similar viscosity behavior to that predicted with the Kreiger-Dougherty equation. The relative viscosities predicted with Eq. (54) at n=0.3 and the Kreiger-Dougherty equation are plotted in Figure 11 vs. the particle volume fraction. Besides the minor discrepancy in low particle volume fractions, a surprising overlap is observed, indicating that the Kreiger-Dougherty equation can be considered as a special case of Eq. (54).

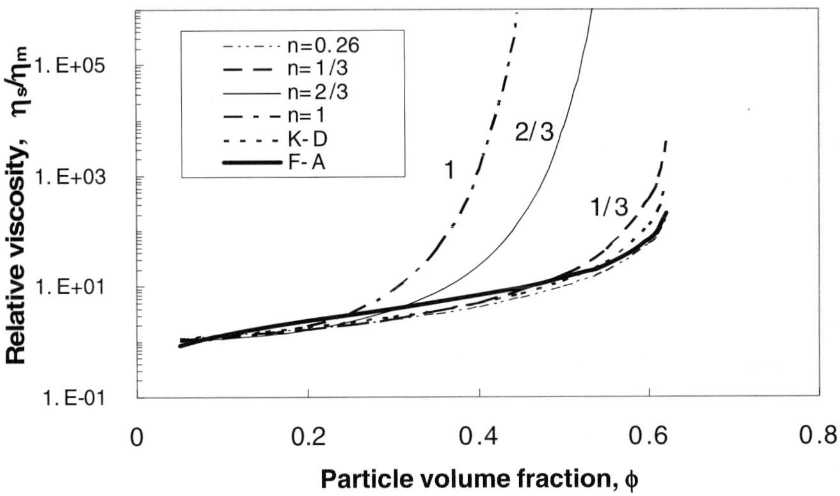

Figure 9. The calculated relative viscosity is plotted against particle volume fraction using Eq. (52-54). A dense random packing structure is assumed and ϕ_m=0.63. K-D represents for the Kreiger-Dougherty equation, and F-A stands for the Frankel-Acrivos equation.

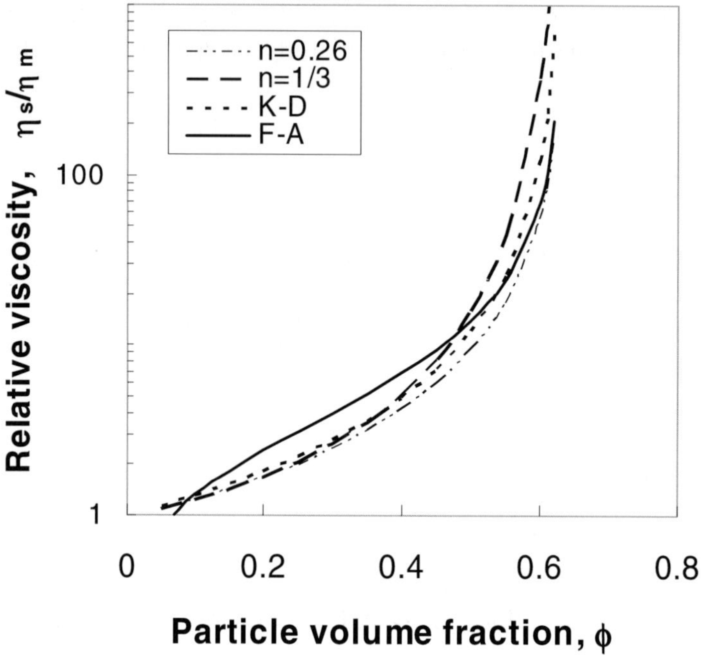

Figure 10 The calculated relative viscosity is plotted against the particle volume fraction using Eq. (40), (41), Eq.(49) at n=0.33, and Eq.(54) at n=0.26. A dense random packing structure is assumed and ϕ_m=0.63. K-D represents for the Kreiger-Dougherty equation, and F-A stands for the Frankel-Acrivos equations.

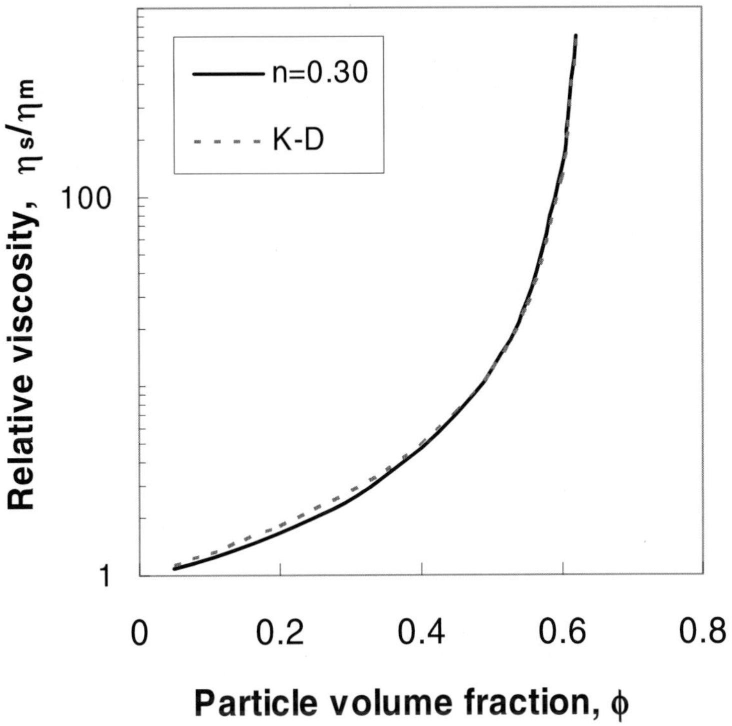

Figure 11. The calculated relative viscosity is plotted against the particle volume fraction using Eq. (40) and Eq. (54) at n=0.30. A dense random packing structure is assumed and ϕ_m=0.63. K-D represents for the Kreiger-Dougherty equation.

Now the question is how to define the parameters n and ϕ_m, the maximum packing fraction, for suspensions in a practical way. Once those two parameters are determined, the viscosity of a suspension then is able to be calculated using the proper equations shown above.

2.1.3 The maximum packing fraction of polydisperse particles

For the monodisperse spherical particle packing, the maximum packing fraction has been reviewed systematically [22]. The six basic cases of sphere packings including both regular and random packing are shown in Table 1.

Table 1 The six basic sphere packing patterns

Packing group	Maximum packing fraction	Coordinate number (points of contact per sphere)
Simple cubic	0.5236	6
Orthorhombic	0.6046	8
Tetragonal	0.6981	10
Rhombohedral	0.7405	12
Dense random	0.63	~9-10
Loose random	0.59	~8

The body-centered and face-centered cubic packing, with the maximum packing fraction 0.68 and 0.74 respectively, were also mentioned in the literature [29] for possible particle packing structures. However, those two packing structures would be very unlikely in practical. The stability of packing structure increases as the voidage decreases. Among the six basic packing patterns, the simple cubic has the greatest energy, while the rhombohedral packing is the most stable packing structure.

In a practical colloidal suspension, the particle always has a size distribution, and cannot be considered as a mono-disperse system. In this case, a method for estimation of the maximum packing fraction on the basis of the particle size distribution information would be very useful. A mathematical expression, originally for binary mixtures, was proposed [30].to estimate the maximum packing fraction, Φ_{max}, for any particle system having a size distribution without significant particle interaction

$$\Phi_{max} = \Phi_{ult} - (\Phi_{ult} - \Phi_{mmono})e^{\alpha[1-\frac{D_5}{D_1}]} \tag{55}$$

$$\Phi_{ult} = 1-(1-\Phi_{mmono})^m \tag{56}$$

where Φ_{ult} = the ultimate packing fraction, which can be calculated from Eq. (56); Φ_{mmono} = the maximum mono-disperse packing fraction, showing in Table 1 and depending on the packing pattern; m = number of different particle diameter classes in suspension (e.g., for binary mixtures, m=2)). α=0.268, a constant. D_1 and D_5 are the particle diameter averages, and can be expressed as:

$$D_x = \frac{\sum_{k=1}^{n} N_k D_k^x}{\sum_{k=1}^{n} N_k D_k^{x-1}} \tag{57}$$

Thus, D_1 and D_5 can be calculated from particle size distribution data obtained with a variety of particle sizing instrumentation. Sudduth developed a mathematical means to computer the ratio of D_5 to D_1, requiring the knowledge of the number of each kind of particle, and the diameter of each particle, or another measure of the composition of particle size diameters in a suspension [30]. Those parameters are not easily determined from the particle size measurement, especially when the system has a very wide size distribution. This rendered Eq. (55) difficult to evaluate. Hao [31] modified Eq.(55) and (56), and developed a simple way for calculating D_5, The maximum packing fraction is thus able to be calculated on the basis of experimental particle size measurement.

For a particulate system of a relatively wide size distribution, the parameter m in Eq. (56) should be a very large number. Since Φ_{mmono} is usually larger than 0.5, thus Φ_{ult} should be very close to 1. Eq. (55) can be approximated as:

$$\Phi_{\max} \cong 1-(1-\Phi_{mmono})e^{\alpha[1-\frac{D_5}{D_1}]} \tag{58}$$

According to the definition, D_1 is the number average particle size that can be directly obtained from particle size measurements. To calculate D_5, a log-normal size distribution has to be assumed, which enables one to determine the number average, D_1, and the number geometric mean, D_{gN}, and the geometric standard deviation, σ_g. If the number distribution of a particulate system is obeying the log-normal law, then other type weighted distributions (such as a volume basis) are also log-normal with the same geometric standard deviation [32]. According to Allen [32], D_1, D_{gn} and σ_g can be related using the following equation:

$$\ln D_1 = \ln D_{gN} + 0.5\ln^2\sigma_g \tag{59}$$

D_{gN} and σ_g thus can be used to calculate D_2, D_3, D_4 using the following transforms:

$$\ln D_2 = \ln \frac{\sum_{k=1}^{n} D_k{}^2 N_k}{\sum_{k=1}^{n} D_k N_k} = \ln D_{LS} = \ln D_{gN} + 1.5 \ln^2 \sigma_g \tag{60}$$

$$\ln D_3 = \ln \frac{\sum_{k=1}^{n} D_k{}^3 N_k}{\sum_{k=1}^{n} D_k^2 N_k} = \ln D_{SV} = \ln D_{gN} + 2.5 \ln^2 \sigma_g \tag{61}$$

$$\ln D_4 = \ln \frac{\sum_{k=1}^{n} D_k{}^4 N_k}{\sum_{k=1}^{n} D_k^3 N_k} = \ln D_{VM} = \ln D_{gN} + 3.5 \ln^2 \sigma_g \tag{62}$$

Where D_{LS} is the length to surface average, D_{SV} is the surface to volume average, D_{VM} is the volume or weight moment average. Using the methodology outlined by Allen, the following relationship for D_5 was derived:

$$\ln D_5 = \ln \frac{\sum_{k=1}^{n} D_k{}^5 N_k}{\sum_{k=1}^{n} D_k^4 N_k} = \ln D_{gN} + 4.5 \ln^2 \sigma_g \tag{63}$$

Since D_{gN} can be calculated using number average or volume average particle size data with Eq. (59) or (62), D_5 can be therefore calculated using Eq. (63). Eq. (59) can be combined with Eq. (63) to yield:

$$\ln \frac{D_5}{D_1} = 4 \ln^2 \sigma_g \tag{64}$$

Eq. (64), together with Eq. (58), indicates that Φ_{max} does not depend on the size but instead the size distribution breadth expressed by σ_g. Experimental results show that the maximum packing fraction calculated from Eq. (58) is relatively higher than that determined with the rheological measurement [31]. The discrepancy is believed to be a result of the constant value of α, which should be a function of the particle size and particle packing structure. An empirical equation was proposed for correlating α with particle size parameters to fit the experimentally determined α data [31]:

$$\alpha = (2c-1)(2+c)^{-c}\Phi_{monom}\sigma_g^{(2c-1)}(D_5/D_1)^{(c^2-1)} \tag{65}$$

Where

$$c = \frac{D_{gN}-0.5}{D_{gN}+0.5} \tag{66}$$

D_{gN} is in μm. When D_{gN} » 0.5 μm, c≈1, and Eq. (65) becomes:

$$\alpha = 0.33\Phi_{monom}\sigma_g \tag{67}$$

For a particulate system with a relatively narrow size distribution and the loose random packing pattern, α approaches to 0.268 at σ_g around 1.38, indicating Sudduth's original equation is only valid for a large particle with a narrow size distribution.

2.1.4 Determining the parameter n

Once the maximum packing fraction can be practically calculated on the basis of the particle size distribution data, the parameter n is the only parameter unknown for calculation of the viscosity of a suspension system. Since the parameter n scales the particle-interaction, one may need to consider two different kinds of interactions. The one is a result of the particle shape. Irregular particle shape may cause the geometric hindering in the space, thus resulting in a strong inter-particle force; the other is a result of the electrostatic interaction due to the particle surface charge. Note that the parameter n is originally associated with the free volume of a suspension, so most likely n is determined by the particle shape. There are many ways to define the particle shape factor [33]. For example, the aspect ratio (also called ellipticity), the surface or volume shape factors [34], the sphericity [35], and circularity (or called roundness) [32], are most commonly used for describing the particle shape. If three perpendicular particle diameters can be defined in such a way that the Length (L)≥ Breadth (B) ≥ Thickness (T), as shown in Figure 12a, then Heywood [36] describes the flakiness (f)

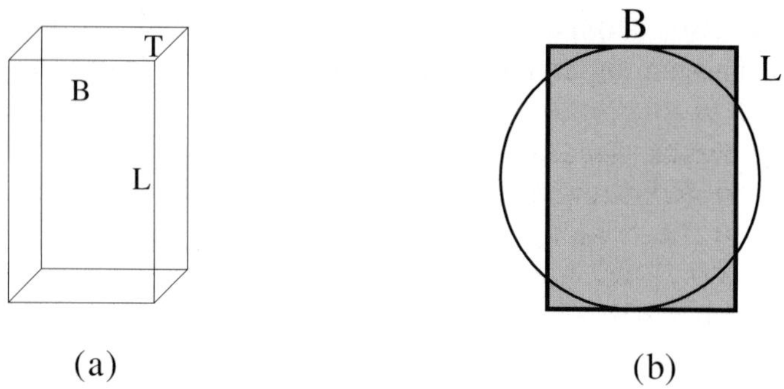

Figure 12 Quantifying the particle shape.

$$f = \frac{B}{T} \tag{68}$$

and elongation e

$$e = \frac{L}{B} \tag{69}$$

Where T is the particle thickness representing the minimum distance between two parallel planes that are tangential to opposite surfaces of the particle; B is the particle breadth representing the minimum distance between two parallel planes that are perpendicular to the planes defining the thickness; and L is the particle length representing the distance between two parallel planes that are perpendicular to the planes defining the thickness and breadth. Under these definitions, the mean diameter d_m is

$$d_m = \frac{L+B+T}{3} \tag{70}$$

If the projected diameter, d_p, is the one of a circle with the same area as the projected area of the particle, the shadow area shown in Figure 12b, then

$$\frac{\pi d_p^2}{4} = L \times B, \qquad d_p = \sqrt{\frac{4LB}{\pi}} \tag{71}$$

The volume shape factor α_V is the average volume of the particle, Vp, divided by the cubic power of the projected diameter, d_p

$$\alpha_V = \frac{V_p}{d_p^3}, \quad V_p = L \times B \times T \tag{72}$$

Similarly the surface shape factor α_s is the average surface area of the particle, S_p, divided by the square of the projected diameter, d_p

$$\alpha_s = \frac{S_p}{d_p^2}, \quad S_p=2(LB+BT+LT) \tag{73}$$

The sphericity is the surface area of a sphere having the same volume as the particle divided by the surface area of the particle. The circularity is the circumference of a circle having the same area as the projected particle divided by the perimeter of the projected particle image. As depicted in Figure 12b,

$$\text{Circularity}=\frac{\sqrt{\pi BL}}{B+L} \tag{74}$$

If the aspect ratio (AR) is defined as:

$$\text{AR–L/B} \tag{75}$$

Eq. (74) can be rewritten as:

$$\text{Circularity}=\frac{\sqrt{\pi AR}}{1 \mid AR} \tag{76}$$

By this definition, the circularity ranges between 0 and 1, and the circularity of the sphere is 1. The circularity of common shapes is listed in Table 2.

Table 2 The circularity of common shapes

Shapes	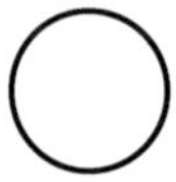 Sphere	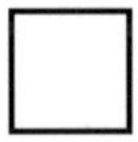 Square	 Equiangular Triangle (ET)	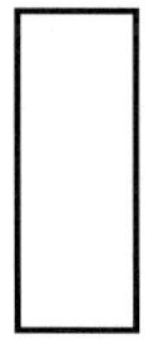	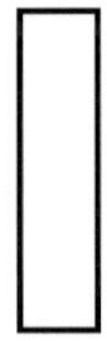	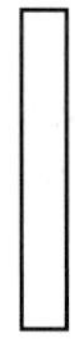
Circularity (CL)	1	0.886	0.777	0.660	0.509	0.429
AR		1		5	10	15
n=0.3/CL	0.30	0.339	0.386	0.455	0.589	0.699
$n=0.3/CL^2$	0.30	0.382	0.497	0.689	1.158	1.630

Table 2 indicates that the circularity gradually decreases as the particle shape changes from a spherical shape to a fiber-like shape. Since when the parameter n=0.3, Eq. (54) predicts an almost identical viscosity behavior as the Kreiger-Dougherty equation does, the parameter n may be defined as 0.30 divided by the circularity of the dispersed particle, or divided by the square of the circularity of the dispersed particle. Another reason for doing this is that the Kreiger-Dougherty equation is usually considered for spherical particle system, the parameter n for this system is already known as 0.30. Note that the circularity is obtained on the basis of two-dimensional shape of particle, it would be more reasonable to define the parameter n is 0.30 divided by the square of the circularity. According to these definitions, the parameter n will increase as the particle shape changes from spherical to fibrous shape, which are shown in Table 2. Eq. (54) thus can be used to predict the viscosity of a suspension containing the particle that the particle shape must be considered. The calculated relative viscosity from Eq. (54) is plotted against the particle volume fraction and shown in Figure 13 for the parameter n defined as 0.3 divided by the circularity and in Figure 14 for the parameter n defined as 0.3 divided by the square of the circularity. The abrupt increase of the relative viscosity occurs at a lower particle volume fraction when the particle shape changes from sphere to fiber. With the increase of the aspect ratio of the fiber, the abrupt viscosity change takes place at a even lower particle volume fraction. This trend becomes much more significant when the parameter n is defined as 0.3 divided by the square of the circularity. The relative viscosity tends to become infinite at the particle volume fraction about 25 vol%, when the aspect ratio of the dispersed particle is 15. There is a plenty of literature dealing with the particle geometry effect on the viscosity of a colloidal suspension, Metzner [37] recommends the following equation:

$$\frac{\eta}{\eta_s} = (1 - \phi / A)^{-2} \tag{77}$$

provided that A can be chosen properly. For an uniform sphere A=0.68, and for fiber with the aspect ratio of 5<AR<30, Kitano [38] suggested:

$$A=0.55-0.013(AR) \tag{78}$$

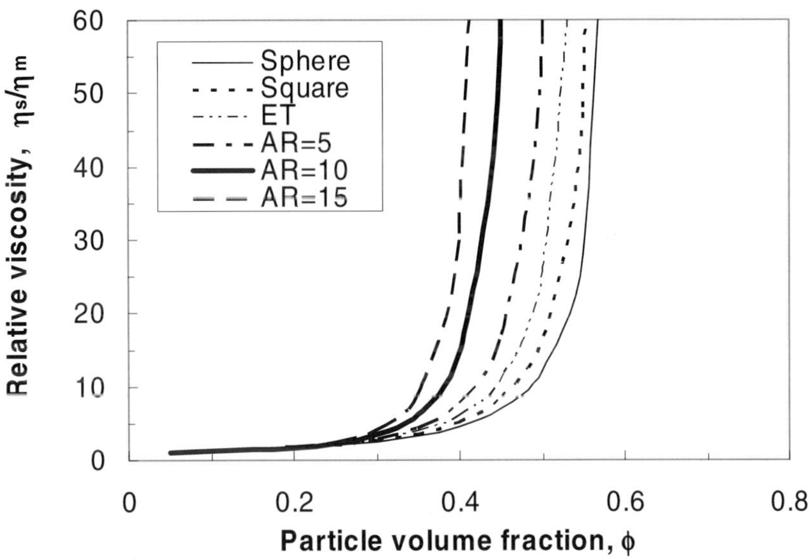

Figure 13 The calculated relative viscosity is plotted against the particle volume fraction using Eq. (54). A dense random packing structure is assumed and ϕ_m=0.63. The parameter n is defined as 0.3 divided by the circularity of the particle.

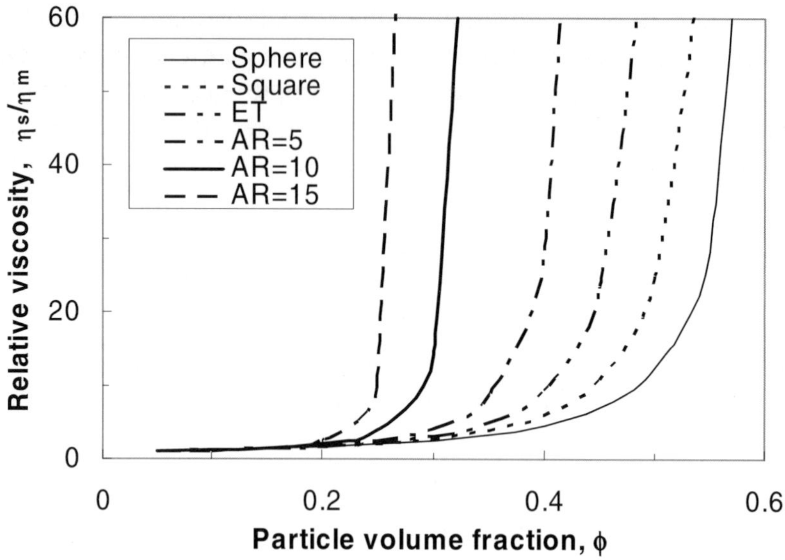

Figure 14 The calculated relative viscosity is plotted against the particle volume fraction using Eq. (54). A dense random packing structure is assumed and ϕ_m=0.63. The parameter n is defined as 0.3 divided by the square of the circularity of particles.

Substituting Eq. (78) into Eq. (77) leads to

$$\frac{\eta}{\eta_s} = \left(1 - \frac{\phi}{0.55 - 0.0139(AR)}\right)^{-2} \tag{79}$$

For comparison, the calculated relative viscosity from the Metzner's equation, Eq. (77), for the spherical particle, A=0.68, from the Kitano's extension equation, Eq.(79), for the particle of the aspect ratio, AR=5, and from Eq. (54) for the particles of the aspect ratio, AR=5, AR=10, and the spherical shape, is plotted against the particle volume fraction in Figure 15. The Kitano's extension equation gives a much higher relative viscosity compared with the result obtained from Eq. (54) at AR=5 for low particle volume fractions. Once the particle volume fraction is larger than 0.48, the relative viscosity predicted with Eq. (79) starts to decrease with the further increase of the particle volume fraction, leading to unreasonable results. Within the medium concentration range between 0.3 and 0.48, Eq.(79)

presents a similar result to that predicted from Eq. (54) at AR=10 rather than AR=5. Metzner's equation gives a similar prediction as Eq. (54) does when the dispersed particle has a spherical shape. For the low particle volume fraction ranging from 0.05 to 0.53, Metzner's equation predicts a slightly higher relative viscosity than Eq. (54) does, while once the particle volume fraction is larger than 0.53, it gives a slightly lower relative viscosity than Eq. (54) does. Clearly, both Metzner's equation and Kitano's extension equation can be considered approximately as special cases of Eq. (54). Note that the Kreiger-Dougherty and Frankel-Acrivos equations are also special cases of Eq. (54), implying that Eq. (54) is much more universal than the currently used equations

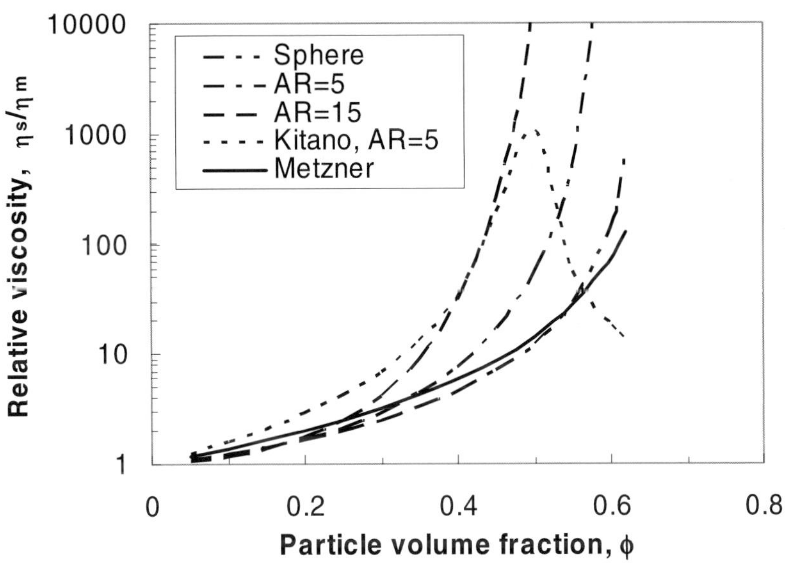

Figure 15 The calculated relative viscosity is plotted against the particle volume fraction using Eq. (54). A dense random packing structure is assumed and ϕ_m=0.63. The parameter n is defined as 0.3 divided by the circularity of the particle. The results calculated from Metzner's equation, Eq. (77), for the spherical particle, and Kitano's extension equation, Eq. (78), for the particle of the aspect ratio, AR=5.

The microstructural theories developed for the rheological properties of fiber suspensions have been reviewed by Huigol and Phan-Thien [39]. Detailed information is out of the scope of this treatise, and interested readers are referred to original literature such as the references [40, 41].

One experimental example from Clarke [42] is shown in Figure 16 for comparison with the prediction shown above. The viscosity of various shaped particles dispersed in water are measured against the particle volume fraction ranging from 5 vol% to 40 vol%. When the particle shape changes from sphere to grain, plate and rod, the critical particle volume fraction, at which the viscosity starts to jump, gradually shifts to lower volume fractions. In comparison with the graph shown in Figure 14, the experimental results are much more closer to the predictions made with Eq. (54) assuming that the parameter n is defined as 0.3 divided by the square of the circularity of the particle. Giesekus's experimental work [43] shows that the relative viscosity of glass fiber suspensions increases substantially with the increase of the particle aspect ratio, which is also in agreement with the prediction given by Eq. (54). As given in Table 2, the parameter n will increase with the circularity or the aspect ratio of the particle, so does the relative viscosity.

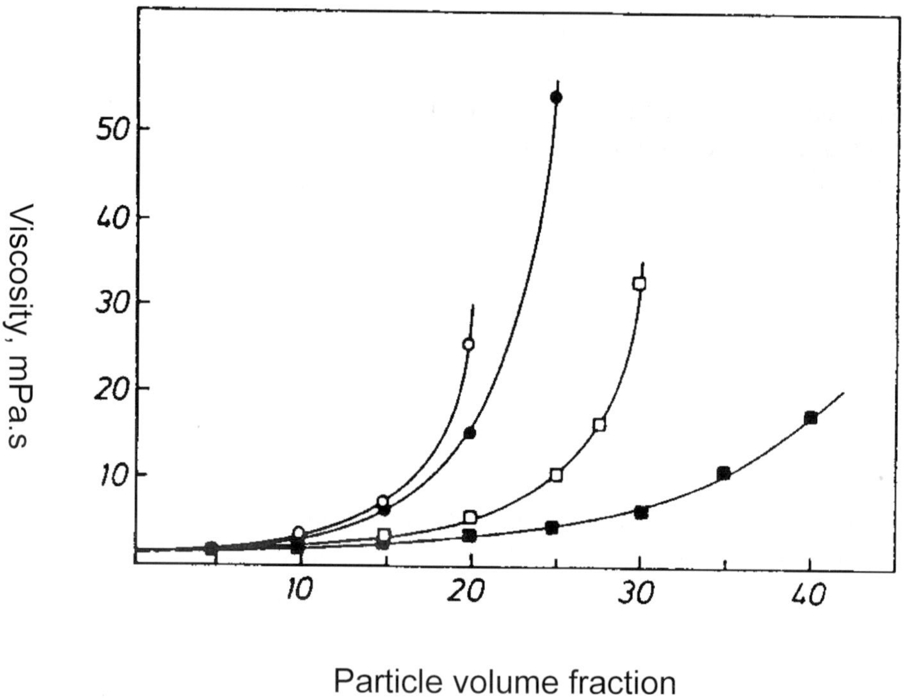

Figure 16 Viscosity of differently shaped particles in water vs. the particle volume fraction at shear rate 300 s^{-1}, (■) spheres; (□) grains; (●) plates; (○) rods. Reproduced with permission from B. Clarke, Trans.Inst.Chem.Eng., 45(1967)251.

2.1.4 Contribution from the particle surface charge

The discussions in the sections from 2.1.1 to 2.1.3 focus on the suspension containing the particle without the surface charge or the negligible interaction between particles. In the case that the particle is strongly charged, the viscosity contributed from the particle interaction must be considered. The interaction between particles will drag the particle to move freely in the system, thus decreasing the free volume of each individual particle and therefore increasing the viscosity of the whole suspension. There are plenty of works for the calculation of the inter-particle potential between two charged colloids [44-46]. Attempts are also made to predict the viscosity of the charged colloids by using the inter-particle potential [21,47] Due to the difficulty and complexity of precise scaling the inter-particle force among a large number of particles in a suspension system, an alternative approach is proposed below for the calculation of the viscosity of the charged colloidal suspensions via continuing to use the free volume concept, and in combination with the effective particle radius concept. Suppose that the viscosity contributed from the particle surface charge doesn't happen until the Debye-length of the double layer is comparable to the particle diameter or the interparticle spacing (IPS, see section 2.1.1), the effective radius, r_{eff}, is thus defined as the particle radius plus the distance between the particle surface and the shear plane, where the Zeta potential is defined. As discussed in the section 1.3, in Chapter 1, the Zeta potential is the potential at the surface $(r + \frac{1}{s\kappa})$ from the center of the particle, so

$$r_{eff} = r + \frac{1}{s\kappa} \tag{80}$$

Where s is an integral number, r is the particle radius, and 1/κ is the Debye length. According to Eq.(9) in Chapter 1

$$r_{eff} = \frac{\zeta \varepsilon_m \varepsilon_0}{d_q} = r + \frac{1}{s\kappa} \tag{81}$$

Where d_q is the particle surface charge density. The effective particle radius should lead to the concepts of the effective particle volume fraction, ϕ_{eff}, the effective inter-particle spacing, IPS_{eff}, the effective free volume of individual particle, V_{feff}, and the total effective free volume, V_{tfeff}. Those parameters can be expressed with the following equations via a simple derivation:

$$\phi_{eff} = \phi \left(\frac{\zeta \varepsilon_m \varepsilon_0}{r d_q} \right)^3 \tag{82}$$

and

$$IPS_{eff} = 2\left(\sqrt[3]{\phi_m / \phi_{eff}} - 1\right) \times \frac{\zeta \varepsilon_m \varepsilon_0}{d_q} = 2\left[r\left(\frac{\phi_m}{\phi} \right)^{1/3} - \frac{\zeta \varepsilon_m \varepsilon_0}{d_q} \right] \tag{83}$$

and

$$V_{feff} = \left(2IPS_{eff}\right)^3 = 64\left[r\left(\frac{\phi_m}{\phi} \right)^{1/3} - \frac{\zeta \varepsilon_m \varepsilon_0}{d_q} \right]^3 \tag{84}$$

and

$$\mathrm{V}_{\mathrm{tfeff}} = V_{feff} \times \left[\phi V_s / \left(\frac{4}{3} \pi r^3 \right) \right] = \frac{48\phi}{\pi} \left[\left(\frac{\phi_m}{\phi} \right)^{1/3} - \frac{\zeta \varepsilon_m \varepsilon_0}{r d_q} \right]^3 \times V_s \tag{85}$$

Again, assuming that the viscosity is related to the unit volume of the free volume, i.e., V_{tfeff}/V_s, thus Eq. (48) can be rewritten as

$$d\eta_s = 2.5\eta_s \left\{ \frac{48\phi}{\pi} \left[\left(\frac{\phi_m}{\phi} \right)^{1/3} - \frac{\zeta \varepsilon_m \varepsilon_0}{r d_q} \right]^3 \right\}^{-n} d\phi \tag{86}$$

or

$$\frac{d\eta_s}{\eta_s} = \frac{2.5}{15.29^n}(\phi_m^{1/3} - \frac{\zeta\varepsilon_m\varepsilon_0\phi^{1/3}}{rd_q})^{-3n} d\phi \qquad (87)$$

Note that the only difference between Eq. (49) and (87) is that there is an additional term $\frac{\zeta\varepsilon_m\varepsilon_0}{rd_q}$ in front of $\phi^{1/3}$. Integrating the Eq. (87) will lead to a generalized viscosity equation for the charged particle with the Debye length comparable to the particle radius or the inter-particle spacing. When n=1, Eq. (87) leads to

$$\ln\frac{\eta_s}{\eta_m} = 0.49(\frac{rd_q}{\zeta\varepsilon_m\varepsilon_0})^3 \left\{ \begin{aligned} &\frac{\sqrt[3]{\phi_m}(4\phi^{1/3} - 3\frac{\zeta\varepsilon_m\varepsilon_0}{rd_q}\phi_m^{1/3})}{(\sqrt[3]{\phi_m} - \frac{\zeta\varepsilon_m\varepsilon_0}{rd_q}\sqrt[3]{\phi})^2} \\ &- \ln[1 - \frac{\zeta\varepsilon_m\varepsilon_0}{rd_q}(\frac{\phi}{\phi_m})^{1/3}] + 1.5 \end{aligned} \right\} \qquad (88)$$

When n=1/3, Eq. (87) becomes

$$\ln\frac{\eta_s}{\eta_m} = -3.02(\frac{rd_q}{\zeta\varepsilon_m\varepsilon_0})^3 \left[\begin{aligned} &\phi_m^{2/3}\ln[1 - \frac{\zeta\varepsilon_m\varepsilon_0}{rd_q}(\phi/\phi_m)^{1/3}] + \\ &\frac{\zeta\varepsilon_m\varepsilon_0}{rd_q}(\phi_m\phi)^{1/3} + 0.5(\frac{\zeta\varepsilon_m\varepsilon_0}{rd_q})^2\phi^{2/3} \end{aligned} \right] \qquad (89)$$

When n=2/3, Eq. (87) becomes

$$\ln\frac{\eta_s}{\eta_m} = -1.22\left(\frac{rd_q}{\zeta\varepsilon_m\varepsilon_0}\right)^3 \left[\begin{array}{c} 2\phi_m^{1/3}\ln\dfrac{1}{1-\dfrac{\zeta\varepsilon_m\varepsilon_0}{rd_q}(\phi/\phi_m)^{1/3}} + \\ \dfrac{(\dfrac{\zeta\varepsilon_m\varepsilon_0}{rd_q})^2\phi^{2/3} - 2\dfrac{\zeta\varepsilon_m\varepsilon_0}{rd_q}\phi^{1/3}\phi_m^{1/3}}{\phi_m^{1/3} - \dfrac{\zeta\varepsilon_m\varepsilon_0}{rd_q}\phi^{1/3}} \end{array}\right] \tag{90}$$

When n≠1, n≠1/3, n≠2/3, Eq. (87) becomes

$$\ln\frac{\eta_s}{\eta_m} = \frac{7.5}{15.29^n}\left(\frac{rd_q}{\zeta\varepsilon_m\varepsilon_0}\right)^3 \left[\begin{array}{c} -\dfrac{(\phi_m^{1/3} - \dfrac{\zeta\varepsilon_m\varepsilon_0}{rd_q}\phi^{1/3})^{3-3n}}{3-3n} + \\ \dfrac{2\phi_m^{1/3}(\phi_m^{1/3} - \dfrac{\zeta\varepsilon_m\varepsilon_0}{rd_q}\phi^{1/3})^{2-3n}}{2-3n} - \\ \dfrac{\phi_m^{2/3}(\phi_m^{1/3} - \dfrac{\zeta\varepsilon_m\varepsilon_0}{rd_q}\phi^{1/3})^{1-3n}}{1-3n} + \\ \dfrac{2\phi_m^{1-n}}{(1-3n)(2-3n)(3-3n)} \end{array}\right] \tag{91}$$

Once the charge density/zeta potential ratio is known, Eq. (88-91) can be used to calculate the viscosity of charged colloidal suspensions. As shown in Chapter one, the relationship between the surface charge density and Zeta potential can be derived using the Poisson's equation under the Debye-Hückel approximation, i.e., the small value of potential, $ze\psi << k_B T$, where ψ is the potential and z is the valency of ion [4].

$$\frac{d_q}{\zeta} = \frac{\varepsilon_m\varepsilon_0(1+\kappa r)}{r} \tag{92}$$

Note that Eq. (92) is only valid when Zeta potential is small, i.e., $ze\zeta << k_B T$, and the electrophoresis retardation and relaxation effects are unimportant. In addition, Eq. (92) is obtained under the assumption that the Zeta potential and surface charge density are independent on the particle volume fraction, which may not be true in a concentrated suspension. Substituting Eq. (92) into Eq. (91), one may obtain an idea on how κr will change the viscosity of the whole system. The calculated relative viscosity with Eq. (91) vs. κr on a wide κr range is depicted in Figure 17. The dispersed particle is assumed as sphere, and three different particle volume fractions are shown in Figure 17. The relative viscosity is extremely sensitive to κr when the double layer is thick, which usually happens in non-aqueous systems. This is easily understood, as the overlapping of double layers may cause the dramatic viscosity increase. When the double layer is thin, i.e., κr is large, the relative viscosity becomes insensitive to κr. For clearly showing the difference between different particle volume fractions, a small scale graph of Figure 17 is plotted in Figure 18. When the double layer becomes thick, the particle volume fraction takes over κr and has a major effect on the viscosity.

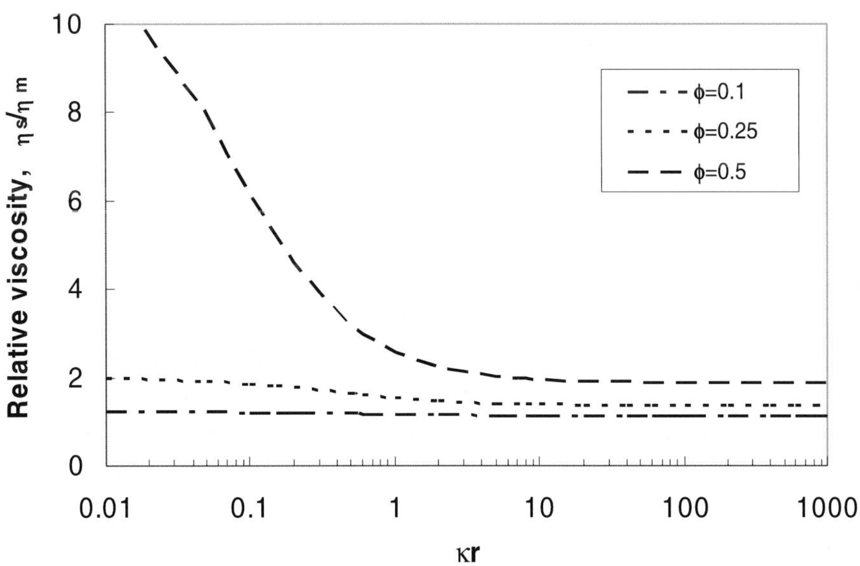

Figure 17 The calculated relative viscosity of a spherical colloidal suspension with Eq. (91). vs. κr at different particle volume fractions (large scale)

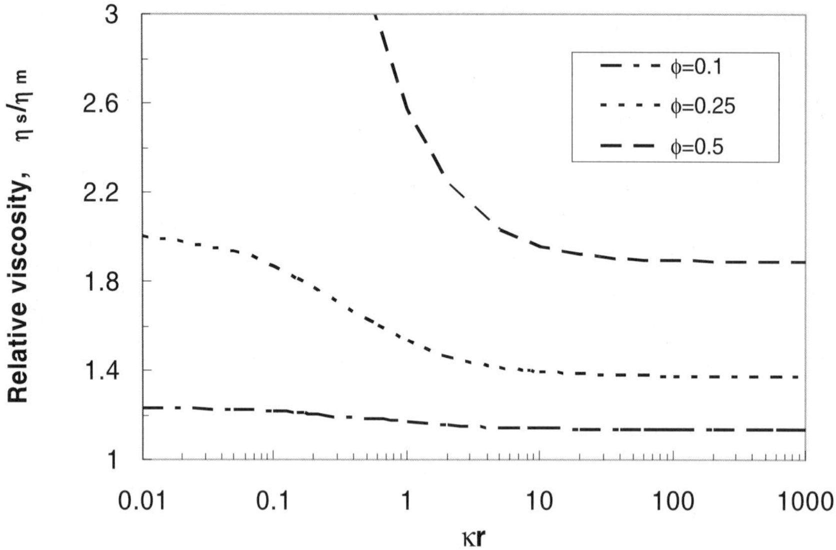

Figure 18 The calculated relative viscosity of a spherical colloidal suspension with Eq. (91). vs. κr at different particle volume fractions (small scale).

The relative viscosity decreases dramatically with the increase of κr when the particle volume fraction is large. When the particle volume fraction is low, the viscosity has less dependence on κr. This is because there are more chances for the double layers to overlap together when the double layers are thick and the particle volume fraction is large. This trend was found experimentally for the nano-sized polystyrene (r=100nm) particle/water suspension [48] with the particle volume fraction 40 vol%, in which the steady shear viscosity increases several orders of magnitude via decreasing the ionic strength from 1.8×10^{-2}M to that of deionized water. Since κr is directly proportional to the ionic strength, decreasing the ionic strength will decrease κr, making the double layer from thin to thick and thus increasing the viscosity of the whole system, according to Figures 17 and 18.

For accurate calculation of the viscosity the particle volume fraction dependence of the Zeta potential and the surface charge density must be taken into account. This means that Eq.(92) must be replaced by a more complicated equation, which can cover a wider range of Zeta potential, a wide range of κr, and a wide range of the particle volume fraction. For non-aqueous systems, the double layer is usually much thicker than that in aqueous systems.

2.2 Electroviscous effect of colloidal suspensions

The viscosity of a liquid containing charged particulate—the colloidal suspension system has been found to increase to a considerable extent even without an external electric field. This can be demonstrated by comparing Eq. (91), the viscosity equation for charged spherical colloidal suspensions, with Eq. (54), the viscosity equation for uncharged spherical colloidal suspensions. The difference is roughly a factor of $(\frac{rd_q}{\zeta\varepsilon_m\varepsilon_0})^3$. If $\frac{rd_q}{\zeta\varepsilon_m\varepsilon_0} > 1$, then a huge viscosity increase is expected due to the charged particle. This effect is historically called the electroviscous (EV) effect, as the electric field within an electric double layer is considerably strong. According to the Gouy-Chapman theory on the electric double layer [49], the surface potential ψ can be related to the ion strength as:

$$\left(\frac{d\psi}{dx}\right)^2 = \left(\frac{2k_B T n_\infty}{\varepsilon_m}\right)\left[\exp\left(-\frac{ze\psi}{2k_B T}\right) - \exp\left(\frac{ze\psi}{2k_B T}\right)\right]^2 \tag{93}$$

where n_∞ is the ion concentration far away from the surface, $d\psi/dx$ is the electric field at the distance x from the surface, z is the absolute value of the valence number. Table 3 shows the values of the electric field E ($=d\psi/dx$) estimated from Eq. (93) for a variety of ψ values and 1:1 electrolytes concentration. With the increase of the surface potential and ion concentration, the induced electric field substantially increases to such an extent that it is almost identical to a typical electric field applied for generating an ER effect. If there is a double layer overlapping between particles, the local electric field should be further strengthened, resulting in a different electroviscous behavior. This may be the reason that the electroviscous effect was initially classified into the primary EV effect, the secondary EV effect, and the tertiary EV effect. The primary EV effect refers to the viscosity increase of the charged colloidal system in which there is no double layer overlapping. The secondary EV effect refers to the viscosity increase of the charged colloidal system in which there is considerable double layer overlapping. The tertiary EV effect refers to the viscosity increase of the polyelectrolyte solution, the coagulated colloidal suspension, or the polymeric particle dispersion in which the polymer chain conformation may change due to the intramolecular double layer overlapping.

Table 3 The electric field (V/mm) calculated using Eq. (93) for various surface potential (close to the Zeta potential) and concentrations of 1:1 electrolytes. Reproduced with permission from P.C. Hiemenz and R. Rajagoplan, Principles of Colloids and Surfaces, Marcel Dekker, 1997, p557

ψ(mV)	$n_\infty=10^{-3}$ (mol/liter)	$n_\infty=10^{-2}$(mol/liter)	$n_\infty=10^{-1}$ (mol/liter)
50	6.36×10^{3} V/mm	2.01×10^{4} V/mm	6.36×10^{4} V/mm
100	1.98×10^{4} V/mm	6.24×10^{4} V/mm	1.98×10^{5} V/mm
150	5.49×10^{4} V/mm	1.74×10^{5} V/mm	5.49×10^{5} V/mm
200	1.51×10^{5} V/mm	4.77×10^{5} V/mm	1.51×10^{6} V/mm

A systematic review on the primary, secondary and tertiary electroviscous effect has been presented by Conway and Dobry-Duclaux [1] in 1960. A brief review on some of those effects has been given by Dukhin [50] and Saville [51], and a more unified review has been presented by Hunter [4]. In a dilute suspension, the apparent viscosity will increase with the particle volume fraction and the surface charge of the particle. A viscosity equation first published by Smoluchowski without proof [52] for describing such a system is

$$\eta = \eta_m \left\{ 1 + 2.5\phi \left[1 + \frac{1}{\sigma \eta_m r^2} \left(\frac{\zeta \varepsilon_m}{2\pi} \right)^2 \right] \right\} \tag{94}$$

where η represents the apparent viscosity of the suspension, η_m is the viscosity of the continuous phase (pure liquid), ϕ is the particle volume fraction, σ is the conductivity of the suspension, r is the radius of the particle, ζ is zeta potential of the particle, ε_m is the dielectric constant of the continuous phase. The derivation of Eq. (94) was subsequently made by Krasny-Ergen [53] with the numeric factor 3/2 in the numerator of the coefficient $(\zeta\varepsilon_m/2\pi)^2$. The Eq. (94) is valid under the presumptions that: a) the suspension is dilute enough (usually the particle volume fraction is less than 10%); b) there is no overlapping between the electric double layers (EDL). This means that Eq. (94) is derived for the primary electroviscous effect, stemming from the distortion of the electric double layer under an electric field and a shear field. If the suspension is concentrated, the electric double layers will probably overlap with each other, and the electric repulsive forces should be dominant. Eq. (94) cannot work for such a system. Note that Eq. (94) is remarkably similar to Einstein's equation. A more useful development has been made by Booth [54] without being limited by the assumption made by Smoluchowsky that the suspension is dilute enough that there is no overlapping between the double layers. Booth's equation is expressed as:

$$\eta = \eta_m \left\{ 1 + 2.5\phi \left[1 + \sum_{n=1}^{\infty} b_n \left(\frac{e\zeta}{k_B T} \right)^n \right] \right\} \quad (95)$$

where b_n is the coefficient of the nth term in $\frac{e\zeta}{k_B T}$. Mathematical development of Eq. (95) leads to [54]

$$\eta = \eta_m \left\{ 1 + 2.5\phi \left[1 + q^* \left(\frac{e\zeta}{k_B T} \right) z(b) \times (1+b)^2 \right] \right\} \quad (96)$$

where

$$q^* = \frac{\varepsilon_m k_B T \sum_1^i n_i z_i^2 u_i^{-1}}{\eta_m e^2 \sum_1^i n_i z_i^2} \quad (97)$$

in which u_i is the mobility of ions of type i, and b=κr. When b is large (thin double layer), z(b)= $\frac{3}{2}\pi b^4$, and Eq. (96) reduces to the Krasny-Ergen relation; When b is small (thick double layer)

$$z(b) = \frac{1}{200\pi b} + \frac{11b}{3200\pi} \quad (98)$$

Note that Eq. (96) is only applicable to the small values of ζ-potential ($\frac{e\zeta}{k_B T}\sim 1$) and small Peclet numbers. The Peclet number, Pe, compares the effect of convection (hydrodynamic) relative to diffusion, and can be defined as

$$\text{Pe}=\left(6\pi\eta_m r^3 \dot{\gamma}/k_B T\right)=r^2\dot{\gamma}/D \tag{99}$$

Where $\dot{\gamma}$ is the shear rate, and D is the diffusion coefficient. Note that other forms of Peclet numbers are also possible, depending on the type of convective flow studied. For example, for oscillatory flows the oscillatory frequency ω should replace the shear rate $\dot{\gamma}$ in Peclet equation (99). When Pe « 1, the distribution of the particle is only slightly disturbed by the flow and the rheological behavior of suspension should be dominated by the diffusional relaxation of the particle. If Pe » 1, the rheological behavior should be dominated by the hydrodynamic effect. If Pe is large, Pe « κr and κr »1, a modified equation of Eq. (96) can be expressed as [55]

$$\eta=\eta_m\left\{1+2.5\phi\left[1+\frac{1}{2\sigma\eta_m r^2}\left(\frac{\zeta\varepsilon_m}{2\pi}\right)^2\times\frac{1}{1+P_e^2}\right]\right\} \tag{100}$$

Eq. (100) is still only valid for small Zeta potentials. Eq. (100) indicates that the relative viscosity (η/η_m) may decrease with the increase of Pe, in other words, a shear thinning behavior is expected.

Direct comparison between theoretical predictions either from the Smoluchowsky-Krasny-Ergen or Booth's equations and experimental data is not possible, as it is difficult to fulfill the requirements of the Smoluchowsky-Krasny-Ergen theory, and the Booth's equation requires the knowledge of the individual mobilities of all ions. Under any approximation, a significant overestimation from both equations were found [56-58]. Note that Eq.(95) and (96) don't contain the electric field strength, which is only built up within the electric double layer. The value of κr thus has a critical effect on the viscosity of suspensions. Many other equations or theories [59-61] are only valid in certain κr ranges and very low particle volume fractions, in which the multiple-particle interaction is negligible.

At high volume fractions the multiple-particle and cooperative effects become significant, and κr has less importance to the viscosity of whole system. In the limit of low shear rates (Pe «1), Russell derived an equation [62]

$$\eta = \eta_m \left\{ \begin{array}{l} 1 + [\eta]\phi + 2.5(\phi[\eta])^2 + \\ \dfrac{3}{40} \ln\left(\dfrac{\alpha}{\ln \alpha}\right)\left[\ln\left(\dfrac{\alpha}{\ln(\alpha / \ln \alpha)}\right)\right]^4 \dfrac{\phi^2}{(\kappa r)^5} \cdots \end{array} \right\} \tag{101}$$

Where α is the ratio of electrostatic energy relative to the thermal energy and can be expressed as:

$$\alpha = \left(\frac{4\pi \varepsilon_m \zeta^2 \kappa r^2}{k_B T} \right) \exp(2\kappa r) \tag{102}$$

[η] is the intrinsic viscosity and ϕ is the particle volume fraction. Comparison with experimental results indicates Eq. (101) is valid at low particle volume fractions.

At high particle volume fractions there are considerable double layer overlappings in the system. Experimental evidences [29,63] suggest that the viscosity or yield stress of such a system decreases with the decrease of the particle volume fraction and with the increase of electrolyte concentration. It doesn't depend on the type and valency of ions.

Generally speaking, the viscosity increase due to the electroviscous effect is not very large, mostly within a factor of two. It is non-comparable to the ER effect, which would gives a ten-thousand times increase in the rheological property. Since the first and secondary electroviscous effects are actually observed without an external electric field in aqueous systems, those two topics will not be explored further.

In early 1939, Winslow began experimental research on the electric-field-induced viscosity increase in a suspension system made from solid semi-conducting particulates dispersed into a low viscosity and very high insulation oil. He first got a patent in 1947 [5] and then in 1949 reported his results in J. Appl. Phys.[64]. He found that a several hundred gram per square centimeter shear force could be obtained under an electric field strength of 3kV/mm. The solid particulate materials he used included starch, flour, gelatin, and limestone. The liquids he used included transformer oil, mineral oil, and silicone oil. He observed that under an electric field a fibrillation structure bridging between two electrodes is formed, and it has a high strength, resulting in the viscosity of the suspension to increase several orders of magnitude. The electric-field-induced effect he observed is much stronger than the so-called electroviscous effect, creating an entirely new research field. This is the reason that the ER effect is also called the "Winslow" effect (in some German literature, the ER effect is also called the Oppermann effect, as G. Oppermann was thought to be a pioneer in this field). Winslow is considered as the pioneer who first found that the fibrillated chain structure was formed in the ER suspensions and attributed the ER effect to the fibrillated structure formation, which becomes a distinctive feature of ER fluids from non-ER fluids. Before an electric field is applied, the particulate in the suspension is randomly distributed, while after an electric field is applied the particulate is orientated along the direction of the applied electric field. The time scale involved in the microstructure change from the disordered state to an ordered state determines the response time of the ER effect. Besides these milestone findings, Winslow also proposed many potential ER devices, such as clutches, brakes, and valves. However, his research did not attract much attention at that time. There were almost no ER related publications until 1960. In the early 1960's, Deneiga [65] investigated how an electric field would change the rheological property of a suspension. Extensive studies were carried out by Klass [66, 67] in 1967, and the dielectric tool was first used to characterize the ER effect, and the polarization of the particle was linked to the ER effect. The distorted electric double layer model was proposed to explain the ER effect. In the early 1970s, the electric double layer model was further extended by Uejima [68] and later in 1984 the proton polarization model, based on the electric double layer model, was proposed by Deinega [69]. Since water is important in the electric double layer formation, Stangroom [70,71] extended the electric double layer model to such an extent that water was regarded as necessary material for the ER effect.

ER research became active at the beginning of 1980, and many publications appeared at that time addressed both basic academic interests and potential industrial applications. ER technology was believed to bring a revolutionary innovation to many industrial areas, especially to the automotive industry. Efforts were made to elucidate the mechanism of the

ER effect and to make high performance ER fluids, which meets the industrial requirements.

3. POLYMERS AND POLYELECTROLYTE SOLUTIONS

This section could have followed section 1 in this chapter. The reason for placing it here is that the concepts and derivations developed in section 2 are needed for deriving the viscosity equation for polymeric electrolytes (polymer solutions) and polymer melts. Similar to the previous two sections, the viscosity of polyelectrolytes and polymer melts without an external electric field are discussed first, and then the viscosity of those materials under an external electric field or having strong surface charges are focused upon thereafter.

3.1 Viscosity of the polyelectrolyte and polymer melt (the free volume of polymeric materials)

The free volume concept has been successfully used for describing the glass transition phenomenon and the Fick's law of diffusion for polymers. One of such successful examples is the Williams-Landel-Ferry(WLF) equation [72,73], which provides the relationship between the time-temperature superposition shift factor and temperature. The free volume is considered to be the "holes" resulting from the packing void or irregularity of polymeric molecules. This concept will be continuously used in this section for deriving the viscosity equation of both polymer melts and polymer solutions.

3.1.1 Viscosity equation of the polymer melt

According to the reptation theory [74,75], the polymer chains are confined laterally to a tube-like region. The chains can only relax by sliding back and forth along the tube like a snake, and cannot cross the wall of the tube. Based on this physical picture of a polymer chain in liquid state, the volume of the reptation tube is presumably considered to be the free volume of a individual chain that can occupy. As illustrated in Figure 19, the volume of the tube, V_{tube},

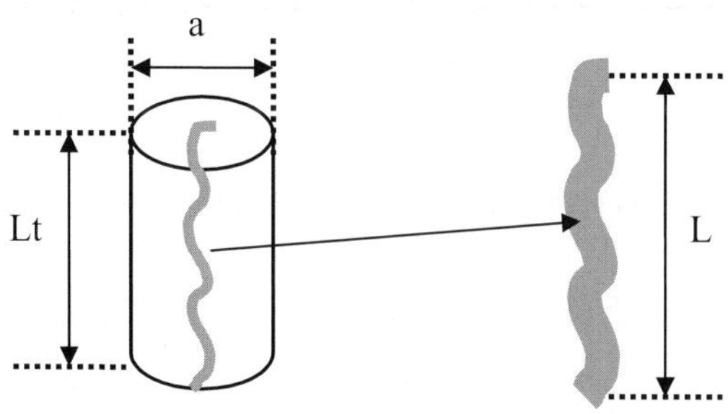

Figure 19 Schematic illustration of the reptation of a polymer chain.

$$V_{tube} = \frac{\pi a^2}{4} \times L_t \tag{103}$$

Where a is the tube diameter, and L_t is the contour length (or primitive path) of the tube. In Doi and Edwards's reptation theory [76], the tube diameter, a, is assumed to be independent of the molecular weight, M, and the contour length is directly proportional to M. A different approach is used here for calculating the free volume of individual polymer chain. Since the confinement tube is formed due to the neighboring entanglement, the more entanglement contacts (or called temporary cross-linked points) in the system will definitely reduce the diameter of such a tube. Also, the squeezing force acts three dimensionally and the diameter reduction occurs two dimensionally. The tube diameter should reduce at a faster pace. Instead of assuming that the diameter decreases linearly with the number of entanglement contacts, the 3/2 power relationship is proposed:

$$a = \frac{a_0}{(M / M_e)^{3/2}} \tag{104}$$

Where M_e is the molecular weight between entanglement (usually one third of the critical molecular weight for the entanglement formation), thus M/M_e roughly represents the number of entanglement steps. The a_0 is the diameter when the molecular weight is identical to the critical molecular weight or M_e, the largest tube diameter. When the entanglement happens, the contour length of a tube will reduce one-dimensionally. Suppose that the entanglement occurs three-dimensionally, the length of the tube should thus be proportional to the cubic power of the number of entanglement steps:

$$L_t = \frac{L_0}{(M/M_e)^3} \tag{105}$$

Where L_0 is the chain length at $M=M_e$. Under those assumptions, Eq. (103) can be rewritten as:

$$V_{tube} = \frac{\pi a_0^2 L_0 M_e^6}{4M^6} \tag{106}$$

The volume of the tube should be the free volume of an individual chain. According to Eq. (30), the viscosity of such an entanglement system, η_s, can be expressed as:

$$\eta_s = \frac{f(2\pi M k_B T)^{1/2}}{2\Delta E} \bullet \left(\frac{\pi a_0^2 L_0 M_e^6}{4M^6} \right)^{-2/3} e^{E_0/k_B T} \tag{107}$$

The shear force, f, multiplied by the molar volume V_{mol} and divided by Avogadro number N_A , should be the energy applied to each chain, ΔE

$$\frac{f}{\Delta E} = \frac{N_A}{V_{mol}} = \frac{N_A \rho}{M} \tag{108}$$

where ρ is the bulk density of the polymer. So Eq. (107) can be re-written as:

$$\eta_s = 0.34 N_A \rho (k_B T)^{1/2} \left(a_0^2 L_0 M_e^6 \right)^{-2/3} M^{3.5} e^{E_0/k_B T} \tag{109}$$

Note that Eq. (109) is only valid when $M>M_e$. Eq. (109) indicates that the viscosity of a polymer melt increases with the 3.5 power of the molecular weight, which has been confirmed with many experimental results [77-80]. Eq. (109) also indicates the viscosity of a polymer melt follows the Arrhenius equation:

$$\eta_m = k e^{E_0/k_B T} \tag{110}$$

with

$$k=0.34N_A\rho(k_BT)^{1/2}\left(a_0^2L_0M_e^6\right)^{-2/3}M^{3.5} \tag{111}$$

k is a constant at given temperature. It is proportional to $T^{1/2}$, containing the information of polymer molecular weight, bulk density, chain diameter, and length at the critical molecular weight, M_c.

When $M<M_c$, there is no entanglement in the system, thus there is no reptation tube formed. However, one may still assume that there is a confinement region, formed by the neighboring chains and similar to the reptation tube that scales the free volume of the chain. The diameter of such a confinement region should continuously increase with the decrease of the molecular weight, as shown in Eq. (104). The only difference should be the contour length of such a confinement. Since there is no entanglement, the length of such a confinement should be proportional to the molecular weight,

$$L = L_0 \bullet \frac{M}{M_e} \tag{112}$$

The volume of such a confinement region, V_c, is

$$V_c = \frac{\pi a_0^2 L_0 M_e^2}{4M^2} \tag{113}$$

Again using the Eq. (30), the viscosity equation of a small molecular weight polymer is

$$\eta_s = 0.34N_A\rho(k_BT)^{1/2}\left(a_0^2L_0M_e^2\right)^{-2/3}M^{0.83}e^{E_0/k_BT} \tag{114}$$

Eq. (114) indicates that the viscosity of a small molecular weight polymer is proportional to the 0.83 power of the molecular weight. Experimental results show the viscosity is roughly proportional to the first power of the molecular weight.

3.1.2 Viscosity equation of the polymer solution

As shown earlier, for the viscosity of two-component systems there are two ways to derive the viscosity equation. One approach is to use Eq. (30), directly correlating the viscosity to the free volume of the dispersed phase on the basis of Eyring's rate theory; Another approach is to use Eq. (48), assuming that the viscosity is proportional to the –n power of the free volume on the basis of Einstein's equation.

3.1.2.1 The viscosity equation derived from Eyring's rate theory

3.1.2.1.1. Theta condition

The viscosity of a polymer solution, a polymer dissolved either in water or an organic solvent, shows a similar rheological behavior to polymer melt when the solution is concentrated. The concentrated solution means that the polymer chains start to interact each other and there are entanglements between chains. By either varying the temperature or concentration, a polymer solution can move from a well-dissolved state (good solvent) to theta state, and finally to a poor solvent (globule) state. In the theta condition or charged polymer chain condition [81], the polymer chains form coils, and can be treated as spherical particles. Flory defined the Theta condition (or called Theta temperature or Flory temperature) as " the critical miscibility temperature in the limit of infinite molecular weight", in which the gyration radius approximates that of the bulk polymer, and the mass-fractal dimension of coil becomes 2. In a good solvent, the polymer chain expands as longer as possible, and the mass-fractal dimension of coil becomes 5/3. Theta condition is treated in this section at first.

Since both the concentration and molecular weight may change the chain entanglement status, the critical molecular weight of a polymer solution, M_{csol}, can be expressed as [80].

$$M_{csol} = \frac{\rho M_c}{c} \tag{115}$$

Where ρ is the bulk density of the polymer, and c is the concentration in g/ml. The number of entanglement contacts, n_e, can be assumed to be proportional to the number of entanglement steps as:

$$n_e = \left[\frac{M}{M_{csol}}\right]^2 = \left[\frac{cM}{\rho M_c}\right]^2 \tag{116}$$

assuming that a two-dimensional entanglement network is formed. The volume fraction concentration, ϕ, has a relationship with c, $\phi = c/\rho$. According to the random-walk statistics theory [82], the time-averaged mean-square distance $<R^2>_0$ is:

$$< R^2 >_0 = N_k b_k^2 \tag{117}$$

Where N_k is the number of effective Kuhn steps, and the b_k is the length of each Kuhn step. The gyration radius of R_g:

$$R_g = \frac{\left\langle R^2 \right\rangle_0^{1/2}}{\sqrt{6}} = \frac{N_k^{1/2} b_k}{\sqrt{6}} \tag{118}$$

For polymer chains, the mass-fractal dimension is 5/3, independent of the goodness of solvent [83]. The length of each Kuhn step should be related to the molecular weight. The more entanglement contacts a polymer chain has, the shorter the Kuhn step size is. Assuming

$$b_k = \frac{b_0}{(M/M_{csol})^{5/3}} \tag{119}$$

and $N_k \approx n_e$, Thus Eq. (118) can be rewritten as

$$R_g = 0.41 b_0 \left(\frac{\rho M_c}{cM} \right)^{2/3} \tag{120}$$

By taking R_g as the radius of spherical coil, the free volume of such a single polymeric coil is:

$$V_{fp} = 64 \left[\left(\frac{\phi_m}{\phi} \right)^{1/3} - 1 \right]^3 \bullet r^3$$
$$= 4.41 b_0^3 (\rho M_c)^2 M^{-2} c^{-3} \left[(\rho \phi_m)^{1/3} - c^{1/3} \right]^3 \tag{121}$$

Eq. (121) comes from the definition of inter-particle spacing, IPS, shown earlier. Now there are two ways to correlate the viscosity with the free volume, as demonstrated in previous sections. Only one way of using Eq. (30) is shown below.

$$\eta_s = \frac{f(2\pi m k_B T)^{1/2}}{2\Delta E} \bullet \left(V_{fp}\right)^{-2/3} e^{E_0/k_B T} \tag{122}$$

where m is the molar mass of the polymer solution. Since the molar mass of a polymer is much larger than that of the solvent, the molar mass of the polymer solution can be approximated as:

$$m \approx \phi M = \frac{cM}{\rho} \tag{123}$$

Assuming that the density of the polymer solution is close to 1, then

$$\frac{f}{\Delta E} = \frac{N_A}{V_{mol}} \approx \frac{N_A \rho}{cM} \tag{124}$$

Thus Eq. (122) becomes:

$$\eta_s = 0.46\left(k_B T\right)^{1/2} \rho^{-1/3} N_A b_0^{-2} M_c^{-4/3} M^{5/6} c^{3/2} \times \left[\left(\rho\phi_m\right)^{1/3} - c^{1/3}\right]^{-2} e^{E_0/k_B T} \tag{125}$$

Eq. (125) clearly indicates that the viscosity is proportional to the 0.83 power of molecular weight in a concentrated polymer solution. The intrinsic viscosity (the relative viscosity minus one divided by the concentration at the dilute limit, $[\eta] = \lim_{c \to 0}(\eta/\eta_m - 1)/c$, where η_m is the viscosity of medium) has been experimentally found to have a relationship with molecular weight of the power ranging between 0.15 (stiff rod) to 0.80 (compact coil), depending on the conformation of the polymer coils. The variation may come from the term $M_c^{-4/3}$, which vary with polymer used in the solution. When the concentration is not very high, $c^{1/3} << (\rho\phi_m)^{1/3}$, the viscosity definitely shows the 3/2 power of the concentration, which was found experimentally [84] and explained with the aid of the dynamic scaling law [85].

For a diluted solution, the conformation of the chain will be that of self-avoiding walk, rather than the random walk discussed above. In other words, the chain must avoid self-intersection from one to another [86]. This is also called long range excluded-volume effect. In this case,

$$R_g = \frac{\langle R^2 \rangle_0^{1/2}}{\sqrt{6}} = \frac{N_k^{3/5} b_k}{\sqrt{6}} \tag{126}$$

Unlike in the concentrated case, the Kuhn number and the Kuhn length should not change with the concentration of polymer. Also, the Kuhn number is assumed to linearly change with the molecular weight in a diluted solution. It is assumed to be the square root of the Kuhn number in the concentrated case. So

$$N_k = \left[\frac{M}{M_e}\right] \tag{127}$$

The Kuhn length should be less dependent on the molecular weight, assuming to be the square root of the Kuhn length in the concentrated case, too,

$$b_k = \frac{b_0^{1/2}}{(M / M_e)^{5/6}} \tag{128}$$

So the gyration radius

$$R_g = 0.41 b_0^{1/2} \left(\frac{M_c}{M}\right)^{0.23} \tag{129}$$

Accordingly the free volume of each polymer coil

$$V_{fp} = 64\left[\left(\frac{\phi_m}{\phi}\right)^{1/3} - 1\right]^3 \bullet r^3$$
$$= 4.41 b_0^{3/2} M_e^{0.7} M^{-0.7} c^{-1} \left[(\rho\phi_m)^{1/3} - c^{1/3}\right]^3 \tag{130}$$

The viscosity of a diluted polymer solution should be

$$\eta_s = 0.46(k_B T)^{1/2} N_A \rho_s M_s^{-1/2} b_0^{-1} M_c^{-0.47} M^{0.47} c^{2/3} \times \left[(\rho\phi_m)^{1/3} - c^{1/3}\right]^{-2} e^{E_0/k_B T} \quad (131)$$

Eq. (131) shows the viscosity is proportional to the 0.67 power of the concentration and 0.47 power of the molecular weight when there is no entanglement in the solution. Experimental data shows the relative viscosity should be proportional to the 0.5 power of concentration [87, 88]. The viscosity (not the intrinsic viscosity) only shows the 0.5 power of the molecular weight when the polymer concentration is low, which is in agreement with the experimental results [89].

3.1.2.1.2 Good solvent

In a good solvent, the polymer chains should expand and form as many contacts as possible with solvent molecules. It would be hard, though it is possible, to continuously treat the polymer chain as a sphere in a good solvent. Two approaches thus can be used for deriving the viscosity equation. The first approach is similar to the way of treating polymer melt shown previously. The only difference from the polymer melt is that the high enough concentration is able to induce the entanglement, as assumed in the theta solvent case. The second approach is to treat the polymer chain as a sphere with the gyration radius depending on the molecular weight and concentration. This approach is obviously not preferable, and only the first approach is shown below, which is exactly analogous to the way of dealing with the polymer melt. According to Eq. (115), Eq. (104) and (105) can be rewritten as:

$$a = \frac{a_0 (\rho M_c)^{3/2}}{(cM)^{3/2}} \quad (132)$$

$$L = \frac{L_0 (\rho M_c)^3}{(cM)^3} \quad (133)$$

Analogous to the polymer melt, the viscosity of the polymer solution in a good solvent should be

$$\eta_s = \frac{f(2\pi M k_B T)^{1/2}}{2\Delta E} \bullet \left(\frac{\pi a_0^2 L_0 \rho^6 M_c^6}{4c^6 M^6} \right)^{-2/3} e^{E_0 / k_B T} \tag{134}$$

Substituting Eq. (124) into Eq. (134) leads to

$$\eta_s = 0.34 N_A \rho (k_B T)^{1/2} \left(a_0^2 L_0 \rho^6 M_c^6 \right)^{-2/3} (cM)^{3.5} e^{E_0 / k_B T} \tag{135}$$

Eq. (135) indicates that in a good solvent and entanglement region, the viscosity has a 3.5 power of cM. Similarly, below the entanglement region,

$$\eta_s = 0.34 N_A \rho (k_B T)^{1/2} \left(a_0^2 L_0 \rho^2 M_c^2 \right)^{-2/3} (cM)^{0.83} e^{E_0 / k_B T} \tag{136}$$

The viscosity has the 0.83 power of cM. All those predictions are in agreement with the experimental results [90]

3.1.2.2 The viscosity equation derived from Einstein's equation

In a polymer solution, no matter whether it is in a Theta solvent or a good solvent condition, the polymer chain could be assumed as a fiber and thus Eq. (54) could be used to predict the viscosity of the solution. The solvent condition changes n value, so does the relationship between the viscosity and the polymer volume fraction. When n≠1, n≠2, and n≠3, Eq. (54) shows

$$\ln \frac{\eta_s}{\eta_m} = \frac{7.5}{15.29^n} \left[\begin{array}{l} -\dfrac{(\phi_m^{1/3} - \phi^{1/3})^{3-3n}}{3-3n} + \dfrac{2\phi_m^{1/3}(\phi_m^{1/3} - \phi^{1/3})^{2-3n}}{2-3n} - \\ \dfrac{\phi_m^{2/3}(\phi_m^{1/3} - \phi^{1/3})^{1-3n}}{1-3n} + \dfrac{2\phi_m^{1-n}}{(1-3n)(2-3n)(3-3n)} \end{array} \right] \tag{137}$$

If n is defined as 0.3 divided by the square of the circularity of the polymer chain, the viscosity of a polymer solution against polymer concentration at different n value is shown in Figure 20. Clearly, the parameter n corresponds to the aspect ratio of polymer chains shown in the table 4. The large n value corresponds to the large aspect ratio of polymer chain. As shown in Figure 12, the abrupt increase of the relative viscosity appears at low volume fraction when the aspect ratio of the polymer chain becomes large. For example, when the polymer chain has the aspect ratio of 50, the abrupt change of the relative viscosity starts at the polymer volume fraction around 15 vol%. While when the polymer chain has the aspect ratio of 500, the abrupt change of the relative viscosity starts at the polymer volume fraction around 10 vol%.

Table 4 The parameter n and the aspect ratio of the polymer chain

n	4.97	9.75	19.3	47.96	95.73
Circularity	0.246	0.175	0.125	0.079	0.056
Aspect ratio	50	100	200	500	1000

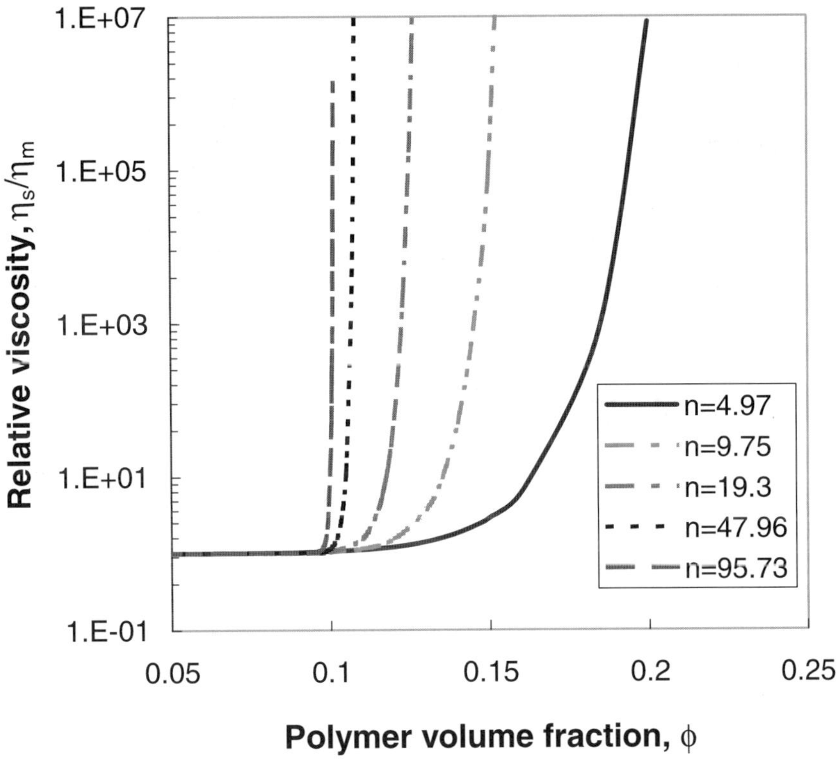

Figure 20 The relative viscosity of a polymer solution vs. polymer volume fraction at different aspect ratio of the polymer chain. A dense random packing structure is assumed and ϕ_m=0.63. The parameter n is defined as 0.3 divided by the square of the circularity of a polymer chain. Calculated with Eq. (137)

Before this "critical" volume fraction value, the relative viscosity gradually increases with the polymer volume fraction. The relative viscosity is also plotted against the parameter n at the polymer volume fraction 10.8 vol% in Figure 21. It increases with the parameter n increasing and the abrupt change of the relative viscosity occurs around n=40. For clearly showing how the relative viscosity changes with the parameter nbelow 40, the enlarged plot is shown in Figure 22. The relative viscosity increases very slowly with n when n is less than 30, which indicates that the aspect ratio of the polymer chain has a very pronounced effect on the relative viscosity once n is larger than 30. Since the aspect ratio relates to the molecular weight and molecular architecture, Figures 21 and 22 present an idea on how the relative viscosity changes with the molecular weight and molecular architecture. Generally speaking, for a linear polymer, the larger the molecular weight is, the longer the polymer chain, and thus the larger the aspect ratio. The relative viscosity will abruptly increase with polymer volume fraction even at a relative low value. For a polymer with a large side chain (long branch or a star-shaped chain), the aspect ratio of such a chain is small, and the abrupt increase of the relative viscosity will occur at a relative high volume fraction. Those derivations are qualitatively consistent with the experimental results. Branches that are long but are still shorter than those required for entanglements were found to decrease the zero-shear viscosity in comparison with a linear polymer of the same molecular weight [91-93]. The zero-shear viscosity of four-branched star polystyrenes is lower than that of a linear polymer of comparable molecular weight, however, it increases more rapidly with the molecular weight as the length of the branches increases [94]. The parameter n should be a function of the molecular weight, the number of entanglement constants, and the molecular architecture.

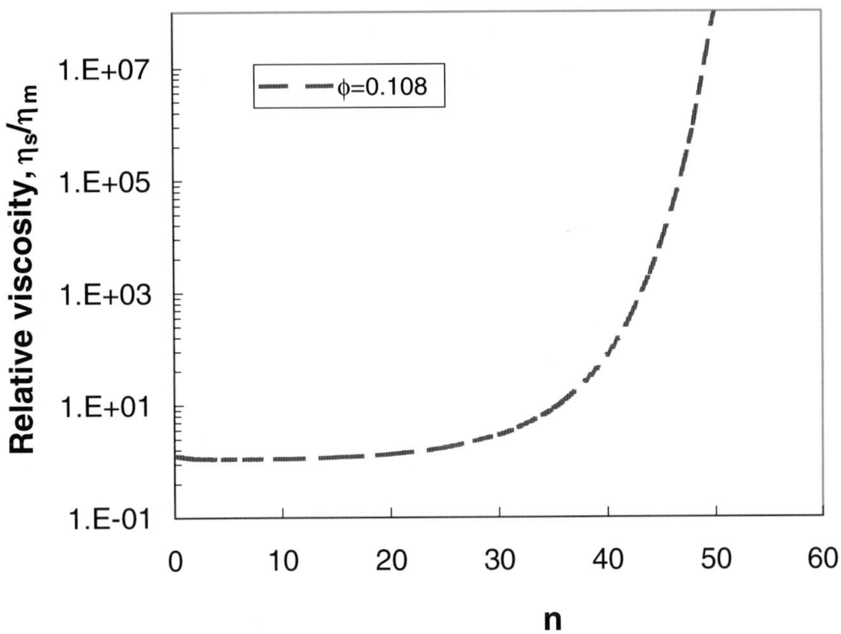

Figure 21 The relative viscosity of a polymer solution vs. the parameter n at polymer volume fraction 10.8 vol%. A dense random packing structure is assumed and ϕ_m=0.63.

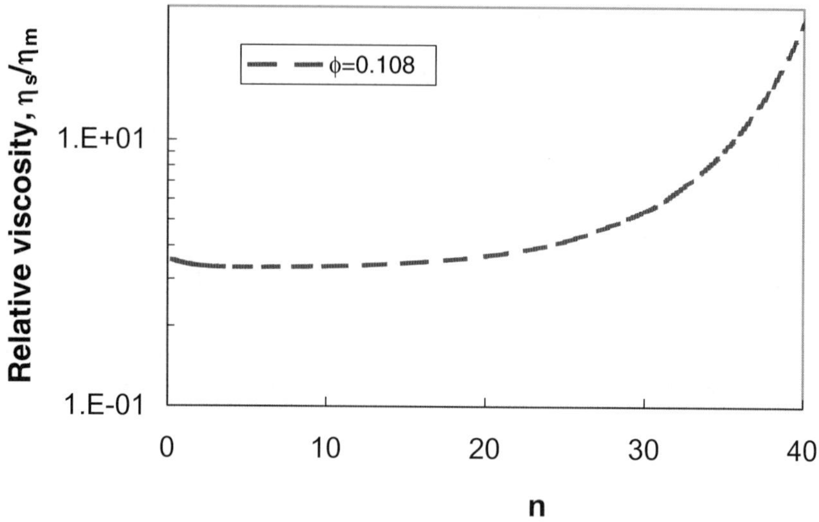

Figure 22 The relative viscosity of a polymer solution vs. the parameter n (n<40) at polymer volume fraction 10.8 vol%. A dense random packing structure is assumed and ϕ_m=0.63.

3.2 The electroviscous effect of polyelectrolytes

After the electroviscous effect was discovered in pure liquids, many solutions containing simple electrolytes were comprehensively investigated. Note that there is no an external electric field applied to the electrolyte systems, as most of such systems are aqueous solutions, which are unable to afford for a high electric field. Poiseuille [95] was the first to observe that the viscosity of electrolytic solutions differs from that of the solvents. Further work was carried out by Jones [96], and Falkenhagen [97]. It was found that the electroviscous effect of an electrolyte solution is much stronger than the effect observed in the pure liquids. According to Jones [98], the viscosity of an electrolytic solution can be represented as:

$$\eta = \eta_m \left[1 + A\sqrt{c} + \left(A^2 - B\right)c + \ldots\right] \tag{138}$$

Where A is zero for nonelectrolytes and positive for all strong electrolytes, and B can be positive or negative but is negative for most salts. Rearranging Eq. (138) leads to

$$\frac{\eta/\eta_m - 1}{c} = cons\tan t + A/\sqrt{c} \tag{139}$$

Experimental data support Eq.(139), as shown in Figure 23. The normalized viscosity decreases with the increase of the concentration, as predicted by Eq. (139).

A general theory of the viscosity of electrolytes was developed by Onsager [99], and the "ion atmosphere" concept was proposed to explain the viscosity increase observed in the electrolyte solutions. Each ion in an ionic solution is surrounded by an atmosphere of ions having a net charge of opposite sign to that of central ion. The distortion of the ion atmosphere due to the overlapping between ion atmospheres may generate more drag force in a shear field, thus leading to the increase of the apparent viscosity.

For the polyelectrolyte solutions, the electroviscous effect actually refers to the viscosity enhancement observed without an external electric field. This issue is already addressed in the section 3.1. Under a strong electric field, a polar polymer dispersed in a non-polar solvent [100] should show a similar behavior as a liquid does, i.e., the apparent viscosity increases proportionally to the square of the applied electric field strength. The rheological properties of polymer-like inverse micelle system, soybean lecithin dispersed in n-decane [101] under an electric field as shown in Figure 24, indicate that the micelle dimension may increase in the presence of an electric field, resulting in the viscosity increase in low frequency shears. However, at relatively high shear frequency, the viscosity under 2 kV/mm is less than that under zero electric field. Detailed discussion on the polyelectrolyte ER systems will be addressed late, included in the homogeneous ER systems that have attracted a huge attention during the late 1990's.

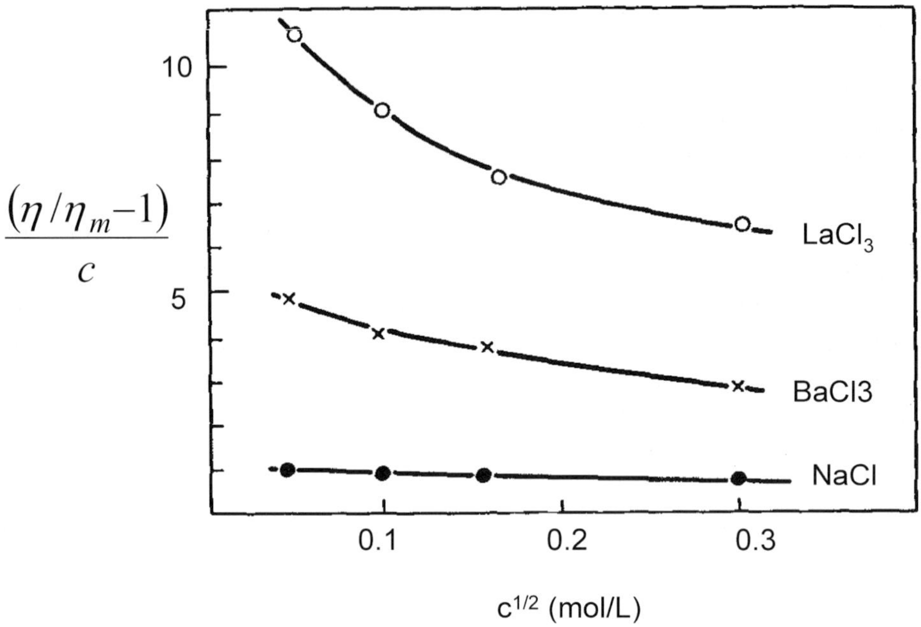

Figure 23 Normalized viscosity $(\eta/\eta_m\text{-}1)/c$ vs. the square root of the electrolyte concentration $c^{1/2}$. Reproduced with permission from G. Jones, and M. Dole, J. Am. Chem. Soc., 51(1929)2950.

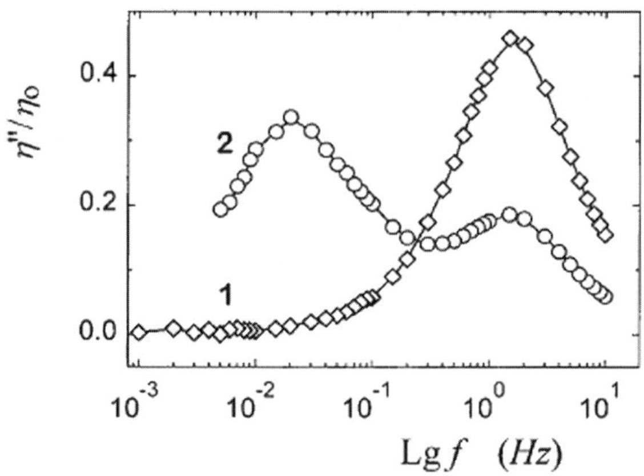

Figure 24 The imaginary part of a complex viscosity, normalized by the zero shear viscosity, vs. the mechanical oscillation frequency without an applied electric field (1) and at 2000 V/mm strength (2) for an organogel contains 250 mg/mL lecithin and 0.9 mol of glycerol per mole of lecithin in *n*-decane. Reproduced with permission from Yu. A. Shchipunov, T. Dürrschmidt, and H. Hoffmann, Langmuir 16(2000)297

4 CONCLUDING REMARKS

The viscosities of liquids, colloidal suspensions and polyelectrolytes or polymeric systems with and without an external electric field can be well described with the free volume concept and the derived viscosity equations are remarkably consistent with experimental results. The main topic of this book is the ER effect of ER fluids, which typically operate under an external electric field. The electroviscous effect that generally doesn't need an external electric field is only briefly covered. Attention is paid to deriving a more universal viscosity equation that can account for colloidal suspensions, pure liquids, and polymeric systems. The free volume concept has proved to be extremely important for such a task.

REFERENCES

[1] B.E. Conway, and A. Dobry-Duclaux, in "Rheology: Theory and applications", F.R. Eirich, ed., Vol.3, page 83, Academic Press, New York, 1960.
[2] J.W. Goodwin, Colloid Science: Specialist Periodical Reports, Vol.2, pp. 246-293, Chem. Soc., London, 1975
[3] D.A. Saville, Ann. Rev. Fluid Mech. 9(1977)321
[4] R.J. Hunter, Zeta Potential in Colloidal Science, Academic Press, New York, 1981
[5] W.M. Winslow, U.S. Patent 2417850, 1947
[6] E.N. da C. Andrade, Phil. Mag.17 (1934)497
[7] S. Glasstone, K.Laidler, and H. Eyring, The Theory of Rate Process, McGraw-Hill, New York and London, 1941.
[8] H. Eyring and J.O. Hirschfelder, J.Phys.Chem., 41(1937)249
[9] A. W. Duff, Phys. Rev., 4(1896) 23
[10] E.N. Andrade, C. Dodd, Nature, 143(1939)26
[11] E.N. Andrade, C. Dodd, Roy. Soc. London Proc. A 187(1946)296
[12] E.N. Andrade, C. Dodd, Roy. Soc. London Proc. A 204(1951)449
[13] E.N. Andrade, J. Hart,, Roy. Soc. London Proc. A 225(1954)463
[14] A.B. Metzner, J. Rheol., 29(1985) 739;
[15] P.M. Adler, A. Nadim, H. Brenner, Adv. Chem. Eng., 15(1990)1
[16] T. Hao, R.E. Riman, J. Colloid and Interface Sci.,2005, in press
[17] S. Kuwabara, J. Phys. Soc. Jpn. 14(1959) 527
[18] M. W., Kozak, and E. J., Davis, J. Colloid Interface Sci. 112 (1986)403
[19] S. Levine, and G. H., Neale, J. Colloid Interface Sci. 47(1974)520
[20] H. Ohshima, J. Colloid Interface Sci. 195(1997)137
[21] A. Ogawa, H. Yamada, S. Matsuda, K. Okajima, and M.Doi, J. Rheol., 41(3)(1997)769
[22] D.J. Cumberland, and R.J. Crawford, The Packing of Particles, Elsevier, Amsterdam, 1987
[23] I.M. Kreiger, and T.J. Dougherty, Trans. Soc Rheol., 3(1959)13

[24] N.A. Frankel, and A. Acrivos, Chem. Eng. Sci., 22(1967)847
[25] A. Einstein, Ann.Phys., 19(1906)289
[26] R.C. Ball and P. Richmond, Phys. Chem. Liq., 9(1980)99
[27] R.D. Sudduth, J. Appl. Polym. Sci., 48(1993)25
[28] M. Mooney, J. Colloid Sci., 6(1952)162
[29] H.A. Barnes, J.F. Hutton, and K. Walters, An Introduction to Rheology, Elsevier, Amsterdam, 1989
[30] R.D. Sudduth, J. Appl.Poly. Sci.,48(1993)37
[31] T. Hao and R.E. Riman, Particle Science and Technology, 21(2003)317
[32] T. Allen, 1997, Particle size measurement, 5th edition, Chapman & Hall, UK
[33] V. Mikli, H. Kaerdi, P. Kulu, amd M. Besterci, Proc. Estonia Acad. Sci. Eng., 7(2001)22
[34] G. Herdan, Small Particle Statistics, 2nd. Edn., Academic, New York, 1960
[35] H. Wadell, J. Geol., 40(1932)250
[36] H. Heywood, J. Pharm. Pharmacol, 15(1963)56
[37] A.B. Metzner, J. Rheol., 29(1985)739
[38] T. Kitano, T. Takaoka, and T. Shirota, Rheol. Acta, 20(1981)207
[39] R.R.Huilgol, and N. Phan-Thien, Fluid Mechanics of Viscoelasticity, Elsevier, Amsterdam, 1997
[40] L.G. Leal, and E.J. Hinch, Rheol.Acta, 12(1973)127
[41] F.P. Folgar, and C.L. Tucker, J. Reinforced Plastics and Composites, 3(1984)98
[42] B. Clarke, Trans.Inst.Chem.Eng., 45(1967)251
[43] H. Giesekus, Disperse systems: Dependence of Rheological Properties on the Type of Flow with Implications for Food Rheology, in" Physical Properties of Foods" R. Jowitt et. al. Ed, Applied Science Publishers, chapter 13, 1983
[44] B.V. Derjaguin, and L Landau, Acta Physicochim URSS, 10(1941)25
[45] E. J. W. Verwey, and J.Th.G. Overbeek, The Theory of the Stability of Lyophobic Colloids, Elsevier, Amsterdam, 1948;
[46] L.N. McCarthney, and S. Levine, J. Colloid Interface Sci., 30(1969)345
[47] W. B. Russell, D. A. Saville, and W.R. Schowalter, Colloidal Dispersion, Cambridge University Press, 1992
[48] I.M. Krieger, and M.Eguiluz, Trans.Soc.Rheol, 20(1976)29
[49] P.C. Hiemenz and R. Rajagoplan, Principles of Colloid and Surface, Marcel Dekker, 1997, p553
[50] S.S. Dukhin, and B.V. Deryaguin, in E. Matijevic. Ed., "Surface and Colloid Science", Vol.7, John Wiley, New York, 1974
[51] D.A.Saville, Ann. Rev. Fluid Mech., 9(1977)321
[52] M. von Smoluchowski, Kolloid-Z, 18(1916)190
[53] W. Krasny-Ergen, Kolloid-Z, 74(1936)74
[54] F. Booth, Proc. Roy. Soc., A203(1950)533
[55] W.B. Russel, J. Fluid Mech., 85(1978)673
[56] H.B. Bull, Trans. Faraday Soc., 36(1940)80
[57] D.R. Briggs, J. Phys. Chem., 45(1941)866
[58] A. Dory, J. Chim. Phys. 52(1955)809

[59] J. Stone-Masui, and A. Watillon, J. Colloid Interface Sci., 28(1968)187
[60] J. Stone-Masui, and A. Watillon, J. Colloid Interface Sci., 34(1970)327
[61] D. Stiger, J. Colloid Interface Sci., 23(1967)379
[62] W.B. Russel, J. Fluid. Mech., 85(1978)209
[63] J. G. Brodnyan, and E.L.Kelley, J. Colloid Sci., 20(1965)7;
[64] W.M. Winslow, J. Appl. Phys. 20(1949)1137
[65] Yu.F. Deinega, G.V. Vinogradov, Colloid J. 24 (1962)570
[66] D.L. Klass, and T.W. Martinek, J. Appl. Phys. 38 (1967) 67;
[67] D.L. Klass, T.W. Martinek, J. Appl. Phys. 38 (1967)75
[68] H. Uejima, Jpn. J. Appl. Phys. 11(1972)319
[69] Yu.F. Deinega, G.V. Vinogradov, Rheol. Acta 23 (1984) 636
[70] J.E. Stangroom, I. Harness, GB Patent 2153372, 1985
[71] J.E. Stangroom, Phys. Technol. 14 (1983)290
[72] M.L.Williams, R.F. Landel, and J.D. Ferry, J. Am. Chem. Soc., 77(1955)3701,
[73] J. D. Ferry, Viscoelastic properties of polymers, 3rd ed, Wiley, 1980, New York
[74] P.G. de Gennes, J. Chem.Phys., 55(1971)572;
[75] S.F. Edwards, Proc. Phys.Soc., 92(1967)9
[76] M.Doi, and S.F. Edwards, The Theory of Polymer Dynamics, Oxford Univ. Press, 1986
[77] T.G. Fox, and P.J.Flory, J.Polym.Sci., 14(1954)315
[78] A. Casale, R.S. Porter, and J.F. Johnson, J. Macromol. Sci-Rev, Macromol Chem., C5(1971)387;
[79] V.R.Rju, G.G. Smith, G.Marin, J.R.Knox, and W.W.Graessley, J. Polym. Sci: Polym. Phys. Ed., 17(1979)1183;
[80] R.K.Gupta, Polymer and Composite Rheology, 2nd Ed., Dekker., New York, 2000
[81] P.J. Flory, Principles of Polymer Chemistry, Cornell Univ. Press, Ithaca, NY 1953
[82] P.J.Flory, Statistical Mechanics of Chain Molecules, Carl Hanser Verlag, New York, 1969
[83] P.G. de Gennes, Scaling Concepts in Polymer Physics, Cornell Univ. Press, Ithaca, New York, 1979
[84] R.F. Prini, and A.E. Lagos, J. Polym. Sci., Part A 2(1964)2917]
[85] A.V.Dobrynin, R.H. Colby, and M.Rubinstein, Macromolecules, 28(1995)1859
[86] R.G. Larson, The structure of complex fluids, Oxford University Press, 1999
[87] R.M. Fuoss, Polyelectrolytes Disc.Faraday Soc., 11(1951)125
[88] R.F. Prini, and A.E. Lagos, J.Polym. Sci., Part A 2(1964)2917
[89] M. Kurata, W.H. Stockmayer, Polymer Handbook, ed. by J.Brandrup and E.H.Immergut, 3rd ed, Wiley, New York, 1989
[90] W.W. Graessley, Adv. Polym.Sci., 16(1974)1
[91] R.A.Mendelson, Polym.Eng.Sci., 9(1969)350
[92] R.A.Mendelson, W.A.Bowles, and F.A.Finger, J.Polym.Sci.: Part A 8(1970)105
[93] J.Miltz, A.Ram, Polym.Eng.Sci.,13(1973)273
[94] T.Masuda, Y.Ohta, and S. Onogi, Macromolecules, 4(1971)763
[95] J.L.M.Poiseuille, J. Chim.Phys., 21(1847)76
[96] G. Jones, and M.Dole, J. Am.Chem.Soc., 51(1929)2950

[97] H. Falkenhagen, Physik Z. 30(1929)611
[98] G.Jones, and M.Dole, J. Am.Chem.Soc., 51(1929)2950
[99] L. Onsager, R. M. Fuoss, J. Phys. Chem. 36(1932)2689
[100] N.Saito, and T.Kato, J. Phys. Soc.Japan, 12(1957)1393
[101] Yu. A. Shchipunov, T. Dürrschmidt, and H. Hoffmann, Langmuir 16(2000)297

Chapter 3

The positive, negative, photo-ER, and electromagnetorheological (EMR) effects

1. POSITIVE ER EFFECT

Since the innovation of the ER effect in 1947 by Winslow, attention has been paid on how to increase the viscosity or yield stress induced by the electric field. The fact that rheological properties increase with the applied electric field strength is termed the positive ER effect. This effect has been comprehensively addressed and is still receiving a large amount of effort. An example of the positive ER effect [1] is shown in Figure 1. The shear stress of zeolite/silicone oil suspension shows a stepwise increase when the applied electric field strength changes from zero to 2.0 kV/mm in a 0.5 kV/mm step. Once the applied electric field is turned off, the shear stress immediately falls to a value that is slightly higher than the original one without an electric field. The reason for this is that the suspended particles usually align along the applied electric field, forming fibrillated bridges that span between two electrodes. Even if the applied electric field is removed, the fibrillated structure remains, resulting in a relatively higher shear stress due to ordered particle structure, in comparison with the randomly distributed particle structure before the electric field is applied. Microstructure images of oxidized polyacrylonitrile particle dispersed in silicone oil obtained with an inverted optical microscope are shown in Figure 2. The fibrillated chains are observed to span from one electrode to another. Under an electric field of 1.0 kV/mm, the chains are thin and entangled, forming the tree-shape structure.

Once the electric field strength is increased from 1.0 to 1.5 kV/mm, two thicker columns build up with much thinner chains parallel to them. They become much closer if the electric field is further increased from 1.5 to 2.0 kV/mm, though there are still small amounts of much thinner chains available. Formation of the mechanically strong chains in ER suspensions is believed to be responsible for the sharp increase of the rheological properties of positive ER suspensions.

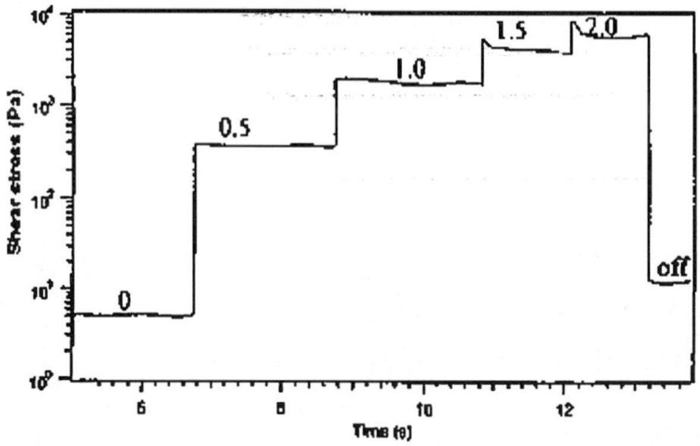

Figure 1 The recorded shear stress of a positive ER suspension (zeolite/silicone oil) against time. The numbers on each step indicate the applied electric field strength (kV/mm). Note that after the electric field is switched off, the shear stress doesn't fully recover to the original value. Reproduced with permission from T. Hao, Adv. Mater., 13(2001)1847

A good positive ER fluid should have: a) a high yield stress preferably equal to or larger than 5 kPa under an electric field of 2kV/mm; b) a low current density passing through the ER fluid preferably less than 20μA/cm^2; c) a wide working temperature range between –30—120°C; d) a short response time. The response time of an ER fluid scales at 10^{-3} second. For some specific purposes, an even faster response is required; e) high stability. The ER fluid should be chemically and physically stable. There should be no particle sedimentation and material degradation problems.

Before 1985, all positive ER fluids contained small amounts of water. Many shortcomings are pertinent to these systems, for example, narrow working temperature range due to water evaporation at high temperatures and icing at low temperatures; high current density due to the large conductivity of water; and device erosion caused by water, etc. A water-free acenequinone radical polymer (PAnQR) ER fluid was developed in 1985 [2]. This sort of polymer has a big-π electron structure as shown in Figure 3 and thus has a relatively high conductivity. The static yield stress (σ_0) of PAnQR (see Figure 3) in partially chlorinated petroleum fraction (Cereclor) [5] against the particle volume fraction is shown in Figure 4. This anhydrous type ER fluid was believed to be much more promising in comparison with hydrous ones. The static yield stress (the minimum stress required to cause the flow [3, 4] can reach close to 15 kPa at 3.6 kV/mm when the particle volume fraction is 35 vol%, as shown in Figure 4.

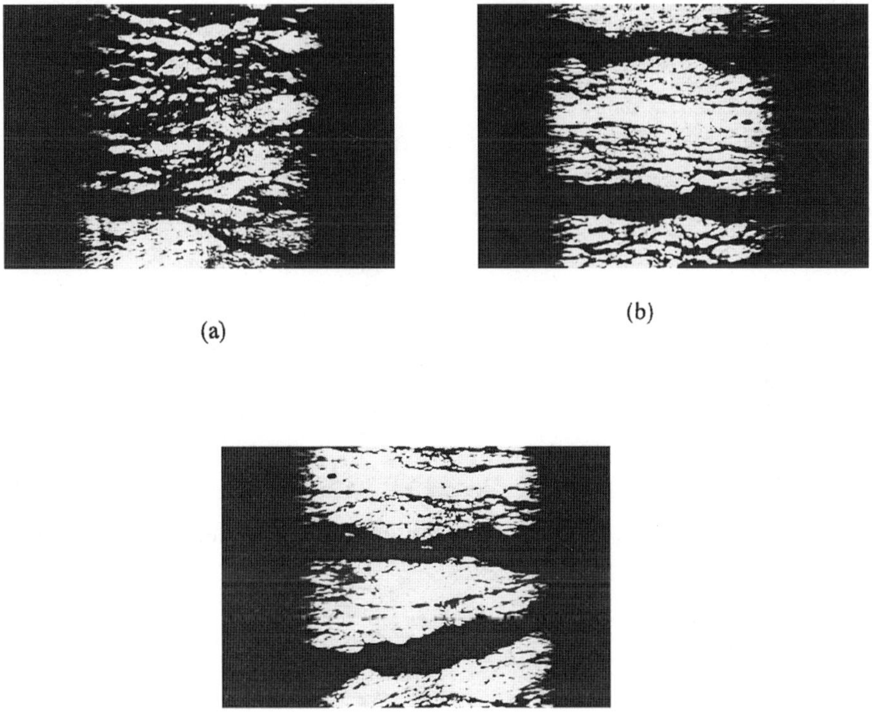

(a) (b) (c)

Figure 2 Microscopic images of oxidized polyacrylonitrile particles dispersed in silicone oil under different electric fields: (a) E=1.0 kV/mm, (b) E=1.5 kV/mm, (c) E=2.5 kV/mm. the left and right side stripes are electrodes and the gap between them are 1 mm. Reproduced with permission from T. Hao, and Y. Xu, J. Colloid Interface Sci., 181(1996)581.

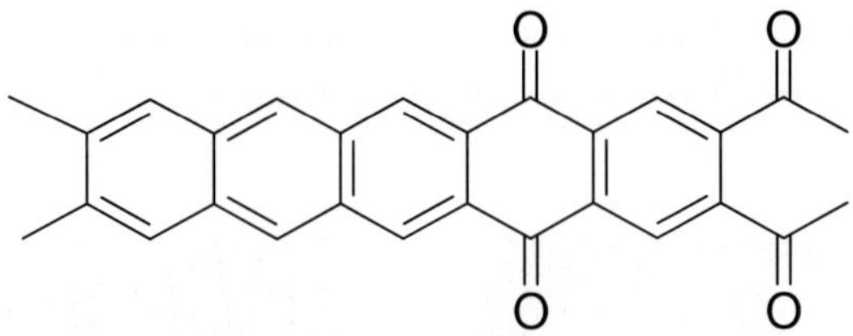

PAnQR

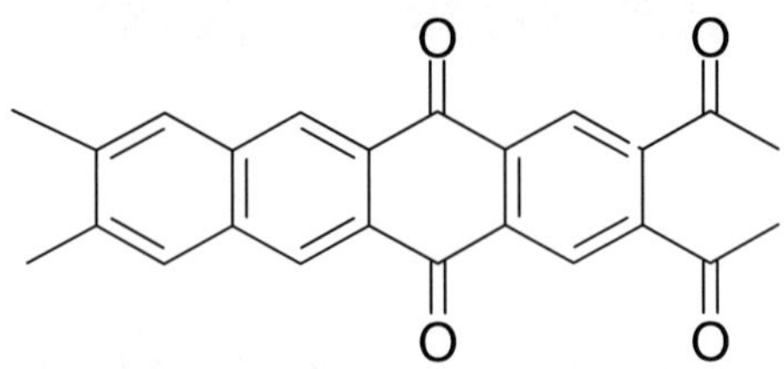

PNQR

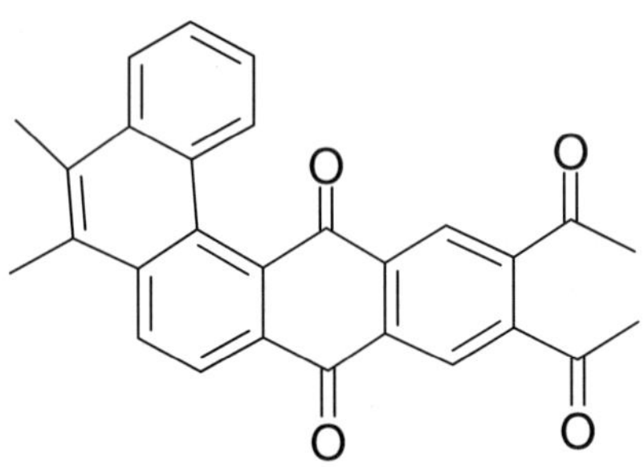

PPhQR

Figure 3 Semiconducting poly(acene quinine) radicals (PAQR) used for water-free ER fluids.

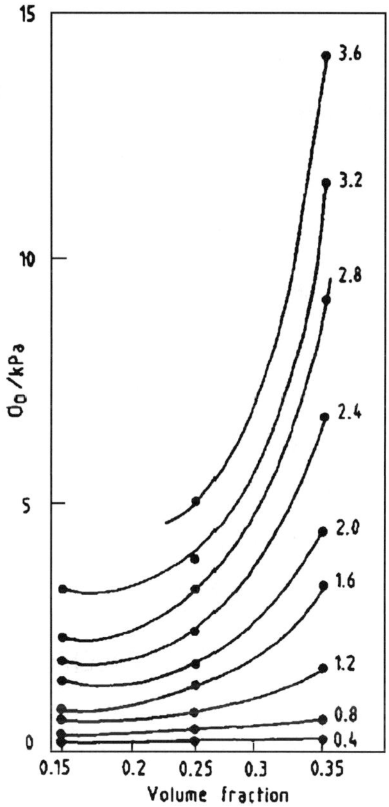

Figure 4 The static yield stress of PAnQR in Cereclor against particle volume fraction. The number indicated on each curve is the electric field (kV/mm). Reproduced with permission from ref. H. Block, J. P. Kelly, A. Qin, and T. Watson, Langmuir, 6(1990)6.

Aluminosilicate materials were found to have a very strong ER effect under water-free condition [6,7]. In 1991, a crystalline alumino-silicate (zeolite, Linde 3A) powder of molecular formula $K_9Na_3[(AlO_2)_{12}(SiO_2)_{12}]$ was dispersed into paraffin oil and the maximum stress of such an ER suspension was found to reach more than 100 kPa at 2.0kV/mm [8]. The yield stress under such a condition reaches 42.6 kPa. The maximum stress against strain amplitude is shown in Figure 5.

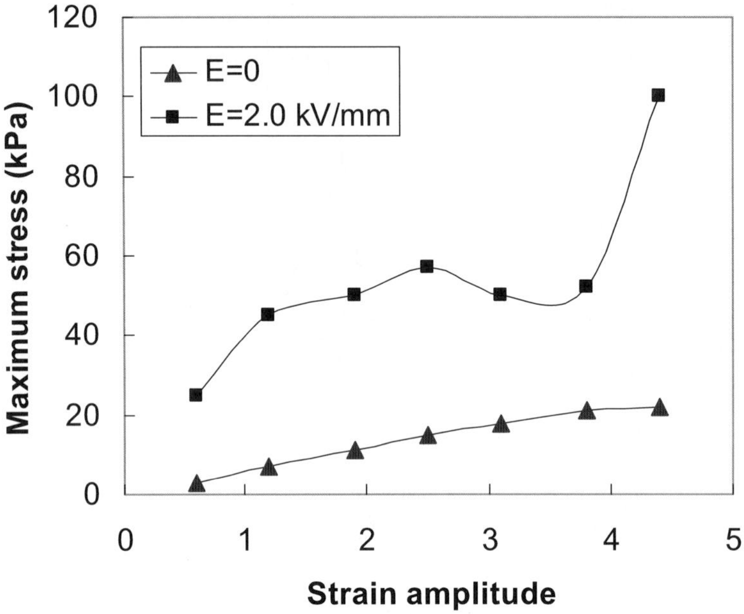

Figure 5 The maximum stress of aluminosilicate/paraffin suspension against the strain amplitude at strain frequency of 10 Hz. Re-plotted from the data given in D.R. Gamota, and F.E. Filisko, J. Rheol. 35(3)(1991)399

If the zeolite particle is dispersed in silicone oil, the yield stress of such a zeolite/silicone oil suspension can reach 27 kPa at the electric field 5 kV/mm [9], as shown in Figure 6. The yield stress strongly depends on the particle volume fraction, The 27 vol% zeolite/silicone oil suspension (ER A in Figure 6) shows a smaller yield stress compared with the 30 vol% suspension (ER B in Figure 6), and the difference becomes more significant at high electric fields.

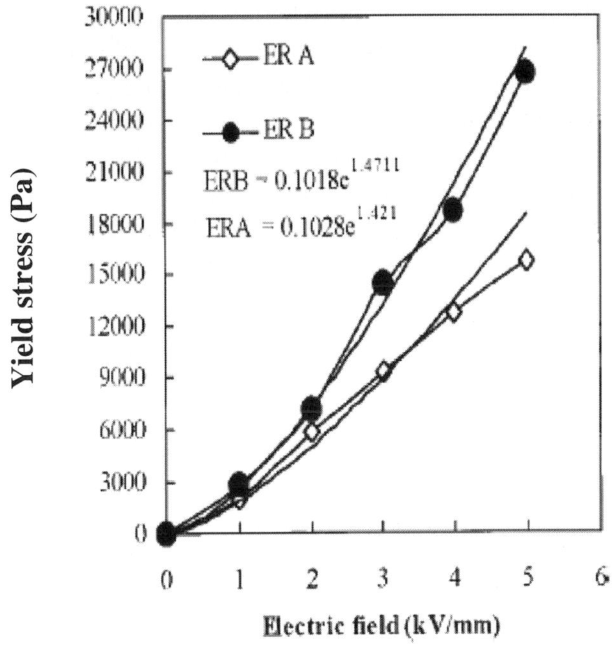

Figure 6 Yield stress of zeolite/silicone oil suspension against the electric field. The particle volume fraction for ER A and ER B is 27% and 30 %, respectively. Reproduced with permission from Y. Tian, Y. Meng, and S. Wen, J.Appl.Phys. 90(2001)493

The strontium titanate synthesized with sol-gel technique [10,11] was dispersed into silicone oil and found to give a yield stress 27 kPa at a dc field of 3 kV/mm, as shown in Figure 7. Without the application of an electric field, the yield stress also increases with the particle volume fraction in a much slower path, and is 7 times smaller than that under 3 kV/mm. However, the urea coated barium titanyl oxalate nanoparticle ($BaTiO(C_2O_4)_2 + NH_2CONH_2$) of average size of 50–70 nm and a surface coating less than 3~10 nm dispersed into silicone oil was found to give a yield stress of 130 kPa at the electric field 5 kV/mm and the particle volume fraction 30 vol% (see Figure 8). After this particle was doped with Rb of fomula [$Ba_{0.8}(Rb)_{0.4}$ $TiO(C_2O_4)_2$+ NH_2CONH_2], it even gave a much stronger ER effect, above 250 kPa at the electric field 5 kV/mm once the particle size was reduced from 50 to 20 nm [12]. Although this paper didn't make clear whether the particle is monodispersed in size (TEM picture available only), it is expected that the shear stress or the viscosity should substantially increase once the particle size decreases, no matter whether the external electric field is applied or not. In contrast, wide particle size distribution does lower the viscosity of the colloidal system and does allow more particle to be added in the system. The limitation is the particle loading. Once the particle size becomes small, the maximum particle loading fraction typically decreases. Based on the inter-particle spacing equation, Eq.(25) in Chapter 2, for 20 nm particle system the inter-particle spacing is estimated as 9.2 nm when the particle volume fraction reaches 20 vol%.

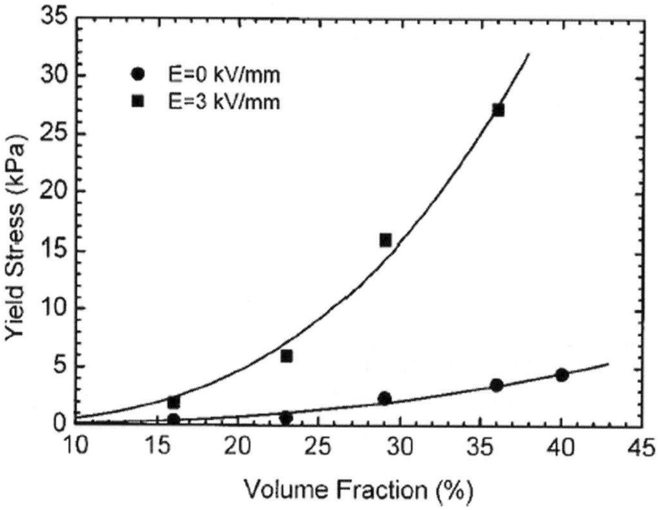

Figure 7 The yield stress of surface modified strontium titanate/silicone oil against the particle volume fraction obtained at 40 °C. Reproduced with permission from Y. Zhang, K. Lu, and G. Rao, Y. Tian, S. Zhang, J. Liang, Appl. Phys. Lett., 80(2002)888

Since the interparticle spacing is almost the half of the individual particle size, there should not be enough space for more particles being dispersed into the system. Such a system should have a very high shear stress already, even without an external electric field. Those suspensions give such a strong ER effect comparable to the magnetorheoleogical effect, while the magnetic field instead of the electric field is used to induce the huge increase of the rheological property. Wen's result definitely breaks the upper limit of the yield stress that the ER fluids can normally reach, which is believed to be much smaller than that of magnetorheological fluids. Note that the theoretical explanation for such a high yield stress observed in urea coated barium titanyl oxalate/silicone oil system has an assumption that the interparticle distance should be on the Angstrom scale, which may be unlikely as this length scale is comparable to that of the chemical bond.

There are many other systems reported to show a large yield stress. For example, surfactant modified polysaccharide and titanium oxide give the static shear stress 37 kPa under dc field 4kV/mm at room temperature [13]. The maximum shear stress value is indeed important, however, the increment before and after an electric field applied is more meaningful in practice.

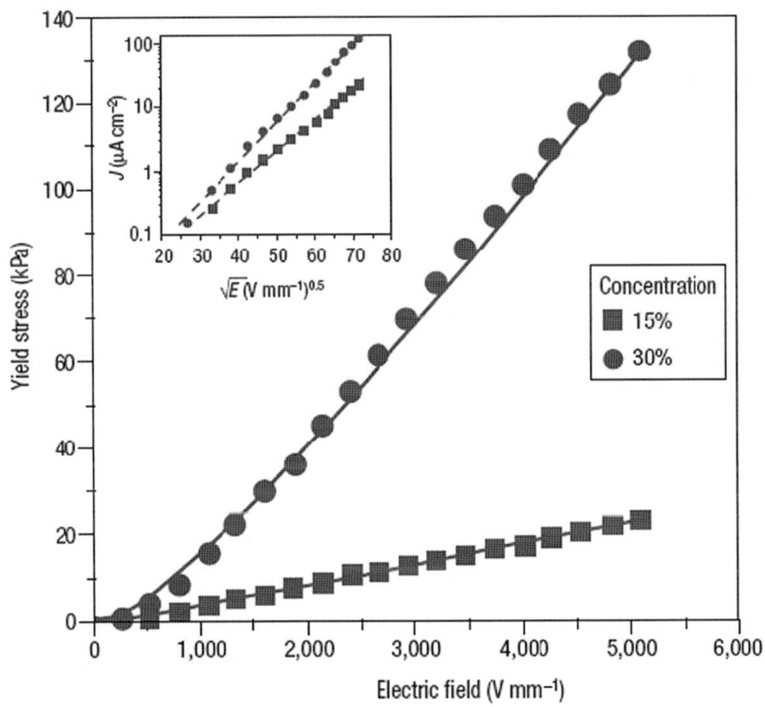

Figure 8 Static yield stress of urea coated barium titanyl oxalate /silicone oil suspension is plotted as a function of applied electric field for two solid concentrations. Inset: logarithm of the current density J plotted as a function of $\sqrt{E}$. Reproduced with permission from W. Wen, X. Huang, S. Yang, K. Lu, and P. Sheng, Nature Materials, 2(2003)727

Note that anhydrous ER fluids encounter another big problem—particle sedimentation, which could cause ER fluids to malfunction totally and severely limit the practical applications. Efforts were then spent to develop homogeneous ER fluids with no particulate material dispersed inside. Liquid crystal polymer/oil and water/oil emulsions ER systems were thus developed, which will be discussed later in detail.

The dispersed particles are polarized under an electric field and re-orientated along the direction of the external electric field, forming the fibrillated chains spanned between two electrodes, as shown in Figure 2. Those particle chains dramatically increase the mechanical strength of ER fluids, resulting in a profound ER effect. The fibrillated chains become thicker and stronger with the increase of the electric field, which is the reason that the shear stress of ER fluids usually increases with the electric field.

Besides the phenomenological mechanism mentioned above, there are many other physical mechanisms proposed to explain the positive ER facts. Those mechanisms will be discussed in more detailed manner in a future chapter.

2. NEGATIVE ER EFFECT

In contrast to the positive ER effect, a negative ER effect refers to a phenomenon where the apparent viscosity of the ER fluid decreases as the external electric field increases. The apparent viscosity of a fumed silica/silicone oil suspension against time under different electric fields is shown in Figure 9 [14]. When the applied electric field changes from 0 to 2.0 kV/mm in a step of 0.5 kV/mm, the apparent viscosity moves downwards instead of upwards as shown in Figure 1. Once the electric field is removed, the apparent viscosity doesn't recover to the original value without an electric field. In 1995, Boissy [15] reported a negative ER effect observed in the suspension containing PMMA [poly(methyl)methacrylate] powder dispersed into a liquid mixture of a mineral oil TF 50 and Ugilec T (a weakly polar solvent from Elf-Atochem co.). He found that the apparent viscosity of the whole suspension decreases as the external electric field increases, as shown in Figure 10. The apparent viscosity decrease becomes more prominent in concentrated suspensions. The apparent viscosity of the liquid mixture without any particle is also plotted against the applied electric field in Figure 10, and it remains flat in the electric field between 0 and 3 kV/mm, indicating that the negative ER effect does not result from the liquid medium.

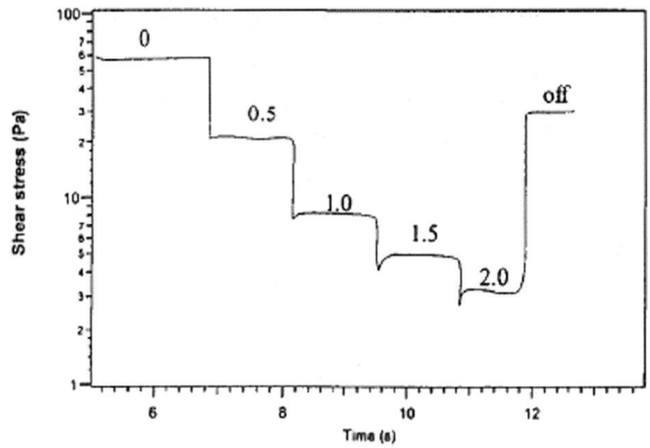

Figure 9 The shear stress of the fumed silica/silicone oil ER fluid with the volume fraction 10 vol% against the time. The number on each step refers to the applied electric field strength (kV/mm). Note that the shear stress doesn't recover to the original value once the external electric field is removed. Reproduced from the T. Hao, The Correlation between the Electric Property of ER fluids and the ER effect, Ph.D. thesis, the Institute of Chemistry, Chinese Academy of Sciences, 1995

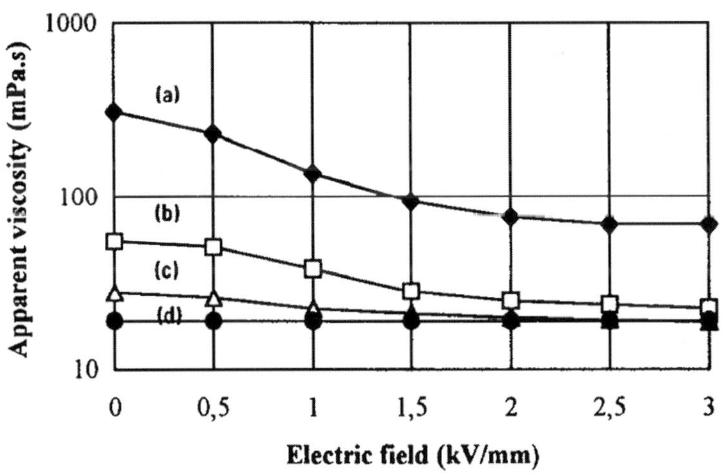

Figure 10 Apparent viscosity of PMMA/Ugilec-mineral oil vs. the electric field strength at various particle concentration, a) 30 vol%; b) 20 vol%; c) 10 vol%; and d) 0. Reproduced with permission from C. Boissy, P. Atten, and J.-N. Foulc, J. Electrostatics, 35(1995)13.

A similar suspension system containing PMMA particle was also found to show the negative ER effect [16], as shown in Figure 11, though the viscosity of those two systems are quite different, probably due to the surfactant AOT(sodium dodecyl sulfate) added in one suspension. Even at a very high shear field, 300 s^{-1}, the negative ER effect is still quite remarkable. In addition to PMMA particle, the Teflon particle was also found to show the negative ER effect once it is dispersed into silicone oil [17]. A clear phase separation was observed for this Teflon/silicone oil suspension under an electric field [19] , as shown in Figure 12. A condensed particle layer was found to form nearby the anode, and this layer was further condensed with the increase of the electric field. Most likely, the Teflon particle was originally carrying negative charges or was negatively charged under an electric field. The more negative charges were generated in high electric fields, resulting in a condensed particle layer nearby the surface of the positive electrode. The negative ER effect was ascribed to lower conductivity of those organic particles compared to that of the dispersing medium. The conductivity of Teflon is in the order of 10^{-14} S/m, while that of silicone oil is 10^{-13} S/m. However, a relatively high conductive inorganic particle, magnesium hydroxide with conductivity $5.8x10^{-9}$ S/m dispersed into silicone oil, was found to also show a negative ER effect [18]. The conductivity of the dispersed particle may not play a critical role in the negative ER effect.

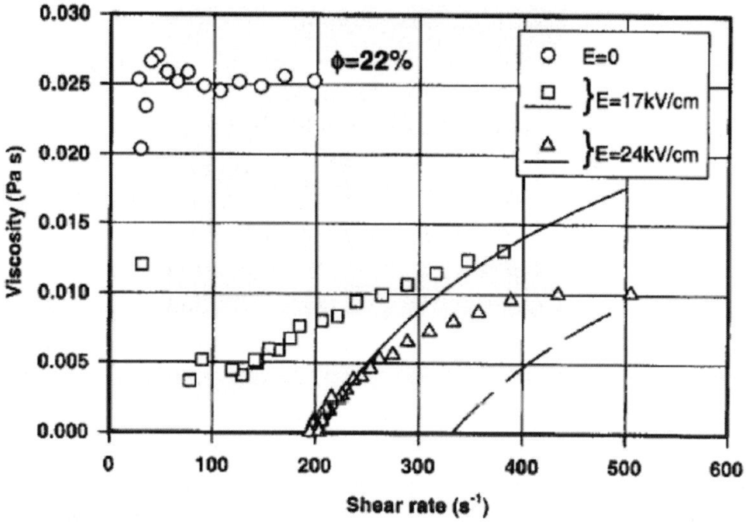

Figure 11 The apparent viscosity of PMMA particles dispersed into a mixture liquid (transformer oil+Ugilec+AOT) vs. shear rate at different dc electric field. The particle volume fraction is 22vol%. Ugilec is a dielectric liquid from Elf Atochem, and AOT is sodium dodecyl sulfate. Reproduced with permission from L. Lobry, and E. Lemaire, J. Electrostatics, 47(1999)61

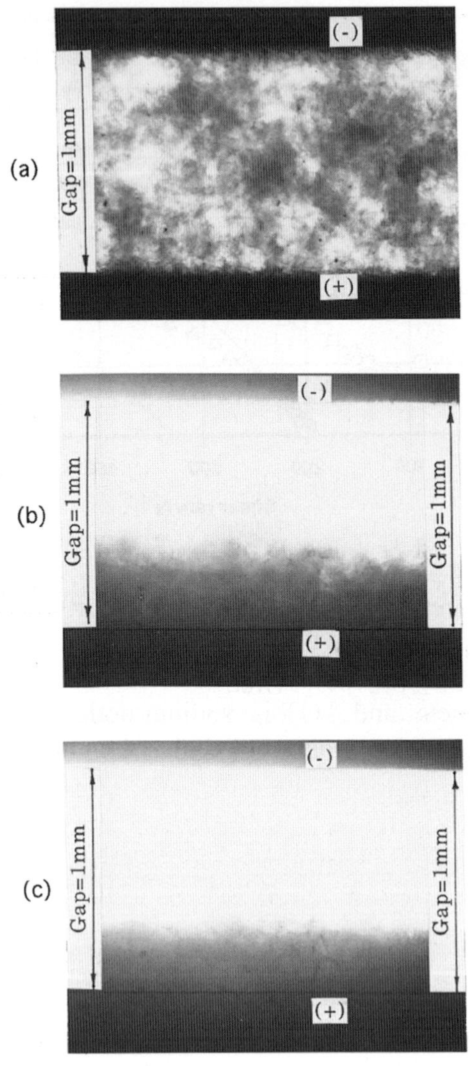

Figure 12 The optical microscopy image of 5 vol% Teflon/silicone oil suspension under the electric field: a) 0; b)1 kV/mm; and c)3 kV/mm. The average particle diameter is 20 μm. Reproduced with permission from C. W. Wu, and H. Conrad, J. Rheol., 41(2)(1997)267.

Besides the particle-type suspensions, the liquid crystalline materials also show the negative ER effect. The apparent viscosity difference (with and without an electric field) of MBBA (N-(4-methoxybenzylidene)-4-butylaniline) solution against the electric field [24] is shown in Figure 13. The viscosity decrease mainly happens when the electric field is less than 1.5 kV/mm. Above 1.5 kV/mm and under 5 kV/mm, the viscosity levels off without changing with the applied electric field. The molecular structure of MBBA is shown in Figure 14. The magnitude of the viscosity decrease in this system is obviously less than that observed in the particle-type suspensions. A castor oil/silicone oil emulsion system also shows a negative ER effect [20].

Another kind of interesting materials can show either positive and negative ER effect, depending on the dispersing medium or the particle concentration. Those materials include side-chain liquid crystalline polysiloxane [21], urethane-modified polypropylene glycol [22], and PMMA stabilized by diblock copolymer [23]. The liquid crystalline polysiloxane was found to show a positive ER effect when it was dissolved in 4'-(pentyloxy)-4-biphenylcarbonitrile. However it showed a negative ER effect when it was dissolved in MBBA(N-(4-methoxybenzylidene)-4-butylaniline) [21], see structure in Figure 14. Urethane-modified polypropylene glycol (UPPG) mixed with dimethylsiloxane (DMS) shows the positive ER effect when the viscosity of UPPG is larger than that of DMS. When the viscosity of UPPG is smaller than that of DMS, a positive ER effect was observed when the DMS concentration is less than 60 wt%, and a negative ER effect was observed when the DMS concentration is higher than 60 wt%. At 80°C, the viscosity of UPPG (with R=ethyl) is larger than that of DMS of viscosity 1 Pa.s, and smaller than that of DMS of viscosity 100 Pa.s. The shear stress of UPPG (ethyl)/DMS (100 Pa.s) mixture against the DMS concentration [25] is shown in Figure 15. The magnitude of positive ER effect at low DMS concentrations is smaller than that of the negative ER effect at high DMS concentrations. At high DMS concentrations the shear stress at 2 kV/mm is much lower than that under zero electric field, indicating that a negative ER effect appears. The concentration seems to play an important role in controlling which type of ER effect will be displayed. A similar trend was also found in a PMMA/decane system stabilized with polystyrene-block-poly(ethylene-co-propylene), in which the PMMA concentration is critical for the system to show the positive or negative ER effect (see Figure 16). When the PMMA particle volume fraction is 0.12, the apparent viscosity starts to decrease at 1.5 kV/mm. However, when the PMMA particle volume fraction is 0.364, the apparent viscosity almost linearly increases with the electric field from 0 to 2.5 kV/mm.

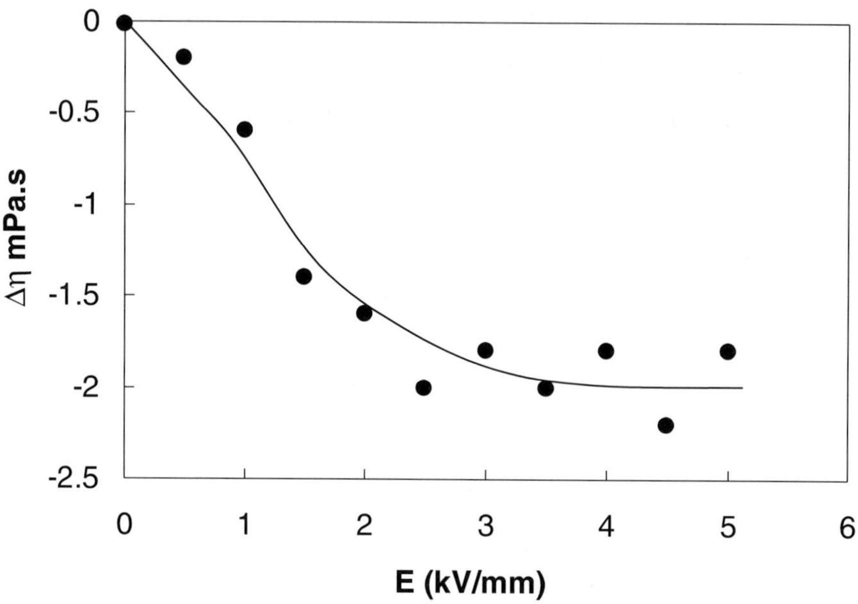

Figure 13 The apparent viscosity difference Δη (with and without an electric field) of MBBA , obtained at 300K and 988.5 s^{-1}, against the electric field strength. The solid line is just for guiding the trend. Redrawn from K. Negita, Chem.Phys.Lett, 246(1995)353.

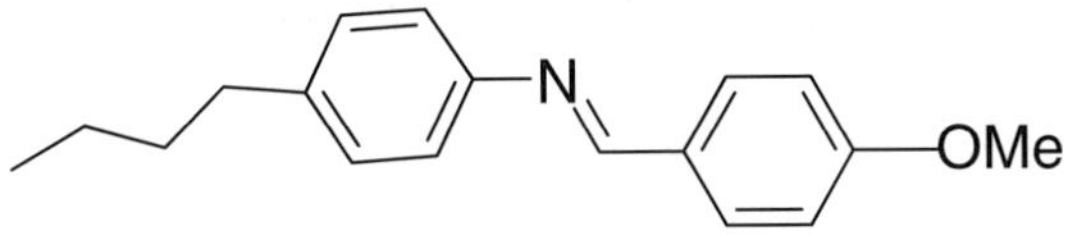

N-(p-methoxybenzylidene)-butylaniline (MBBA)

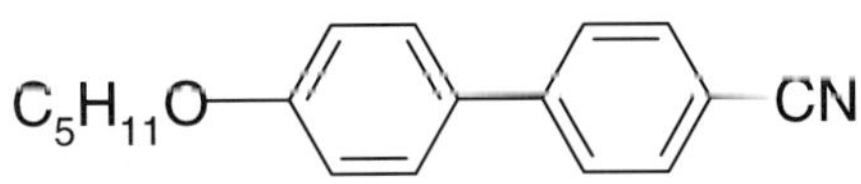

4'-(pentyloxy)-4-biphenylcarbonitrile (5OCB)

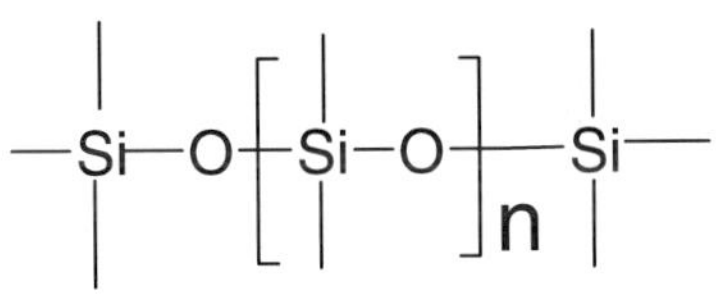

Dimethylsiloxane (DMS)

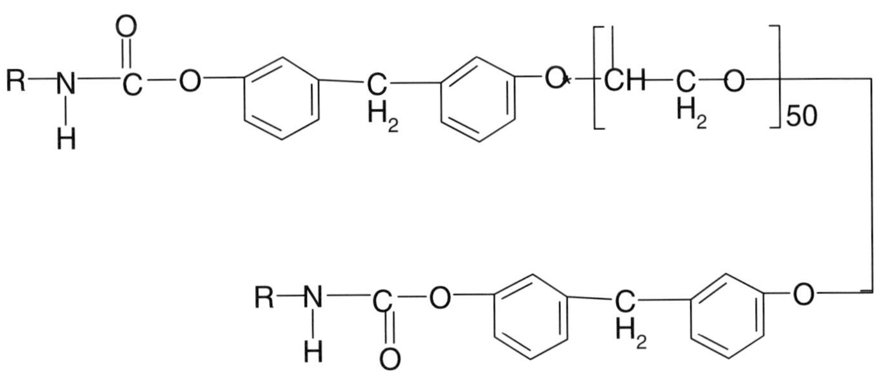

R=CH_3-CH_2-, FCH_2-CH_2-, CN-CH_2-CH_2-

Urethane-modified-polypropylene glycol (UPPG)

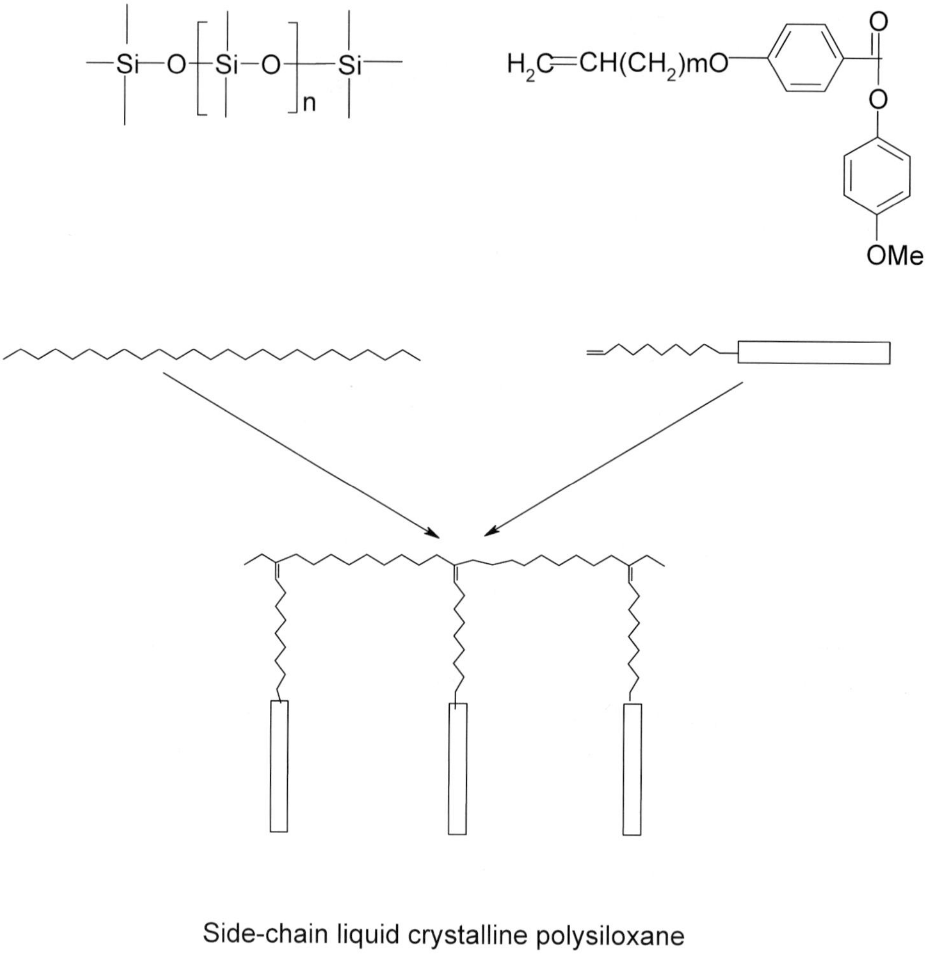

Figure 14 The molecular structure of negative and positive ER materials

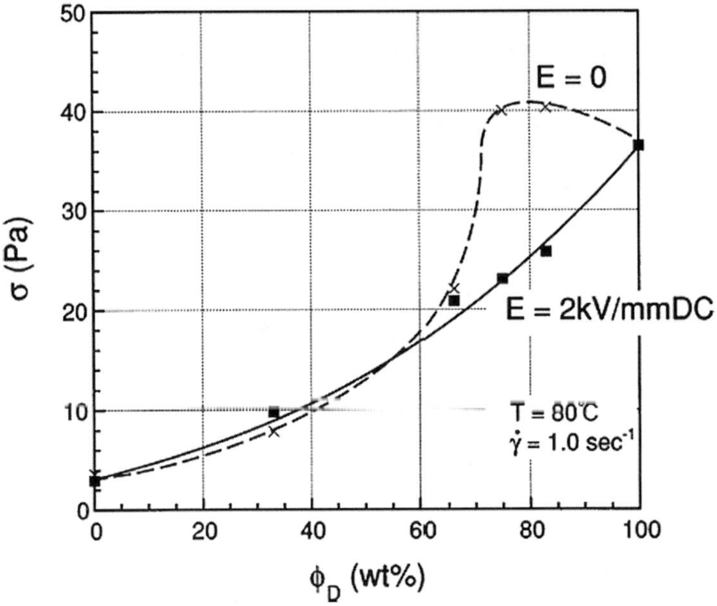

Figure 15 Shear stress of UPPG(ethyl)/DMS(100 Pa.s) mixture vs. DMS concentration obtained at 80 °C. Reproduced with permission from H. Kimura, K. Aikawa, Y. Masubuchi, J. Takimoto, K.Koyama, and T. Uemura, J. Non-Newtonian Fluid Mech., 76(1998)199

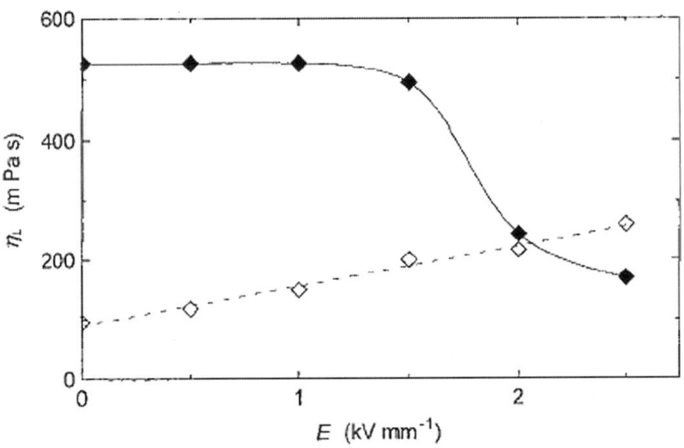

Figure 16. The apparent viscosity η_L at low shear rate (15.5 s^{-1}) against the electric field strength. The particle volume fraction (♦) 0.12; (◊) 0.364. Reproduced with permission from V. Pavlinek, P. Saha, O. Quadrat, and J.Stejskal, Langmuir, 16(2000)1447

There are two reasons believed to be responsible for the negative ER effect. The first is that the particles are either charged nonuniformly (mixed charges) or uniformly (single sign charge). For the nonuniformly charged case, some particles are negatively charged and some are positively charged. Under an electric field, the particles migrate towards both electrodes due to electrophoresis, leading to the phase separation in the suspension. For uniformly charged particles, the particles also migrate towards one electrode under an electric field. In either case two phases exist in the system, as shown in Figure 17. The first phase locates in the surface of one or both electrodes and has a high particle concentration; The second one is only the dispersing liquid without the particle. Another mechanism is that the Quincke rotation was thought to be responsible for the viscosity reduction [16]. Particles are polarized under an electric field. The charges (ions or other charge species) from the liquid may accumulate around the particle surface after the particles are polarized, as shown in Figure 18. The charges on the top of the particle surface have the same sign as the upper electrode does, and the charges at the bottom of the particle have the same sign as the lower electrode does. This situation is unstable. The repulsive force between charges around the particle surface and the electrodes creates a rotational force upon the particle, causing the particle to turn around under an electric field.

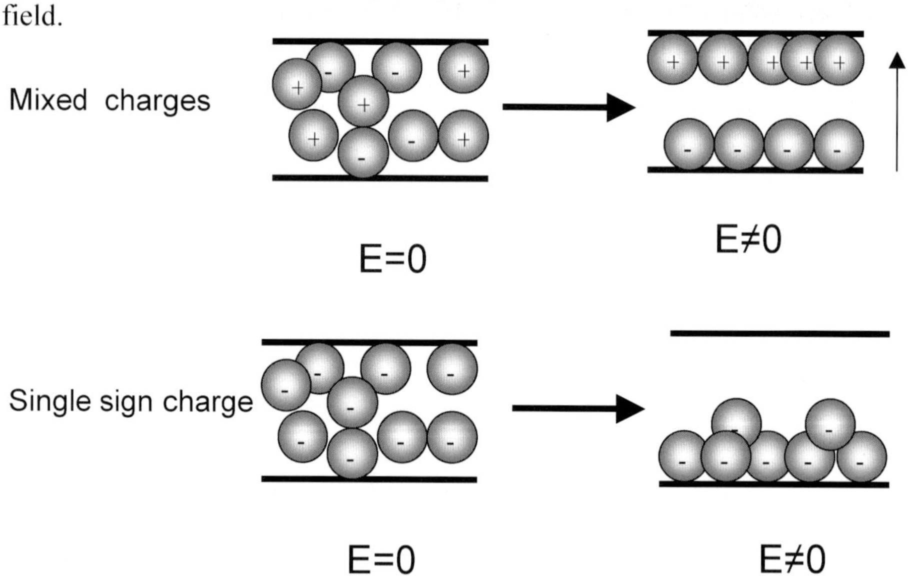

Figure 17 Phase separation under an electric field due to electrophoresis, which leads to the negative ER effect.

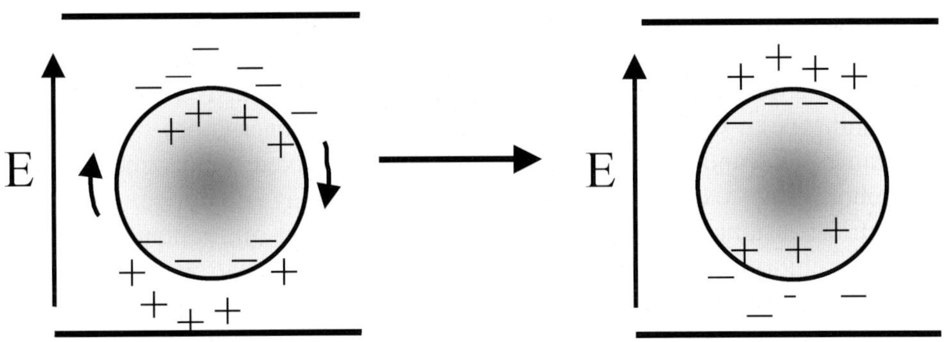

Figure 18 Quincke rotation of each individual particle resulted from the particle polarization and charge distribution around the polarized particle.

After the rotation of the particle, the interparticle force becomes weaker as the external surface charges on the particle create a local electric field just opposite to the external electric field, which leads to the negative ER effect.

For a two-liquid mixture system, one liquid may form droplets, dispersed into another continuous liquid phase. Under an electric field, the liquid droplet may behave like a solid particulate, polarized and aligned along the direction of the electric field. A liquid-droplet bridge may form and connect two electrodes. Therefore, the viscosity of the whole system may be determined by the viscosity of the fibrillated liquid droplet. If the viscosity of this liquid droplet phase is lower than that of the continuous phase, a negative ER effect is thus expected. Which liquid would form the droplet is dependent on the relative concentration of those two liquids. This is the reason that the viscosity and the concentration of two liquids are important for the system to show a positive or a negative ER effect.

A criterion for evaluating the negative effect and the positive effect based on the physical parameters of both the dispersed and dispersing materials was proposed [26].

3. PHOTIC(PHOTO-) ER EFFECT

When the dispersed particle is photoactive, the ER effect could be enhanced by the UV illumination no matter whether the ER effect is positive or negative. This phenomenon is termed the photoelectrorheological (photo-ER) effect. In a US patent published in 1971 [27], photosensitive electrorheological fluids were disclosed as ink materials for image recording. Dispersed particulate materials include the organic colorants such as Monastral Green B (phthalocyanine pigment), Indofast Yellow Toner (Flavanthrone), Quido Magenta (substituted quinacridone), etc., and inorganic colorants such as cadmium sulfide mixed with zinc sulfide phosphor, zinc oxide, etc.. Dispersing liquid may comprise any low dielectric constant oil materials, for example, mineral oil, heptane, hexane, paraffin, chlorinated or fluorinated hydrocarbons. Both ac and dc electric field can be used to control the photo-ER ink. However, this patent doesn't provide any photo-ER effect data to show how photonic radiation can change the viscosity of those ER fluids. A phenothiazine particle suspension [28] was also reported to show the photo-ER effect. A series of research works on photo-ER effect have been done using TiO_2 nanoparticle (average particle size about 29nm) as dispersed phase [29-32]. Water content on the particle surface was found to play an important role. Low water content (for example, 1.2wt%) makes the suspension display a positive photo-induced ER effect, while high water content (for example 10 wt%) gives a negative photo-induced ER effect. Also, the shear rate can switch the ER effect from positive to negative. Photo-generated carriers, either electrons or holes, were thought to change the electric property of TiO_2 and then to enhance the ER performance. The possible mechanism includes the following steps [32]:

$$TI^{IV}O_2+e^-+H^+\leftrightarrow Ti^{III}O_2\cdot H \quad (1)$$

$$2H^++2e^-\leftrightarrow H_2 \quad (2)$$

$$O_2+e^-\leftrightarrow (O^-{}_2)_{aq} \quad (3)$$

$$H_2O+h^+\leftrightarrow \cdot OH+H^+ \quad (4)$$

$$\cdot OH+ e^-+H^+\leftrightarrow H_2O \quad (5)$$

$$2\cdot OH\leftrightarrow H_2O_2 \quad (6)$$

The photo-generated charge carriers (electron e-, and hole h+) react with water, producing either OH- or H+, which are released to the bulk solution and leaving excess charges on the particle. Since the TiO_2 particle is dispersed into silicone oil, the particle may acquire charge by colliding with electrodes or other particles. When water content is high, the particle mainly acquires the charge from the water, and the particle behaves like self-charged species. The repulsive force should be strong enough to make the negative ER effect happen. Whereas when water content is low, the particle mainly acquires charge from other particles via contact, the bridged structure is favorable and the positive ER effect is expected. The shear rate definitely changes how particles are in contact with each other, thus the ER effect induced by the photo irradiation is also dependent on shear rate. An illustration graph of photo-induced positive and negative ER effect is shown in Figure 19. The experimental data of the TiO_2 P25 dispersed into the silicone oil under light illumination was plotted in Figure 20 as the flow rate (in gram per minute) vs. time [32]. The flow rate decrease indicates a strong flow resistance, i.e., a positive ER effect.

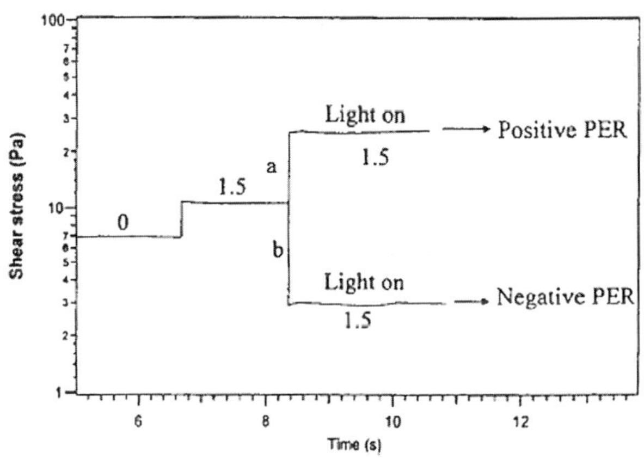

Figure 19 The illustration of photo-induced positive and negative ER effect plotted as the shear stress vs. time. The number on each step is the electric field, kV/mm. Reproduced with permission from T. Hao, Adv. Mater., 13(2001)1847.

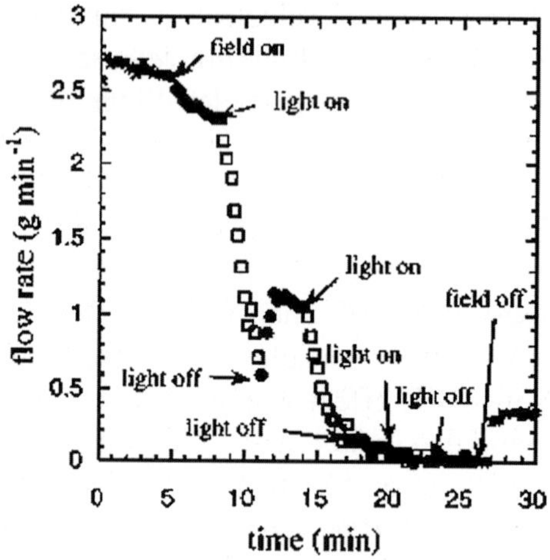

Figure 20 The flow rate (gram per minute) of the TiO_2 P25-based fluid under light on and light off conditions measured as a function of time under an applied electric field, 1.4kV/mm. Reproduced with permission from Y. Komoda, N. Sakai, T.N. Rao, D.A. Tryk, A. Fujishima, Langmuir 14 (1998) 1081

A few cycles of light on and off are shown in Figure 20 and a constant positive ER effect is observed. Note that the photo-induced ER effect is relatively small in comparison with the effect induced by an external electric field.

4. ELECTROMAGNETORHEOLOGICAL (EMR) EFFECT

The electromagnetorheological effect (EMR effect) is the combination of the electrorheological(ER) effect and magnetorheological(MR) effect. Before proceeding to the EMR effect, one may need to take a brief look at the MR effect

4.1 Magnetorheological effect (MRE)

Similar to the ER effect, the magnetorheological effect (MRE) is termed the remarkable increase of the rheological properties of magnetorheological fluids (MRF) induced by an external magnetic field. The magnetic dipole and multipole moments induced on each particle interact with each other, leading to the formation of fibrillated particle structures along the direction of the magnetic field [33]. So for both positive

ER and MR effects, the fibrillated chain structures induced by the external field and formed by the randomly dispersed particles are believed to be responsible for the dramatic rheological property increase. In this sense, the MR fluid is quite similar to the ER fluid, except that the particle constituting the MR fluid must be ferromagnetic or a magnetic nonlinear particle. Iron, nickel, cobalt, and ceramic ferrites, etc., are good candidates for making MR fluids. Ferrofluids, on the other hand, are stable dispersions made from superparamagnetic particle of size 5-10 nm, such as iron oxide. The conventional ferrofluids usually don't show a yield stress and only give a rather small viscosity change in a magnetic field [34-36]. There is no report of a negative MR effect to date, in parallel with the negative ER effect.

The MR effect was first documented in the late 1940s [37, 38], almost at the same time as the ER effect was discovered by Winslow. Normally, the MR fluids are made of magnetizable (iron or cobalt) particles dispersed in either insulating oil like mineral or silicone oils, or conducting medium like water or glycol [39]. A variety of surfactants or lubricants are usually added to MR suspensions for preventing particle settling and enhancing the MR effect, as the typical diameter of magnetic particle is about 3 to 5 microns. Nanometer-sized particles definitely give a better stability. However, they are limited by a low mechanical strength under a magnetic field, typically 5 kPa, comparing to more than 100 kPa for a micrometer sized particle [40]. Figure 21 show the yield stress of a carbonyl iron/silicone oil system vs. the applied magnetic field [41]. The average particle size is 5 μm and the particle volume fraction is 46 vol%. Without a normal compression pressure, this MR fluid can easily reach 120 kPa at H=514 kA/m. The compression pressure can increase the yield stress to 600 kPa under a magnetic field 514 kA/m. The mechanical strength of a MR fluid is controlled by the saturation magnetization of the suspended particle, and thus a particle with a large saturation magnetization is preferred [42].

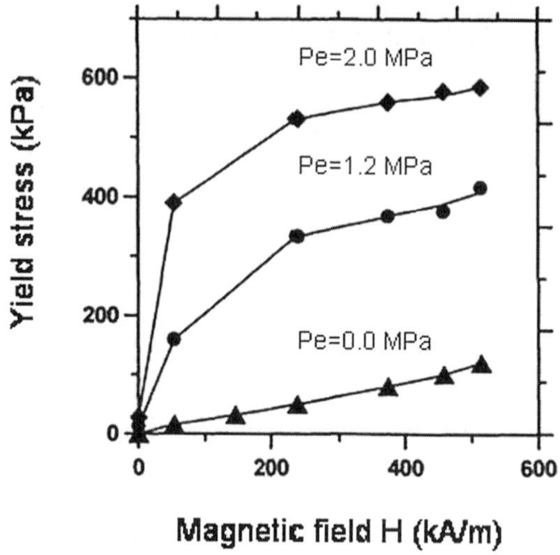

Figure 21 The yield stress of carbonyl iron/silicone oil with particle volume fraction 46 vol% against the applied magnetic field. The average particle size is 5 μm. Reproduced with permission from X. Tang, X. Zhang, and R. Tao, Int. J. Modern Phys. B, 15(2001)549.

Since the typical solid phase of MR fluids are metal or metal oxide ferrites and have a mass density larger than that of most suspending fluids, stabilizing a micrometer-sized particle against sedimentation is a challenging issue facing the technological application of these materials [33]. Small magnetic particle sizes [43], high solid loadings, and high viscosities of dispersing oil are helpful in reducing the rate of sedimentation. However, these preferences can have obvious disadvantages such as low mechanical strength and high zero-field viscosity. Adding surfactant (stearates) or nanosilica particles into the MR fluids was found to be the best way to improve particle stability [44, 45]. There are two mechanisms proposed for explaining why silica particles can influence the stability of MRFs. First, silica–silica interparticle hydrogen bonding facilitates the formation of a thixotropic network that prevents the sedimentation and eventually reduces redispersion difficulties. Secondly, a silica particle may adhere to an iron particle by acid-base reaction in non-polar media [46]. In the presence of the magnetic field, the shell of adsorbed silica will partially screen the magnetic interactions between iron particles, thus reducing the magnetorheological effect. A rheological study of the stabilization of microsized iron/silicone oil suspension by addition of silica nanoparticle was made by Vicente [45]. Figure 22 shows the normalized yield stress (σ_y / B^2) of an iron particle (~1 μm in diameter) dispersed in silicone oil containing nano-sized silica (~7 nm in diameter) vs. silica concentration, where B is the magnetic flux density. When the silica concentration is low, below 20 g/L, the presence of nanosized silica particles hinders the MR effect, while once the silica concentration is larger than 20 g/L, the normalized yield stress seems to increase with the silica concentration. As the authors claimed in the paper, this is a manifestation of the fact that in such high silica concentrations the magnetic interaction between silica-covered iron particles are almost fully shielded. The yield stress increase results from the crowding effect of the nanosilica particles, as at such high silica concentration region the yield stress is independent of the iron concentration and the applied magnetic field.

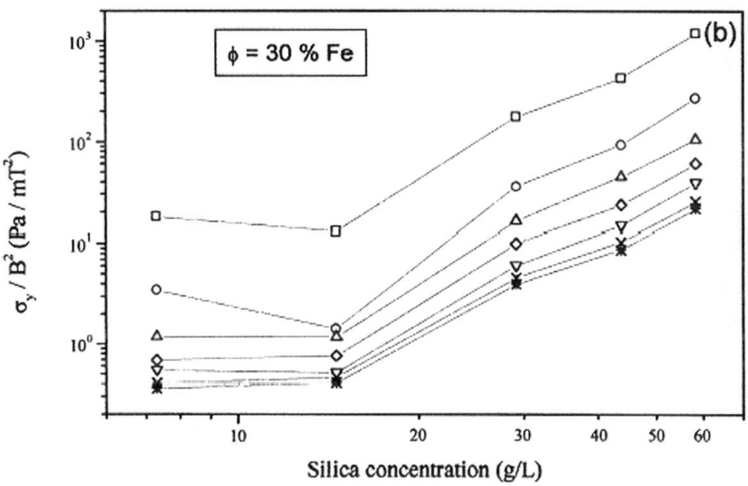

Figure 22 Normalized yield stress (σ_y / B^2) of an iron particle (~1μm in diameter) dispersed in silicone oil containing nano-sized silica (~7 nm in diameter) vs. silica concentration. Reproduced with permission from J. de Vicente, M. T. Lopez-Lopez, F. Gonzalez-Caballero, and J. D. G. Duran, J. Rheol, 47(2003)1093. The symbols correspond to the following magnetic flux densities(mT): □ 0.345; ○ 0.725,△ 1.085; ◇ 1.42; ▽ 1.77; × 2.17, and * 2.38.

The typical features of both MR and ER fluids are shown in Table 1. MR fluids have a relatively high yield stress with small voltage in comparison with ER fluids. However, MR fluids require a huge power consumption.

Table 1 Typical property of both MR and ER fluids.

	MR fluid	ER fluid
Maximum yield stress	50~200 kPa	10—130 kPa
Particle	Iron or cobalt metal or oxide	Semi-conductive particle
Dispersing medium	Insulating oil or polar liquid (water, glycol)	Insulating oil
Stability	Less stable due to large density of particle	More stable than MR fluid
Power input	50 V, 2A	3000 V, less than 1 mA

4.2 The EMR effect

The combination of both electric and magnetic fields can have an appreciable impact on the rheological property of some fluids in comparison with the situation where only one field, either the electric or the magnetic field, is applied [47-50]. The EMR effect is termed the synergistic effect of viscosity increase under the combination of applied electric and magnetic fields [51]. The direction of the applied magnetic field can be either parallel or perpendicular to the applied electric field (see Figure 23). Koyama shows that [52] the ER effect of the TiO_2-coated iron particle/silicone oil suspension was further enhanced by the external magnetic field if the direction of this magnetic field parallels that of the electric field. If the magnetic and electric fields are perpendicular to each other, the ER effect was also enhanced, however, the magnitude is less than that when those two fields are parallel. An example of 11 vol% TiO_2-coated iron/silicone oil suspension under the reaction of both electric and magnetic fields is shown in Figure 24. The magnetic field further increases the shear stress of this suspension no matter whether the magnetic field is parallel or perpendicular to the electric field.

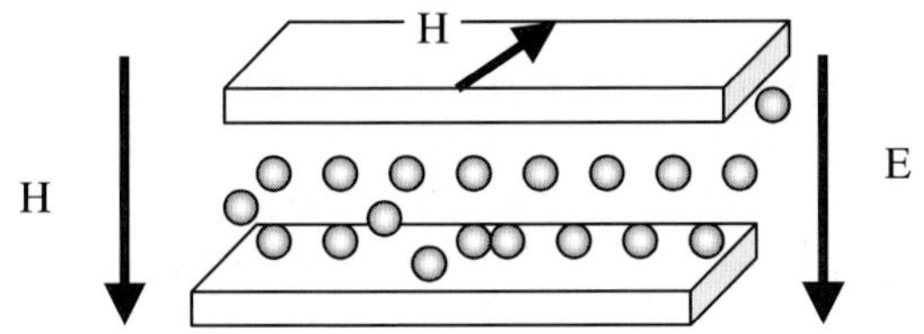

Figure 23 The schematic illustration of EMR effect setup. H represents the applied magnetic field and E represents the electric field.

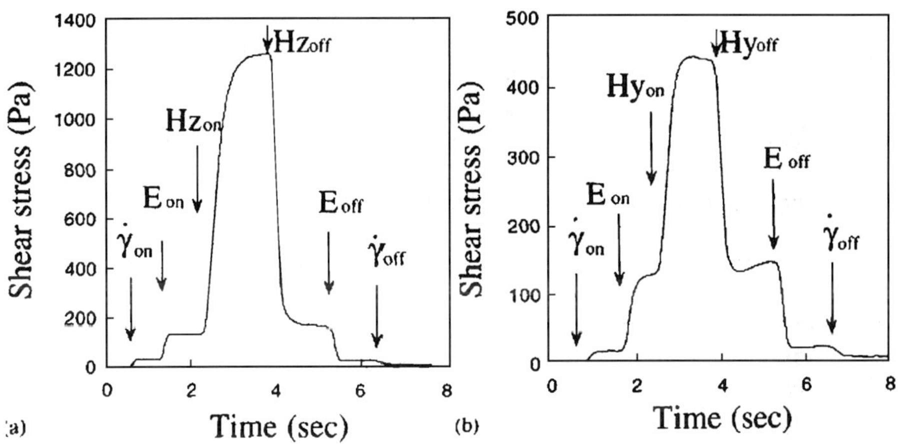

Figure 24 The shear stress of 11 vol% TiO_2 coated iron/silicone oil suspension vs. time. (a) the electric and magnetic fields parallel to each other and (b) the electric and magnetic fields are perpendicular to each other. Shear rate=2.8 s^{-1}, E=1.0 kV/mm, and H=2 kOe. Reproduced with permission from K. Koyama, K. Minagawa, T. Watanabe, Y. Kimakura, and J. Takimoto, J. Non-Newtonian Fluid Mech., 58(1995)195

REFERENCES

[1] T. Hao, Adv. Mater., 13(2001)1847
[2] H. Block, and J.P. Kelly, GB Patent 2170510, 1985
[3] R.T. Bonnecaze, and J.F. Brady, J. Rheol., 36(1)(1992)73
[4] A.M. Kraynik, Proc. 2nd Int.Conf. on ER fluids, J. D. Carlson, A.F. Sprecher, and H. Conrad, Ed., Technomic, Lancaster, PA, 1990
[5] H. Block, J. P. Kelly, A. Qin, and T. Watson, Langmuir, 6(1990)6
[6] D.G. Bytt, GB Paten 2189803, 1987
[7] F.E. Filisko and W.F. Armstrong, European Patent 0313351, 1989
[8] D.R. Gamota, and F.E. Filisko, J. Rheol. 35(3)(1991)399
[9] Y. Tian,Y. Meng, and S. Wen, J. Appl. Phys. 90(2001)493
[10] Y. L. Zhang, Y. Ma, Y. Ch. Lan, K. Q. Lu, and W. Liu, Appl. Phys. Lett. 73(1998)1326;
[11] Y. Ma, Y. L. Zhang, and K. Q. Lu, J. Appl. Phys. 83 (1998) 5522
[12] W. Wen, X. Huang, and P. Sheng, Appl. Phys. Lett., 85(2004)299
[13] X.P. Zhao, and X. Duan, Materials Lett., 54(2002)348
[14] T. Hao, The Correlation between the Electric Property of ER fluids and the ER effect, Ph.D. thesis, the Institute of Chemistry, Chinese Academy of Sciences, 1995
[15] C. Boissy, P.Atten,and J.-N. Foulc, J. Electrostatics, 35(1995)13
[16] L. Lobry, and E. Lemaire, J. Electrostatics, 47(1999)61.
[17] C. W. Wu, and H. Conrad, J. Rheol., 41(2)(1997)267
[18] J. Trlica, O. Quadrat, P. Bradna, V. Pavlinek, and P. Saha, J. Rheol., 40(5)(1996)943
[19] C.W. Wu, and H. Conrad, J. Rheol., 41(2)(1997)267
[20] J. Ha, and S. Yang, J. Rheol., 44(2000)235
[21] N. Yao, and A. Jamieson, Macromolecules, 30(1997)5822
[22] H. Kimura, K. Aikawa, Y. Masubuchi, J. Takimoto, K.Koyama, and T. Uemura, J. Non-Newtonian Fluid Mech., 76(1998)199
[23] V. Pavlinek, P. Saha, O. Quadrat, and J. Stejskal, Langmuir, 16(2000)1447
[24] K. Negita, Chem.Phys.Lett, 246(1995)353
[25] H. Kimura,K.Aikawa, Y. Masubuchi, J. Takimoto, K.Koyama, and T. Uemura, J. Non-Newtonian Fluid Mech., 76(1998)199
[26] T. Hao, A. Kawai, and F. Ikazaki, Langmuir, 16(2000)3058
[27] L. Carreira, and V. Mihajlov, US Patent 3553708, 1971
[28] F.E. Filisko, Proc. Int..Conf.ER Fluids, R.Tao Ed., World Scientific, 1991, p116
[29] Y. Komoda, T.N. Rao, A. Fujishima, Langmuir 13 (1997) 1371.
[30] T.N. Rao, Y. Komoda, N. Sakai, A. Fujishima, Chem. Lett. (1997) 307.
[31] N. Sakai, Y. Komoda, T.N. Rao, D.A. Tryk, A. Fujishima, J.Electroanal. Chem. 445 (1998)1.
[32] Y. Komoda, N. Sakai, T.N. Rao, D.A. Tryk, A. Fujishima, Langmuir 14 (1998) 1081
[33] J.M. Ginder, MRS Bull. 23(1998)26
[34] P.P. Phule, and J.M. Ginder, MRS Bulletin, 23(1998)19
[35] R.E. Rosensweig, Ferrohydrodynamics, Cambridge Univ. Press, Cambridge, 1985
[36] H.M. Laun, C. Kormann, and N. Willenbacher, Rheol. Acta, 35(1996)417
[37] J. Rabinow, AIEE Transactions, 67(1948)1308
[38] J. Rabinow, U.S.Patent 2,575,360, 1951

[39] J. de Vicente, A.V. Delgado, R.C. Plaza, J.D.G. Duran, and F. Gonzalez-Caballero, Langmuir, 16(2000)7954
[40] K.D. Weiss, and T.G. Duclos, Proc. 4th Int. Conference on ER & MR Fluids, R. Tao and G.D. Roy, Ed., World Scientific, Singapore, 1994, pp. 43–59
[41] X. Tang, X. Zhang, and R. Tao, Int. J. Modern Phys. B, 15(2001)549
[42] J.D. Carlson, and B.F. Spencer Jr., Proc. 2nd Workshop on Structural Control: Next Generation of Intelligent Structures, Hong Kong, China, 1996, pp. 99–109
[43] C.Kormann, H. M. Laun, and H. J. Richter, Int. J. Mod. Phys.B 10(1996)3167
[44] G. Bossis, O. Volkova, S. Lacis, and A. Meunier, ''Magnetorheology: Fluids, structures and rheology,'' in Ferrofluids, S. Odenbach, Ed., Springer, Bremen, 2002, pp. 202–230;
[45] J. de Vicente, M. T. Lopez-Lopez, F. Gonzalez-Caballero, and J. D. G. Duran, J. Rheol, 47(2003)1093
[46] G.A. van Ewijk, and A. P. Philipse, Langmuir, 17(2001)7204
[47] T. Fujita, J. Mochizuki, and I.J.Lin, J.Magn.Mater., 122(1993)29
[48] W.I. Kordonsky, S.R. Gorodkin, and E.V. Medvedeva, Proc. 4th Int.Conf. on ER fluids, R. Tao and G.D. Roy Ed., World Scientific, Singapore, 1994, pp22-36;
[49] K. Minagawa, T. Watanabe, K. Koyama, and M. Sasaki, Langmuir, 10(11)(1994)3926;
[50] K. Minagawa, T. Watanabe, M. Kudo, M. Sasaki, and K. Koyama, Proc. Pacific Conference on Rheology and Polym. Proc, Kyoto, 1994, pp68-69
[51] K. Minagawa, T. Watanabe, M. Munakata, and K. Koyoma, J. Non-Newtonian Fluid Mechn., 52(1994)59
[52] K. Koyama, K. Minagawa, T. Watanabe, Y. Kimakura, and J. Takimoto, J. Non-Newtonian Fluid Mech., 58(1995)195

Chapter 4

The electrorheological materials

1 GENERAL FEATURES OF ER FLUIDS

Most ER fluids are made of an insulating oil embodying particulate of the volume fraction between 0.05 and 0.50. The applied electric field usually ranges from 0.5 to 3 kV/mm. Those ER fluids are called the heterogeneous ER fluids in contrast with the homogeneous ER fluids that are composed of a liquid (for example liquid crystal polymer) dispersed in oil. If a small amount of adsorbed water exists in the systems, those ER fluids are normally called the hydrous ER fluids, in parallel with water-free or anhydrous ER fluids, in which no detectable water residue exists. Generally speaking, ER fluids contain three components: The dispersed phase that is either a liquid or a solid particulate, the dispersing phase that is an insulating oil, and the unavoidable additives that could be any polar material or surfactant with the function to enhance the ER effect and/or the stability of the whole suspension. The additives include inorganic salts, water, as well as the surfactant intentionally added in. The typical characteristics of ER fluids are shown in Table 1.

Table 1 The Experimental Characteristics of an ER fluid

	Liquid medium	Dispersing phase	ER Suspension
Dielectric Constant	~2	2—10000	
Conductivity (S/m)	10^{-10}—10^{-16}	~10^{-7}	10^{-9}—10^{-16}
Viscosity (Pa.s) at zero electric field	0.01—10		0.1—100
Yield stress (Pa) at 3 kV/mm	0		15 kPa

ER fluids usually contain small amounts of water. In 1985 Block [1] developed a water-free acenequinone radical polymer ER fluid, which was termed anhydrous ER fluid. Anhydrous ER fluid was believed to be much more promising from the standpoint of industrial applications and thus received much attention. Many other kinds of water-free ER fluids were developed later in the 1980s [2, 3]. Since these very promising water-free ER fluids have a particle sedimentation problem, homogeneous ER fluids with no particulate material inside were intensively investigated, especially the low molecular weight liquid crystalline (LC) materials. Starting from the later 1960s, the viscosity of LC materials was found to increase under an electric field, though the increment was only a few times [4, 5]. In 1992, an extremely large ER effect was found by Yang [6] and Inoue [7] in certain LC materials. Other LC ER systems were developed thereafter for the purpose of mechanism study [8-10]. A chlorinated paraffin oil/silicone oil emulsion system [11] and polymeric glycol/silicone oil emulsion system [12] were also found to exhibit an ER effect. The mechanism of the rheological response of the oil-in-oil emulsion system under an electric field was addressed by Ha [13] . Homogeneous ER fluids are believed to offer a new type of ER fluids. However, high viscosity under zero electric field and liquid–liquid segregation problem are obvious obstacles for such systems.

The ER fluids are either heterogeneous or homogeneous depending on whether the dispersed material is in liquid or solid form. In the heterogeneous category, the dispersed phase could be inorganic, organic or polymeric particulate materials. In the inorganic group, oxide and non-oxide materials give a quite different ER effect, and are therefore discussed separately. Figure 1 shows the preliminary classification of the ER fluids. Commonly used preparation approaches for fabricating ER fluids are introduced first in the following section and the general features of the above-mentioned three components are summarized thereafter, followed by the introduction of important positive, negative, photo-ER, and EMR materials.

1.1 Preparations of ER fluids

Preparations of ER fluids basically need the same techniques as fabrications of colloidal suspensions or emulsions. A very important issue in preparing ER fluids is that the water content should be strictly controlled, as ER fluids should withstand a high electric field, several kilovolts per millimeter in most cases. Heat treatment of powder is obviously necessary and thermal analysis tools are frequently used to determine the heat treatment condition. The dispersed particle of size between 0.1 and 100 μm is preferred in the viewpoints of both the ER effect and the fluid stability. Fine powder with narrow particle size distribution is always desired since even few large particles can substantially reduce the breakdown strength of an ER fluid. Milling of powder is another critical step for refining. The steel and iron ball mill with steel cylinder is not suitable for fabricating the ER fluids, as attrited metal fine powder will largely increase the current densities of ER fluids, causing serious arc sparking even in a relatively low electric field. Mixing of the powder and an oil material needs high shear mills. Colloidal mills and ultrasonic activators are frequently used to make homogeneously dispersed suspensions. Attention should always be on how to avoid bringing moisture or other highly conductive impurities into the system. Surfactants or other kinds of additives may be added at this stage to further enhance the particle sedimentation stability. Prepared ER fluids should be sealed and stored in a dry place for characterizations. Figure 2 shows a typical preparation procedure for making an ER fluid.

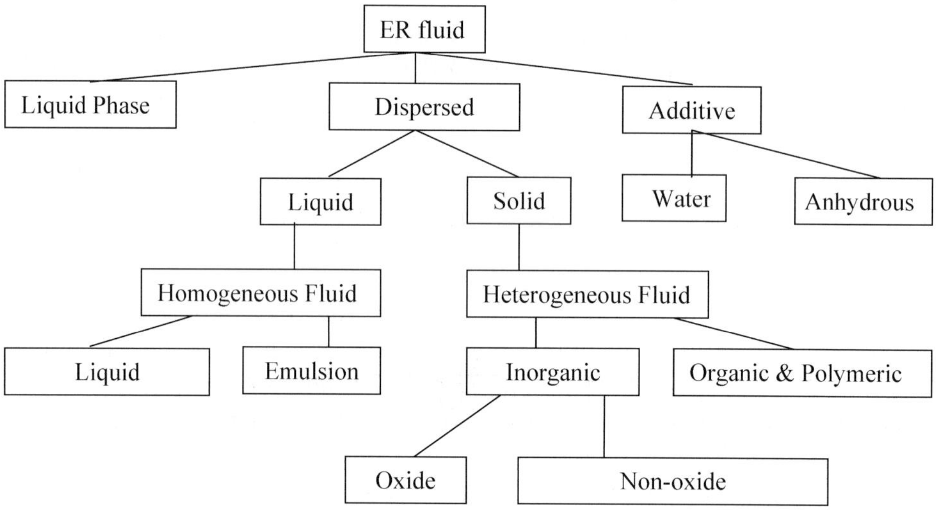

Figure 1 The classification of ER materials

1.2 Liquid continuous phase

The liquid continuous phase of an ER fluid is usually an insulating oil. Polydimethylsiloxane is a commonly used liquid for many ER systems. Other oil materials such as vegetable oil and mineral oil are also frequently used. Stangroom [14] concluded that an ideal dispersing liquid material should: a) have a high boiling point and low solidifying point. In other words, it should not easily evaporate within whole working temperature range; b) have a low viscosity for keeping the viscosity of whole suspension at a low level at zero electric field; c) have high resistance and high breakdown strength, and withstand a high electric field; d) have a high density (larger than $1.2 g/cm^3$ is better).

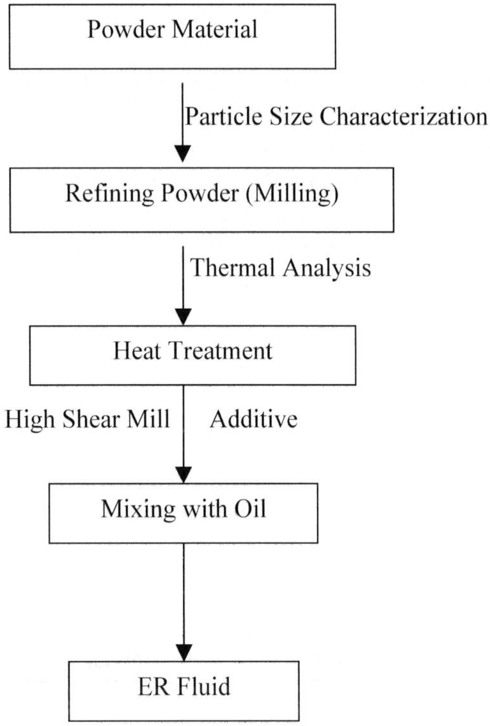

Figure 2 Preparation procedures of fabricating ER fluids.

The particle sedimentation problem might not occur until the densities of both the liquid and the solid match each other; e) have a high chemical stability. The liquid medium would not degrade or chemically react with other material once the ER fluid is prepared; f) have an obvious hydrophobicity, and would not adsorb too much moisture from environment; g) have low toxicity and a low cost. Currently used oil material includes silicone oil, vegetable oil, mineral oil, paraffin, kerosene, chlorinated hydrocarbon, transformer oil etc. Large density oil, such as fluoro or phenyl silicone oil, is also used.

1.3 Dispersed phase

1.3.1 Solid particulate–heterogeneous ER materials

1.3.1.1 Inorganic oxide materials

Some metallic oxides or ceramic materials sintered from several oxides were found to give a good ER effect. The compositions of those ER fluids are summarized in Table 2. Since the oxide materials are hydrophilic, most oxide ER fluids contain water even when special dehydration precaution is taken, which is the big shortcoming of this kind of fluids.

Table 2 Oxide ER fluids

Dispersed phase	Dispersing phase	Additive	Ref.
Piezoelectric Ceramic	Mineral oil or xylene	Water or glycerol oleates	[15]
Iron(II/III) Oxide	Petroleum fractions or dibutyl sebacate	Water or Surfactant	[16]
Silica	Kerosene or dibutyl sebacate Mineral oil, silicone oil	Water and soaps water and glycerol oleates	[16] [17]
Tin(II) oxide	Petroleum fractions	water or surfactant	[20]
Titanium Dioxide	Mineral oil or p-xylene or polyphenylmethylsiloxane	Water and glycerol oleates	[19]
Cr_2O_3 doped TiO_2	Dimethylsilicone	No	[18]
Al_2O_3, Cu_2O, MgO, ZnO, La_2O_3, ZrO_2, Ta_2O_3, MnO_2, CoO, Nb_2O_3 etc.	Mineral oil	Polybutylsuccimide	[19]
$AgI/Ag_2O/V_2O_5/P_2O_5$	Silicone oil		[20]

1.3.1.2 Non-oxide inorganic materials

Non-oxide inorganic ER fluids were mainly developed in the later 1980s and early 1990s. They could give an extremely strong ER effect without any amount of water, though water could also enhance the ER effect substantially. The innovation of these kinds of ER fluids was very inspiring at that time. Among them, aluminosilicates, especially zeolite family materials, received a great deal of attention. Table 3 shows some non-oxide inorganic ER fluids reported to date.

Table 3 Non-oxide inorganic ER fluids

Dispersed phase	Dispersing phase	Additive	Ref.
Aluminosilicate(Si/Al ratio from8:1 to 175:1)	Silicone oil or hydrocarbon	Surfactant	[21]
Crystalline Zeolite $(M_{(x/n)}[(AlO_2)_x(SiO_2)_y]\bullet wH_2O$ M is metal cation	Silicone oil or high dielectric constant hydrocarbon oil		[22]
Zeolite	Silicone oil or hydrocarbon oil		[23]
Micro-glass beads	transformer oil		[2]
Aluminosilicate with 1-25% crystallized water, Si/Al ratio between 0.15-0.80	silicone oil	polyhydroxyl siloxane	[24]
Silicate , Silica- alumina	mineral oil, polyalkylene, paraffin mineral oil, polyphenyl, phosphoric acid ester, etc	sulfonates phenates phosphonates succinic acid, etc	[25]
$LiN_2H_5SO_4$	silicone oil	Block copolymer	[26]
BN, AlN, B_4C	silicone oil	Succimide	[27]
$[Mg_3(Si_{3.67}, Al_{0.33})O_{10}(OH)_2]$ $Na_{0.33} \cdot nH_2O$	*n*-hexadecane	cationic surfactants	[28]

1.3.1.3 Organic and polymeric materials

Although the non-oxide inorganic ER fluids do not contain water and have a strong ER effect, the conductivity of those ER fluids is relatively too high, especially in a high temperature environment. The density of the dispersed phase is much higher than that of the dispersing medium, causing instability of the ER suspension for a long period of time. The inorganic particle is usually too hard for the ER devices, potentially creating scratch and abrasive to the device surface. The organic and polymeric materials are softer than inorganic materials, and they may not have such a disadvantage as the inorganic particle. Some organic and polymeric ER fluids are shown in Table 4.

Organic and polymeric ER fluids can be classified into two categories. The first is the material having a conjugated π bond. Two methods are generally used to control the conductivity of the conjugated material: doping the metallic ion or metallic oxide and controlling the carbonating temperature. Acene-quinone radical polymers and polyacrylonitrile are polymers of conjugated π bond. This kind of material could be highly polarized under an electric field, thus they may have a large dielectric constant. The second category is the material having a highly polarizable group on the molecular chain, such as hydroxyl, cyano, or amido groups, etc.. Polymethyacrylic acid, starch, cellulose, chitosan, dextran, cyclodextrin, etc., belong to this group. Cyclodextrins (CDs) are cyclic molecules that consist of six to eight glucose units: α, β, γ-cyclodextrins with six, seven, and eight glucose units, respectively. Their cylindrical structures with cavities of about 0.7 nm deep and 0.5– 0.8 nm inside diameter yield various derivatives via the formation of host-guest complex. The chemical structure of some polymeric materials is shown in Figure 3. Those materials are also called polyelectrolytes and have a high molecular weight and a high charge density. Almost all polyelectrolytes, including both natural and synthetic, can display a remarkable ER effect, which is probably related to their strong moisture adsorptivity. However, the ER effect of the organic and polymeric ER fluids is usually weaker than that of the non-oxide inorganic materials.

Table 4 The organic and polymeric ER fluids

Dispersed phase	Dispersing phase	Additive	Ref.
Acene-quinone radicals polymer	Chlorinated hydrocarbon		[29]
Crosslinked polyvinylsilane	Fluorinated silicone oil	Electrolytes	[30]
Cellulosic material	Transformer oil, vegetable oil, etc silicone oil	Water or other electrolyte	[31] [32]
Chitosan	Silicone oil		[33-35]

Polyphenyl Polyvinylideneshalides	Mixture of carbon-based oil	Aromatic hydroxyl compound	[36]
Polypyrroles	Transformer oil, vegetable oil Chlorinated paraffins, etc		[37]
Polyanilines	Silicone oil		[38] [39]
Ionic Dye Material	Mineral oil, silicone oil, White oil, etc	Molecule containing hydroxyl carboxyl	[40]
Sodium carboxylmethyl Dextran	Polychlorinated biphenyls or o-dichlorobenzene, etc	Sorbitan monooleate	[41]
Polymeththacrylic acid cross linked with divinylbenzene	Hydrocarbons	Water	[41]
Starch	Mineral oil, transformer Hydrocarbon oil	Water Water	[42] [43]
Oxidized Polyacrylonitrile	Silicone oil		[44]
Carbonated aromatic sulfonic acid or a salt	Fluorosilicone oil or other modified silicone oil		[45]
Cyclodextrin–Epichlorohydrin–Starch	Silicone oil		[46]
CD-PAN	silicone oil		[47]

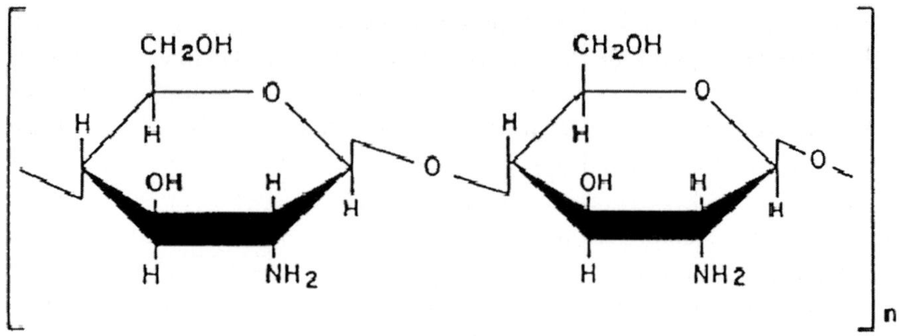

(a) Chitosan

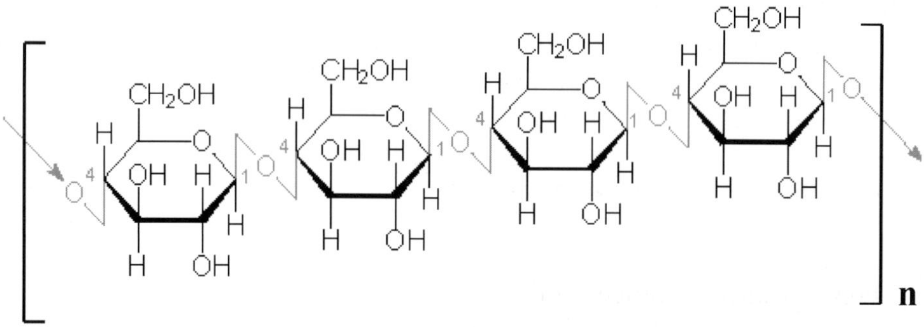

(b) Cellulose

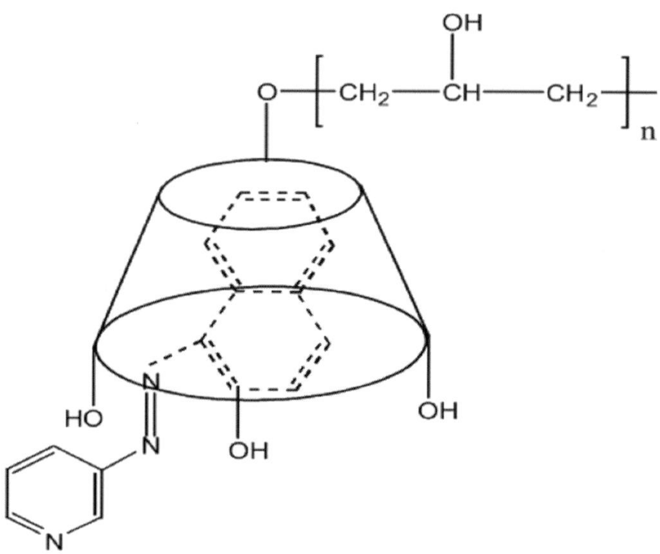

(c) β-Cyclodetrin polymer- 1-(2-pyridylazo)-2-naphthol

Figure 3. Chemical structure of some polymeric ER materials.

1.3.2 Liquid material—homogeneous ER fluids

The homogeneous ER fluids, a liquid dispersed into an insulating oil, were thought to be the best ER fluids, because this kind of ER fluid does not have the particle sedimentation problem, which is the big disadvantage of the heterogeneous ER fluids. However, this kind of ER fluid gives a relatively weak ER effect: phase segregation frequently happens. They also have a large viscosity at zero electric field, which is unfavorable for practical applications. Liquid crystalline polymer materials constitute a large portion of homogeneous ER fluids. Some homogeneous ER fluids are shown in Table 5.

Table 5 The homogeneous ER fluids

Dispersed phase	Dispersing phase	Additive	Ref.
Aluminum soap	mineral oil, silicone oil Polyalpha-olefins, etc	2,6-ditertbuthylphenol	[48]
Poly(γ-glutamate)	Cyclic Ketone		[49]
Poly(n-hexyl isocyanate)	p-xylene		[6]
LC Polysiloxanes	Silicone oil		[9]
LC Polysiloxanes	4'-(pentyloxy)-4-biphenylcarbonitrile		[9]
4-n-pentyl-4'-cyanobiphenyl			[8]
Chlorinated paraffin/silicone oil emulsion	Silicone oil		[11]
Urethane-modified polypropylene/silicone oil emulsion	Silicone oil		[12]
Caster oil/silicone oil emulsion	Silicone oil		[13]

1.4 Additives

Additives are polar materials that can adhere to the surface of the dispersed particle. Block [50] and Tomizawa [51] presented a comprehensive review on ER additives. Water, acid (inorganic and organic), alkali, salt, and surfactants are the most commonly used additives, and are listed in Table 6. The amount of the additive is very important. Less than 0.01wt% would not give any enhancement and larger than 5 wt% would generate a large electric current [51]. When water is an additive, no matter whether it is intentionally added in the system or it unavoidably comes with the dispersed particle, the ER effect of such a system is strongly dependent on how water binds with the particle. The shear stress of ER fluids is unambiguously found to go through a maximum with the increase of water content [41, 52, 53]. An example is shown in Figure 4, the relative shear

Table 6 ER additives

Additives	Examples
Acids	**Inorganic**: Sulfuric, hydrochloric, nitric, perchloric, chromic, phosphoric, and boric acids. **Organic**: Acetic, formic, propionic, butyric, isobutyric, valeric, oxalic, and malonic acids.
Bases	Hydroxides of alkali metals and alkaline earth metals, carbonates of alkali metals, and amines, polyhydric alcohol, for examples, NaOH, KOH, $Ca(OH)_2$, Alkylamine,and ethanolamine, etc
Salts	Inorganic salts such as LiCl, NaCl, NaBr, KI, $AgNO_3$, NH_4NO_3, K_2SO_4, Na_2CO_3, K_3PO_4, and alkali metal Salts of formic, acetic, oxalic, and succinic acids
Surfactant/other additives	Water, glycerol oleates, soaps, polybutylsuccimide, polyhydroxylsiloxane, sulfonates, phenates, block copolymer, proteins, liquid crystal, rare earth electrolytes, etc.

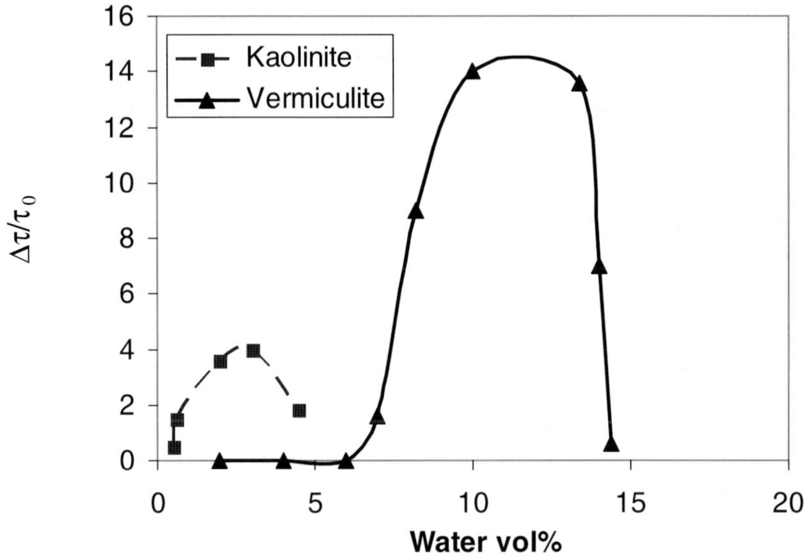

Figure 4 Relative shear stress of clay mineral suspensions against the water content. The particle volume fraction is 0.33, and the electric field is 2.8 kV/mm. Replotted from the data of Yu F. Deinega, G.V. Vinogradov, Rheol. Acta 23 (1984)636

stress, $(\tau_E - \tau_0)/\tau_0 = \Delta\tau/\tau_0$, of two types of clay mineral ER suspensions is plotted against the water concentration, where τ_E is the shear stress under an electric field, and τ_0 is the shear stress under zero electric field. The relative shear stress goes through a maximum for both suspensions but at different water content, which was believed to result from the different adsorptive capability of particle to water. A different water content dependence was found by Uejima [31]. The apparent viscosity of the crystalline cellulose/chlorine insulator oil suspension increases with water content and saturates when the water content exceeds a certain value, as shown in Figure 5.

Once water is removed from an ER system, it may loose ER effect. An example is the wet TiO_2/paraffin oil suspension as shown in Figure 6, the torques of the wet and dried TiO_2/paraffin oil suspensions vs. the applied electric field is presented [54]. The wet TiO_2 was used as received and the dried TiO_2 was maintained at 160°C for 5 hours under a liquid N_2 trapped vacuum. A remarkable ER effect was observed in the wet TiO_2/paraffin suspension and there is almost no ER effect once the wet TiO_2 powder was dried and the water was removed.

The function of water in the ER effect was primarily assumed to increase the dielectric constant of the particle, which in turn results in an enhancement of particle interaction [31; 52]. Another hypothesis is that water would bind the particles together due to the high surface tension of water [55]. Detailed mechanism on how water enhances the ER effect will be addressed in a future chapter.

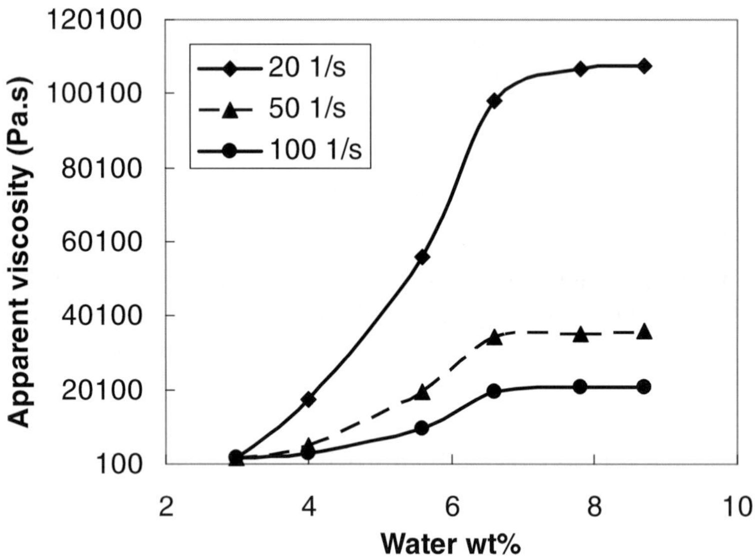

Figure 5 Apparent viscosity of crystalline cellulose/chlorine insulator oil against the water content measured at 30 °C. The particle weight fraction is 10 wt% and the electric field 1000 V. Replotted from the data of H. Uejima, Jpn. J. Appl. Phys. 11 (1972) 319.

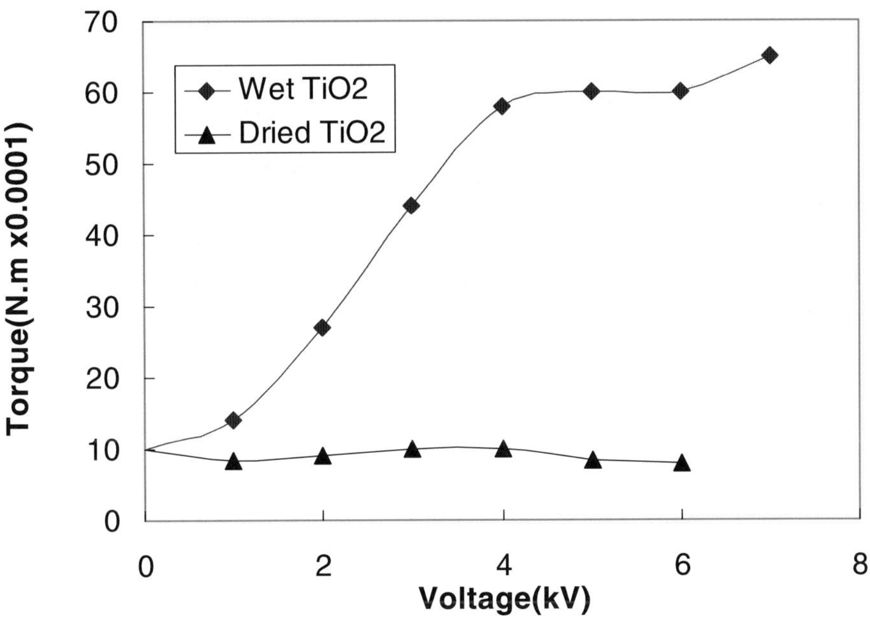

Figure 6 The torques of the wet and dried TiO_2/paraffin oil suspensions vs. the applied electric field, Replotted with permission from F. E. Filisko, Electrorheolgical materials, Encyclopedia of Smart Materials, Mel Schwarz, Ed., John Wiley & Sons, Inc, 2002, p376.

Surfactant has two roles for an ER suspension: Improving the particle sedimentation property and enhancing the ER effect [56]. A maximum yield stress is observed when the surfactant concentration varies from 0 to 7 wt% for alumina/silicone oil suspensions with different water content (see Figure 7). When the surfactant concentration is low, the yield stress increases with the concentration increase of the surfactant. This may result from the surfactant-enhanced particle polarization, in which proton transportation rate

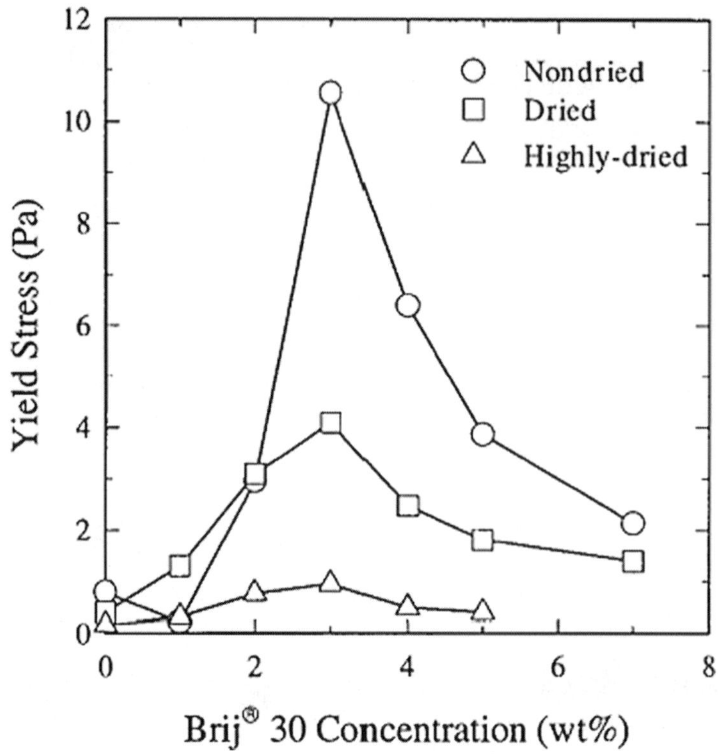

Figure 7 Yield stress as a function of Brij 30 concentration for 20 wt% neutral alumina suspensions in silicone oil, with varying water contents. Nondried, 3.47 wt% (g H_2O/100 g dried alumina) ; dried, 1.20 wt%; highly dried, 0.53 wt% (E =1.5 kV/mm, f_E = 500 Hz). Brij 30 is $C_{12}H_{25}(OCH_2CH_2)_4OH$. Reproduced with permission from Y.D. Kim and D. Klingenberg, J. Colloid Interface Sci. 183(1996) 568

may increase via neighboring hydrogen bonds at relative high surfactant concentrations. Once the surfactant concentration exceeds a certain value, the decreased yield stress may result from the electric field-induced phase separation of surfactant-rich phase that forms conductive interparticle bridges, as shown in Figure 8. The microscopic structure of two glass beads with different surfactant concentrations under different electric fields is directly observed under a microscope. The left side three images in Figure 8 are taken under a fixed electric field (E=513 V/mm, and f_E=500 Hz) with the surfactant concentration varying from 1 wt% to 7 wt%.

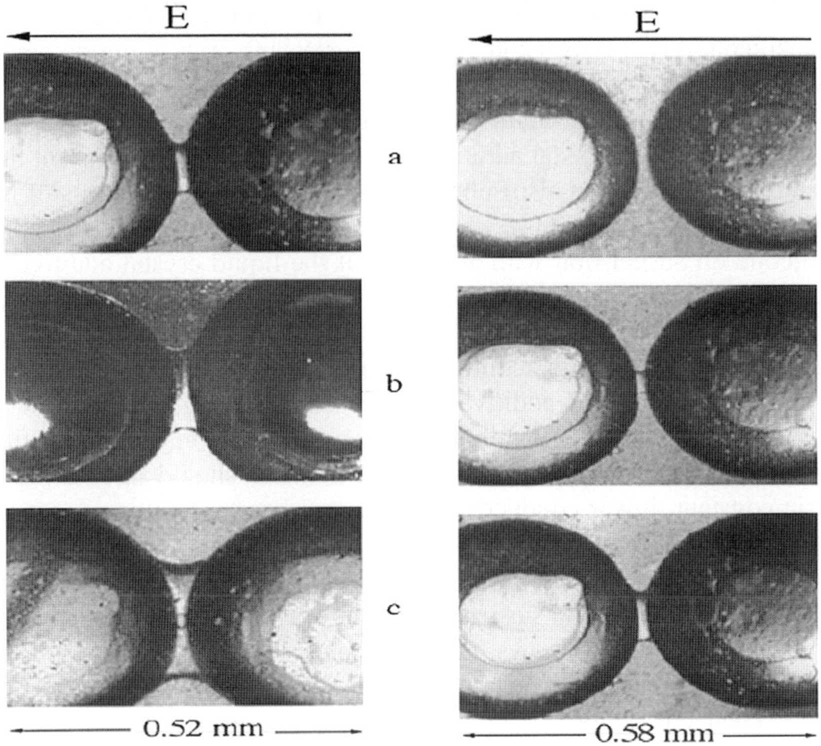

Figure 8. Surfactant structure formation between two glass beads. In surfactant solutions under an applied electric field of 513 V/mm (f_E=500 Hz); (a) 1 wt%, (b) 3 wt%, and (c) 7 wt% Brij 30. In a 1 wt% Brij 30 solution for different electric field strengths (f_E =500 Hz); (d) E = 0 V/mm, (e) E =256 V/mm, and (f) E= 641 V/mm. Reproduced with permission from Y.D. Kim and D. Klingenberg, J. Colloid Interface Sci. 183(1996) 568.

Clearly, a big surfactant bridge between two glass beads builds up with the increase of the surfactant concentration. The right side three images in Figure 8 are taken under a fixed surfactant concentration of 1 wt% but at different electric fields, increasing from 0 to 641 V/mm. Without an external electric field, there is no a surfactant bridge between two glass beads. Once a small electric field is applied, a small surfactant bridge comes up and gradually becomes bigger and bigger with the further increase of the applied electric field. Those bridges may reduce the local electric field strength between particles, thus reducing the interparticle force and hence the yield stress of whole ER suspension. This surfactant bridge model is further developed for quantitatively predicting the yield stress of the alumina/silicone oil suspension at the surfactant concentration range between 0 and 7 wt% [57,58].

Polypeptides such as proteins are found to markedly enhance the ER effect [59]. The liquid crystal additive with a higher dielectric constant than that of the carrier fluid can also substantially increase the yield stress of Na-type zeolite/oil ER fluid [60]. The yield stress of a 30 vol% zeolite NaY/silicone oil suspension with and without the liquid crystal additive, 4-heptyl-4′-cyance-biphenyl, is plotted as a function of the applied electric field in Figure 9. The addition of 1 vol% 4-heptyl-4′-cyance-biphenyl can increase the yield stress of zeolite NaY/silicone oil suspension almost 4 times. A theory on the ER additive is proposed by taking into account the surface tension, dielectric and conduction effects on the ER fluid performance [61]. Based on this theory, an additive should have a higher dielectric constant, lower conductivity, larger surface tension as opposed to that of the carrier fluid.

Rare earth (RE) electrolyte additive, such as $RE(ClO_4)_3$, $RE(NO_3)_3$, and $RECl_3$, even can control the suspension to show a positive or a negative ER effect [62]. Figure 10 shows the apparent viscosity of diatom earth particle/silicone oil in the presence of $Nd(ClO_4)_3$ against the electric field strength. When the diatom earth particle concentration is 28.6 wt%, this suspension shows a weak positive ER effect when the applied electric field is less than 1.5 kV/mm. Once the applied electric field is larger than 1.5 kV/mm, this suspension show a negative ER effect. In contrast, when the diatom earth particle concentration is 24.1 wt%, this suspension shows a relatively strong positive ER effect, and the apparent viscosity jumps to a high value at 1.5 kV/mm. Note that Nd^{3+} content is the same, 2.46 wt%, for both suspensions, and the ratio of the rare earth electrolyte to the diatom earth particle is low in 28.6 wt% suspension than that in 24.1wt% suspension. The different ER performances could result from the different ratio of the rare earth electrolyte to the diatom earth particle.

Besides water, other polar liquids such as alcohol, dimethylamine, acetamide, diethylamine, glycerol, etc., can enhance the ER effect substantially. The small amount of polar liquid may dramatically increase the dielectric constant of the dispersed particle, which was regarded as the possible reason that the ER effect was manifested. Anionic, cationic, and nonionic surfactants are commonly used in the ER systems, while amphoteric surfactant is rarely used.

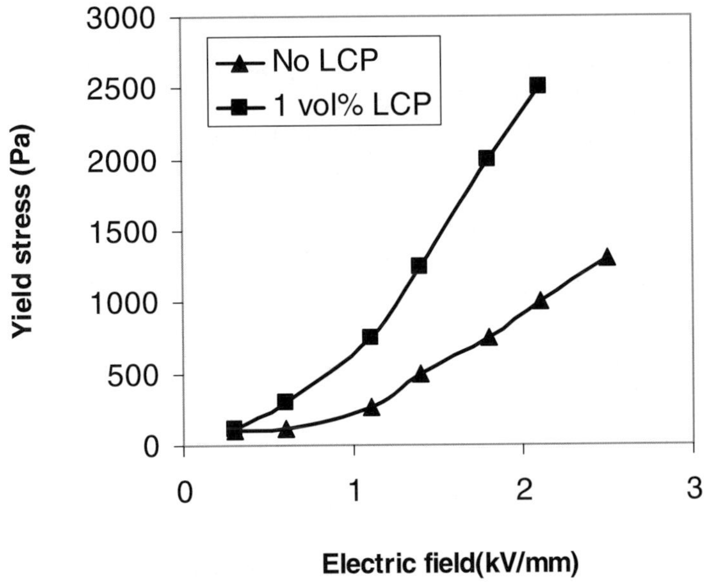

Figure 9 The electric field dependence of the yield stress of a 30 vol% Zeolite NaY/silicone oil suspension with and without the liquid crystal additive, 4-heptyl-4'-cyance-biphenyl. Replotted from the data of X. Duan, H. Chen, Y. He, W. Luo, J. Phys. D: Appl. Phys. 33(2000)696

1.5 Stability of ER suspensions

Particle sedimentation is often a problem in ER fluids containing the solid particle. As mentioned above, additive and surfactant are frequently used for enhancing both the stability of the ER suspension and the ER effect finally. One way to resolve this problem is to make a polymer coated microballoon particle, matching the density between the particle and the carrier liquid and thus reducing the particle sedimentation. An example is the poly(vinyl alcohol) (PVA) coated silica microballoon dispersed in the mixture of heptane and toluene [63]. The shear stress against the electric field is shown in Figure 11 for such a system. In both particle concentrations, 10 wt% and 30 wt%, the coated samples show a much better sedimentation property and much stronger ER effect, also. However, it may be hard to solely attribute the enhanced ER performance to the improved sedimentation property, as the PVA may act as an additive to enhance the ER effect.

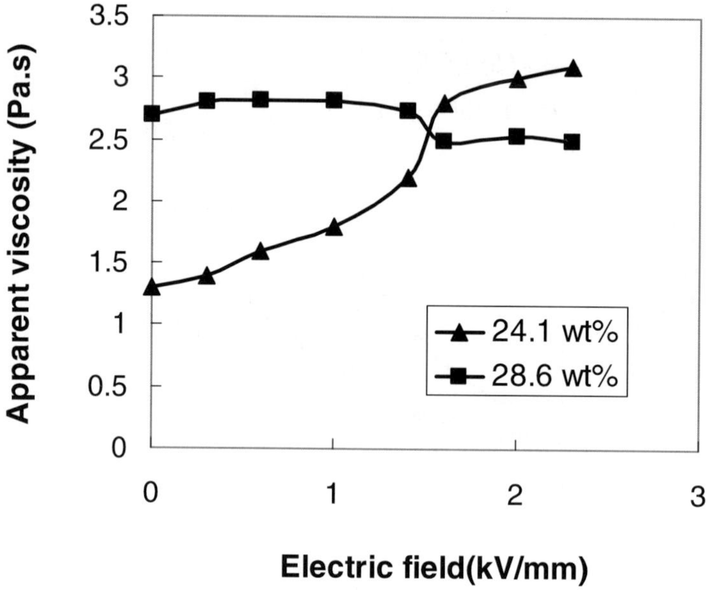

Figure 10 Apparent viscosity of diatom earth particle/silicone oil in the presence of $Nd(ClO_4)_3$ against the electric field strength. Nd^{3+} content in both suspensions is 2.46 wt%, and the shear rate is 16.49 s^{-1}. Replotted from the data of J. Li, L. Zhao, H. Liu, Solid State Commun. 107(1998)561

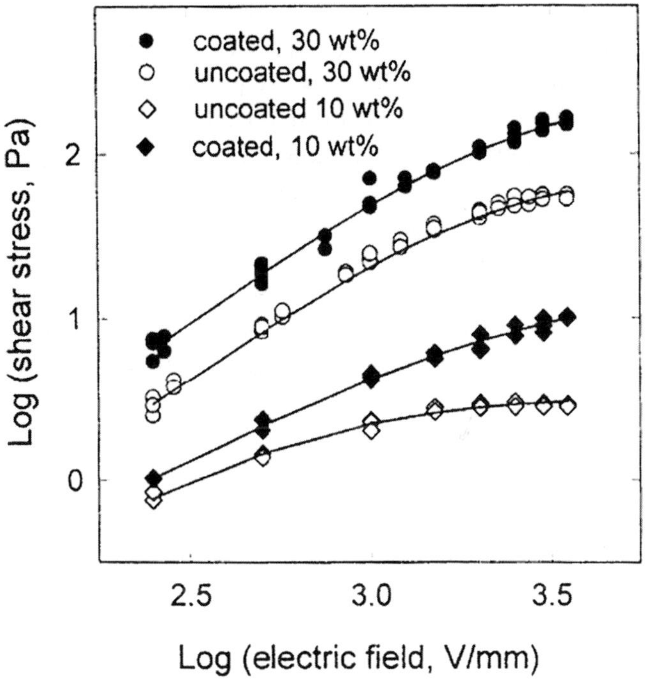

Figure 11 The shear stress of PVA coated silica microballoons dispersed in the mixture of heptane and toluene. Reproduced with permission from M. Qi, and M.T. Shaw, J. Appl. Polym. Sci., 65(1997)539

A claimed direct correlation between the stability of the suspensions against sedimentation and the magnitude of ER effect was presented by Dürrschmidt [28]. The ER properties of synthetic clay mineral (saponite)/n-hexadecane containing cationic surfactants and various additives were investigated. A summary of all the cationic surfactants and additives used in this study is shown in Table 7. The cationic surfactants are chlorinated or brominated ammoniums/pyridiniums, and the additives are alcohols. The sedimentation ratio, defined as the volume of upper clear liquid divided by the volume of whole suspension, is plotted against the yield stress of suspensions containing different surfactants and the same amount of decanol at 2000 V/mm in Figure 12. The low sedimentation ratio corresponds to the high stability of the suspension, which shows the high yield stress, implying that the more stable the suspension is, the stronger ER effect this suspension has. If the surfactant is kept the same, and the additives are changed for the suspensions, a similar trend is also found. Figure 13 shows the sedimentation ratio against the yield stress of the suspensions containing various additives and the same surfactant, $(C_{16})_2$DMABr in Table 7. Without any additive, the suspension is the most unstable and gives a weakest ER effect. For the same additive, for example, GMO in Table 7, the suspension becomes more stable with the concentration increase of GMO, and presents a much stronger ER effect. Again, the general trend is that the most stable suspension shows the strongest ER effect. Since similar surfactants or additives were used, they may modify the dielectric constant of the dispersed particle in a similar way after they were added into the suspensions. It is thus reasonable to attribute the enhanced ER effect to improved particle sedimentation stability. Particle settling may compete with the formation of chain-like or column structure in an electric field. So unstable ER suspensions should obviously give a weak ER effect.

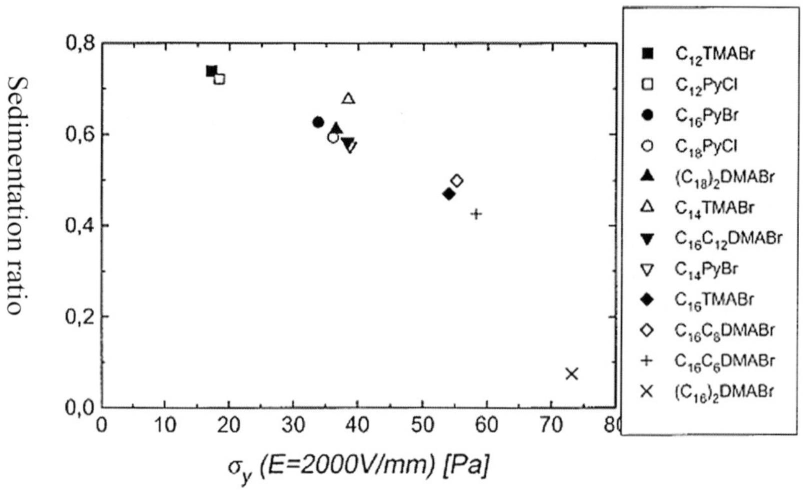

Figure 12 Sedimentation ratio ($V_{\text{clear liquid}}/V_{\text{total}}$) vs. yield stress at a field strength of 2000 V/mm for samples containing 0.05 g/ml of hydrophobically modified saponite and 0.47 mol/l of 1-decanol in *n*-hexadecane. The surfactants adsorbed on the saponite particles are given in the legend. All samples stood 1 day at rest before the sedimentation ratio was measured. Reproduced with permission from T. Dürrschmidt, and H. Hoffmann, Colloids and Surfaces A, 156(1999)257

Table 7 Cationic surfactants and additives used in the clay mineral (saponite)/n-hexadecane ER suspensions, Reproduced with permission from T. Dürrschmidt, and H. Hoffmann, Colloids and Surfaces A, 156(1999)257

Surfactant	Formula	Supplier
Cationic[a]		
Dodecyltrimethylammonium bromide (C_{12}TMABr) puriss quality	$C_{12}H_{25}N^+(CH_3)_3Br^-$	Eastman Kodak
Tetradecyltrimethylammonium bromide (C_{14}TMABr) puriss quality	$C_{14}H_{29}N^+(CH_3)_3Br^-$	Aldrich
Hexadecyltrimethylammonium bromide (C_{14}TMABr) puriss quality	$C_{16}H_{33}N^+(CH_3)_3Br^-$	Merck
Dodecylpyridinium chloride (C_{12}PyCl) puriss quality	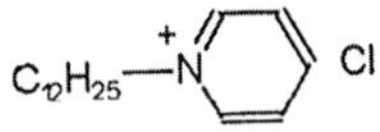	Merck
Tetradecylpyridinium bromide (C_{14}PyBr) puriss quality	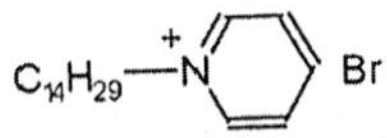	Aldrich
Hexadecylpyridinium bromide (C_{16}PyBr) puriss quality	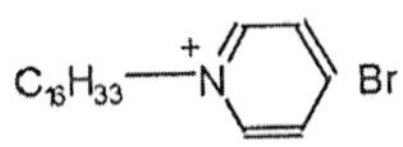	Aldrich
Octadecylpyridinium chloride (C_{18}PyCl) puriss quality	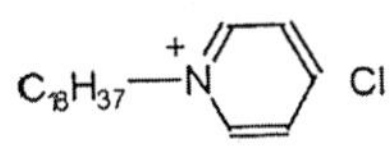	Henkel
Hexadecylhexyldimethylammonium bromide ($C_{16}C_8$DMABr) puriss quality	$(C_{16}H_{33})(C_6H_{13})N^+(CH_3)_2Br^-$	Hoechst
Hexadecyloctyldimetylammonium bromide (($C_{16}C_8$)DMABr) puriss quality	$(C_{16}H_{33})(C_8H_{17})N^+(CH_3)_2Br^-$	
Dodecylhexadecyldimetylammonium bromide (($C_{16}C_{12}$)DMABr) puriss quality	$(C_{16}H_{33})(C_{12}H_{25})N^+(CH_3)_2Br^-$	Hoechst
Dihexadecyldimethylammonium bromide (($C_{16})_2$DMABr) puriss quality	$(C_{16}H_{33})_2N^+(CH_3)_2Br^-$	c
Dioctadecyldimethylammonirum bromide ($C_{18})_2$DMABr) puriss quality	$(C_{18}H_{37})_2N^+(CH_3)_2Br^-$	Hoechst
[b]Additive		
Decanol (C_{10}OH) purum quality	$C_{10}H_{21}OH$	Fluka
Octadecanol (C_{18}OH) puriss quality	$C_{18}H_{37}OH$	Fluka
Glycerol monooleate (GMO) Technical product containing 38–40% diglyceride, 16–18% triglyceride and 4–5% glycerol	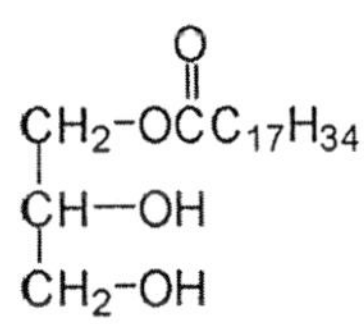	Fluka
Polyethyleneglycol ethers of alcohols (technical products):		
$C_{12}E_9$	$C_{12}H_{25}O(CH_2CH_2O)_9H$	Henkel
$C_{18}E_{2.5}$	$C_{18}H_{37}O(CH_2CH_2O)_2H$/ $C_{18}H_{37}O(CH_2CH_2O)_3H$ 1:1	
Pluriol P 2000, poly(propylene oxide), MW ≈ 2000	$H[CH(CH_3)CH_2O]_mH$, ($m \approx 35$)	BASF

[a] Cationic surfactants used for modifying saponite hydrophobically.
[b] Additives used to stabilize suspensions of hydrophobically modified saponite in *n*-hexadecane.
[c] The substance was synthesised by the reaction of hexadecyldimethylamine and hexadecyl bromide in acetonitrile.

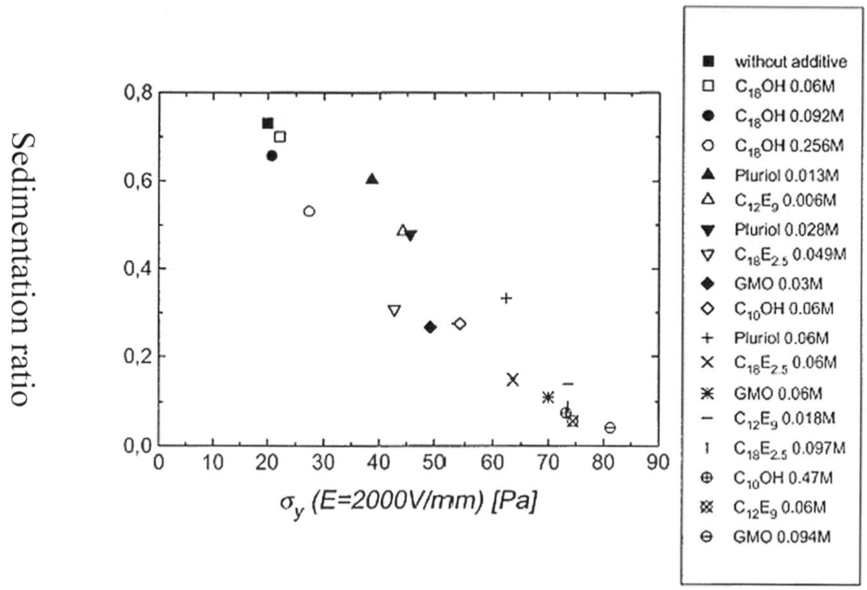

Figure 13 Sedimentation ratio ($V_{clear\ liquid}/V_{total}$) vs. yield stress at a field strength of 2000 V/mm for samples containing 0.05g/ml of hydrophobically modified saponite (adsorbed surfactant: $(C_{16})_2$DMABr) and different additives in *n*-hexadecane. The type and concentration of additives are listed in the legend. All samples stood 1 day at rest before the sedimentation ratio was measured. Reproduced with permission from T. Dürrschmidt, and H. Hoffmann, Colloids and Surfaces A, 156(1999)257.

2 POSITIVE ER MATERIALS

Positive ER materials are the ones whose rheological properties dramatically increase with the applied electric field. A plenty of efforts have been spent to develop high performance positive ER materials. Particle-type positive ER materials give a strong ER effect but have particle sedimentation problem, as mentioned earlier. Efforts are spent on developing liquid/oil emulsion ER systems and liquid/liquid miscible homogeneous ER fluids without the particulate inside. Composite core/shell particulates, mostly inorganic core shelled or modified with polymer, are developed for improving both the ER effect and the fluid stability. In this section, we will selectively discuss some positive ER materials, which in our opinion are important and intriguing in the family of the ER materials.

2.1 Aluminosilicates

Amorphous silicate ceramics are very important ER materials [64]. Among them, aluminosilicates are popularly used to make the hydrous [65] and anhydrous [22, 66-69] ER fluids of strong ER effect. The yield stress of such an ER material can easily reach 10 kPa at 2.5 kV/mm and particle loading 45 wt%. They encompass a family of zeolite materials of a general formula $M_{(x/n)}[(AlO_2)_x(SiO_2)_y]\bullet wH_2O$ (where M is a metal cation or a mixture of metal cations of average valence charge n, x, y and w are integers), including clay (saponite or montmorillonite) [28], zeolite 3A, 5A, and X–type [70,71], and various molecular sieves [72-74]. Plenty of these materials are commercially available and can be modified in various aspects with regard to the crystal structure, cation composition, and particle size and shape, etc., constituting a large number of ER fluids.

The aluminosilicates usually have a structure depicted in Figure 14, composed of tetrahedral AlO_4 and SiO_4 linked through oxygen atoms to form open frameworks. The negative charges accompanied with each aluminum atom in the frameworks are balanced by the extra-framework metal cations. The mobility of metal cations is found to play a critic role for the ER response in these systems [66, 71]. The pore size of the molecular sieve materials is also important to the ER activity [70].

The aluminosilicates materials are really attractive due to the strong ER effect they have. Since the ER effect is associated with mobility of metal cations in this type material, the current density is relatively high, contributed from the movement of metal cations, adsorbed moisture or the crystallized water, especially in high temperatures. Another shortcoming is particle sedimentation problem, which can be improved using suitable surfactants [28]. The aluminosilicate particles are hard and abrasive to the ER device.

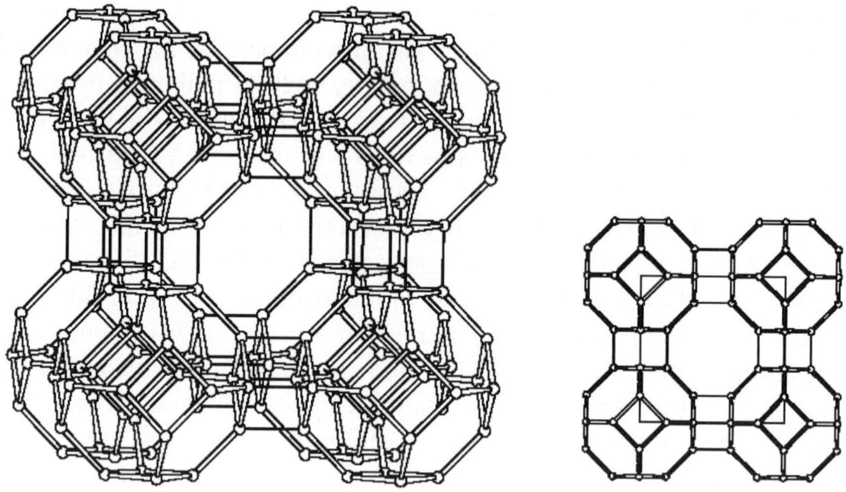

Figure 14 The perspective view of the unit cell of Linda type A zeolite (LTA, Na^{+}_{12} $(H_2O)_{27/8}$ $[Al_{12}Si_{12}\ O_{48}]_8$) is shown in the left and the parallel projection is in the right.

2.2 Conductive organics and polymers

Organic and polymeric semi-conductive materials are found to show a strong ER effect. Examples are oxidized polyacrylonitrile [44,75], polyanilines [38, 39,76], poly(p-phenylene) [77,78], ionic dye materials [79], polypyrroles and their derivatives [37, 80], polythiophenes [80] etc.. They generally are electronic conductive materials with a π conjugated bond structure. They are believed to have a better dispersing property as opposed to that of inorganic materials. However, the ER effect of the organic and polymeric ER fluids is relatively weaker compared with that of the aluminosilicate materials.

2.2.1 Oxidized polyacrylonitrile

Oxidized polyacrylonitrile is one kind of semi-conductive polymeric materials with a strong ER effect [44,75,81]. Polyacrylonitrile has strong polar cyano groups, thus attaining a high dielectric constant. When polyacrylonitrile is heated at 200~300°C under oxygen atmosphere, the cyano groups will be cyclized, as shown in Figure 15. If continuously heated from 600 to 1300 °C, the adjacent chains will join together, forming ribbon-like fused ring polymers with nitrogen atoms along their edges. The newly formed ribbons can merge to form even wider ones while more nitrogen is expelled from the system. An almost pure carbon material of graphite structure is formed in the end if the polyacrylonitrile is carbonated thoroughly. By controlling the carbonating temperature and time, various oxidized polyacrylonitrile materials of different nitrogen and hydrogen contents thus different conductivities and dielectric constants are obtained. They are ideal candidates to be used for investigating the ER mechanism [67] and for making high performance ER fluids [82]. The ER effect of oxidized polyacrylonitrile materials still has much room for improving via chemical doping and surface modification methods.

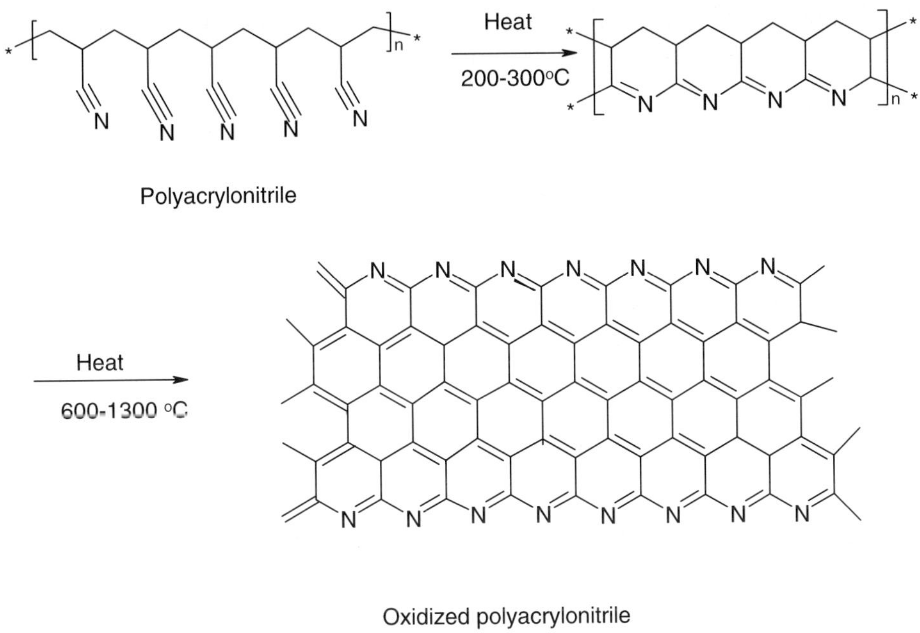

Figure 15 Carbonating process from the polyacrylonitrile to oxidized polyacrylonitrile

2.2.2 Polyanilines and polypyrroles

Polyanilines and polypyrroles, including their derivatives, are popularly used as positive ER materials, as they can be easily prepared with a controllable conductivity via changing the pH value (see Figure 16) using both the conventional chemical and electrochemical methods [83,84]. The electrochemical method is believed to be better for controlling the oxidized state of polyaniline [85]. The polyaniline and polypyrrole particulates are soft and nonabrasive to the ER devices, thus attracting a large attention once they were found to have an ER effect. A large number of publications address those conductive polymer ER fluids [37, 84, 85-90]. However, several problems are associated with these systems such as poor dispersion stability due to strong particle aggregation and high current density. Various approaches are thus proposed to improve the performance through coating polyaniline surfaces with non-conductive polymer [91], synthesizing N-substituted polyaniline or polypyrrole copolymers [86, 92-94], doping inorganic or organic materials into polyanilines [84, 90], using polymeric stabilizer [76] mixing polyanilines or polypyrroles with organic semiconductors such as tetrathiafulvalene (TTF) or tetracyanoquinodimethane (TCNQ) [80] etc. Polyanilines can be hybridized with inorganic materials such as montmorillonite [95] and coated on the surface of inorganic particles [96] for improving the ER performance. Approaches for fabricating polyaniline-coated [97] or polypyrrole-coated [98] ceramic composite materials are available, and many polymer/inorganic composite ER materials are expected to have the advantages of both polymeric and inorganic ER materials.

2.2.3 Carbonaceous materials and fullerenes

Carbonaceous materials are obtained via heat treatment from various sources, including coal, liquefied coal, coke, petroleum, resins, carbon blacks, paraffins, olefins, pitch, tar, polycyclic aromatic compounds (naphthalene, biphenyl, naphthalene sulfonic acid, anthracene sulfonic acid, phenanthrene sulfonic acid, etc.), polymers (polyethylene, polymethylacrylate, polyvinyl chloride, phenol resin, polyacrylonitrile, etc.) [99-101]. This kind of fluids is claimed to show a strong ER effect, low electric power consumption and excellent durability [101]. Several publications addressed the ER effect and physical properties of carbonaceous ER fluids [102-104].

Fullerene-type materials are also found to show remarkable ER effect. Fullerene enriched soot and fullerenes mixture, particularly C_{60} mixed with C_{70} with trace amount of C_{84} and C_{92}, display an electrorheological behavior [105]. The ER properties of fullerene type materials can be tailored by appropriate encapsulation of ions within the hollow sphere or adsorbed on the surface.

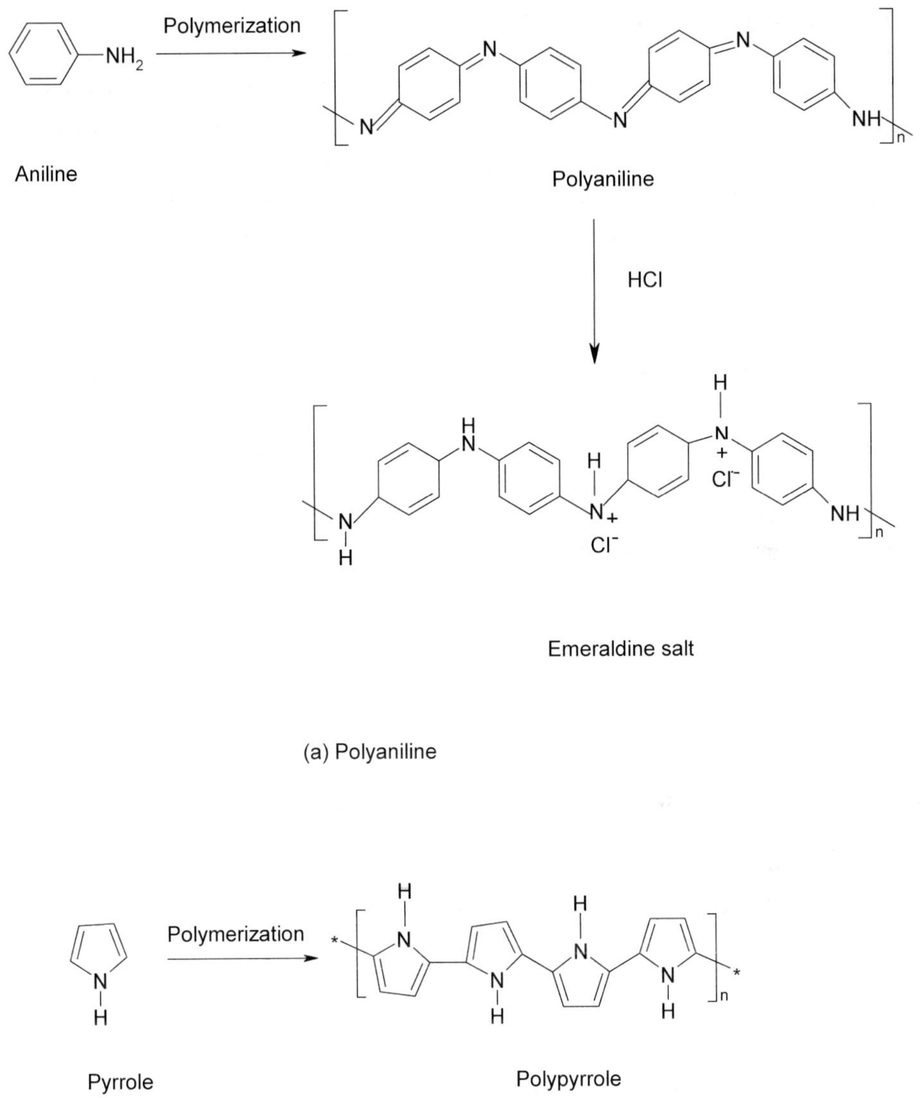

(b) Polypyrrole

Figure 16 Molecular structures of polyaniline (a) and polypyrrole (b). Emeraldine salt is mostly used for making the ER fluid.

2.3 Superconductive materials

The high-Tc superconducting particles, such as $YBa_2Cu_3O_{7-x}$, $NdBa_2Cu_3O_x$ and $YbBa_2Cu_3O_x$ and $Bi_2Sr_2CaCu_2O_{8+x}$, are found to form the fibrillated structure at room temperature in silicone oil [106], indicating that they also have the ER effect. A macroscopic ball of diameter 0.25mm is formed in millisecond if superconducting particles of the size 1-2 μm are dispersed in the liquid nitrogen under an electric field 0.8kV/mm.

2.4 Liquid materials

Since particulate-type ER suspensions have particle sedimentation problem, liquid materials are therefore used as dispersed phases. The liquid may be dispersed (immiscible) or dissolved (miscible) into the insulating oil. Various liquids are reported to show a remarkable ER effect, particularly the liquid crystal materials.

2.4.1 Immiscible with the dispersing phase

Oil-in-oil emulsion systems display a relatively strong ER effect. Examples of such ER active emulsions are chlorinated paraffin/polydimethylsiloxane [11], castor oil/polydimethylsiloxane [13], urethane-modified polypropylene glycol/dimethylsiloxane [12] etc.. The ER effect in emulsions is attributed to the stretched droplets that form fibrillation chains along the direction of the electric field. This is a typical feature for any emulsion system in which the two liquids have a quite different dielectric constant and conductivity. Figure 17 shows the water droplet chains formed in a supercritical fluid carbon dioxide medium under a 60 Hz ac field of a very low field strength, E_{max}=10 V/mm [115]. A synergetic effect is observed in an system composing of polyanilines dispersed in a chlorinated paraffin/silicone oil emulsion [107].

A strong ER effect is also found in the liquid crystal polymer/oil immiscible system. The liquid crystal polymers include various main-chain and side-chain polysiloxane [9, 108,109], poly (6-(4′-cyanobiphenl-4-yloxy) hexylacrylate) [110], poly(γ-benzyl-L-glutamate) [111] etc.. The molecular structures of these polymers are shown in Figure 18. There are many papers addressed the ER mechanism of those immiscible liquid crystal systems [10, 112, 113]. Molecular weight distribution of the liquid crystal polymer seems to play an important role [114].

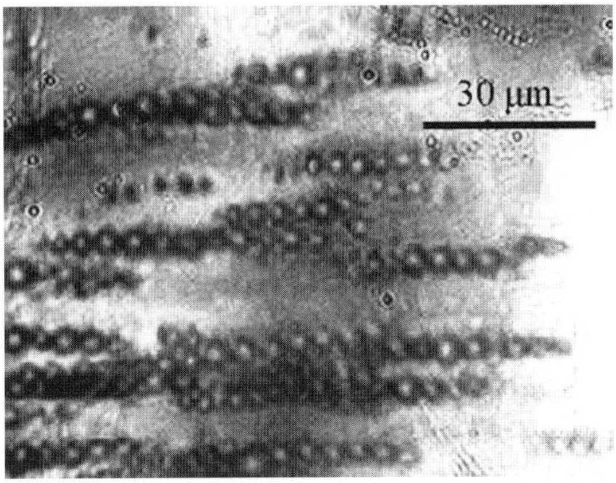

Figure 17 Photomicrograph of water droplet chains in supercritical fluid CO_2 formed in a 60Hz ac electric field (E_{max}=10 V/mm) for 3wt% water and 3 wt % 5b-$C_{12}E_8$ at 55 °C and 386 bar. 5b-$C_{12}E_8$ is The nonionic methylated branched hydrocarbon surfactant, Tergitol TMN-6. Reproduced with permission from W. Ryoo, J. L. Dickson, V. V. Dhanuka, S. E. Webber, R. T. Bonnecaze, and K. P. Johnston, Langmuir 21(2005)5914

2.4.2 Miscible with the dispersing phase

Low molecular weight liquid-crystal polymers could show the ER effect by themselves or dissolving into another liquid medium, forming miscible homogeneous ER fluids. Those liquid crystal polymers include 4-n-pentyl-4′-cyanophenyl [8], poly(n-hexyl isocyanate) [6, 114], poly(1-(4-hydroxy-4′-biphenyl)-2-(4-hydroxyphenyl)-butane [115], poly((γ-glutamate) [116]. Many other polymers can be found in the reference [117]. Maximum transmitted stresses of those systems range from 1.7 to 3.0 kPa [118]. The poly(n-hexylisocyanate) is claimed to show a transmitted shear stress 16 kPa [118]. Some organics also show the ER effect. Examples are aluminum soaps [119], and lecithin/water/glycerol organogels [120].

Homogeneous ER fluids are very promising, as they do not have particle sedimentation problem in contrast with the heterogeneous ER fluids. However, they usually have a high viscosity under zero electric field and mostly they only show the ER effect in the nematic phase. Sometimes liquid–liquid segregation problem occurs.

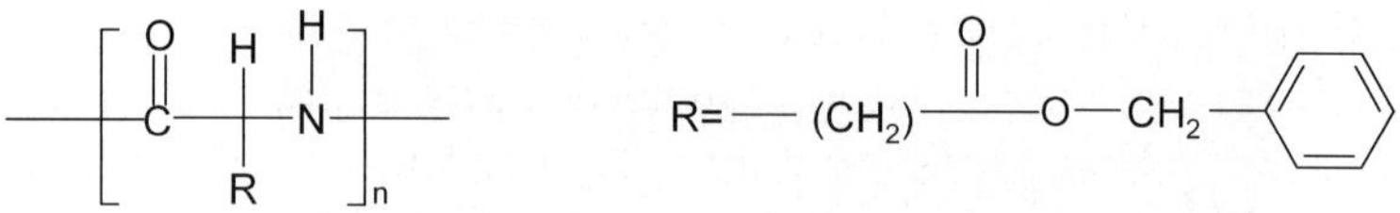

(a) Poly(γ–benzyl-L-glutamate)

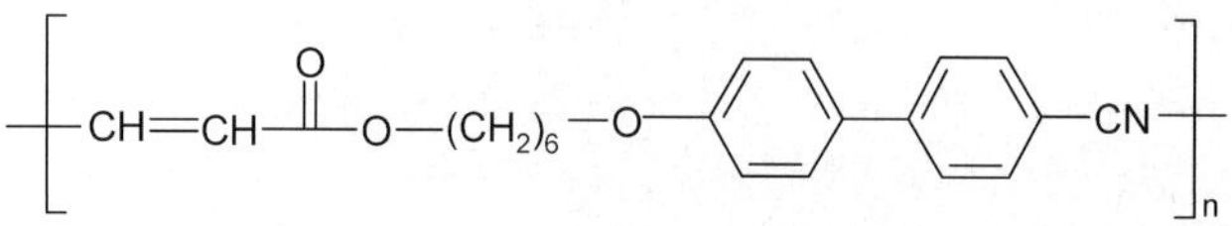

(b) poly(6-(4'-cyanobiphenl-4-yloxy) hexylacrylate

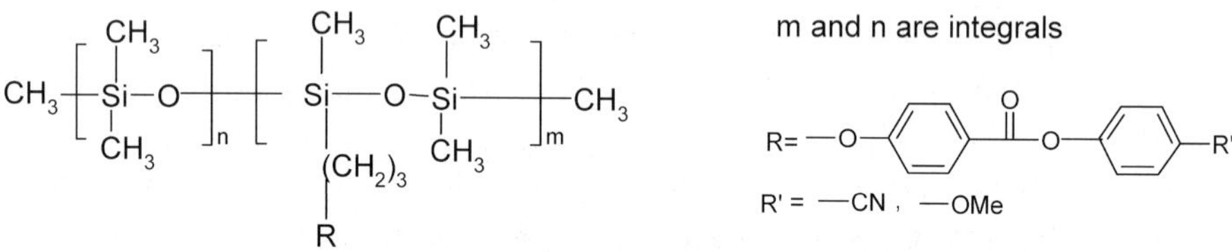

(c) Side chain liquid-crystal polysiloxane

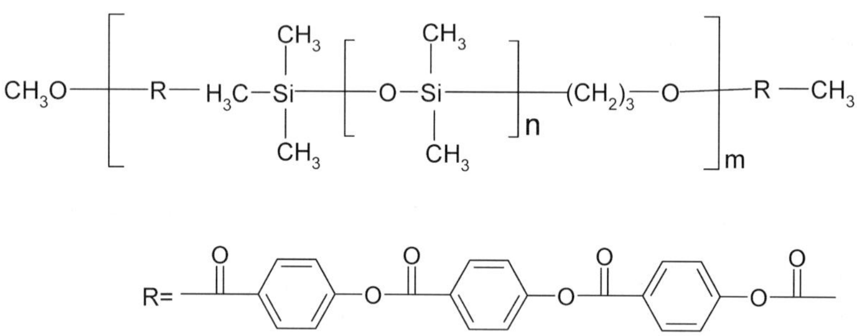

(d) Main chain liquid-crystal polysiloxane

Figure 18 Molecular structures of ER liquid crystal polymers

2.5 Core-shell composite particulates

Neither inorganics nor polymeric materials are perfect ER materials, as discussed above. Core-shell inorganic/polymer or inorganic/inorganic composite particulates may combine the advantages of both the inorganic and polymeric materials, showing enhanced ER effect and dispersing stability. Theoretical investigations show that the outermost coating should be a material of high dielectric constant (greater than 1000), high electrical breakdown strength (greater than 7 kV/mm) and a reasonable level of conductivity (about 10^{-8} Sm^{-1}) that is used to increase the density of electrostatic energy [121,122]. The multi-coated particulates are preferable for making high shear stress ER fluids. Experimental results show that the particle sedimentation property is improved once the particles are coated with a polymer [63]. The double-coated glass spheres show superior static yield stress to that of uncoated ones [123]. Styrene-acrylonitrile-clay composite particles show high shear stresses in the presence of an electric field [124]. However, the silica particles coated with polyaniline give a reduced ER effect [96]. The multi-coated powder, an acrylic resin core covered with several layers of films of different refractive index (titanium dioxide film, polystyrene film, silver metal film, etc.), is claimed to be suitable for color ink materials [125]. Polymeric particles shelled with much smaller inorganic microparticles on their surfaces are reported [126,127]. Butyl acrylate and 1,3-butanediol dimethacrylate crosslinked copolymer sphere coated titanium oxide and tin oxide shows the remarkable ER effect [128].

2.6 Design of high performance positive ER fluids

There are many variables that can influence the ER effect tremendously, making the design of a desired ER fluid difficult. Those variables are the applied electric field strength, the frequency of the electric field, the particle conductivity, the particle size and shape, the particle dielectric properties, the particle volume fraction, temperature, water content, the liquid medium, etc., which are already summarized in the references [129, 130] and will be discussed in more detailed manner in a future chapter. Drawing a guideline for designing high performance positive ER fluids requires a deep understanding of the ER mechanism. Recently, the criteria for screening materials to make high performance ER fluids based on experimental and theoretical results were proposed [67, 131,132]. According to the criteria, the dielectric loss tangent of the dispersed material should be larger than 0.1, and the dielectric constant ratio of the dispersed material to the dispersing medium should be between 50 and 60. Detailed information can be found in future chapter.

3 NEGATIVE ER MATERIALS

Fumed silica (white carbon black) [81], segmented polyurethane [133, 134], PMMA (polymethylmethacrylate) [135], magnesium hydroxide [136], poly(tetrafluoroethylene)[137], 4-methoxybenzylidene-4′-n-butylaniline [138], were found to show the negative ER effect. Liquid crystalline polysiloxane was found to show a positive ER effect when it was dissolved in the 4'-(pentyloxy)-4-biphenylcarbonitrile, however it showed a negative ER effect when it was dissolved in N-(4-methoxybenzylidene)-4-butylaniline [9]. Urethane-modified polypropylene glycol showed a negative ER effect in case that its viscosity is lower than that of the dispersing phase, while it showed a positive ER effect when its viscosity is higher than that of dispersing phase [12]. PMMA particles stabilized by polystyrene-block-poly(ethylene-co-propylene) could show either a negative or a positive ER effect, depending on the particle concentration and the content of diblock copolymer stabilizer [139]. Caster oil/silicone oil emulsion system showed a negative ER effect when the shear field and the electric field are competitive, and the viscosity reduction magnitude increases with both the viscosity and conduction ratios of the dispersed to the dispersing phases [13].

4. PHOTO-ER MATERIALS

For some ER materials the ER effect can be modified either positively or negativety via light illumination. Those materials are called photo-ER active materials, as defined previously. TiO_2 nano-particle is one of such materials that are photo-active under an electric field [140-143]. The water was found to play an important role: Low water content makes the suspension display a positive photo-induced ER effect, while high water content results in a negative photo-induced ER effect. Photo-generated carriers were thought to change the electric properties of the TiO_2 particles and then the ER performances. The dye particles (monastral Green B, copper phthalocyanine) [144] and phenothiazine [145] were also reported to show photo-ER effect.

5. ELECTRO-MAGNETO-RHEOLOGICAL MATERIALS

As defined previously, materials that can react positively with both an external electric field and/or a magnetic field separately or simultaneously are called the electro-magneto-rheological (EMR) materials. Synergic effect usually takes place and the observed EMR effect is stronger than the electrorheological effect if the electric field is only applied or the magnetorheological effect if the magnetic field is only applied. Iron particles [146], iron particles coated with titania [147], glass spheres coated with nickel and titania [148], and various ferromagnetic particles (such as magnetite, manganese ferrite, barium ferrite, iron, cobalt, nickel, permalloy, iron nitride) coated with metallic oxides (such as SiO_2, Al_2O_3, TiO_2, BaO, etc.) [149] were reported to show the EMR effect. Zeolite particle containing iron was also found to show EMR effect [150]. Composite particle SiO_2/Ni/TiO_2 was also found to show EMR effect [151].

REFERENCES

[1] H. Block, J.P. Kelly, GB Patent 2170510, 1985
[2] D.G. Bytt, GB Patent 2189803, 1987
[3] F.E. Filisko, W.E. Armstrong, U.S. Patent 4744914, 1988
[4] J.J. Wysocki, J. Adama, W. Hass, J. Appl. Phys. 40(1969)3865
[5] T. Honda, T. Sasada, K. Kurokawa, Jpn. J. Appl. Phys. 17(1978)1525
[6] I. K. Yang, and A. D. Shine, J. Rheol. 36(1992)1079
[7] A. Inoue, S. Maniwa, European Patent 478034A1, 1992
[8] K. Negita, J. Phys. Chem., 105(1996)7837;
[9] N. Yao, A.M. Jamieson, Macromolecules, 30(1997)5822
[10] K. Tajiri, H. Orihara, Y. Ishibashi, M. Doi, A. Inoue, J. Rheol. 41(1997)335
[11] X. Pan, G.H. McKinley, J. Colloid Interface Sci. 195(1997)101
[12] H. Kimura, A. Aikawa, Y. Masubuchi, J. Takimoto, K. Koyama, T. Uemura, J.Non-Newtonian Fluid Mech., 76(1998)199
[13] J. Ha, S. Yang, J.Rheol. 44(2000)235
[14] J. E. Stangroom, I. Harness, GB Patent 2153372, 1985
[15] G.G. Petrzhik, O.A. Chertkova, A.A. Trapeznikov, Dokl. Akad. Nauk USSR, 253(1980)173
[16] W.M. Winslow, U.S. Patent 2661596, 1953
[17] W.M. Winslow, U.S. Patent 3047507, 1962
[18] J. Yin, and X. Zhao, Chem. Mater.16(2004) 321
[19] M. Kanbara, H. Tomizawa, European Patent 0361931, 1990
[20] M. Osuchowski, and J. Plocharski, Int. J. Modern Phys. B, 16(2002)2378
[21] D. Gillies, L. Sutcliffe, P. Bailey, GB Patent 2219598, 1989
[22] F.E. Filisko, W.F. Armstrong, European Patent 0313351, 1989
[23] F.E. Filisko, W.F. Armstrong, European Patent 0265252, 1988
[24] J. Goossens, G. Oppermann US Patent 4702855, 1987

[25] H. Tomizawa, M. Kanbara, N. Yoshimura, J. Mitsui, H. Hirano, European Patent 0342041, 1989
[26] J.D. Carlson, US Patent 4772407, 1988
[27] M.Kanbara, and H. Tomizawa, European Patent 0361931, 1990
[28] T. Dürrschmidt, H. Hoffmann, Colloids Surfaces A: 156 (1999) 257
[29] H. Block, J.P. Kelley, US Patent 1501635, 1987
[30] W. Podszun, R. Bloodworth, G. Oppermann, US Patent 5503763, 1996
[31] H. Uejima, Jpn. J. Appl.Phys. 11(1972)319
[32] U. Choi, B. Ahn, Colloids Surfaces A: 168(2000)71
[33] U. Choi, Colloids Surfaces A: 157 (1999)193
[34] U. Choi, Y. Park, S. Lee, Colloids and Surfaces A: 211(2002)85
[35] S. Wu, F. Zeng, and J. Shen, J Appl Polym Sci 67(1998)2077
[36] J. W. Pialet, US Patent 5558811, 1996
[37] C.P. Bryant, K. Lal, J.W. Pialet, US Patent 5435932, 1995
[38] C.J. Gow, C. F. Zukoski, J. Colloid Interface Sci., 126(1990)175
[39] J.W. Pialet, C.P. Bryant, K. Lal, European Patent 0562067, 1997
[40] J.D. Carlson, K.D.Weiss, J. E. Bares, WO Patent 9312192, 1993
[41] J.E. Stangroom, GB Patent 1570234, 1980
[42] W. M. Winslow, U.S. Patent 2417850, 1947
[43] J.E. Westhaver, US Patent 3970573, 1976
[44] Y. Xu, and R. Liang, J. Rheol., 35(1991)1355
[45] S. Endo, European Patent 0964053, 1999
[46] Z.Gao, and X. Zhao, J. Appl. Polym. Sci. 93(2004)1681
[47] Z.Gao, and X. Zhao, Materials Lett. 57 (2002) 615
[48] D. Prick, H. Grasshoff, H. Kohnz, P. Finmans, T. Carstensen, D. Jakubik, W. Weber, D. Winkler, US Patent 5800731, 1998
[49] K. Haji, M. Sasaki, M.Matsuno, European Patent 0798368, 1997
[50] H. Block, J. P. Kelley, J. Phys. D: Appl. Phys. 21(1988) 1661
[51] H. Tomizawa, M. Kanbara, N. Yoshimura, J. Mitsui, H. Hirano, European Patent 0 342 041, 1989
[52] Yu. F. Deinega, G.V. Vinogradov, Rheol. Acta 23 (1984)636
[53] A.V. Lykov, Z.P. Shulman, R.G. Gorodkin, A.D. Matsepuro, J. Eng. Phys. 18 (1970)979
[54] F. E. Filisko, Electrorheolgical materials, Encyclopedia of Smart Materials, Mel Schwarz, Ed., John Wiley & Sons, Inc, 2002, p376
[55] J.E. Stangroom, Phys. Technol. 14(1983)290
[56] Y. Kim, and D. Klingenberg, J. Colloid Interface Sci. 183(1996)568
[57] Y. Kim, J. Colloid and Interface Sci., 236(2004) 225
[58] Y. Kim, and S. Nam, J. Colloid and Interface Sci., 269 (2004) 205
[59] Y. Kim, D. Klingenberg, US Patent 5 435 931, 1995
[60] X. Duan, H. Chen, Y. He, W. Luo, J. Phys. D: Appl. Phys. 33(2000)696
[61] X. Duan, W. Luo,W.Wu, J. Phys. D: Appl. Phys. 33(2000)3102
[62] J. Li, L. Zhao, H. Liu, Solid State Commun. 107(1998)561
[63] M. Qi, and M.T. Shaw, J. Appl. Polym. Sci., 65(1997)539
[64] D. Gamota, A. Shubring, B. Mueller, F. Filisko, J. Mater. Res. 11(1996)144
[65] H. Conrad, A. Sprecher, Y. Choi, Y. Chen, J. Rheol. 35(1991)1393

[66] F. Filisko, L. Radzilowski, J. Rheol. 34(1990)539
[67] T. Hao, Z. Xu, Y. Xu, J. Colloid Interface Sci. 190(1997)334
[68] R. Baranwal, and R. Laine, J. Am. Ceram. Soc. 80(1997)1436
[69] T. Hao, A. Kawai, F. Ikazaki, Langmuir 14(1998)1256
[70] M. Cho, H. Choi, I. Chin, W. Ahn, Microporous Mesoporous Mater. 32(1999)233
[71] H. Bose, in Proc. Int. Conf. ER Fluids, M. Nakano and K. Koyama Eds, World Scientific, Singapore 1998, p. 240.
[72] T. Hao, Y. Hao, Y. Xu, J. Colloid Interface Sci. 184(1996)542.
[73] T. Hao, Y. Xu, J. Colloid Interface Sci. 185(1997)324.
[74] H. Choi, M. Cho, K. Kang, W. Ahn, Microporous Mesoporous Mater. 39(2000)19.
[75] T. Hao, Y. Chen, Z. Xu, Y. Xu, Y. Huang, Chin. J. Polym. Sci. 12(1994)97
[76] B. Chin, O. Park, J. Colloid Interface Sci. 234(2001)344.
[77] J. Plocharski, M. Rozanski, H. Wycislik, Synth. Met. 102(1999)1354.
[78] J. Trlica, P. Saha, O. Quadrat, J. Stejskal, Physica A 283(2000)337.
[79] D. Carlson, D. Weiss, E. Bares, World Patent WO9 312 192, 1993
[80] W. Pialet, O. Havelka, US Patent 5 879 582, 1999
[81] T. Hao, Ph.D. Thesis, the Institute of Chemistry, Chinese Academy of Sciences, Beijing, 1995.
[82] R. Liang, and Y. Xu, in Proc. Int. Conf. on Eletrorheological Fluids (Ed: R. Tao), World Scientific, Singapore 1994, p. 233
[83] A. MacDiarmid, Synth. Met. 84(1997)27.
[84] P. Romero, N. Pastor, M. Cantu, Solid State Ionics 101-103(1997)875.
[85] K. Slack, J. Akhavan, H. Block, in Proc. Int. Conf. ER Fluids, Eds: M. Nakano, K. Koyama, World Scientific, Singapore 1998, p. 314.
[84] J. Plocharski, M. Rozanski, H. Wycislik, Synth. Met. 102(1999)1354
[85] W. Podszun, R. Bloodworth, G. Oppermann, US Patent 5 503 763, 1996.
[86] J. Goodwin, G. Markham, B. Vincent, J. Phys. Chem. 101(1997)1961.
[87] H. Xie, J. Guan, J. Guo, J. Appl. Polym. Sci. 64(1997)1641.
[88] O. Quadrat, J. Stejskal, J. Rheol. 42(1998)13.
[89] J. Lee, M. Cho, H. Choi, M. Jhon, Colloid Polym. Sci. 277(1999)73.
[90] S. Kim, J. Kim, H. Choi, M. Suh, M. Shin, M. Jhon, Colloid Polym. Sci. 278(2000)894.
[91] J. Akhavan, K. Slack, V. Wise, H. Block, in Proc. Int. Conf. ER Fluids Eds: M. Nakano, K. Koyama, World Scientific, Singapore 1998, p. 322
[92] M. Cho, J. Kim, H. Choi, M. Suh, R. Webber, M. Jhon, Colloid Polym. Sci. 278(2000)61.
[93] H. Choi, J. Kim, K. To, Polymer 40(1999)2163.
[94] J. Trlica, P. Saha, O. Quadrat, J. Stejskal, Eur. Polym. J. 36(2000)2313
[95] J. Kim, S. Kim, H. Choi, M. Jhon, Macromol. Rapid Commun. 20(1999)450.
[96] J. Trlica, P. Saha, O. Quadrat, J. Stejskal, J. Phys. D: Appl. Phys. 2000, 33, 1773.
[97] L. Frisch, B. Xi, Y. Qin, M. Rafailovich, N. Yang, X. Yan, High Perform. Polym. 12(2000)543
[98] G. Cho, B. Fung, D. Glatzhofer, J. Lee, and Y.Shul, Langmuir, 17(2001)456
[99] M. Sasaki, Y. Kobayashi, K. Haji, US Patent 5 705 088, 1998.

[100] T. Maruyama, T. Ogino, Y. Ishino, T. Saito, T. Haraoka, K. Takagi, H. Hatano, US Patent 5 779 880, 1998.
[101] S. Endo, H. Wada, European Patent 0 964 053, 1999
[102] H. Choi, J. Kim, S. Yoon, R. Fujura, M. Komatsu, J. Kim, M. Jhon, J. Mater. Sci. Lett. 18(1999)1445.
[103] X. Pan, G. McKinley, Langmuir 14(1998)985.
[104] A. Bezryadin, R. Westervelt, M. Tinkham, Phys. Rev. E 59(1999)6896.
[105] B. Powell, US Patent 5 445 759, 1995
[106] R. Tao, X. Zhang, X. Tang, P. Anderson, Phys. Rev. Lett. 83(1999)5575
[107] B. Chin, O. O. Park, J. Rheol. 44(2000)397
[108] A. Inoue, Y. Ide, H. Oda, J. Appl. Polym. Sci. 64(1997)1319.
[109] A. Inoue, S. Miniwa, J. Appl. Polym. Sci. 55 (1995)113.
[110] R. Kishi, T. Kitano, H. Ichijo, Mol. Cryst. Liq. Cryst. 280(1996)109.
[111] K. Tanaka, R. Akiyama, K. Takada, J. Appl. Polym. Sci. 66(1997)1079
[112] A. Inoue, Y. Ide, S. Maniwa, H. Yamada, H. Oda, MRS Bull. 23(1998)43.
[113] K. Tajiri, H. Orihara, Y. Ishibashi, M. Doi, A. Inoue, J. Rheol. 43(1999)137
[114] K. Tanaka, A. Takahashi, R. Akiyama, and N. Kuramoto, Phys. Rev. E 59(1999)5693.
[115] W. Ryoo, J. L. Dickson, V. V. Dhanuka, S. E. Webber, R. T. Bonnecaze, and K. P. Johnston, Langmuir 21(2005)5914
[114] I. Yang, and I. Huang, J. Polym. Sci., Part B: Polym. Phys. 35(1997)1217
[115] Y. Chiang, A. Jamieson, M. Kawasumi, V. Percec, Macromolecules 30(1997)1992.
[116] K. Hji, M. Sasaki, M. Matsuno, European Patent 0 798 368, 1997
[117] S. Miniwa, A. Inoue, US Patent 5 891356, 1999
[118] K. Tse, A. Shine, Macromolecules 33(2000)3134
[119] D. Pirck, H. Grasshoff, H. Kohnz, P. Finmans, T. Carstensen, D. Jakubik, W. Weber, D.Winkler, US Patent 5 800 731, 1998.
[120] Y. Shchipunov, T. Durrschmidt, H. Hoffmann, J. Colloid Interface Sci. 212(1999)390.
[121] C. Wu, H. Conrad, J. Phys. D: Appl. Phys. 31(1998)3312.
[122] H. Zhao, Z. Liu, Y. Liu, Solid State Commun. 116(2000)321
[123] W. Tam, G. Yi,W. Wen, H. Ma, M. Loy, P. Sheng, Phys. Rev. Lett. 78(1997)2987.
[124] J. Kim, M. Noh, H. Choi, D. Lee, M. Jhon, Polymer, 41(2000)1229
[125] K. Nakatsuka, T. Atarashi, European Patent 0 913432, 1999
[126] K. Edamura, Y. Otsubo, M. Mizoguchi, US Patent 5 736 064, 1998.
[127] S. Katayana, N. Yamada, I. Yoshinaga, Japanese Patent 10 140 171, 1998.
[128] K. Hanaoka, S. Tanaka, Y. Nakazawa, T. Fukami, and K. Sakurai, Electrical Engineering in Japan, 142(2) (2003)1
[129] T. C. Jordan, M. T. Shaw, IEEE Trans. Electr. Insul. 24(1989)849
[130] H. Conrad, MRS Bull. 23(1998)35
[131] T. Hao, A. Kawai, F. Ikazaki, Langmuir, 15(1999)918.
[132] T. Hao, A. Kawai, F. Ikazaki, Langmuir,16(2000)3058
[133] T. Uemura, K. Minagawa, K. Koyama, Chem. Lett. Chem. Soc. Jpn. 1995-1996(1994)
[134] T. Uemura, K. Minagawa, K. Koyama, Polym. Prepr. 35(1994)360.
[135] C. Boissy, P. Atten, J.-N. Foulc, J. Electrostatics, 35(1995)13

[136] J. Trlica, O. Quadrat, P. Bradna, V. Pavlinek, P. Saha, J. Rheol. 40(1996)943
[137] C. Wu, and H. Conrad, J. Rheol. 41(1997)267
[138] K. Negita, Chem. Phys. Lett. 246(1995)353
[139] V. Pavlinek, P. Saha, O. Quadrat, J. Stejskal, Langmuir, 16(2000)1447.
[140] Y. Komoda, T. N. Rao, A. Fujishima, Langmuir, 13(1997)1371.
[141] Y. Komoda, N. Sakai, T. N. Rao, D. A. Tryk, A. Fujishima, Langmuir 14(1998)1081.
[142] Y. Komoda, T. N. Rao, N. Sakai, D. A. Tryk, A. Fujishima, J. Electroanal. Chem. 459(1998)155.
[143] Y. N. Sakai, Y. Komoda, T. N. Rao, D. A. Tryk, A. Fujishima, J. Electroanal. Chem. 445(1998)1
[144] L. Carreira, V.S. Mihajlov, U.S. Patent 3553708, 1971
[145] F. E. Filisko, Proc. Intern. Conf. Electrorheological Fluids , R.Tao, Ed., World Scientific, 1992, p116
[146] K. Koyama, K. Minagawa, T. Watanabe, Y. Kumakura, J. Takimoto, J.Non-Newtonian Fluid Mech., 58 (1995)195
[147] H. Takeda, K. Matsushita, Y. Masubuchi, J. Takimoto, K. Koyama, Proc. Int. Conf. ER fluids, M. Nakano and K. Koyama Ed., World Scientific, Singapore, 1998, p571
[148] P. Sheng, W. Wen, N. Wang, H. Ma, Z. Lin, W. Tam, C. Chan, Physica B, 279(2000)168
[149] M. Sasaki and H. Sato, US Patent 5702630, 1997
[150] A. Shibayama, T. Miyazaki, K. Yamaguchi, K. Murakami, and T. Fujita, Int. J. Modern Phys. B, 16(2002)2405
[151] H. Guo, X. Zhao, H. Guo, and Q. Zhao, Langmuir 19(2003)9799

Chapter 5

Critical parameters to the electrorheological effect

The ER effect depends on the applied electric field strength, the frequency of the electric field, the particle conductivity, the particle dielectric properties, the particle volume fraction, temperature, water content, the liquid medium, even the electrode pattern, etc. Those parameters are reviewed in this chapter.

1 THE ELECTRIC FIELD STRENGTH

Since the ER effect is induced by the electric field, the impact of the electric field strength on the ER effect has been extensively studied. A critical electric field strength, E_c, exists, and the ER fluid could not show any ER effect until the applied electric field strength exceeds the critical one [1,2]. Theoretical treatment [3] shows that the critical electric field may be expressed as:

$$E_c = \frac{\rho - c}{|\alpha|}\sqrt{8\pi k_B T/(c\rho\bar{v}\varepsilon_m)} \quad (1)$$

where ρ is the density of particle, c is the concentration of particle in g/ml. $\bar{v}$ is the average volume of a particle. $k_B T$ is the thermal energy, ε_m is the dielectric constant of liquid medium. α is a constant

$$\alpha = \frac{g}{g + \varepsilon_m/(\varepsilon_p - \varepsilon_m)} \quad (2)$$

where ε_p is the dielectric constant of particle, and g is a numerical constant related to the particle geometry.

$$g \begin{cases} = 1/3 \text{ for spherical particle} \\ > 1/3 \text{ for rod particle} \\ < 1/3 \text{ for disk particle} \end{cases} \quad (3)$$

Eq. (1) indicates that the critical electric field strength decreases with the increase of particle concentration. The critical electric field concept was also used in several other theoretical treatments [4-6]. However, the critical electric field was assumed to be a constant only depending on the electric property of the dispersing medium. The liquid-solid phase transition induced by the external electric field is supposed to happen at the critical electric field that is in the order of several hundred to one thousand volts per millimeter. In consideration of electric-field-induced particle aggregation in ER fluids, Khusid [7] suggested that the critical electric field is the function of particle volume fraction, the dielectric constant and conductivity of both the particle and dispersing medium. It should be as low as 14 V/mm, as confirmed experimentally [8]

The relationship between the shear stress (or yield stress) and the electric field is complicated. In early literature, the yield stress was reported to linearly increase with the electric field [1]:

$$\tau_y=k(E-E_c) \tag{4}$$

Where τ_y represents the yield stress of an ER fluid, k a constant, E the applied electric field strength. However, many other researchers [9-12] found that the yield stress should be directly proportional to the square of the electric field strength. An example is shown in Figure 1, in which the yield stress of silica/silicone oil vs. the product of the square of applied electric field and the particle volume fraction is plotted. When the particle volume fraction is less than 30 vol%, a good linear relationship between the yield stress and the square of the applied electric field is obtained. When the electric field is high enough, the yield stress and the apparent viscosity tend to become saturated [13-16] . A schematic diagram of such a saturation phenomenon is shown in Figure 2. There are many other examples, both theoretical and experimental, showing that at low electric fields the yield stress is proportional to the square of the electric field strength and at high electric fields it is proportional to the 3/2 power of the electric field strength [17-19]. The yield stress of microencapsulated polyaniline with the melamine-formaldehyde (MCPA) dispersed in silicone oil is plotted against the applied electric field in Figure 3, showing that two regimes exist for this system. When the applied electric field is less than 1 kV/mm, the yield stress scales as $\tau_y \propto E^2$, while once the applied electric field is above 1 kV/mm, the yield stress scales as $\tau_y \propto E^{1.5}$. Directly measured interaction force between two polyamide semispheres with radius 7 mm immersed in silicone oil is shown in Figure 4 [20]. Clearly two regimes were found: at low electric field, the attraction force is proportional to the square of the electric field strength, and at high electric field, the attraction force is linearly proportional to the electric field. The relationship between yield stress and the electric field should follow the trend of the dielectric property of the suspension on the electric field, which will be discussed later in more detail.

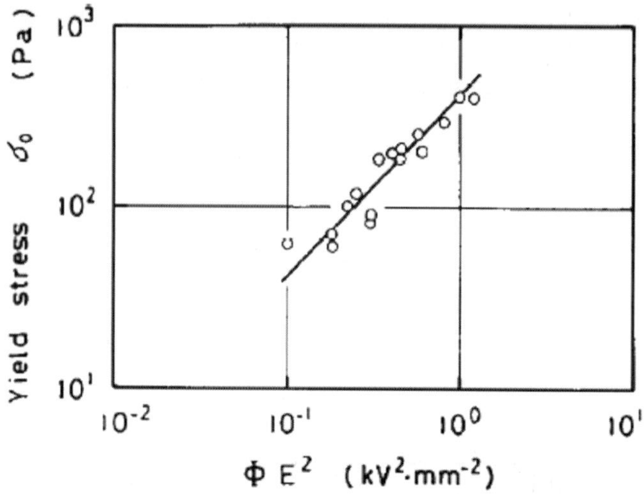

Figure 1 The yield stress of silica /silicone oil suspension against the product of the square of applied electric field and the particle volume fraction. The particle volume fraction is below 30 vol% and the particle size is 0.45 μm in diameter. Reproduced with permission from Y. Otsubo, J. Rheol. 36 (1992) 479.

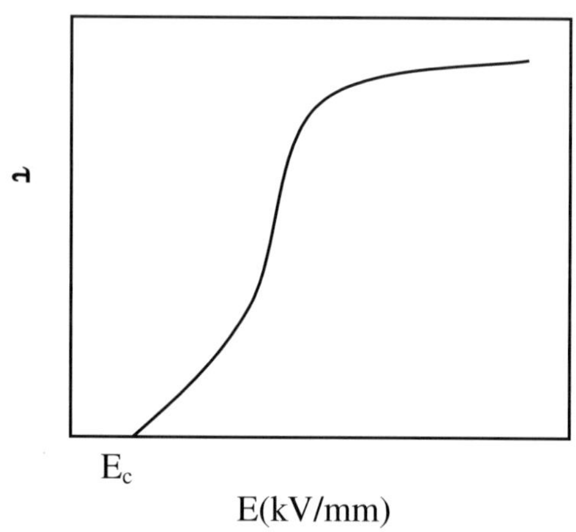

Figure 2 Schematic illustration of shear stress vs. the applied electric field.

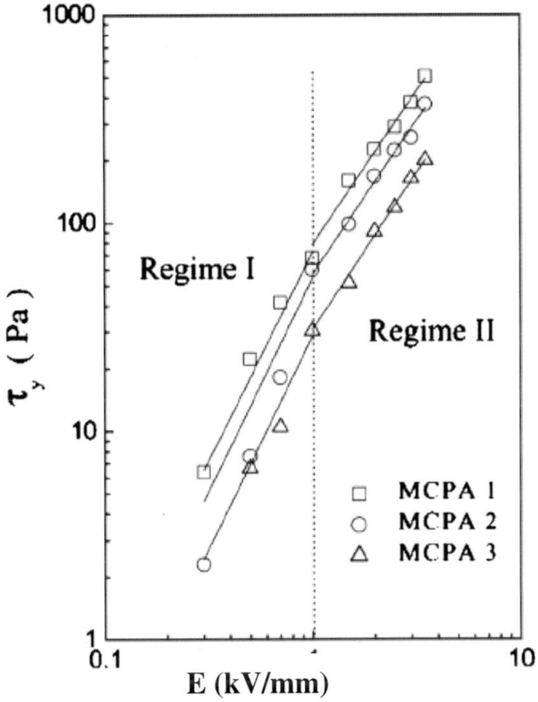

Figure 3 The yield stress τ_y vs. *E* for 20 wt% suspension of MCPA (microencapsulated polyaniline with the melamine-formaldehyde) particle dispersed in silicone oil. The ratios by weight of polyaniline to melamine-formaldehyde resin are 10/114 for MCPA1, 10/152 for MCPA2 , and 10/190 for MCPA3 , respectively. Regime I $\tau_y \propto E^2$, and in Regime II, $\tau_y \propto E^{1.5}$. Reproduced with permission from H.J. Choi, M.S. Cho, and J.W. Kim, Appl. Phys. Lett.,78(2001)3806.

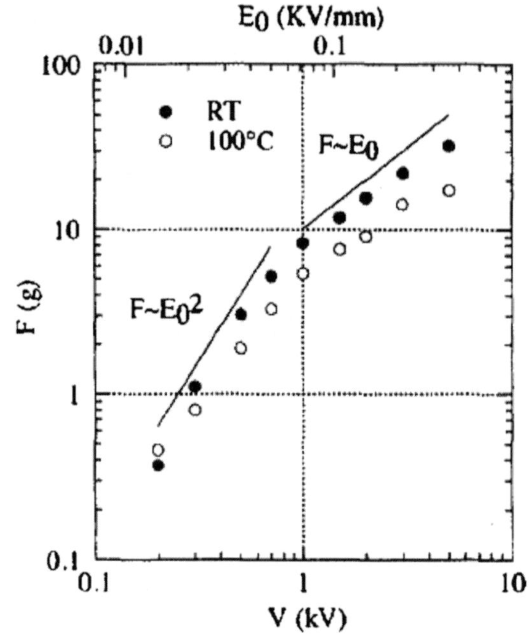

Figure 4 Attraction force F vs. applied voltage V (or mean electric field E_0) at room temperature and at 100 °C. F is the attraction force between two hemispheres(radius R=7mm) made of polyamide and immersed in silicone oil. Reproduced with permission from P. Gonon, and J-N Foulc, J. Appl. Phys. 87(2000)3563

2 FREQUENCY OF THE ELECTRIC FIELD

A dc electric field is mostly used to generate a detectable ER effect. An ac electric field is very useful to study the mechanism of the ER effect and to determine the response time of an ER fluid. Since an ER fluid has a response time around 1 millisecond, its viscosity and yield stress are expected to decrease with the increase of frequency, as it would be unable to catch up with the oscillation of the electric field at high frequencies. Klass [21,22] first addressed this issue, and found that the apparent viscosity of a silica/naphthentic system with nonionic surfactant decreases with the increase of frequency, as shown in Figure 5.

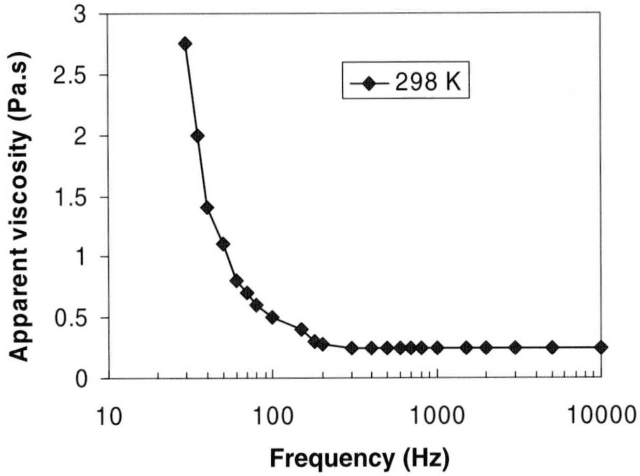

Figure 5 Apparent viscosity of silica/naphthenic system with nonionic surfactant against the frequency of ac field obtained at room temperature. The field strength is 930 V/mm, shear rate 23.6 s^{-1}, and particle volume fraction is 0.38. Redrawn from D.L. Klass, T.W. Martinek, J. Appl. Phys. 38 (1967) 75.

If the ratio of the surfactant concentration to particle concentration became lower, a quite different dependence of the apparent viscosity on frequency was obtained, as shown in Figure 6. A sharp decrease followed by recovery of the apparent viscosity was found to occur at 200 Hz. A so-called Flow-Modified-Polarization (FMP) was proposed to explain this sharp discontinuity by Block [23, 24] A resonance between the applied electric field and mechanical field takes place when the shear rate is 4π times the frequency of the applied electric field.

$$\dot{\gamma} = 4\pi f_E \qquad (5)$$

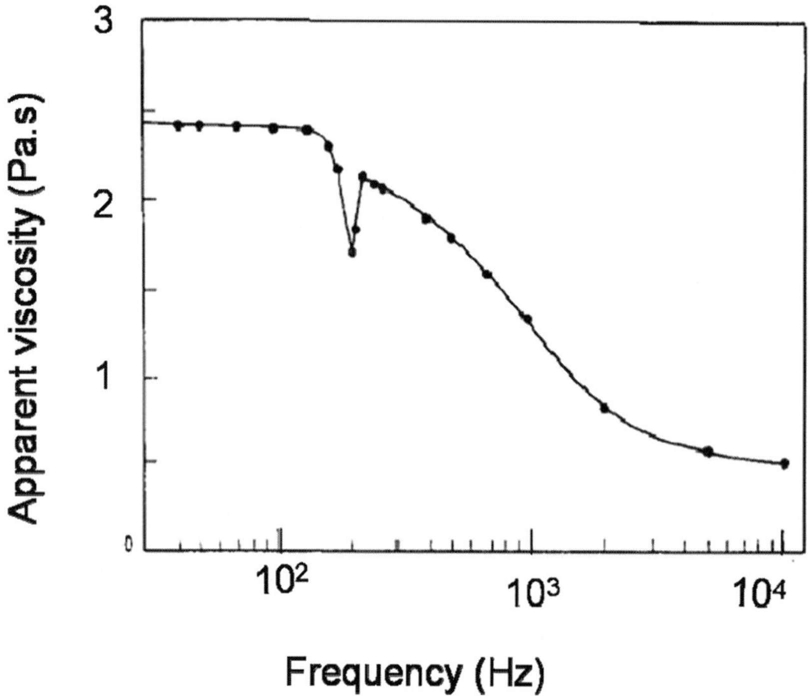

Figure 6 Apparent viscosity of silica/naphthenic against the frequency of ac field. The field strength is 930 V/mm, shear rate 753 s^{-1}, and particle volume fraction 0.46. Replot of the data from D.L. Klass, T.W. Martinek, J. Appl. Phys. 38 (1967) 67.

The dielectric constant of poly(lithium methacrylate) dispersed in Cereclor oil at a volume fraction of 0.3 vs. the frequency of electric field under different shear rates is shown in Figure 7. The dielectric constant reaches a maximum value when Eq. (5) is satisfied. This phenomenon is called the FMP resonance between a shear field and an electric field. The flow field definitely has an influence on the particle polarization and hence on the dielectric constant. The dielectric constant thus becomes a function of both the shear rate and the frequency of the applied electric field. Especially when the flow field is rotational and strong enough to such an extent that the particle is able to spin, it may compete with the applied electric field for particle polarization. In other words, the particles or particle clusters can be orientated not only under an electric field, but also under a shear flow field. FMP was also observed in rigid or flexible polymer solutions [25-27].

Theoretical treatments on FMP [28-30] confirmed the resonance condition as expressed in Eq. (5).

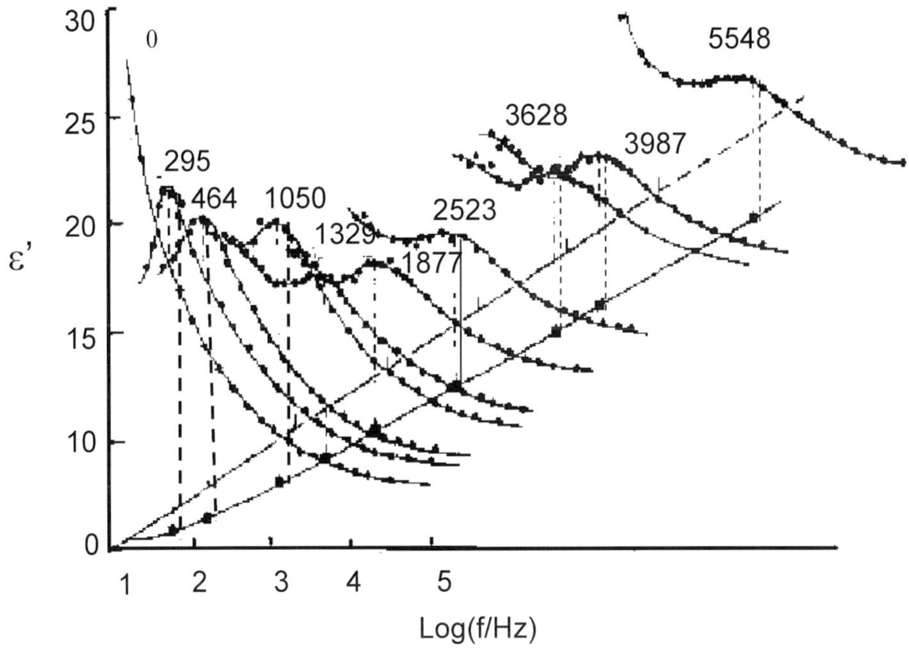

Figure 7. The dielectric constant of poly(lithium methacrylate) dispersed in Cereclor oil at the particle volume fraction of 0.3 vs. the frequency of electric field under different shear rates that are indicated adjacent to individual curves. Reproduced with permission from H. Block, J.P. Kelley, A. Qin, T. Watson, Langmuir 6 (1990)6

A weak particle interaction is expected when the resonance between an electric field and a flow field occurs. This is the reason that a sharp decrease of the apparent viscosity is observed in Figure 6. However, another experimental result showed that the shear stress of an ER fluid, 400 nm TiO_2 particle/silicone oil, increases rather than decreases when the FMP resonance occurs, as shown in Figure 8 [31]. Computer simulation indicated that the FMP plays a crucial role in producing ER effect at high shear rates [32]. Note that the FMP resonance would not always occur in all ER fluids, and is also hard to detect even if Eq.(5) is satisfied. In most cases the shear stress or apparent viscosity would follow the trend shown in Figure 5, which was confirmed in many ER fluids [33-36]

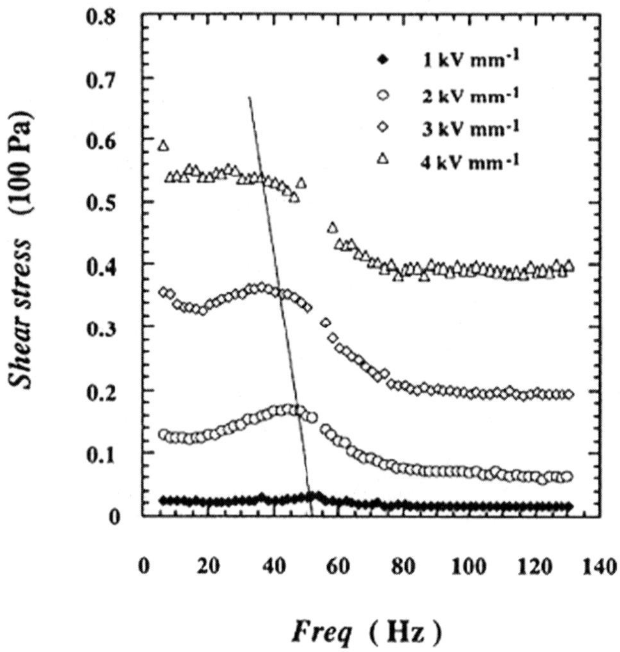

Figure 8 Shear stress of TiO_2/silicone oil vs. the frequency of electric field at a shear rate 659 s^{-1} and particle volume fraction 5.5 vol%. The shear stress peak shifts with the amplitude of the field. At the limit of zero ac field, the peak frequency converges to $659/4\pi$ =52.4Hz, which is the value at which the resonance is supposed to happen. Reproduced with permission from K. Negita and Y. Ohsawa, Phys. Rev. E, 52(1995)1934.

The Wagner-Maxwell polarization [37-39] model was used to explain the frequency dependence of the yield stress [40] It was found that the yield stress drop corresponds to the decrease of dielectric constant of the whole suspension when the frequency increases. The particle conductivity determines whether the yield stress drop appears at a high frequency or a low frequency. Highly conductive particle gives a marked ER effect even in very high frequency fields, indicating that this suspension has a short response time. An oxidized polyacrylonitrile/silicone oil ER fluid was investigated under an ac field, and a good agreement between the experiment result and the prediction based on the Wagner-Maxwell polarization was obtained, as shown in Figure 9. The experimentally determined shear stress under 300 V/mm and the calculated dielectric constant from the Wagner-Maxwell model are displayed in Figure 9, and both of them follow a similar frequency dependence.

An anomalous frequency dependence of the interaction force between two identical $SrTiO_3$ spheres (6.30 ±0.01 mm in diameter) was found when those two spheres almost touch each other (gap 10 μm) [41] (see Figure 10). The interaction force increases with the increase of the frequency of applied electric field, which is opposite to the trend mentioned above. The reason may be that such big spheres are unable to move with the applied ac electric field, and thus the frequency of the electric field functions differently in this experiment.

The response time of an ER fluid can be extracted from the relationship between shear stress and frequency of applied electric field [42]. A time-regulated high-voltage power supply was designed to [43] investigate the time dependence of particle aggregation and corresponded ER effect induced by the interaction among aggregated particles.

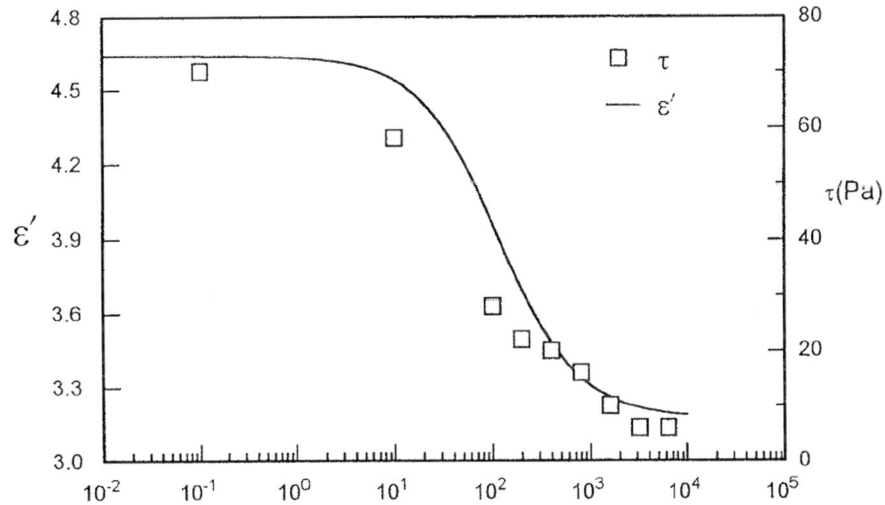

Figure 9. The shear stress obtained at 300 V/mm and dielectric constant predicted using Wagner model (solid line) are shown against the frequency of applied electric field. Reproduced with permission from T. Hao, J. Colloid Interface Sci. 206 (1998)240.

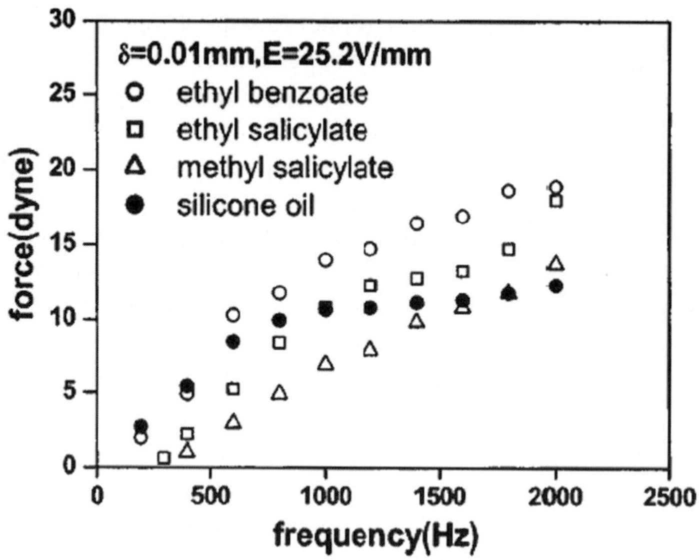

Figure 10 Frequency dependence of the force between two $SrTiO_3$ spheres in various liquid media. Reproduced with permission from Z. Wang, R. Shen, X. Niu, K. Lu, and W. Wen, J. Appl. Phys., 94(2003)7832.

3 PARTICLE SIZE AND SHAPE

The particulate size and shape have an impact on the ER effect. The influence of particle size on the ER effect is quite diversified in the literature. The particles of size from 0.1 to 100 μm are commonly used in preparation of ER fluids. The ER effect is expected to be weak if the particle is too small as the Brownian motion tends to compete with the particle fibrillation. Very large particle is expected to display a weak ER effect too, as the sedimentation would prevent the particle to form fibrillation bridges. However, the experiment results didn't show a monotonic relationship between the ER effect and the particles size. The yield stress of two ER suspensions made from $BaTiO_3$ powders of different sizes dispersed in silicone oil was found to be quite different: At dc field, the suspension containing the particle of a mean size 0.35 μm shows much stronger ER effect than another one with a mean size 70 nm [34]. However, under ac field the situation changed. The yield stress of those two suspensions against ac electric field (60 Hz) are shown in Figure 11. At low electric fields, 70 nm $BaTiO_3$ suspension shows a relative higher yield stress than that of 350 nm $BaTiO_3$ suspension. However, both of them give an identical yield stress at high electric field.

Three fumed silica of primary particle sizes 16, 18, and 19 nm were also examined for the influence of the particle size on the ER behavior after they were dispersed into silicone oil [44]. The silica of the smallest particle size shows the largest apparent viscosity under an electric field. This result is consistent with the data presented in Figure 11, which is also consistent with the result observed in colloidal suspensions without an applied electric field. There is a large body of literature on how particle size affects the viscosity of colloidal suspensions [45]. The different ER effects should come from the viscosity difference as a result of the particle size difference even without an external electric field. For dried glass beads of 6-100 μm in diameter dispersed in silicone oil, the yield stress was found to increase with particle size at low shear rate (2 s^{-1}) [46]. However, once the shear rate continuously increases up to 100 s^{-1}, the particle size effect gradually diminishes. A similar phenomenon was also observed in humidified glass beads [47], as shown in Figure 12. When Φ_1/Φ is zero, i.e., there is no small particle in the system, and all particles are of the average diameter 100 μm, the suspension has the maximum yield stress. Note that those experimental results were obtained by using the particle with size distribution other than monodispersed. Theoretical investigation [48] indicated that the ER fluid consisting of a single-sized particle gives the large yield stress.

The computer simulation results obtained by using molecular-dynamics method show that the shear stress of an ER fluid should be proportional to the cube of the particle diameter [49]. When one size of particle mixed with another size forming a bimodal suspension system, the shear stress was found to decrease both theoretically[49] and experimentally [47] with the volume fraction of smaller particle, reaching a minimum at a certain point that is dependent on the ratio of two particle sizes. These are consistent with the experimental results obtained in bimodal systems that are ER inactive [50]. However, an unusually very large enhancement of static yield stress was observed by adding nanoparticles of lead zirconate or lead titanate to an ER fluid containing 50 μm glass sphere [51,52]. A similar effect was observed in mixing two monodispersed sulfonated poly(styrene-co-divinylbenzene) particles, one is 50 μm and the other is 15 μm in diameter (see Figure 13).

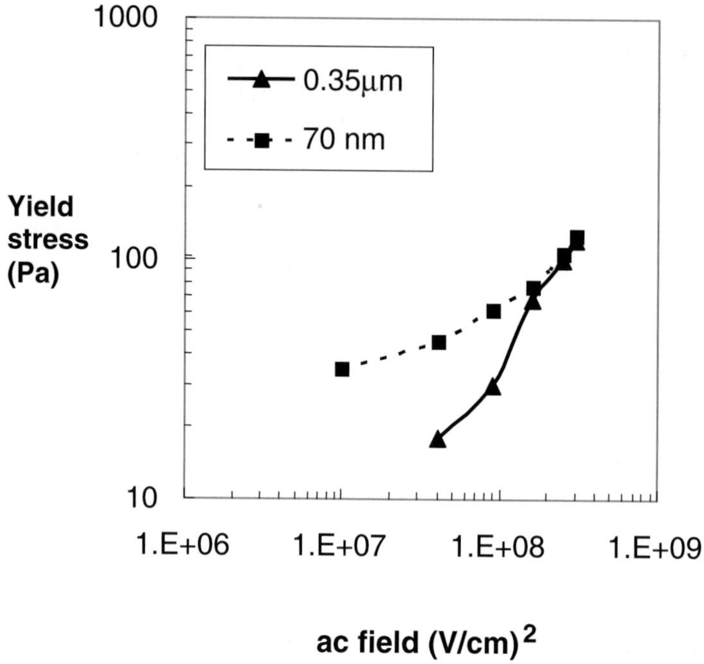

Figure 11 Yield stress of $BaTiO_3$/silicone oil suspensions with a same particle volume fraction 10 vol% but a different mean particle sizes, 350 and 70 nm.. Reproduced with permission from [D.V. Miller, C.A. Randall, A. Bhalla, R.E. Newnham, and J.H. Adair, Ferroelectric Letters, 15(1993)141

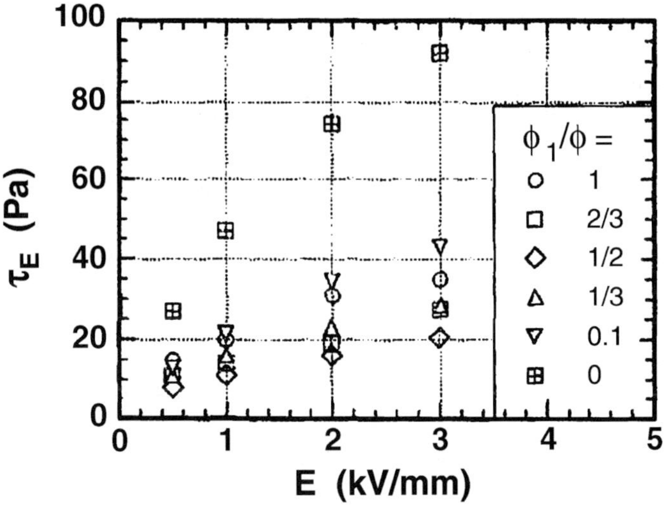

Figure 12 Yield stress against the electric field for various mixed suspensions containing humidified glass beads. Φ_1 is the volume fraction of small particles (average diameter 6 μm) and Φ is the total volume fraction (the small particles plus the large particle of average diameter 100 μm). Reproduced with permission from C.W Wu, and H. Conrad, J.Appl. Phys., 83(1998)3880]

The maximum ER effect was obtained when the mixing ratio is 0.5. When highly uniform monodispersed styrene-acrylonitrile copolymer (SAN) particles with different diameters (about 5, 10, and 15 μm) were mixed together, the bimodal ER fluid exhibited lower yield stress than the ER fluid of single-sized large particle [53]. SEM picture of SAN particle is shown in Figure 14. A perfectly spherical and monodispersed particle is used for the study. Figure 15 shows the yield stress against the mixing ratio of small particle to the total volume fraction of particle at three different electric fields. A synergistic effect was observed at the mixing ratio of 0.2. The ER behaviors of four inorganic powders (talc, magnesium hydroxide, aluminum hydroxide, and titanium dioxide) of different particle sizes and shapes dispersed in silicone oil were compared with that of polyaniline/silicone oil suspension [54]. The non-spherical or irregular particle shape can considerably increase rigidity of the structure formed in the electric field and enhance the ER performance. The dielectric properties of those suspensions were measured and particle polarizability was attributed to account for the

ER behaviors. The suspensions with mixing particle sizes definitely have different particle packing structure, depending on the size ratio used for mixing. A more compact packing structure may be formed at certain size ratio, and thus an enhanced ER effect should be expected; A more loose packing structure may be formed if the size ratio is not right. In this case, a weakened ER effect should be expected. This is the major reason that diversified results were reported in literature. Regarding how the particle packs together for monodispersed and polydispersed particles, the interested readers are recommended to read the book written by Cumberland and Cawford [55].

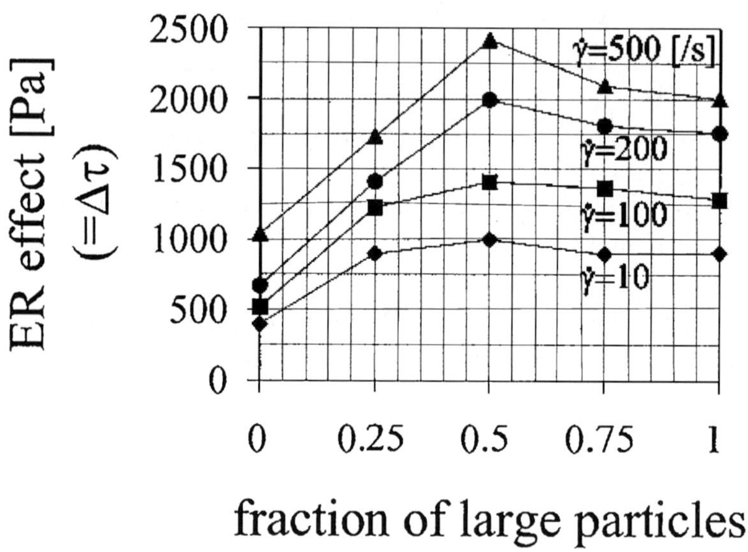

Figure 13 The ER effect $\Delta\tau$, defined as the increment in the shear stress induced under 2.5 kV/mm over that at zero field, is plotted against the fraction of the large particle (50 μm) in the mixture, formed by mixing with the small particle (15μm). The suspension was made of sulfonated poly(styrene-co-divinylbenzene) particle dispersed in silicone oil. Reproduced with permission from H. See, A. Kawai, and F. Ikazaki, Rheo. Acta, 41(2002)

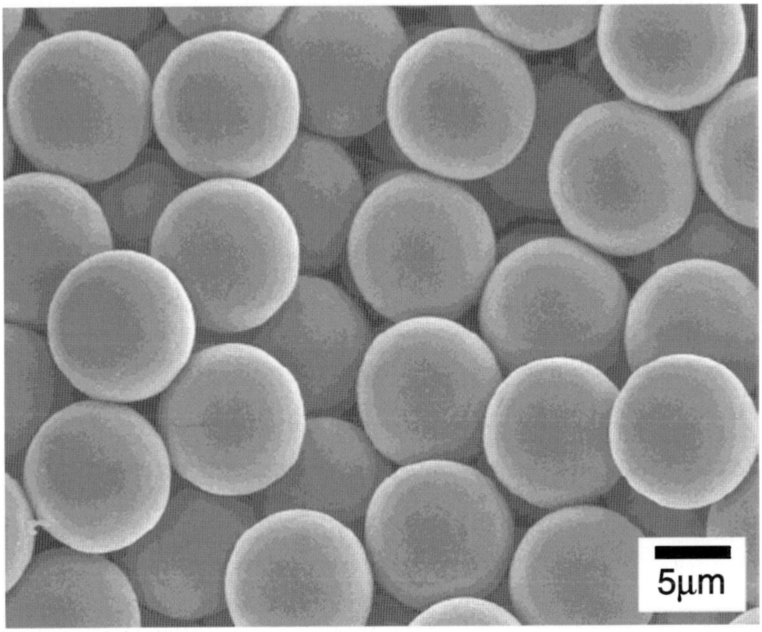

Figure 14. SEM image of monodisperse hydrolyzed SAN particle. Reproduced with permission from J. Jun, S. Uhm, S. Cho, and K. Suh, Langmuir 20(2004)2429

Few research works addressed how the particle shape influences the ER effect. It is well known that the dielectric properties of a heterogeneous system largely depend on the geometry of dispersed particles [39,56]. Since the ER effect is induced by an external electric field, the dielectric property of a suspension is believed to play a significant role in the ER effect, so does the geometry of the dispersed particle. Ellipsoidal particle is expected to give a stronger ER effect than the spherical particle as the ellipsoidal particle strengthen the particle chain formation due to a greater electric-field-induced moment. Experimental results [57] show that the dynamic modulus almost increases linearly with the particle geometric aspect ratio (length-to-diameter) of poly-*p*-phenylene-2,6-benzobisthiazole particle/mineral oil suspensions (see Figure 16). This result also confirms that the ER effect of the elongated particle is larger than that of spherical one when their sizes are comparable [58].

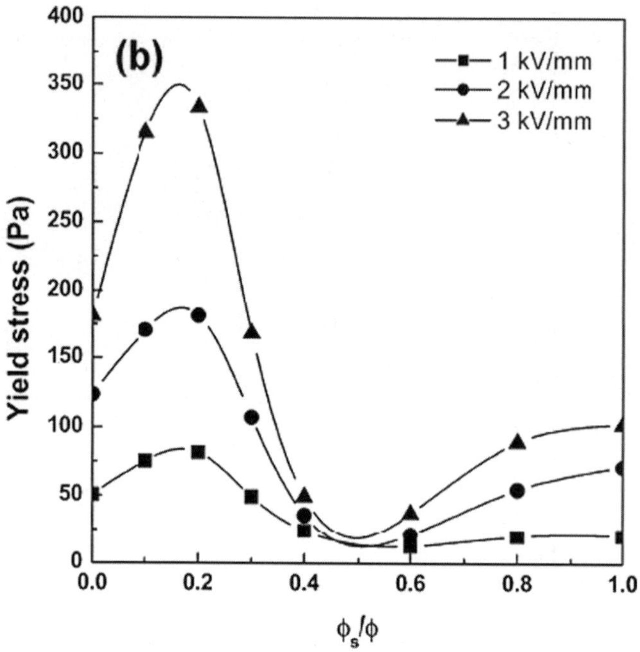

Figure 15 The shear yield stress vs the ratio of volume fractions of small SAN particle to the total particle volume fraction at the electric fields $E = 1$, 2, and 3 kV/mm. Reproduced with permission from J. Jun, S. Uhm, S. Cho, and K.Suh, Langmuir 20(2004)2429

An ellipsoidal/spherical particles blending system shows much stronger ER effect than one-component system [59]. A contradictory result was obtained in hydroxyl-zinc compounds such as plate shape particle of $Zn_5(OH)_8(NO_3)_2 \cdot 2H_2O$ and rod-like particle of $Zn_5(OH)_8(CH_3COO)_2 \cdot 2H_2O$ /silicone oil suspensions. The rod particle suspension showed weak ER effect.[60], which was attributed to the low dielectric constant of the rod particle suspension. It may be hard to draw such a conclusion as the particle materials are different.

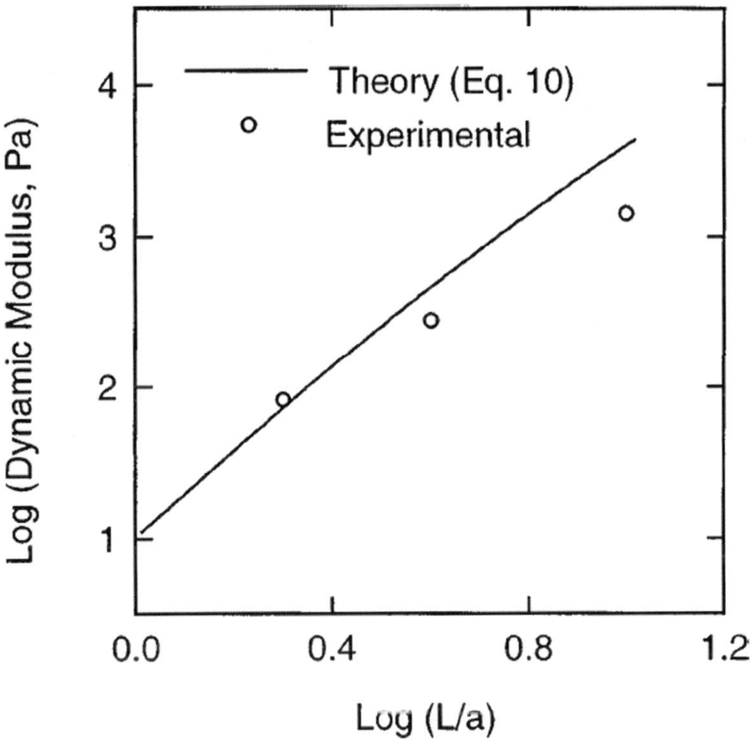

Figure 16 Dynamic modulus vs. aspect ratio of the dispersed particle for poly-*p*-phenylene-2,6-benzobisthiazole particle/mineral oil suspensions. Data obtained at frequency 6.28 rad/s and strain amplitude 0.01, electric field 2 kV/mm and particle volume fraction 15 vol%. Reproduced with permission from R. Kanu, M. Shaw, J. Rheol. 42(1998) 657

4 PARTICLE CONDUCTIVITY

Particle conductivity has a strong influence on ER performance. Block [24] studied how ER effect is influenced by the particle conductivity using the acene-quinone radical polymer/silicone oil, and found that the static yield stress peaks at a particle conductivity around 10^{-5} S/m (see Figure 17). The molecular structure of the acene-quinone radical polymers is shown in Figure 18 and Figure 3 in Chapter 3. A similar tendency was found in oxidized polyacrylonitrile/silicone oil ER system [61]. However, the yield stress peaks at the particle conductivity of about 10^{-7} S/m, rather than 10^{-5}

S/m (see Figure 19). The ER effect of the polyaniline particle of different conductivity dispersed into silicone oil was studied and the largest ER effect was found to occur in the suspension of polyaniline particle of conductivity 10^{-5} S/m [62]. Besides the influence on the ER effect, the particle conductivity also determines the current density of the whole suspension and the response time of the ER fluid. The current density of the oxidized polyacrylonitrile(OP)/silicone oil suspensions obtained at 2.5 kV/mm as a function of particle conductivity is shown in Figure 20 [61]. The current density almost linearly increases with the conductivity of particle. The response time was found to be inversely proportional to the particle conductivity both experimentally [63] and theoretically [64]. The response time can be determined from the relationship between the shear stress and the frequency of applied electric field. Such an example is shown in Figure 21, in which the shear stress of two aluminosilicate/silicone oil suspensions is plotted vs. frequency. The suspension with particle of conductivity 6.0 x 10^{-7} S/m displays a response time 0.6 ms, much shorter than that of the suspension of the particle conductivity 8.4 x10^{-10} S/m, 0.22s [42].

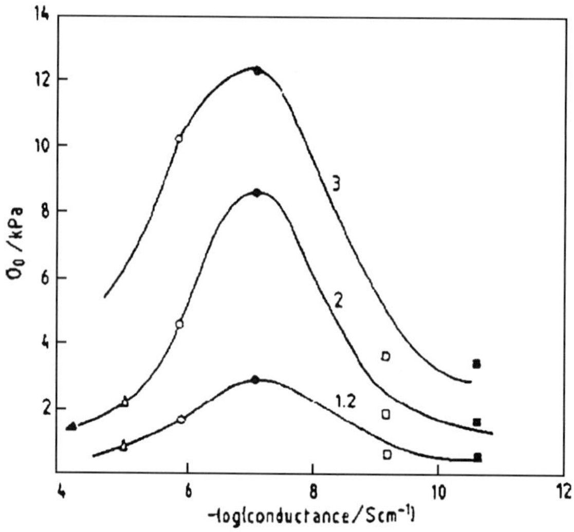

Figure 17. Variation of electric field induced static yield stresses shown by the poly(acenequinones) dispersed in Cereclor (volume fractions of 0.35) as a function of the bulk conductances of the poly(acenequinones): ▲ PPQR; Δ, PPhQR; ○, PAnQR; ●, PNQR, □, PFQR; ■ PTQR. Field strengths (dc) in **kV** mm^{-1} indicated on the figure. Reproduced with permission from H. Block, J.P. Kelley, A. Qin, T. Watson, Langmuir 6 (1990)6

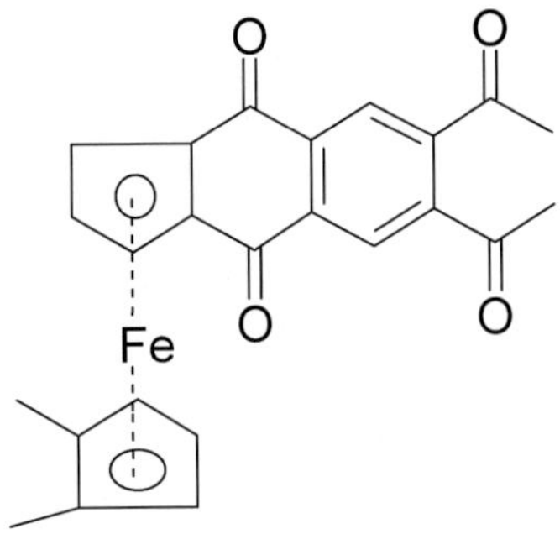

PFQR

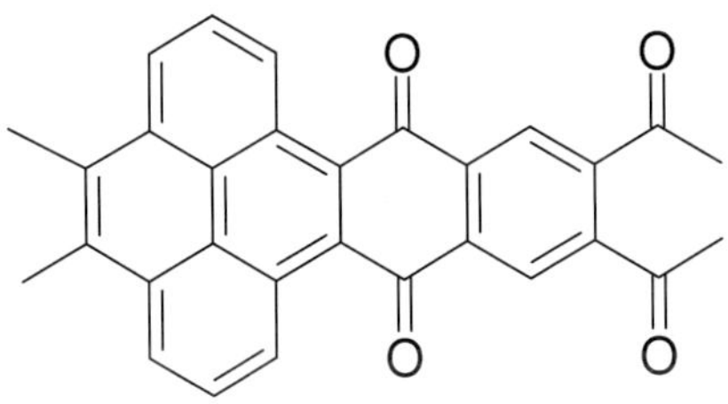

PPQR

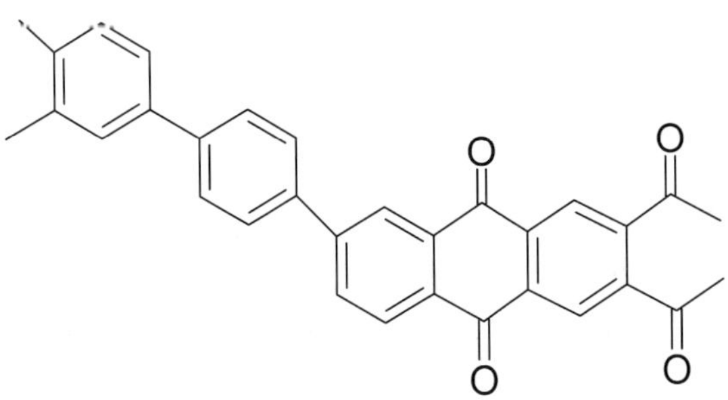

PTQR

Figure 18 Molecular structure of poly(acene quinones) used in Figure 8.

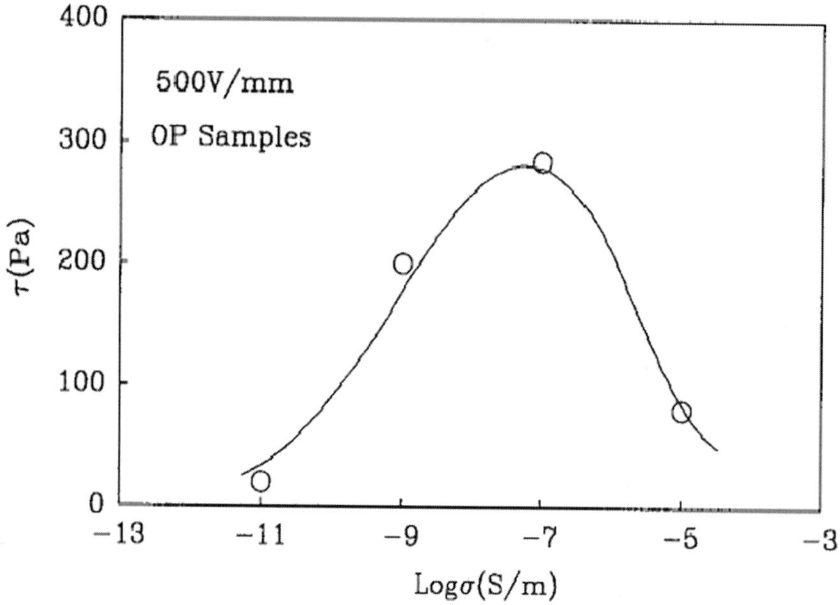

Figure 19 The shear stress of the oxidized polyacrylonitrile(OP)/silicone oil suspensions obtained at 0.5 kV/mm and shear rate 0.3 s^{-1} as a function of particle conductivity. The particle volume fraction of all suspensions is 35 vol%. Reproduced with permission from T. Hao, Z. Xu, Y. Xu, J. Colloid Interface Sci. 190 (1997) 334

If particles with different conductivity were blended together, the ER effect would be dependent on the ratio of two particles. Carbonaceous particles, both with a mean size 3.5 μm and almost identical size distribution but different conductivities, one is in the order of 10^{-6} S/m and another is in the order of 10^{-5} S/m, were blended in a high insulation silicone oil. The shear stress of this blended suspension vs. the ratio of high conductive to low conductive particles is shown in Figure 22. The shear stress goes through a minimum value at high conductive particle ratio 20 %, and then increases with the increase of the high conductive particle ratio.

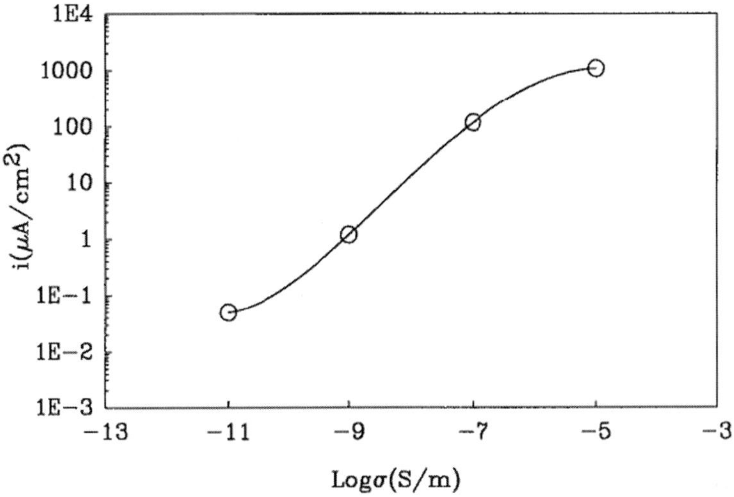

Figure 20 The current density of the oxidized polyacrylonitrile(OP)/silicone oil suspensions obtained at 2.5 kV/mm as a function of particle conductivity. The particle volume fraction of all suspensions is 35 vol%. Reproduced with permission from T. Hao, Z. Xu, Y. Xu, J. Colloid Interface Sci. 190 (1997) 334

Since particle conductivity is so important to the ER effect, there is much literature on the conductivity dependence of the ER effect. Forces between particles in chains induced by an electric field were calculated using the finite-element technique and a dipole-approximation method by Davis [65]. It was found that in an electric field of relatively high-frequency (> 0.1–1 kHz), the forces and the shear modulus would depend upon the dielectric constants of both the suspending liquid and the dispersed particle. In low-frequency or dc electric fields, the forces and the shear modulus would depend upon the conductivities of the components. If the ratio of particle-to-liquid conductivity substantially exceeds the ratio of dielectric constant, a large enhancement of the modulus would be expected. Furthermore, the conductivity model was proposed to explain the ER effect [66, 67]. The ER effect was thought to be determined by the particle-to-liquid conductivity ratio when the conductivity of particle is higher than that of liquid. A negative ER effect would be expected if the conductivity of the particle is lower than that of the liquid. Further expansion of the conductivity model was made by Tang [5], and detailed information on the conductivity model will be discussed in Chapter 9.

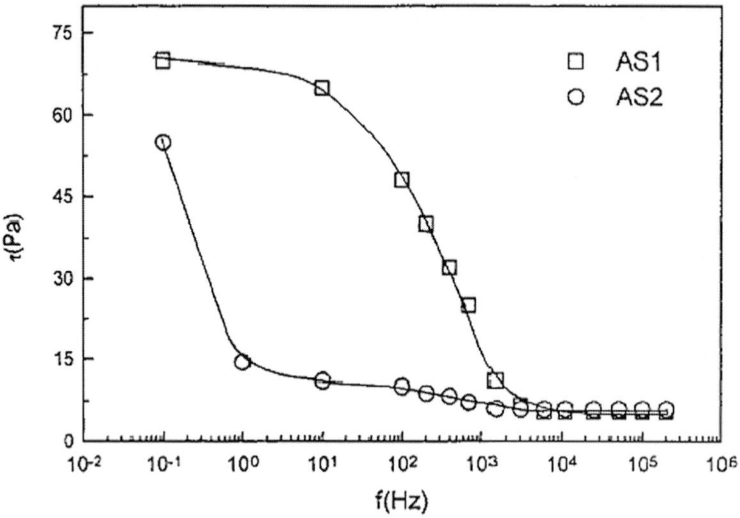

Figure 21 Shear stresses of the aluminosilicate/silicone oil suspensions vs the frequency of applied electric field (0.3 kV/mm).. The particle conductivity of AS1 and AS2 suspensions are 6.0 x 10^{-7} and 8.4 x10^{-10} S/m. the response time determined on this figure is 0.6 ms for AS1 and 0.22 s for AS2, respectively. Reproduced with permission from T. Hao, A. Kawai, and F. Ikazaki, Langmuir 14(1998)1256

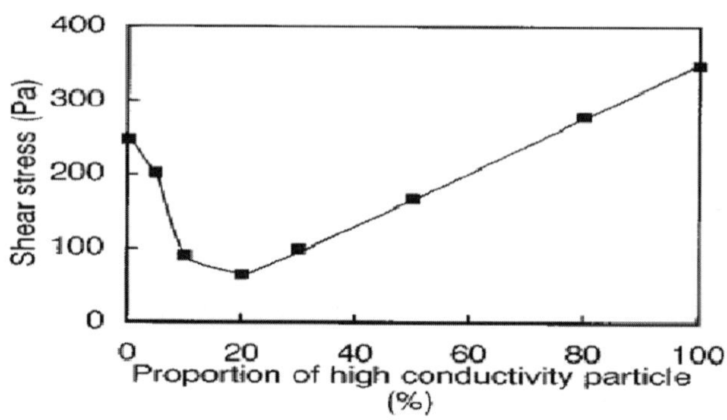

Figure 22 The shear stress of blended carbonaceous particles with same size distribution but different conductivity in silicone oil vs. the ratio of the high conductive particle to the lower conductive particle. The total particle volume fraction was kept constant, 30 vol%. Data were obtained under 2 kV/mm, and shear rate 100 s^{-1}. Reproduced with permission from R. Sakurai, H. See and T. Saito, J. Rheol., 40(3)(1996)395

5 PARTICLE DIELECTRIC PROPERTY

The ER effect is obviously induced by an external electric field, and the polarization definitely plays an important role. The dielectric properties of both the dispersed particle and the dispersing medium should be critical to the ER effect. The dielectric tool thus is frequently used for investigating how dielectric property influences the ER effect.

The dielectric property of an ER suspension would change with the particle volume fraction, the applied electric field, and the water content adsorbed on the particle surface. The dielectric properties of ER fluids at high electric fields or high shear fields are specially important, as ER fluids most likely operate under these conditions, and the microstructure of an ER fluid at high electric or shear field should be quite different from the microstructure at low electric field or quiescent state. A linear increase relationship between the dielectric constant of the whole ER suspension and the particle volume fraction was found in the silica/silicone oil system [22], as shown in Figure 23. The electric double layer overlapping was thought to be the main reason for this linear relationship. This linear relationship was also found in crystalline cellulose/chlorine insulator oil system [33]. The dielectric constant of silica/silicone oil system against the applied electric field strength at different particle volume fraction is shown in Figure 24. According to Klass [22], the dielectric constant of whole suspension increases with the applied electric field only when the particle volume fraction is less than 10 vol%. When the particle volume fraction is between 10 and 46 vol%, the dielectric constant doesn't change with the applied electric field strength at low field and obviously decreases with further increase of the electric field. Once the particle volume fraction reaches 46 vol%, the dielectric constant decreases with the increase of the applied electric field strength. Note that the ER effect does increase with the applied electric field strength, though the dielectric constant of whole suspension decreases at high electric field. The apparent viscosity of this suspension against the applied electric field is shown in Figure 25. In the whole concentration range, the apparent viscosity increases with the electric field, indicating that the ER effect doesn't follow the trend of the dielectric constant. A similar trend was found by Deinega [70] that the dielectric constant increases with the electric field strength (0.4-4kV/mm), and levels off at high electric field. If the particle surface adsorbs the moisture, the dielectric constant of crystallized cellulose/insulator oil suspension linearly increases with the moisture content, as shown in Figure 26, and reported in ref. [33]. This result may shed light on why the water or other polar liquid could enhance the ER effect. After comparing the dielectric properties of ER suspensions with their ER effects, Block [24] postulated that the polarization rate and its magnitude may be very important for the ER effect, and Filisko [71,72] even assumed the ER effect may be associated with the low-field dielectric dispersion. However, there was no clear relationship between the dielectric property and the ER effect at that time.

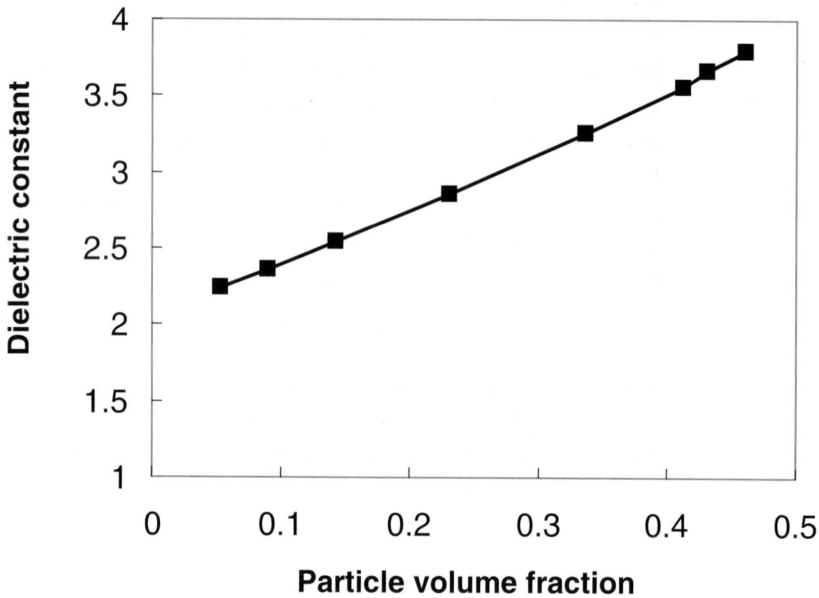

Figure 23 The dielectric constant of the silica/silicone oil suspension vs. the particle volume fraction. The data were obtained at 5 MHz and 25°C. Redrawn from D.L. Klass, T.W. Martinek, J. Appl. Phys. 38 (1967) 75.

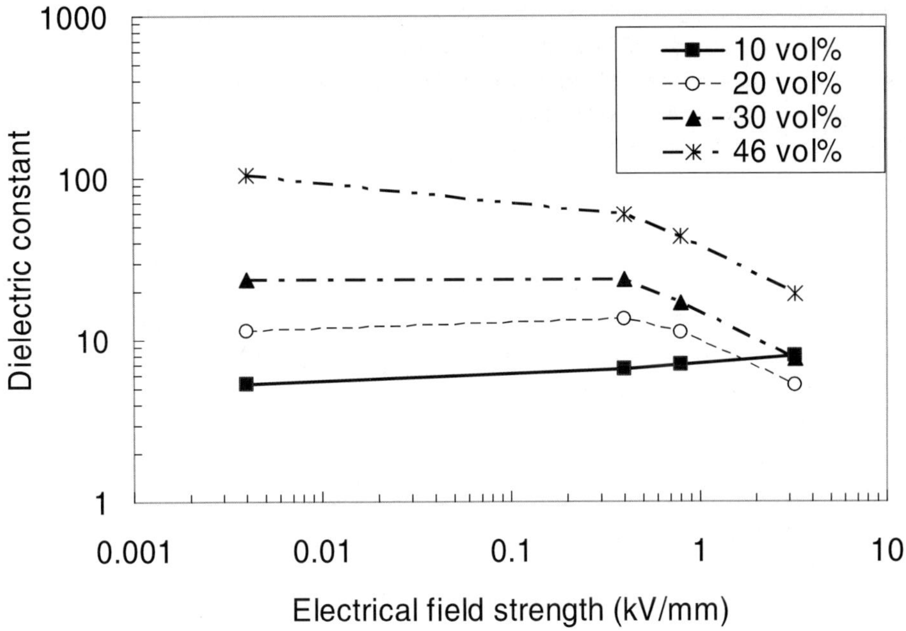

Figure 24 The dielectric constant of the silica/silicone oil suspension vs. the applied electrical field strength at different particle volume fraction. The data were obtained at 65 Hz and 65°C. Redrawn from D.L. Klass, T.W. Martinek, J. Appl. Phys. 38 (1967) 75.

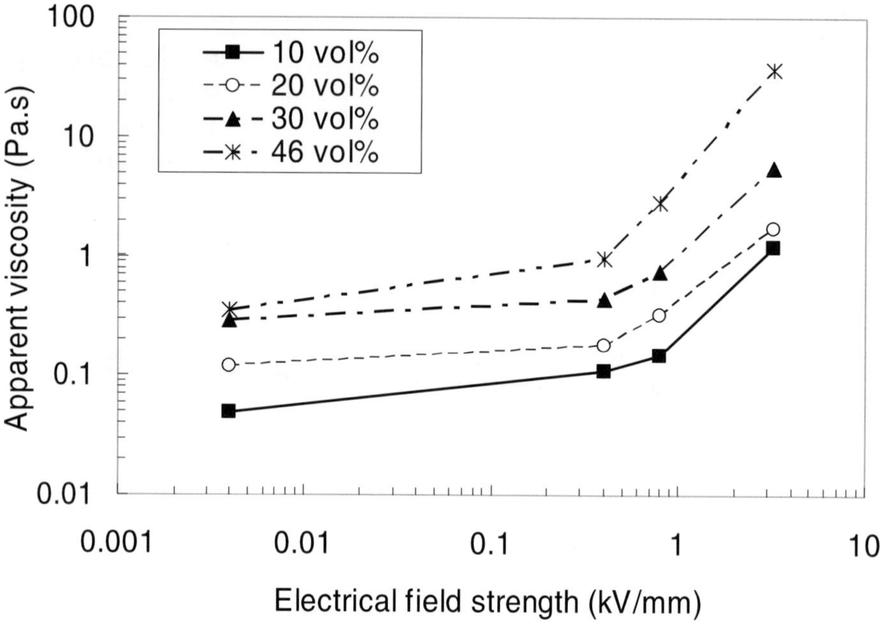

Figure 25 The apparent viscosity of the silica/silicone oil suspension vs. the applied electrical field strength at different particle volume fraction. The data were obtained at shear rate 23.6 s^{-1} and 65°C. Redrawn from the data in D.L. Klass, T.W. Martinek, J. Appl. Phys. 38 (1967) 75.

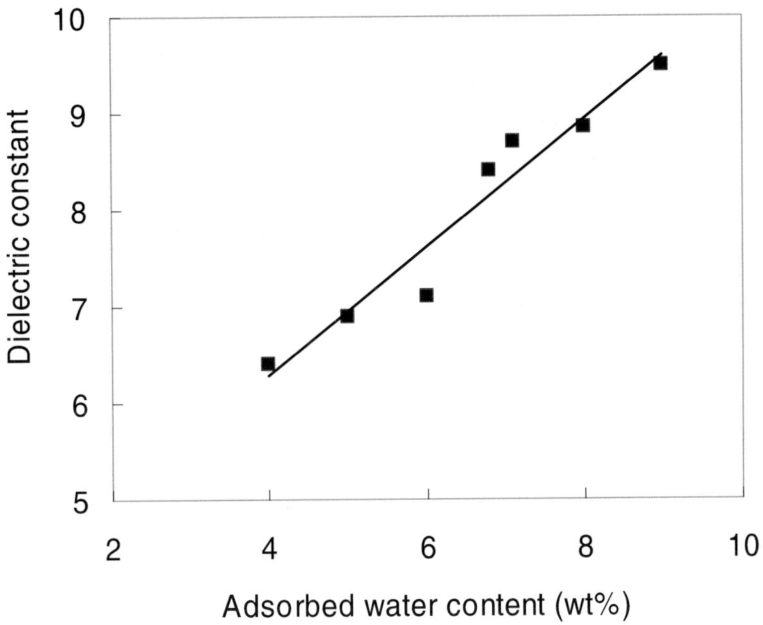

Figure 26 The dielectric constant of crystallized cellulose/chlorinated insulator oil against the adsorbed water content on the particle surface. The particle concentration is 10 wt%, the electric field frequency is 1000 Hz, and the temperature is 20 °C. Redrawn from H. Uejima, Jpn. J. Appl. Phys., 11(1972)319.

How the dielectric properties of both the dispersed particle and dispersing medium affect the ER effect is an important question being addressed for a long time. Since the dispersing medium is usually an insulating oil of dielectric constant about 2, much attention has been paid to the dielectric property of the dispersed particle. Mismatch of the solid and liquid phase dielectric constants (at high frequency field) or conductivities (at dc or low frequency field) was proposed for interpreting ER phenomena [65, 73,74]. However, the obvious differences between the experimental results and the theoretical predictions [75] suggest that this polarization model does not adequately reflect the physical fundamentals of the ER effect. The conductance model [4, 5, 66], in which the particle interaction is thought to be determined by the ratio of particle-to-fluid conductivity rather than dielectric constant, was also proposed, and good agreement between the experimental results and the predictions was obtained. Nevertheless, the conductance model can only be applied to the situation in which the ER suspension microstructure has been fully formed. It does not consider the microstructure transition process of the ER suspension from a randomly distributed state to an ordered chain state. So the conductance model does not present a clear physical picture of the field-induced response process and therefore is unable to explain, for example, why most ER fluids respond in milliseconds, why some ER fluids show a quite weak ER effect in a dc field but become considerably active under an ac field, or why an optimal field induced shear stress is found to occur at dispersed particle conductivity around 10^{-5} S/m [24]. A successful theoretical treatment was made by Khusid [7]; In his work the electric field-induced aggregation process and the interfacial polarization process are both considered. This theory can provide a reasonable interpretation for most ER phenomena, especially the ones that cannot be understood by virtue of the polarization and conductance models. However, quantitative discrepancies still pertain to this theory. For example, according to this theory, the shear stress of $KNbO_3$ /silicone oil system should decrease monotonically with increasing frequency; however, experimentally it increases with frequency [76]. Also, the shear stress maximum of semiconducting poly(acenequinones) dispersed in a chlorinated hydrocarbon oil [7,24] should have appeared at the particle conductivity around 10^{-9} S/m, but experimentally it occurred at 10^{-5} S/m. Those disagreement imply that other important processes or parameters relevant to the polarization have been ignored. Hao [61,77] studied how the particle dielectric property affects the ER effect experimentally. The oxidized polyacrylonitrile (OP) and aluminosilicate (AS) powder of different dielectric constant (from 2—10) and conductivities ranging from 10^{-5}—10^{-12} S/m were used. It was found that there is a complicated relationship between the ER effect and the particle dielectric constant/ dielectric loss, as shown in Figure 27 and 28.

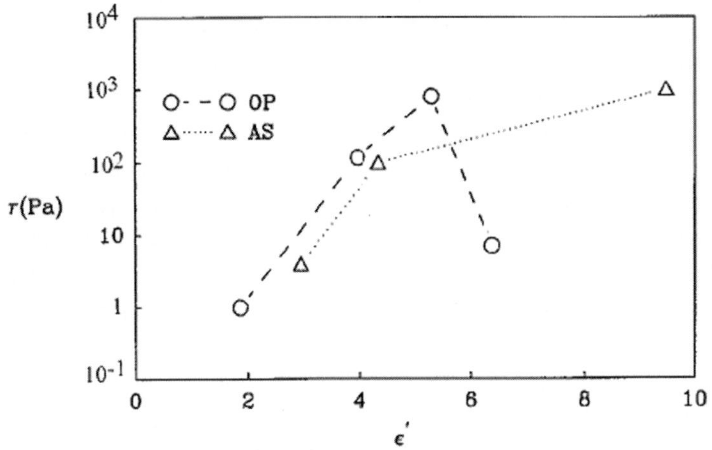

Figure 27. The shear stresses of oxidized polyacrylonitrile (OP) and aluminosilicate (AS)/silicone oil suspensions obtained at 0.5 kV/mm against the dielectric constants of dispersed particle obtained at 1000 Hz. The particle volume fraction is 35 vol%. Reproduced with permission from T. Hao, Z. Xu, Y. Xu, J. Colloid Interface Sci. 190 (1997) 334.

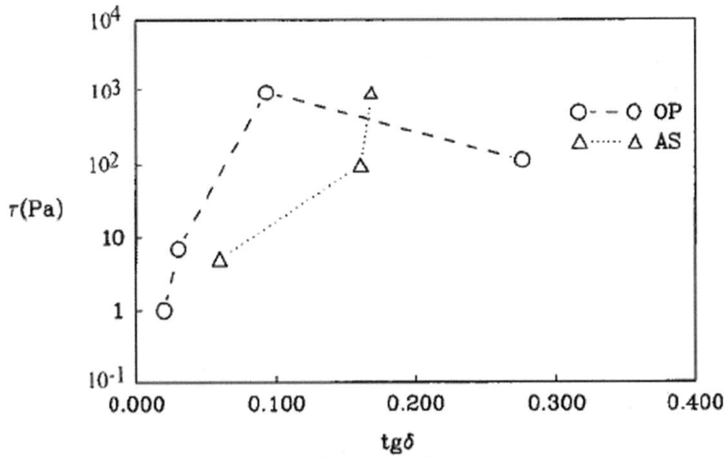

Figure 28 The shear stresses of oxidized polyacrylonitrile (OP) and aluminosilicate (AS)/silicone oil suspensions obtained at 0.5 kV/mm against the dielectric loss tangent of dispersed particle obtained at 1000 Hz. The particle volume fraction is 35 vol%. Reproduced with permission from T. Hao, Z. Xu, Y. Xu, J. Colloid Interface Sci. 190 (1997) 334.

Note that the suspensions with a strong ER effect do have a high dielectric constant and loss tangent at the frequency 1000 Hz. The shear stress doesn't change monotonically with the dielectric constant or dielectric loss even within the same series of samples, and the particle dielectric loss is found to play an important role in the ER response. By analyzing the dielectric data of currently used ER materials, an empirical criterion was proposed for selecting ER material: The particle dielectric loss tangent should be around 0.10 at 1000 Hz; Then the larger the particle dielectric constant is, the stronger the ER effect will be. The experimental results evidently show that suspensions with a dielectric relaxation peak at a frequency around 1000 Hz may exhibit a remarkable ER effect at room temperature, and among them the higher the dielectric constant of the dispersed particle is, the stronger the ER effect of the suspension is. The dielectric constant and dielectric loss tangent data, cited from Hippel [78], for certain patented or reported ER solid materials are shown in Table 1. All materials have a relatively high dielectric loss tangent value, agreeing with the conclusion derived from the experimental data in Ref.[61].

This conclusion could be well understood by virtue of the Wagner-Maxwell polarization model, which will be discussed in detail in the future chapter. The dielectric loss tangent and dielectric constant against the particle conductivity, predicted with the Wagner-Maxwell model, is shown in Figure 29. The presumption made for such a calculation is that the applied electric field frequency is 1000 Hz, the particle volume fraction 35%, and the dielectric constant of insulating oil is 2. The dielectric loss tangent of the whole ER suspension was found to peak at the particle conductivity of 10^{-7} S/m and the peak value is coincidentally around 0.10 [79]. The predicted dielectric constant was found to increase abruptly at the particle conductivity of 10^{-8} S/m and reaches a plateau at the particle conductivity 10^{-6} S/m. If a high dielectric loss tangent, about 0.10, is necessary for the strong ER activity, the strongest ER effect would be found at the particle conductivity about 10^{-7} S/m, which has already been confirmed experimentally. Frequency dependence of the shear stress can be explained qualitatively on the basis of these findings. If an ER suspension has a large dielectric loss in low frequency fields, it will display a strong ER activity in low frequency fields but will become ER inactive in high frequency fields because its dielectric loss will decrease with frequency to a value too small to generate an ER effect. Thus the shear stress of the ER suspension should decrease with the increase of frequency, which has been witnessed in many ER suspensions [21, 22,33, 80-82] However, if an ER suspension has a low dielectric loss in low frequency fields, it is unable to exhibit a clear ER activity in low frequency fields but the shear stress may gradually increase with the increase of frequency because its dielectric loss goes up in high frequency fields, which has been found in the suspensions such as $KNbO_3$ [76] and $BaTiO_3$ systems [83]. Obviously, too large a dielectric loss is inappropriate for ER activity since too strong thermal motion may compete with the fibrillation process. Determination of a suitable dielectric loss tangent range would be of special significance. The "water" mechanism, ever generally accepted by the electrorheology community, also can be easily understood with this empirical criterion. A small amount of water can increase the dielectric loss tangent of the solid material to a considerable high level compared to the dry state. However, once the water content exceeds the material absorptivity, the free water would cause too large a dielectric loss or much stronger thermal energy, unfavorable to fibrillation processes. A maximum shear stress would be found to occur at a specific water content value, which has already been observed experimentally [14, 33, 83-85].

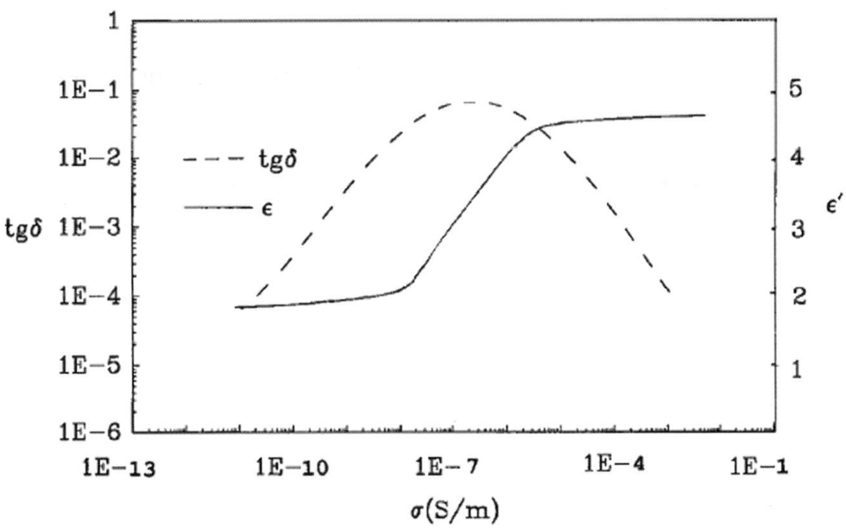

Figure 29 The conductivity dependence of the dielectric loss tangent and the dielectric constant of an ER fluid under the assumptions of $\varepsilon'=2$, $\varepsilon''= 10$, $\Phi= 0.35$, and $f=1000$ Hz predicted with Wagner-Maxwell model. Reproduced with permission from T. Hao, J. Colloid Interface Sci., 206(1998)240.

Table 1 Dielectric constant ε' and dielectric loss tangent (tanδ) of certain ER solid materials (20°C)

Material	ε'	tanδ	frequency (Hz)
Silica-glasses (25% Na_2O+75%SiO_2)	9.7	0.24	1000
Sandy soil (dry)	3.42	0.08	1000
Sandy soil (2-18% moisture)	2.72	0.13	1000
TiO_2	200	0.35	1000
Phenol-formaldehyde	7	0.11	1000
Nitrate cellulose	8.4	0.10	100
Methyl cellulose	7.6	0.128	100
	6.8	0.057	1000
Polyvinyl alcoholacetate			
25°C	7.8	0.05	1000
85°C	100	1.3	1000
Polyethyl methacrylate			
25°C	2.75	0.042	1000
85°C	3.36	0.11	1000
Oxidized polyacrylonitrile	5.5	0.085	1000

This empirical criterion could be qualitatively interpreted if such an assumption could be made: There are two processes, particle turning and particle polarization, happening in an ER system under an electric field. A large dielectric loss should be responsible for the particle turning process, while a large dielectric constant is responsible for the thermal stability and the mechanical strength of fibrillated chains. Thus, a good ER solid material should have a large dielectric loss and a large dielectric constant. Some important ER phenomena could be qualitatively explained using this criterion. Examples include: Why the shear stresses of some ER fluids decrease with frequency while the shear stresses of others increase with it; why the strongest ER effect takes place at the particle conductivity about 10^{-7} S/m; and why undetectable water can enhance the ER effect considerably. Although large dielectric loss was thought to be needed for a strong ER response, too large a dielectric loss is of course inadequate; otherwise the suspension would give off a great deal of heat, resulting in too strong particle thermal motion, which would destroy fibrillated chains and lead to a weak ER effect finally.

There is a large body of literature on the relationship between the dielectric property and the ER effect. Another different criterion of the similar physical essential to Hao's was also proposed [86, 87]. This criterion argued that for a good ER fluid, its dielectric relaxation frequency should be between 100 and 10^5 Hz and the difference of the dielectric constant below and above the relaxation frequency must be large. The microcrystalline cellulose, polymer–inorganic composite particle, and sulphonated poly(styrene-co-divinylbenzene) (SSD) particle, dispersed in silicone oil, were used for experimental study. The ER effect, scaling as the viscosity under an electric field divided by the viscosity at zero electric field, η_E/η_0, was plotted against the suspension dielectric constant difference at 10^2 and 10^5 Hz, $\Delta\varepsilon' = \varepsilon'_{10^2} - \varepsilon'_{10^5}$, in Figures 30-32 for microcrystalline cellulose, polymer–inorganic composite particle, and SSD, respectively. For those materials, there is a clear trend showing that the ER effect increases with the suspension dielectric constant difference between 10^2 and 10^5 Hz. A huge dielectric constant difference in this frequency range indicates that there should be a dispersion peak, corresponding to the Maxwell-Wagner polarization that will be discussed in more detail in the future. The larger the suspension dielectric constant difference, the larger the dielectric loss tangent. This is the reason that in Hao's criterion, the larger dielectric loss tangent is emphasized. Essentially, those two criteria are physically the same, but expressed in a different way. Many other ER systems show the similar relationship between the dielectric property and the ER effect. Examples are the silica particle coated with polyaniline [88], asphaltene [89], polyaniline coated with polymethyl methacrylate [90], and polyaniline/molecular sieve composite [91].

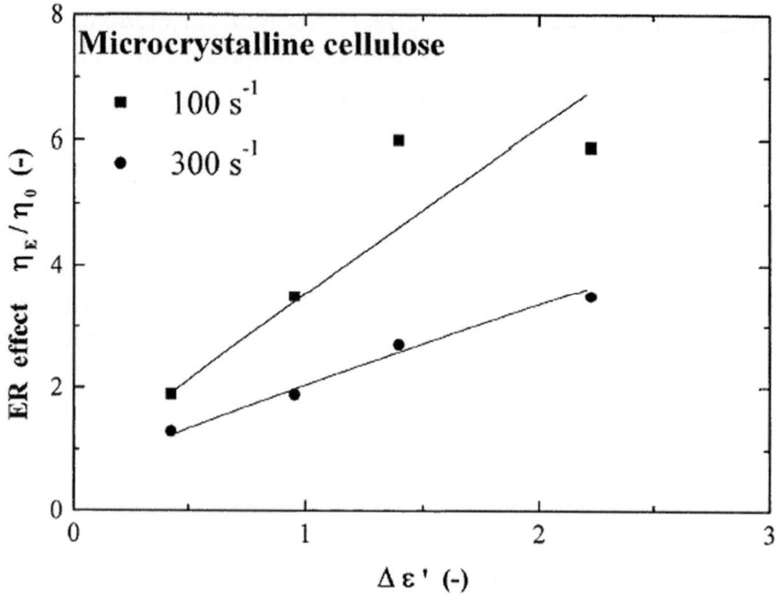

Figure 30 The ER effect scaled as the viscosity under an electric field divided by the viscosity at zero electric field, η_E/η_0, vs. the suspension dielectric constant difference at 10^2 and 10^5 Hz, $\Delta\varepsilon' = \varepsilon'_{10^2} - \varepsilon'_{10^5}$ for microcrystalline cellulose. The particle concentration is 10 wt%, and the electric field is 2.4 kV/mm. Reproduced with permission from F. Ikazaki, A. Kawai, T. Kawakami, K. Edamura, K. Sakuri, H. Anzai et al., J. Phys. D: Appl..Phys. 31 (1998)336

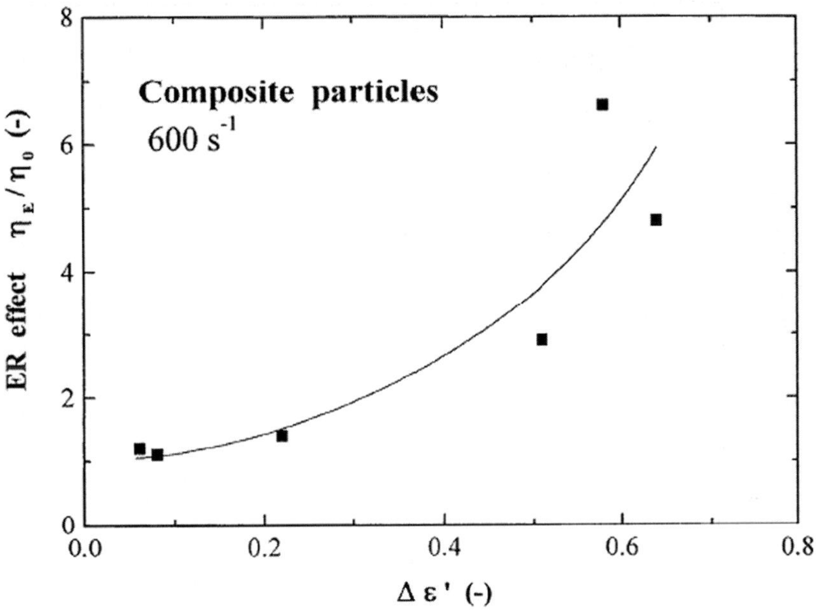

Figure 31 The ER effect scaled as the viscosity under an electric field divided by the viscosity at zero electric field, η_E/η_0, vs. the suspension dielectric constant difference at 10^2 and 10^5 Hz, $\Delta\varepsilon' = \varepsilon'_{10^2} - \varepsilon'_{10^5}$ for the polymer-inorganic composite particle. The particle concentration is 8.7 vol%, and the electric field is 2.4 kV/mm. Reproduced with permission from F. Ikazaki, A. Kawai, T. Kawakami, K. Edamura, K. Sakuri, H. Anzai et al., J. Phys. D: Appl. Phys. 31 (1998)336

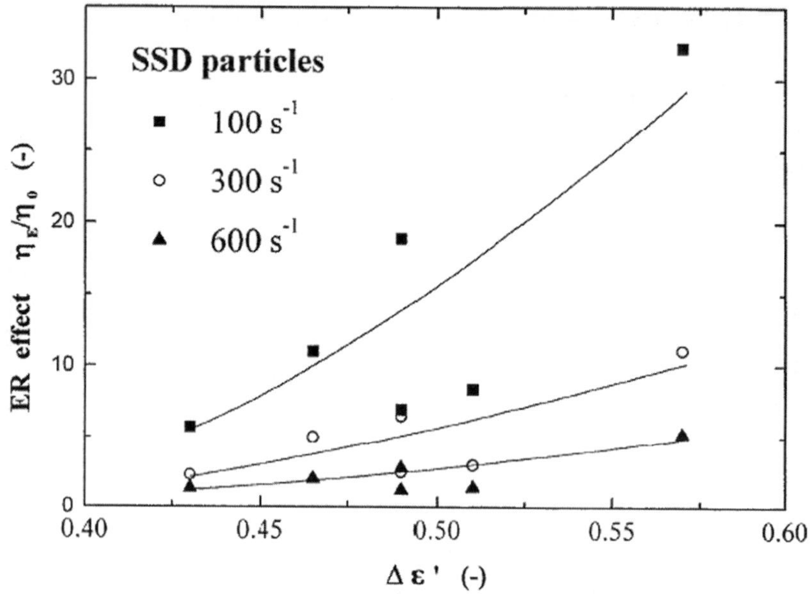

Figure 32 The ER effect scaled as the viscosity under an electric field divided by the viscosity at zero electric field, η_E/η_0, vs. the suspension dielectric constant difference at 10^2 and 10^5 Hz, $\Delta\varepsilon' = \varepsilon'_{10^2} - \varepsilon'_{10^5}$ for sulphonated poly(styrene-co-divinylbenzene) (SSD) particle. The particle concentration is 6.8 vol%, and the electric field is 2 0 kV/mm. Reproduced with permission from F. Ikazaki, A. Kawai, T. Kawakami, K. Edamura, K. Sakuri, H. Anzai et al., J. Phys. D: Appl..Phys. 31 (1998)336

6. PARTICLE SURFACE PROPERTY

Particle surface property has an impact on both the particle dispersion stability and the ER effect. An excellent example is the nano-sized carbon black particle with a primary particle size 22 nm [92]. This kind of carbon black particle can easily form aggregates in any oil medium, resulting in a quick sedimentation. After the carbon black surface was grafted with a polymer, as schematically shown in Figure 33, the grafted particle/silicone oil suspension showed a remarkably long dispersion stability. The particle didn't settle at least for two months. Even under centrifuging at 9000 G for 1 hour, there is no obvious particle settling or aggregation observed. Under ac electric field of 1000 Hz, the shear stress 145 Pa was obtained at shear rate 150 s^{-1}, the application of an electric field 3 kV/mm, and the particle weight fraction 30 wt%, as shown in Figure 34. Surface modification seems to be very important for the ER effect.

The relative shear stress, the shear stress under an electric field divided by that without an electric field, of nanosized silica particle (150 nm)/silicone oil suspension was found to increase with the number of adsorbed lithium cations per square nm of particle surface, as shown in Figure 35 [93]. For the purpose of reducing the particle density and then improving the particle stability, a polymer core particle with conductive inner layer and a non-conductive outer layer was designed by Saito [94], as depicted in Figure 36. The ER effect was found to increase with the conductive layer (silver) thickness and decrease with the non-conductive layer (silica) thickness [94]. Figure 37 shows the increment of shear stress, $\Delta\tau$, obtained at shear rate 800 s^{-1} and 2 kV/mm, vs. the thickness of silica layer. The ER effect linearly decreases with the silica thickness. The increment of shear stress vs. the silver content is shown in Figure 38. S-shape relationship between the increment of shear stress and the silver content is obtained. A jump of the shear stress increment occurs at silver content about 0.25 g/m^2, and a saturation begins at silver content about 0.3 g/m^2.

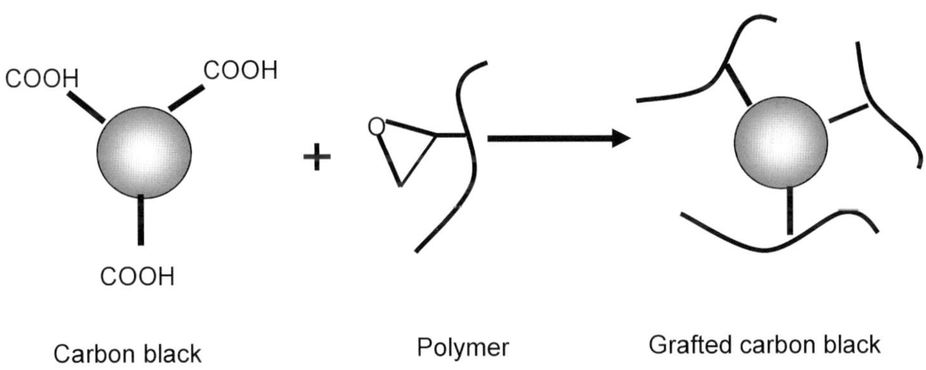

Figure 33 Schematic illustration of polymer grafting process of carbon black. Redrawn from M. Konishi, T. Nagashima, and Y. Asako, Proc. 6th intern.conf. on ER and MR suspensions and their applications, M. Nakano and K. Koyama ed., World Scientific, Singapore, 1998, p12

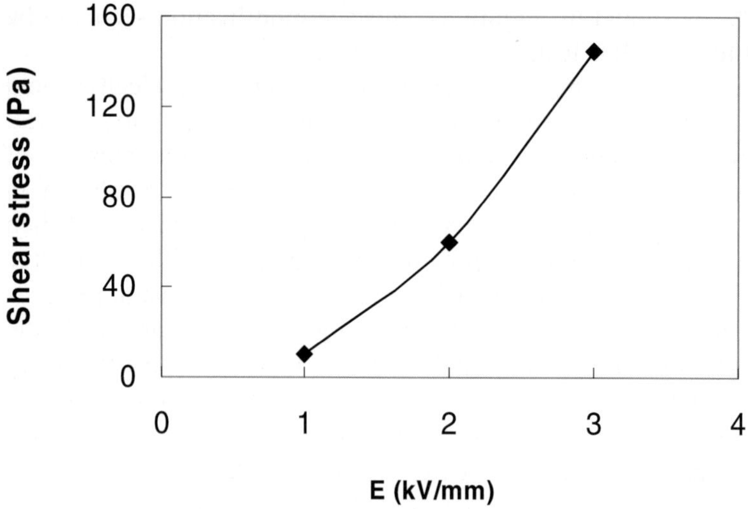

Figure 34 Shear stress of grafted carbon black/silicone oil vs. electric field of 1000 Hz. Shear rate 150 s^{-1}, and particle weight fraction is 30 wt%. Reproduced with permission from M. Konishi, T. Nagashima, and Y. Asako, Proc. 6th Int. Conf. on ER and MR suspensions and their applications, M.Nakano and K.Koyama ed., World Scientific, Singapore, 1998, p12

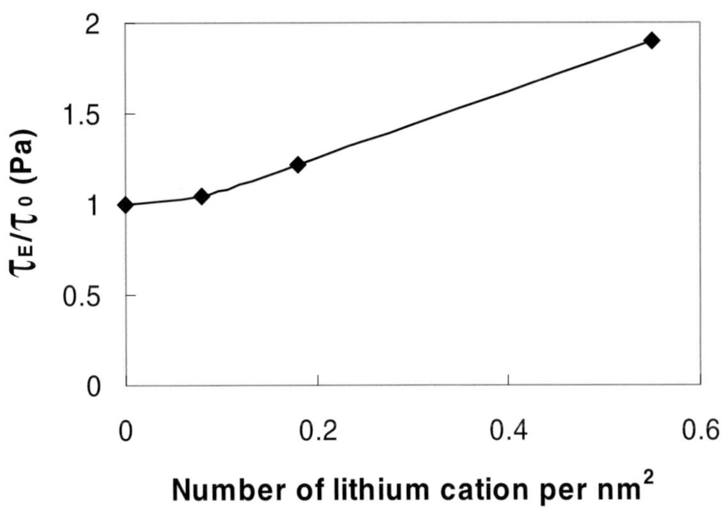

Figure 35 the relative shear stress ($\tau_{1kV/mm}$ divided by τ_0) vs. the number of adsorbed lithium cations per square nanometer of silica surface. The shear rate is 0.045 s^{-1}, and the particle volume fraction is 18.24 vol%. Redrawn from C. Gehin, and J.Persello, Int. J. Modern Phys. B, 16(2002)2494

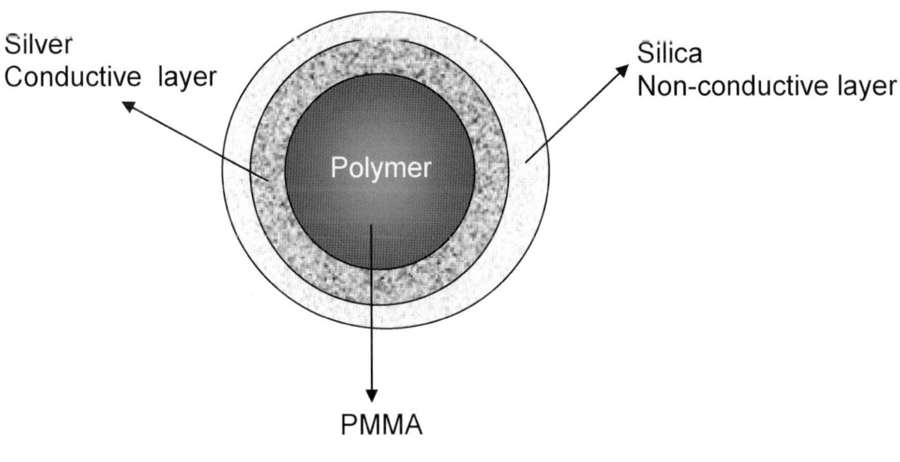

Figure 36 A Poly(methyl methacrylate) (PMMA) polymer core particle with a silver conducting inner layer and a silica outer layer.

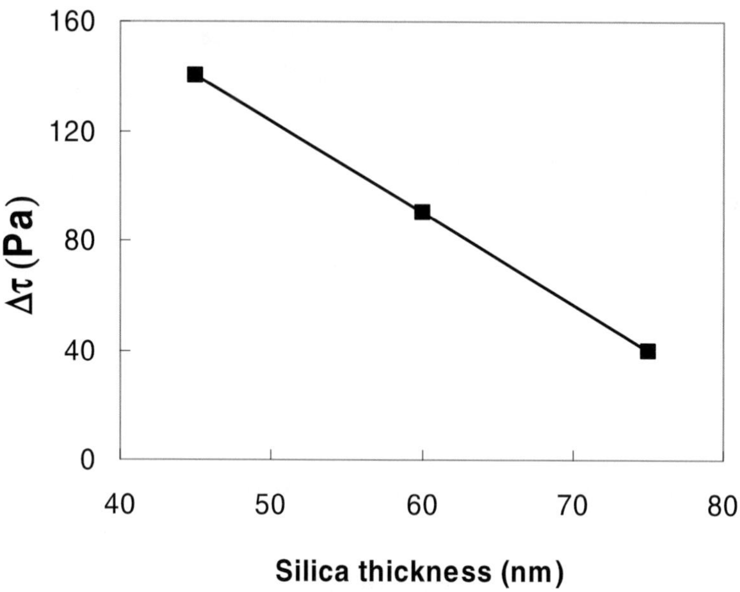

Figure 37 The increment of shear stress of PMMA/Ag/Silica particle/silicone oil vs. the thickness of silica outer layer. The applied electric field is 2 kV/mm, the particle volume fraction is 20 vol%, and shear rate is 800 s^{-1}. Redrawn from T. Saito, H. Anzai, S, Kuroda, and Z. Osawa, Proc. 6th Int. Conf. on ER and MR suspensions and their applications, M.Nakano and K.Koyama ed., World Scientific, Singapore, 1998, p19.

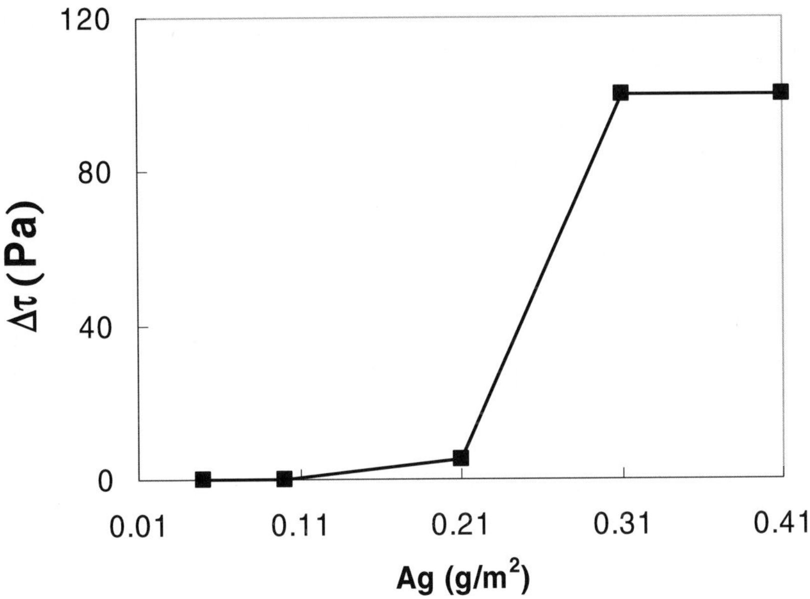

Figure 38 The increment of shear stress of PMMA/Ag/Silica particle/silicone oil vs. the amount of silver deposited on PMMA surface. The applied electric field is 2 kV/mm, the particle volume fraction is 20 vol%, and shear rate is 800 s^{-1}. Redrawn from T. Saito, H. Anzai, S. Kuroda, and Z. Osawa, Proc. 6th Int. Conf. on ER and MR suspensions and their applications, M. Nakano and K. Koyama ed., World Scientific, Singapore, 1998, p19.

Particle surface modification could be characterized with the surface energy measurement. How the surface energy of dispersed particle affects the ER effect was systematically addressed by Hao [95]. A set of oxidized polyacrylonitrile (OP) materials of different surface properties were employed for such a purpose. The five kinds of water-free ER fluids composed of oxidized polyacrylonitriles (OP) particle dispersed in a low viscosity silicone oil were used for correlating the ER effect with the particle surface energy. The powdered OP materials with average particle size 0.1-10 μm were treated at 150 °C for 8 h, and then dispersed in silicone oil immediately at the particle volume fraction of 35 vol%. The surface energy was measured by means of the dynamic wicking method [96]. Water and ethylene glycol were used as wicking liquids. The geometric mean method [97] was employed to determine the dispersion component and the polar component of the surface energy. For a low surface energy material, the dispersion component of the surface energy would be large if a nonpolar liquid can easily spread on its surface, whereas its polar component would be large if a polar liquid can easily spread. The shear stress of the OP suspension vs. the surface polarity is shown in Figure 39. The surface polarity is defined as the percentage of polar component in surface energy. The shear stress was found to peak at the polarity 40% or so. Evidently, the highly polar material does not always result in a high ER response. Furthermore, the shear stress of the OP suspension obtained at shear rate 0.3 s^{-1} and 2.0 kV/mm was also experimentally found to peak for the surface energy of the dispersed particle, as shown in Figure 40. A shear stress peak was observed at the surface energy around 45 mN/m.

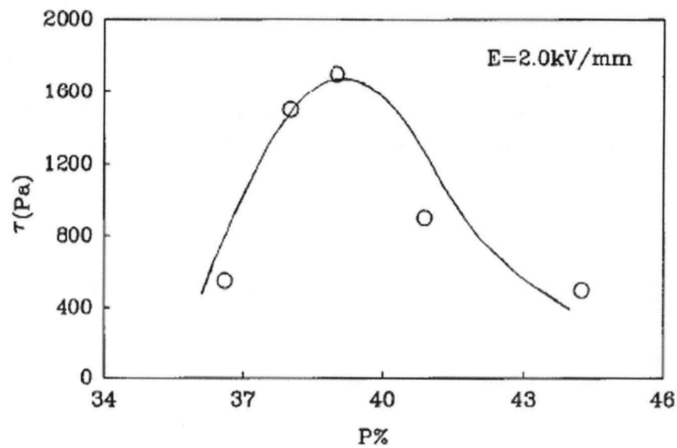

Figure 39 The shear stress vs. the polarity of dispersed particle surface. The data were obtained at the electric field 2.0 kV/mm, shear rate 0.3 s^{-1}, and particle volume fraction 35 vol%. Reproduced with permission from T. Hao, and Y. Xu, Appl. Phys. Lett., 69(1996)2668.

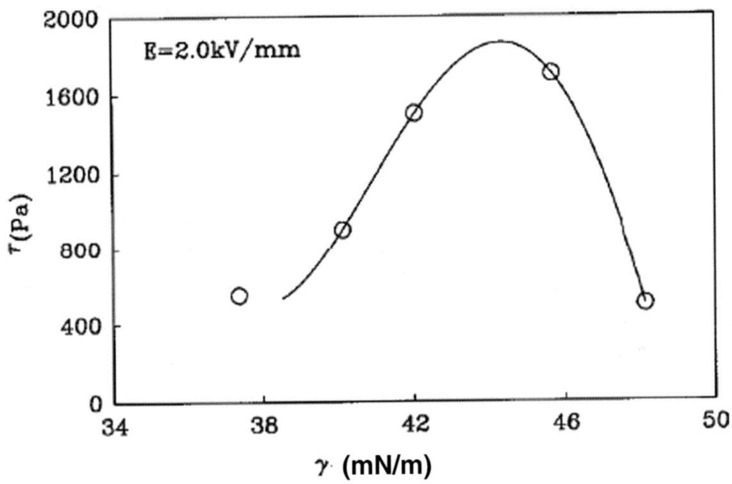

Figure 40 The shear stress vs. the surface energy of dispersed particle. The data were obtained at the electric field 2.0 kV/mm, shear rate 0.3 s^{-1}, and particle volume fraction 35 vol%. Reproduced with permission from T. Hao, and Y. Xu, Appl. Phys. Lett., 69(1996)2668.

The material surface energy essentially arises from the van der Waals forces between two bodies, and would strongly correlate with its dielectric properties. The dispersion energy U_{12}^{d} between a molecular type 1 and molecular type 2 was given by [98]

$$U_{12}^{d} = -\frac{23hc\alpha_1\alpha_2}{8\pi^2 r_{12}^{7}} \qquad (r_{12}>10\ \text{nm}) \tag{6}$$

and [F. London, Trans. Faraday Soc. 33 (1937)57]

$$U_{12}^{d} = -\frac{3\alpha_1\alpha_2 I_1 I_2}{2r_{12}^{6}(I_1+I_2)} \qquad (r_{12}<10\ \text{nm}) \tag{7}$$

where h is Planck's constant, c is the light speed, α_1, α_2, I_1, I_2 are the polarizability and ionization potential of molecules 1 and 2, respectively. r_{12} is the intermolecular distance. The dipole-dipole energy was given as [99]

$$U_{12}^{p} = -\frac{2\mu_1^2\mu_2^2}{3k_BTr_{12}^6} \tag{8}$$

where μ is the dipole moment, r_{12} is the center to center distance between two dipoles, and k_BT is the thermal energy. In most cases, induction energy (dipole-induced dipole) is very small compared with the dispersion energy and the dipole energy. For solid materials, hydrogen bond energy could also be neglected. Thus the total attraction energy can be expressed:

$$U_{12} = U_{12}^{d} + U_{12}^{p} = -\frac{A_{12}}{r_{12}^6} \tag{9}$$

where A_{12} is the total attraction constant given by:

$$A_{12} = A_{12}^{d} + A_{12}^{p} \tag{10}$$

A_{12}^{d} is the attraction constant for dispersion interaction and A_{12}^{p} is the attraction constant for dipole interaction. They can be expressed as:

$$A_{12}^{d} = \frac{3\alpha_1\alpha_2 I_1 I_2}{2(I_1 + I_2)} \tag{11}$$

$$A_{12}^{p} = \frac{2\mu_1^2\mu_2^2}{3k_BT} \tag{12}$$

Based on the quasi-continuum model [100], the surface energy γ could be expressed as:

$$\gamma = \frac{\pi n^2 A_{12}}{16z_0^2} \tag{13}$$

where n is the number of molecules per unit volume, z_0 is the equilibrium separation between two bodies, where the attraction force is zero. Eq.(10) for the attraction constant separates the surface energy into two terms:

$$\gamma = \gamma^d + \gamma^p \tag{14}$$

where

$$\gamma^d = \frac{3\pi n^2 \alpha_1 \alpha_2 I_1 I_2}{32 z_0^2 (I_1 + I_2)} \tag{15}$$

$$\gamma^p = \frac{\pi n^2 \mu_1^2 \mu_2^2}{24 z_0^2 k_B T} \tag{16}$$

Eq. (15) and (16) indicate that the dispersion component of the surface energy is determined by the polarizability of the molecule, while the polar component is determined by the molecule dipole moment. It is known that the dielectric constant of a material can be expressed as follows if the local field is assumed to be the mean field [101].

$$\varepsilon = \varepsilon_\infty + \frac{4\pi n \alpha}{(1 + \varpi^2 \tau_t^2)} \tag{17}$$

where ε_∞ is infinite frequency dielectric constant, α is the effective polarizability, ϖ is angular frequency and τ_t is the relaxation time. Eqs. (15) and (17) clearly show that the surface energy directly correlates with the dielectric constant, and a large dielectric constant means a large dispersion component. Once the leakage conductivity is neglected, the dielectric loss in the electric field is given by [101]

$$tg\delta = \frac{4\varpi t \pi n \mu^2}{3 k_B T \left(\varepsilon_0 + \varepsilon_\infty \varpi^2 t^2\right)} \tag{18}$$

where μ is dipole moment of molecule, ε_0 is dc dielectric constant of molecule (frequency is zero), t is a time constant related to polarization

$t = \tau_t[(2+\varepsilon_0)/(2+\varepsilon_\infty)]$. Equations (16) and (18) show that the dielectric loss tangent correlates with the polar component of the surface energy, and a large dielectric loss tangent means a large polar component. Experimental results [15,102] showed that the suspension would exhibit a strong ER effect if the dispersed particle has both a large dielectric constant and dielectric loss. According to Eqs. (15)–(18), the large dispersion component and polar component of the surface energy appear to be both needed for a strong ER effect. In other words, there should be an appropriate proportion of the polar component to the dispersion component of surface energy, which accords well with the experimental results presented in Figs. 39 and 40. The results clearly imply that the surface modification would be an effective means for improving ER activity.

The findings above could be used to explain why water or other polar liquids can obviously enhance ER effect. The yield stress of aluminosilicate based or polyvinyl alcohol based ER suspensions were experimentally found to go through a maximum within the water content range from 0% to 16% [14]. It is well known that the small amount of water or other polar liquids could easily bind onto the surface of the dispersed particle, resulting in the reduction of the polar component of surface energy. The proportion of the dispersion component to the polar component should also change [97]. Evidently, the shear stress of ER fluids goes through a maximum against water content is confined by the correlation of the shear stress with the surface polarity or surface energy shown above.

7 PARTICLE VOLUME FRACTION

The yield stress and the apparent viscosity of an ER suspension are largely dependent on the particle volume fraction. A linear relationship between the yield stress and the particle volume fraction was derived on the basis of the fibrillation model [103] and compared with the experimental data of the hydrated poly(methacrylate) particle in a chlorinated hydrocarbon suspension obtained by Marshall [104]. The theoretical prediction was only valid in high particle volume fractions, and failed in low particle volume fractions, as stated in this paper and shown in Figure 41.

However, other theoretical works showed that the yield stress or viscosity goes through a maximum as the particle volume fraction increases, and the maximum appears at a very high particle volume fraction [105-107]. Under the reaction of an electric field, the spherical particle was assumed to form the body-centered tetragonal (bct) structure, and the yield stress could be estimated using the many-body electrostatic interaction method proposed in the ref. [106, 108]. The calculated yield stress, normalized by $2\,\varepsilon_m\,(\beta_1 E)^2$, is plotted as a function of particle volume fraction ϕ for several values of α in Figure 42, where ε_m is the dielectric constant of dispersing medium, $\beta_1 = (\alpha - 1)/(\alpha + 2)$, and α is the dielectric constant ratio of the particle to

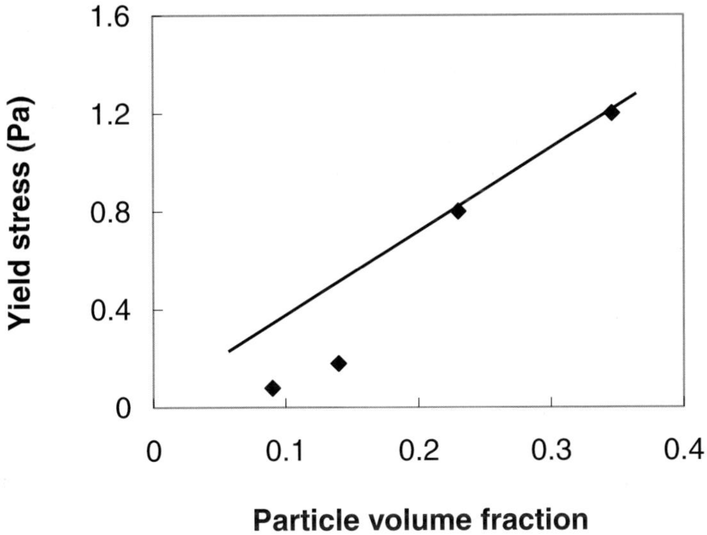

Figure 41 The yield stress of the hydrated poly(methacrylate) particle in a chlorinated hydrocarbon vs. the particle volume fraction. Data was obtained in an electric field 400V/mm. Redrawn from L. Marshall, C.F. Zukoski, and J. Goodwin, J. Chem. Soc., Faraday Trans. 1, 85(1989)2785.

the dispersing medium, $\alpha = \varepsilon_p / \varepsilon_m$ and E is the electric field strength. The yield stress peaks around 45 vol%, and the exact position depends on the dielectric constant ratio of the particle to the dispersing medium. The yield stress peak moves to the high particle volume fraction side in the case that the dielectric constant ratio of the particle to the dispersing medium is large. A similar trend was found between the shear modulus of the bct structure and the particle volume fraction as shown in Figure 43. A microscopic model that takes into account only the shape anisotropy (such as chain or cylinder) of the particle aggregates and a macroscopic model that takes into account only the inter-particle force inside the aggregates were further proposed for calculating the yield stress of either a MR or an ER fluid [107].

The normalized yield stress ($\tau_y / \mu H^2$, where μ is the permeability of the suspension and H is the magnetic field strength) vs. the particle volume fraction is shown in Figure 44 for comparing the difference between the macroscopic and microscopic models. $\alpha = \mu_p / \mu_m$=0, corresponds to the case of magnetic holes in a MR fluid. For both models, the yield stress peaks at the particle volume fraction 30 vol%. It appears that those two models agree relatively well. Another yield stress equation was derived on the basis of electric-field-induced phase transition and the formation of the bct lattice structure [109]. The normalized yield stress τ_y/E^2 x10^2 was found to gradually increase with particle volume fraction when the volume fraction is low and to abruptly jump at the particle volume fraction around 35 vol%.

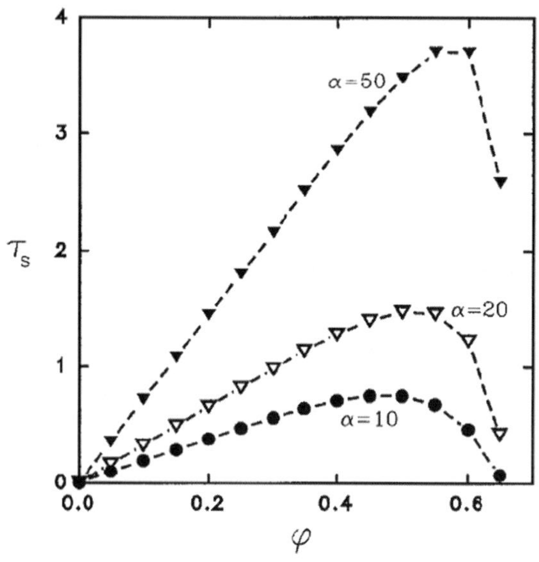

Figure 42. The static yield stress τ_s of the bct structure, normalized by 2 ε_m (β_1E)^2, plotted as a function of particle volume fraction φ for several values of α. ε_m is the dielectric constant of dispersing medium, $\beta_1 = (\alpha - 1)/(\alpha + 2)$, and α is the dielectric constant ratio of the particle to the dispersing medium, $\alpha = \varepsilon_p / \varepsilon_m$, and E is the electric field strength. Reproduced with permission from H.J.H. Clercx, and G. Bossis, J. Chem. Phys., 103(1995)9426.

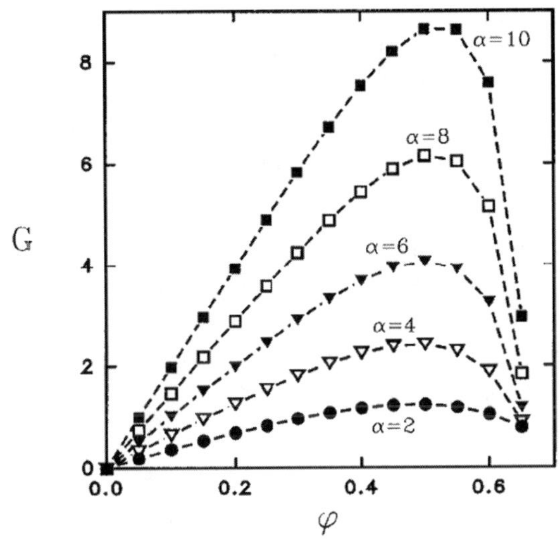

Figure 43. The shear modulus *G* of the bct structure, normalized by 2 ε_m $(\beta_1 E)^2$, plotted as a function of particle volume fraction ϕ for several values of α. ε_m is the dielectric constant of dispersing medium, $\beta_1 = (\alpha - 1)/(\alpha + 2)$, and α is the dielectric constant ratio of the particle to the dispersing medium, $\alpha = \varepsilon_p / \varepsilon_m$, and E is the electric field strength. Reproduced with permission from H.J.H. Clercx, and G. Bossis, J. Chem. Phys., 103(1995)9426.

Experimental results don't exactly follow the theoretical prediction. Uejima [H. Uejima, Jpn. J. Appl. Phys. 11(1972)319] found experimentally that the relative viscosity of crystalline cellulose/chlorine suspension peaks at particle weight percentage 10% (see Figure 45). However, Block [24] and Xu [10] found the yield stress or complex viscosity parabolically increases with the particle volume fraction, as shown in Figure 4 in Chapter 3 and Figure 46. Those are contradictory to Uejima's result and to Bossis' theoretical prediction. Hao [110] found that there was a critical particle volume fraction for the ER suspension. The complex viscosity of oxidized polyacrylonitrile/silicone oil suspension vs. the particle volume fraction at the electric field 1.5 kV/mm is shown in Figure 47.

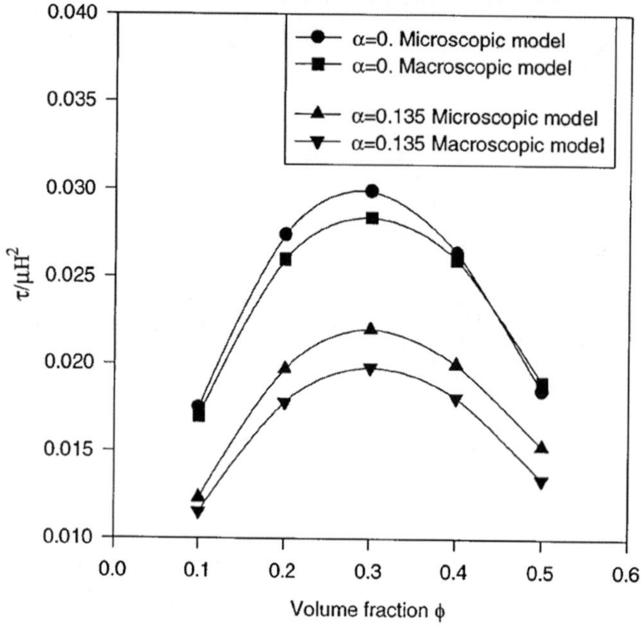

Figure 44 Normalized yield stress ($\tau / \mu H^2$, where μ is the permeability of the suspension and H is the magnetic field strength) vs. the particle volume fraction for different α values ($\alpha = \mu_p / \mu_m$, the permeability ratio of the particle to the medium). Reproduced with permission from G. Bossis, E. Lemaire, O. Volkova, H. Clercx, J. Rheol. 41(1997)687.

The data was obtained at the shear strain of 200% and frequency 2 s^{-1}. There is a clear critical particle volume fraction in this suspension system, about 38 vol%. When the particle volume fraction exceeds this critical value, a sharp increment of the rheological property is observed. Percolation theory was used to explain this phenomenon, which will be discussed in detail in later on. Note that Figure 46 is consistent with Figure 47, and Figure 46 may only show the low particle volume fraction portion of Figure 47.

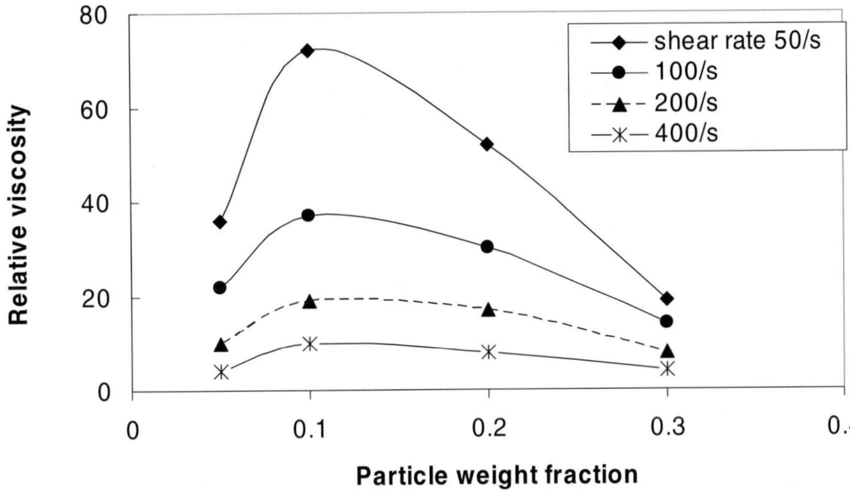

Figure 45 The relative viscosity, the apparent viscosity at 1000 V devided by the viscosity at zero electric field, vs. the particle volume fraction of crystalline cellulose/chlorine suspension obtained at different shear rates. The adsorbed water content is 8.3 wt%. Redrawn from H. Uejima, Jpn. J. Appl. Phys. 11(1972)319.

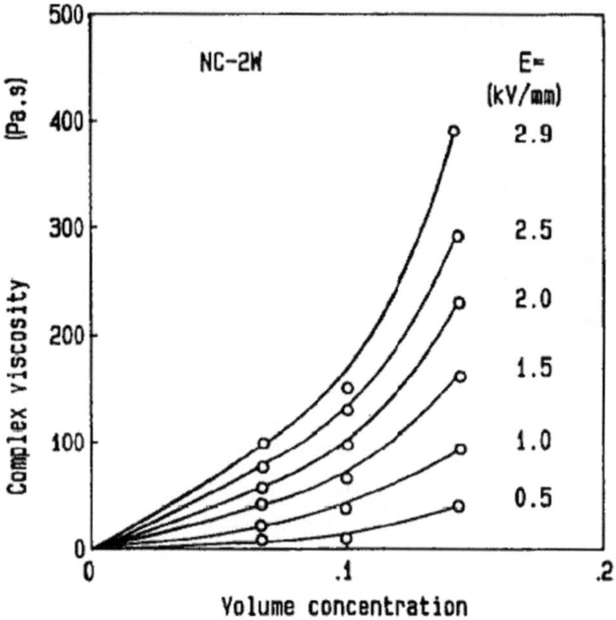

Figure 46 Complex viscosity of oxidized polyacrylonitrile/silicone oil suspension vs. the particle volume fraction at different electric fields. The mechanical strain is 1 and frequency is 5 s^{-1}. Reproduced with permission from Y. Xu, R. Liang, J. Rheol. 35 (1991) 1355.

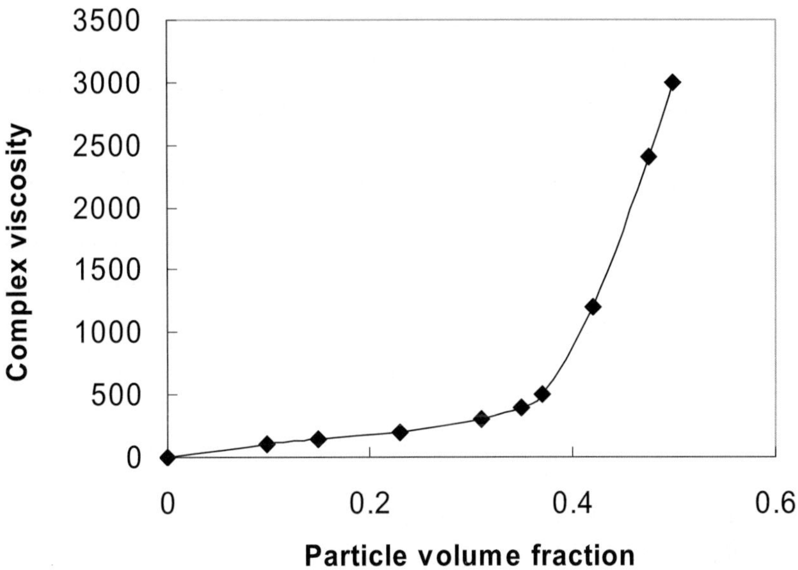

Figure 47. Complex viscosity of oxidized polyacrylonitrile/silicone oil suspension vs. the particle volume fraction at the electric field 1,5 kV/mm.The shear strain is 200% and frequency 2 s^{-1}. Redawn from the data of T. Hao, Y. Chen, Z. Xu, Y. Xu, Y. Huang, Chin. J. Polym. Sci. 12(1994)97.

An interesting approach for deriving the relationship between the yield stress and particle volume fraction was proposed by Mezzasalma [111] on the basis of Helmholtz free energy change under an electric field [109]. The yield stress equation was derived as:

$$\tau_y(\phi) = A_\tau \theta^2(\phi)\phi^m \tag{19}$$

and

$$\theta\left(\bar{V}, \psi_m, \psi_p, \phi\right) = \frac{\phi + \bar{V}(1-\phi)}{\psi_m \phi + \bar{V}\psi_p(1-\phi)} \tag{20}$$

where A_τ is a heuristic yield stress parameter, m is a positive heuristic coefficient, $\psi_{m,p} \propto h\overline{\omega}_{m,p}$, an average energy related to the characteristic frequency $\overline{\omega}_{m,p}$ either in a liquid medium or a solid particle state , h is the Planck constant, and $\bar{V}$ is the molar volume ratio of the solid particle to the liquid medium. $\psi_{m,p}$ and $\bar{V}$ should be considered as phenomenological parameters. If $\psi_p \geq \psi_m$, the yield stress should always increase as particle volume fraction increases; However, if $\psi_p \leq \psi_m$, as soon as the particle volume fraction exeeds a critical value, the yield stress should decerase with the increase of the particle volume fraction. In other word, the yield stress goes through a maximum value with the volume fraction increase, as indicated earlier both theoretically and experimentally. Eq. (19) was applied to several ER systems and the predicted yield stress was plotted with the experimental data points vs. the particle volume fraction in Figure 48. In Figure 48 (a) and (b), hydrated poly(methacrylate) particle was dispersed in a chlorinated hydrocarbon [104]; in Figure 48(c), the polyaniline particle was dispersed in silicone oil [62]; in Figure 48 (d) chlorinated paraffin/silicone oil emulsions [112]; in Figure 48 (e) a numerical simulation [113]. In Figure 48(d), the yield stress τ^* was normalized to the one without the external electric field. In case of Figure 48(e), the data were obtained with several different dielectric constant ratio between the solid and the liquid medium, and the yield stress was normalized by the factor of $2\varepsilon_m(\beta E)^2, where \beta = \dfrac{\varepsilon_p - \varepsilon_m}{\varepsilon_p - 2\varepsilon_m}$, ε_p and ε_m are dielectric constant of solid particle and liquid medium, respectively. As one may see, all parameters are heuristic and would be hard to determine for each individual system, seriously limiting the applicability of the Eq. (19).

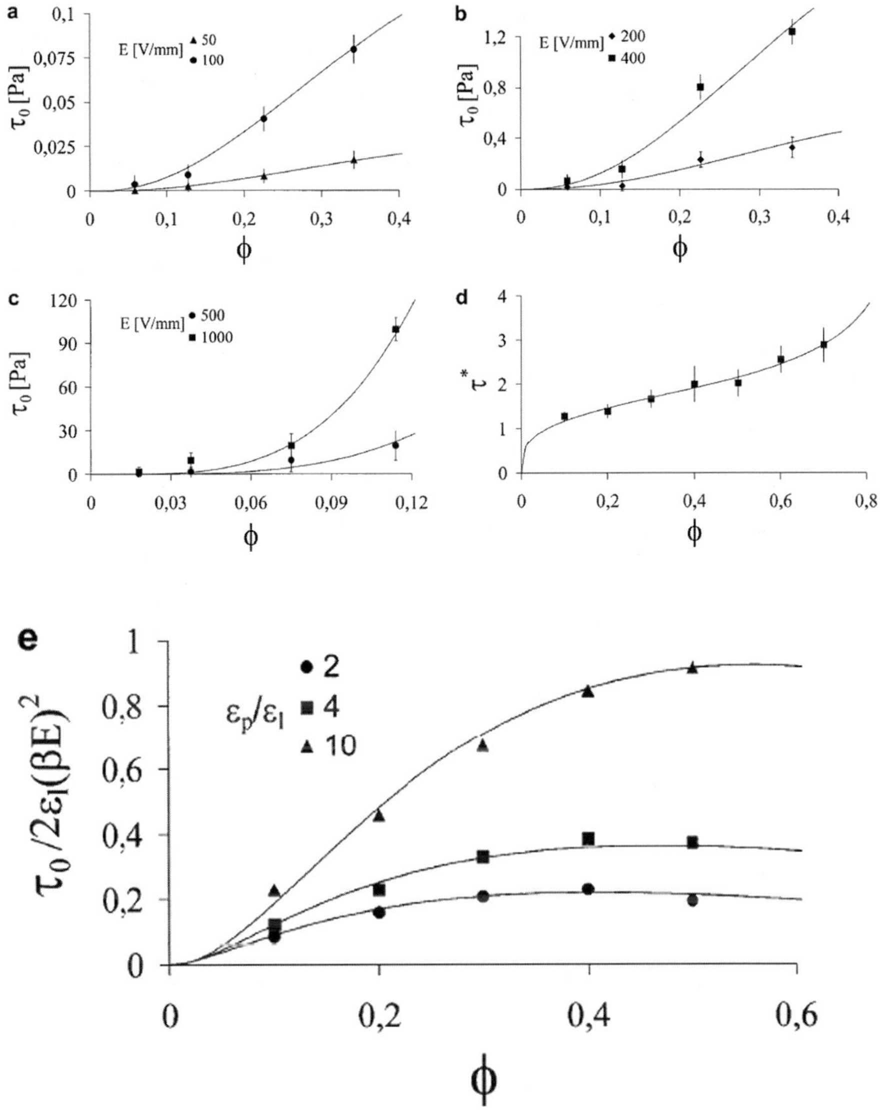

Figure 48 Yield stress τ_0 vs. the particle volume fraction ϕ fitted with Eq. (19) for various ER systems. The data shown in a) and b) for hydrated poly(methacrylate) particle in a chlorinated hydrocarbon (Ref. L. Marshall, C.F. Zukoski, and J. Goodwin, J. J. Chem. Soc., Faraday Trans. 1, 85(1989)2785; c) for polyaniline/silicone oil [C.J. Gow, and C.F. Zukoski, J. Colloid Interface Sci., 136(1990)175]; d) for chlorinated paraffin/silicone oil emulsions [X.Pan, and G.H. McKinley, J.Colloid and Interface Sci., 195(1997)101]; e) for a numerical simulation [R.T. Bonnecaze, and J.F. Brandy, J.rheol., 36(1992)73]. The extrapolated coefficients were: a) and b)

$\bar{V} = 5.2, m = 2.6, \psi_p / \psi_m = 0.1, and$ ψ_m =10.5, 4.8, 2.3, and 1.2 with increasing E; c) $\bar{V} = 0.3, m = 4.8, \psi_p / \psi_m = 0.1, and$ ψ_m=1.5 x10^{-3}, 3.1x10^{-3}; d) $\bar{V} = 10.2, m = 0.29, \psi_p / \psi_m = 0.1, and$ ψ_m =6.7 x10^{-2}; e) $\bar{V} = m = 2.2, \psi_p / \psi_m = 0.1, and$ ψ_m==3.1, 2.2, 1.3 with increasing dielectric constant ratio. A_τ=1 in all cases.

8 TEMPERATURE

There are two reasons why temperature might change the ER effect substantially. The first is that temperature can definitely change the polarizability of the ER suspension, as the particle conductivity and dielectric property vary with temperature. The second is that temperature would directly impact particle thermal motion. If the Brownian motion is intensified at high temperatures and becomes strong enough to compete with the particle fibrillation, and the ER effect should become weak. Whether the elevated temperature would intensify or weaken the ER effect is really dependent on which factor would become dominant at that temperature. Of course, the physical property of the dispersing medium should also vary with temperature, which may modify the ER effect in a considerable scale. But in most cases, it is negligible compared with the two factors mentioned above.

The effect of temperature on ER activity has been addressed in both anhydrous and hydrous ER systems. An obviously improved ER response was found at high temperatures in several ER suspensions, including both the inorganic and polymeric systems [21, 22, 114, 115]. Improved ER response was regarded to be a consequence of readily facilitated polarization of the double layer at high temperatures [21,22]. A similar temperature enhancement was found in a humidified zeolite-based ER fluid [115]. The shear stress of zeolite/silicone oil suspension with particle concentration 34 wt% vs. temperature is shown in Figure 49. The shear stress clearly increases with temperature and levels off at a relatively high temperature, around 120 °C. There is no shear stress decrease even when temperature goes up to 150 °C.The improvement, however, was attributed to the dielectric mismatch between the particle and the carrier fluid and a local concentration of the electric field. The diffusion of Na^+ in the zeolite particle is considered to be responsible for the temperature dependence of the electric conductance of the whole ER fluid. For anhydrous polyelectrolyte ER systems, some of these materials were found to work much better at 100 °C than at room temperature [114]. The locally mobile ions, mobiling within chain coils or along chains but not free to move between chains, were thought to be associated with ER activity [114]. In a word, the readily facilitated polarization of the electric double layer in a hydrous ER system was thought to contribute to the ER effect improvement at high temperatures, while the change of the particle intrinsic property (for example ion mobility inside the particle) was thought to be responsible for the ER effect enhancement in anhydrous ER systems. Hydrous ER suspensions are believed to have a narrow temperature range somewhat between –20°C and +70 °C, possibly due to the water solidification and evaporation. An anhydrous ER suspension could work in a wide temperature range. However, it is limited by the large conductance at high temperature, as most anhydrous ER fluid is made from ionic materials.

The mechanism of how temperature affects the ER activity has received a huge attention. The direct attraction force was measured betweem two semispheres made from semicrystalline polyamide with a radius 7 mm immersed in silicone oil [20]. The percentage variation of the force ΔF/F (compared to the force at room temperature) between those two semispheres is plotted as a function of temperature in Figure 50.

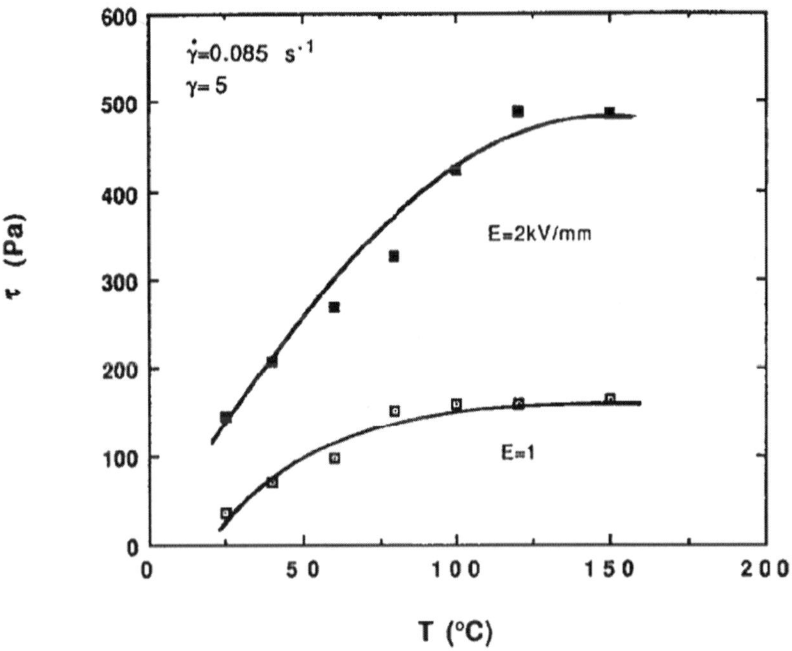

Figure 49 The shear stress of zeolite/silicone oil suspension vs. temperature. The particle concentration is 34 wt%. Reproduced with permission from H. Conrad, A. F. Sprecher, Y. Choi, and Y. Chen, 35(1991)1393.

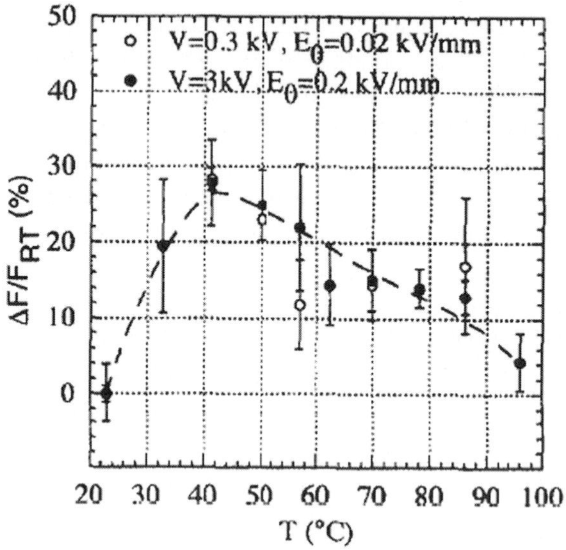

Figure 50 Percentage variation of the force ΔF/F (compared to the force at room temperature) between two semispheres with radius 7mm as a function of temperature T. Measurments were given for two electric fields, E_0=0.02 kV/mm and E_0=0.2 kV/mm, corresponding to the regime $F \propto E0$ and $F \propto E_0^2$, respectively. Error bars are representative of the maximum and minimum force values obtained for four successive measurements performed in about 5 min. Reproduced with permission from P. Gonon, and J-N. Foulc, J. Appl. Phys. 87(2000)3563.

The current density of an ER suspension should be a direct indicator on how temperature affects the ER performance, as it closely relates to the conductivity of the dispersed particle and the temperature. The conductivity of the insulating oil is usually very low and can be neglected. Since the order is increased when an ER fluid fibrillate under an electric field, and the entropy of whole suspension thereby reduces, and an excessive work must be spent to maintain this state. The magnitude of the current density likely scales the power demand, and the minimum current density demanded in an ER fluid is an important parameter. The current density versus temperature was found to fit to the power law at an electric field [117], increasing exponentially with the elevated temperature. However, any increase of the current density would result in excessive power demands with possible serious implications in terms of power supply and energy dissipation in ER devices. For these reasons, determination of the conductive mechanism of ER fluids and the demanded current level seems to be significant for both the design of the high performance ER fluids and the application devices. The Quasi-one-dimensional Variable Range Hopping model was found to govern the conductance of the semiconductive polymer-based binary ER system [118], which will be discussed in a more detailed way in future chapter. An approximate linear relation was phenomenologically found between the current density and the shear stress [117].

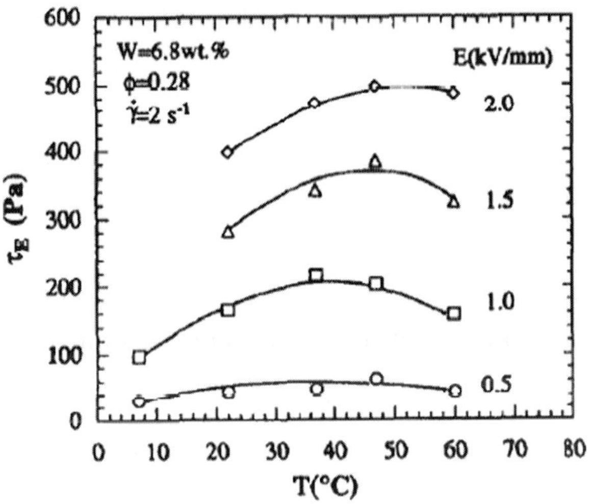

Figure 51 Shear stress of corn starch/corn oil vs. temperature on at different electric fields. The water content is 6.8 wt% and particle volume fraction is 28 vol%. Reproduced with permission from H. Conrad, Y. Li, and Y. Chen, J. Rheol. 39(1995) 1041.

The influence of temperature on the relationship between the shear stress and the current density was comprehensively addressed using the molecular sieve and permutite particles of a similar molecular structure dispersed into silicone oil [79]. The molecular structure of those two materials can be expressed as $(MO)_n(Al_2O_3)_x(SiO_2)_y(H_2O)_z$, where n, x, y, z are integral numbers and M represents the metallic atom. The particle was heated under at least 500 °C for several hours before it was mixed with silicone oil. The shear stress of the molecular sieve/silicone oil suspension versus the elevated and reduced temperature at an electric field 1.5 kV/mm and particle volume fraction 35 vol% is shown in Figure 52. The temperature was raised up from room temperature around 20 °C and then cooled down back to the room temperature. With the increase of temperature, the shear stress of the molecular sieve/silicone oil suspension slowly decreases, whereas the current density increases with temperature, as shown in Figure 53. When the suspension gradually cools from a high temperature to room temperature, the shear stress goes up at low temperatures. Note that an anomalous shear stress peak was found to appear at the temperature about 80°C. As expected, the current density decreases with reduced temperature and increases with elevated temperature. Ideally, the shear stress of an ER fluid measured in the elevated temperature process and then in the reduced temperature process should coincide with each other. However, experimentally the shear stress in the reduced temperature process is much higher than in the elevated temperature process, and the current density in the reduced temperature process is less than that in the elevated temperature process. If the lower current density in the reduced temperature process resulted from the water evaporation during the elevated temperature process, the shear stress in the reduced temperature process would be less than in the elevated temperature process, however, this is just opposite to the experimental result. Thus some other mechanisms rather than a water-activating mechanism would be responsible for the increment of the shear stress after an elevated temperature process. The shear stress of the molecular sieve/silicone oil suspension as a function of the current density in the elevated and reduced temperature processes is shown in Figure 54. Note that the shape of the curve shown in Figure 54 is somewhat similar to that in Figure 52, implying that an optimal current density and then an optimal energy demand may exist for ER systems.

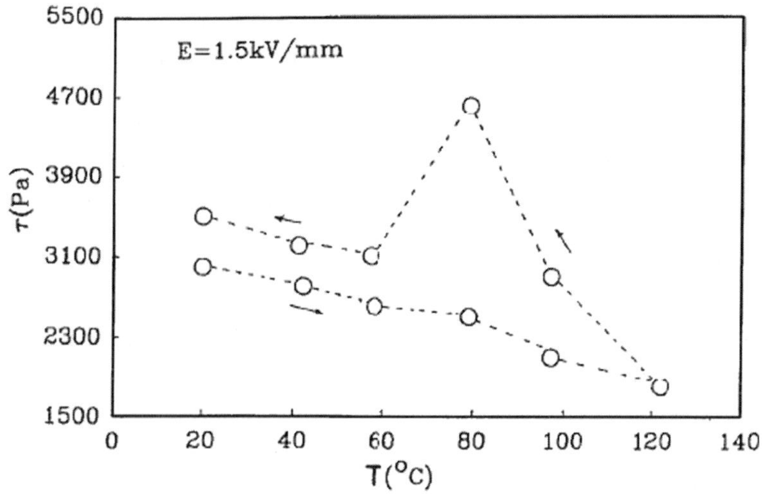

Figure 52. Shear stress of the molecular sieve/silicone oil suspension versus the elevated and reduced temperature at an electric field 1.5 kV/mm. Particle volume fraction is 35 vol%, shear rate is 0.3 s^{-1}, and electric field is 1.5 kV/mm. Reproduced with permission from T. Hao, H. Yu, and Y. Xu, J. Colloid Interface Sci., 184(1996)542.

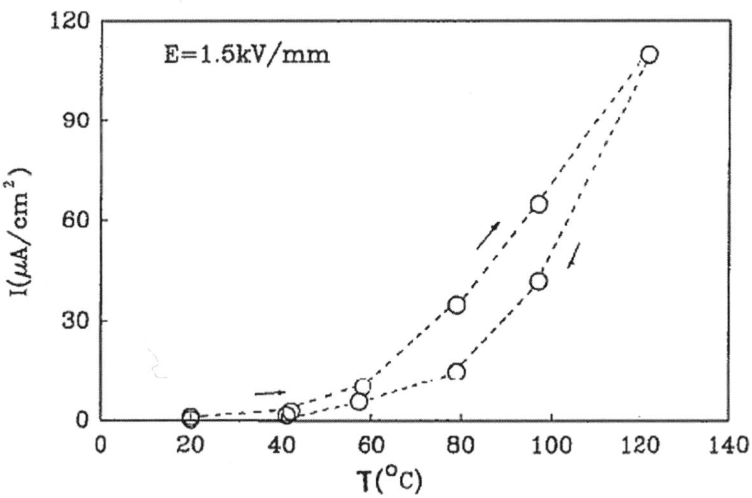

Figure 53 Current density of the molecular sieve/silicone oil suspension versus temperature at an electric field 1.5 kV/mm. Particle volume fraction is 35 vol%. Reproduced with permission from T. Hao, H. Yu, and Y. Xu, J. Colloid Interface Sci., 184(1996)542.

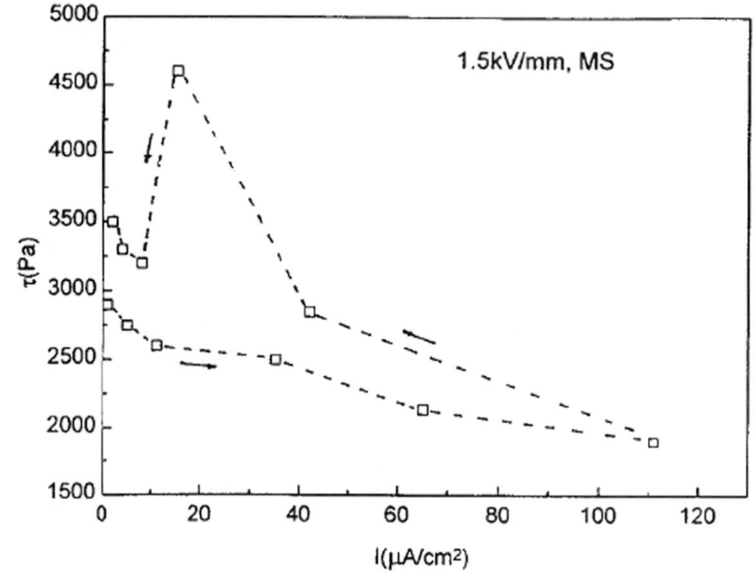

Figure 54 Shear stress of the molecular sieve/silicone oil suspension versus current density at an electric field 1.5 kV/mm. Particle volume fraction is 35 vol%, and shear rate is 0.3 s^{-1}. Reproduced with permission from T. Hao, H. Yu, and Y. Xu, J. Colloid Interface Sci., 184(1996)542.

For the permutite/silicone oil suspension, the shear stress vs. temperature is shown in Figure 55, obtained at the particle volume volume fraction 35 vol% and electric field 1.5 kV/mm. In this case, the shear stress peaks at 40 ^{o}C, different from the molecular sieve/silicone oil suspension, the shear stress of which peaks at 80 ^{o}C. il suspension. Also, the shear stress data of the permutite/silicone oil suspension obtained in the elevated temperature and reduced temperature don't overlap, similar to the molecular sieve/silicone oil. The difference between those two suspensions are that the shear stress of permutite/silicone oil suspension obtained in the reduced temperature process is less than that in the elevated temperature process, contrary to the situation in the molecualar sieve/silicone oil suspension. The current density of permutite/silicone oil suspension vs. temperature is shown in Figure 56. The current density values of those two ER fluids obtained in the reduced temperature process are all less than that in the elevated temperature process. The shear stress of permutite/silicone oil vs. the current density is shown in

Figure 57. Approximately, a similar shear stress dependence on the current density, especially the shear stress peaks at certain current density, was found for both the permutite and molecular sieve type suspensions. Those experimental results could be well explained using the Wagner-Maxwell model, which will be discussed extensively in future chapters. The Wagner-Maxwell polarization is strongly related to the conductivity of the dispersed particle. The conductance of an ER fluid might be determined by the conductivity of the dispersed particle, the number of the particles in the fibrillated chains or columns, and the contact resistances between the neighboring particles. The conductivity of the carrier medium is negligible as it is comparatively small, on the magnitude of 1×10^{-14} S/m. The current density is therefore mainly confined by the conductivity of the dispersed particle. As shown in the section 4 in this chapter, the strongest ER effect would take place at the particle conductivity about 1×10^{-7} S/m, which can be schematically shown in Figure 58. Supposing that the conductivity of the molecular sieve/silicone oil suspension is at point D, and that of the permutite/silicone oil suspension is at point B. During reduced temperature process, their conductivities all decrease, thus one may assume that the particle conductivity of the molecular sieve/silicone oil suspension moves from the point D to C, and that of the permutite/silicone oil suspension moves from the point B to A. Clearly, $\tau_A < \tau_B$, $\tau_C > \tau_D$. In other word, the shear stress of the molecular sieve/silicone oil suspension would increase when the particle conductivity decreases, whereas that of permutite/silicone oil the suspension would go to the opposite side. Generally speaking, a decrease in shear stress with the elevated temperature would be found in a suspension with the particle conductivity larger than the optimal value, while a increase in the shear stress with temperature would be found if the particle conductivity is less than the optimal value. Nevertheless, this interpretation cannot explain the fact that a shear stress peak of the molecular sieve/silicone oil suspension is found to occur in the reduced temperature process and a similar peak of the permutite/silicone oil suspension is found in the elevated temperature process. The possible reason might be that the optimum particle conductivity is achieved at those temperatures. However, it is not clear why this peak was not observed in the reversed process. Unlike the curve shown in Figure 58, the curves of the shear stress versus the current density depicted in Figs. 54 and 57 are not smooth, indicating that a more complicated process or mechanism is involved.

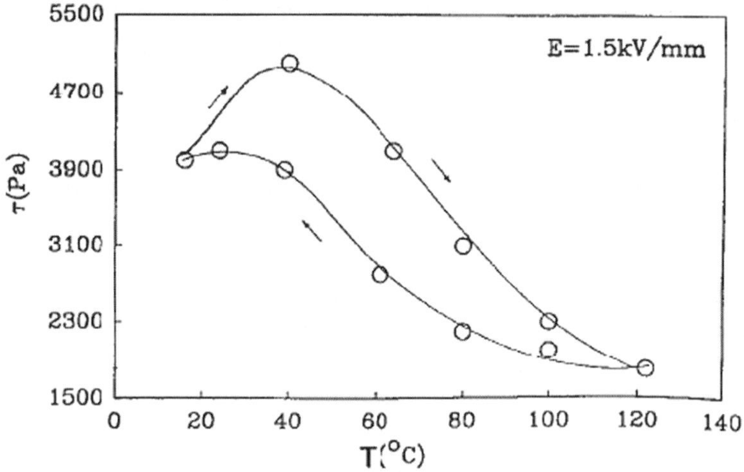

Figure 55 Shear stress of the permutite sphere (PS)/silicone suspension vs the elevated and reduced temperature. Particle volume fraction is 35 vol%, shear rate is 0.3 s^{-1}, and electric field is 1.5 kV/mm. Reproduced with permission from T. Hao, H. Yu, and Y. Xu, J. Colloid Interface Sci., 184(1996)542.

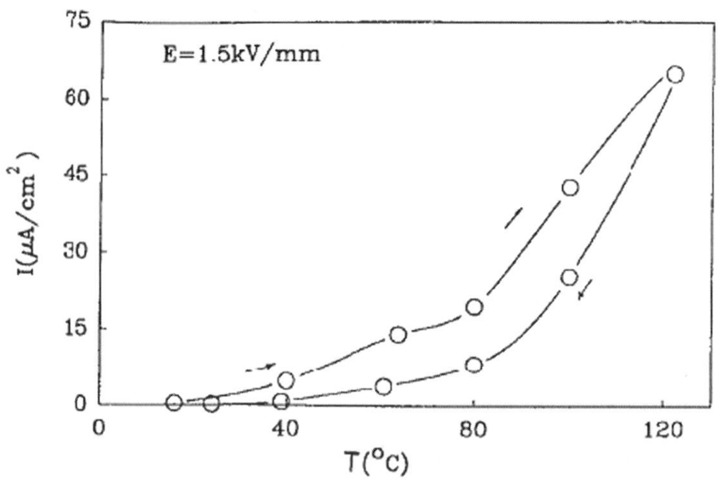

Figure 56 Current density of the permutite sphere (PS)/silicone suspension versus the elevated and reduced temperature. Particle volume fraction is 35 vol%, shear rate is 0.3 s^{-1}, and electric field is 1.5 kV/mm. Reproduced with permission from T. Hao, H. Yu, and Y. Xu, J. Colloid Interface Sci., 184(1996)542.

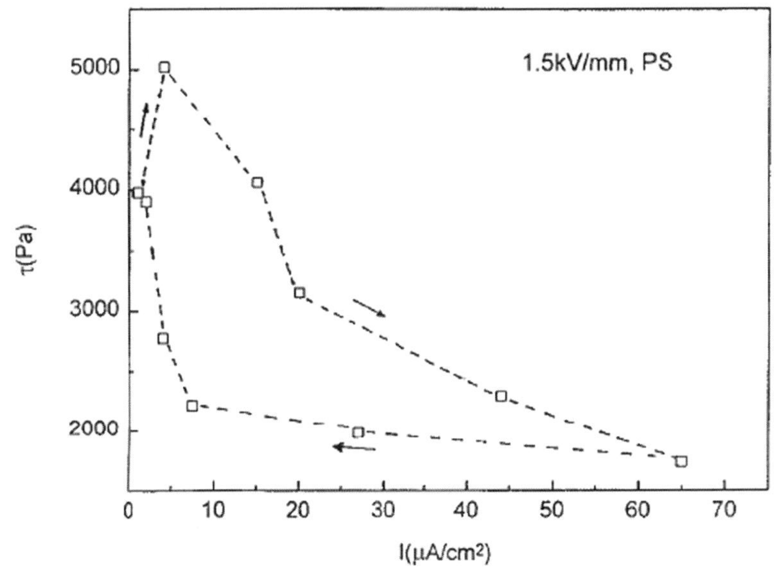

Figure 57 shear stress of the permutite sphere (PS)/silicone suspension versus the current density during the elevated and reduced temperature proceses. Particle volume fraction is 35 vol%, shear rate is 0.3 s^{-1}, and electric field is 1.5 kV/mm. Reproduced with permission from T. Hao, H. Yu, and Y. Xu, J. Colloid Interface Sci., 184(1996)542.

There are many other examples to further confirm that temperature dependence of ER effect is confined by the conductivity of the dispersed particle. The shear stress of KNO_3 particle/silicone oil vs temperature at different dc electric fields is shown in Figure 59 [119,120]. The conductivity of KNO_3 varies exponentially with the increase of temperature below and above the Curie temperature (*Tc* ~129 °C), respectively. Interestingly, after it was mixed with silicone oil, the measured shear stress of this suspension at shear rate 3.18 s^{-1} was found to have a similar temperature dependence as the conductivity of ferroelectric single crystal KNO_3. The upper dash line in Figure 59 shows the conductivity of KNO_3 against temperature. There is a steep increase of conductivity at Curie temperature 129 °C [121]. The shear stress of KNO_3 particle/silicone oil suspension abruptly increases at the Curie temperature, and then decreases rapidly to a level that is still higher than that below the Curie temperature, clearly indicating that the shear stress is strongly dependent on the conductivity of the dispersed particle, and the largest shear stress appears at the particle conductivity close to $3x10^{-6}$ S/m.

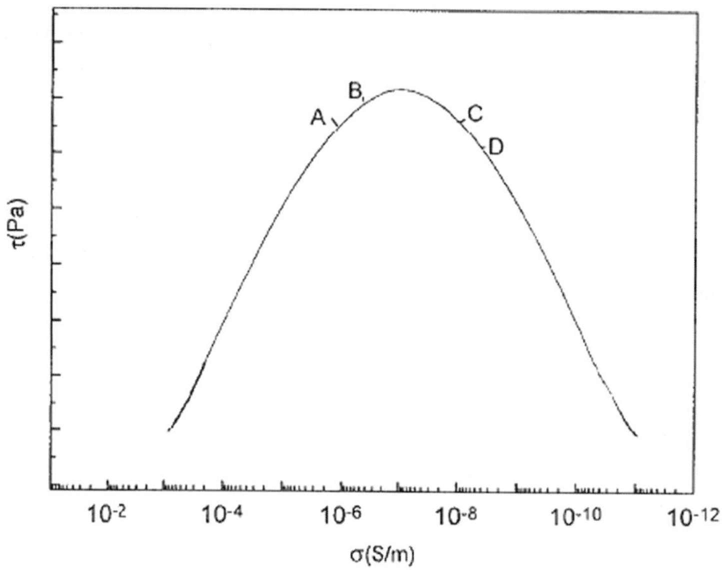

Figure 58 Schematic illustration of the shear stress versus particle conductivity. Reproduced with permission from T. Hao, H. Yu, and Y. Xu, J. Colloid Interface Sci., 184(1996)542.

For a given material, the dielectric constant should be directly correlated with its conductivity. Once it is dispersed into a insulating liquid, the dielectric property of this suspension should be governed by the Wagner-Maxwell polarization, or called the interfacial polarization [37-39, 125-129], which is strongly dependent on the conductivity of dispersed particle and will be discussed in future chapter. Experimental evidence shows that the shear stress of an ER suspension follows the same trend as the dielectric constant of the dispersed particle when temperature varies [130]. An good example is the ferroelectric single crystal triglycine sulphate TGS ($(NH_2CH_2COOH)_3 \cdot H_2SO_4$). The dielectric constant of ferroelectric single crystal TGS ($(NH_2CH_2COOH)_3 \cdot H_2SO_4$) versus temperature at different frequencies is plotted in Figure 60. Along the ferroelectric direction (*b*-axis), the dielectric constant of TGS shows a dramatic change. Those along other directions are relatively independent of temperature in the tested range. The average dielectric constant of TGS could be calculated using the following formula [131]

$$\varepsilon_p = \theta\varepsilon_b + (1-\theta)\frac{(\varepsilon_a + \varepsilon_b + \varepsilon_c)}{3} \tag{21}$$

where ε_p is the average dielectric constant of particle, θ is the preferred orientation degree of particle under an electric field, *a*, *b*, **c** refers to three crystal axes of the particle and *b* is the polarization axis (direction of the particle chains). The shear stress of TGS particle/silicone oil suspension and the average dielectric constant of TGS particle are plotted againt temperature in Figure 61. Interestingly, both the shear stress and the average dielectric constant varies continuously with temperature and peak at the Curie temperature (Tc = 50 °C). This result clearly indicates that temperature changes the conductivity and dielectric constant of dispersed particle, leading to the change of the shear stress of whole ER suspension.

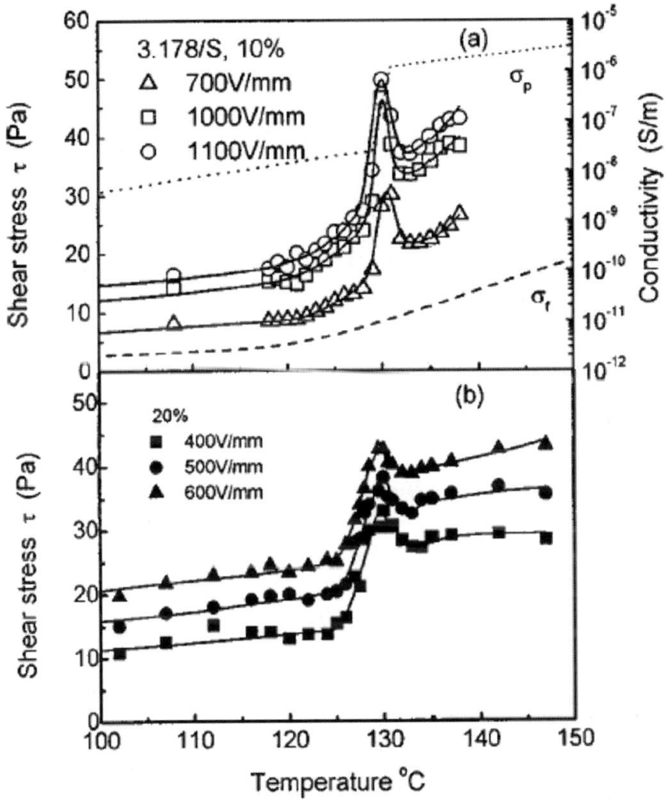

Figure 59. The shear stress of KNO_3 particle/silicone oil vs temperature at different dc electric fields. a) particle volume fraction is 10%, and b) 20%. Dashed lines in (a) represent the conductivity of particles and that of oil vs temperature, respectively, measured at 1000 V/mm under dc electric fields. Reproduced with permission from Y. Lan, X. Xu, S. Men, and K. Lu, Appl.

Phys. Lett., 73(1998)2908; K.Lu, Y. Lan, S. Men, X., Xu, X.,Zhao, S. Xu, Int. J. Modern Phys. B, 15(2001)938

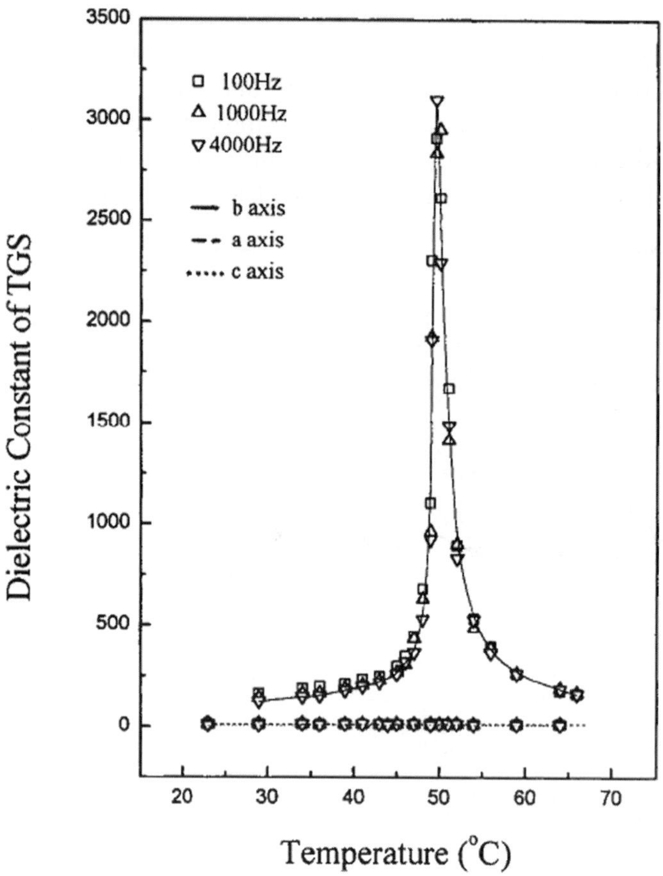

Figure 60 Dielectric constant of single crystal TGS ($(NH_2CH_2COOH)_3 \cdot H_2SO_4$) versus temperature at different frequencies. The full curve is the fitting curve of the dielectric constant along the ***b***-axis, the dashed and dotted lines those along the ***a***-axis and ***c***-axis respectively. Reproduced with permission from Y.C. Lan, S. Q. Men, X. Y. Xu, and K. Q. Lu, J. Phys. D: Appl. Phys. 33(2000)1239

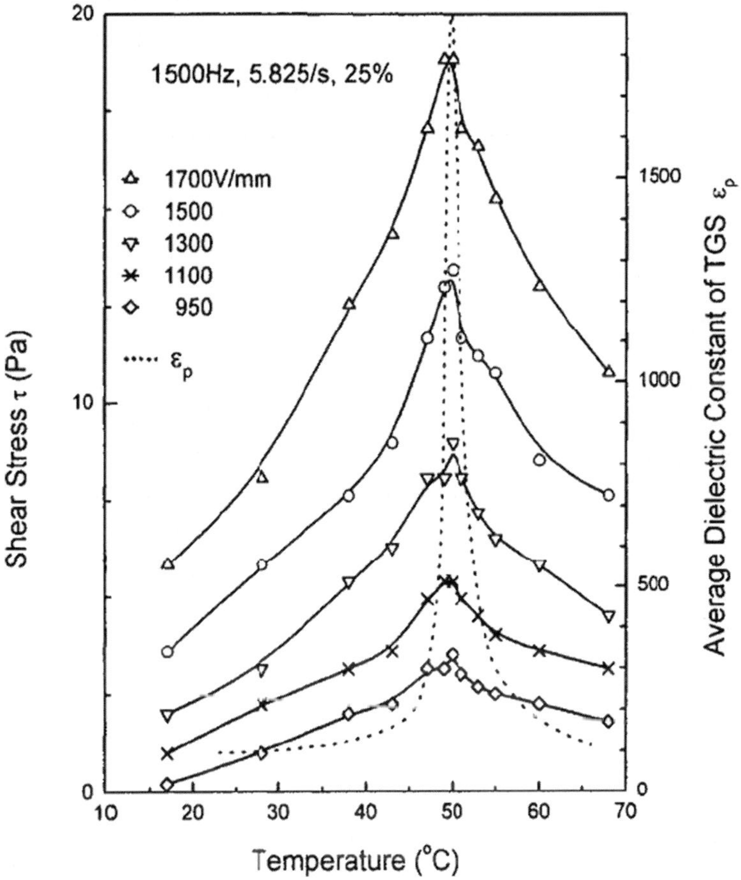

Figure 61. The shear stress of TGS ($(NH_2CH_2COOH)_3 \cdot H_2SO_4$) particle dispersed in silicone oil with the particle volume fraction of 25% vs.temperature at 1500 Hz ac fields. The dashed curve represents the average dielectric constant of the particle at 1500 Hz. Reproduced with permission from Y.C. Lan, S. Q. Men, X. Y. Xu, and K. Q. Lu, J. Phys. D: Appl. Phys. **33**(2000)1239.

9 LIQUID MEDIUM

Generally, the dispersing phase has a low dielectric constant and does not have a strong impact on the ER activity, apart from an influence on the response time of the ER fluid due to its viscosity, conductivity, and dielectric constant. Theoretically, a particulate material should display a similar ER response regardless whether dispersed in silicone oil or mineral oil. However, in some cases the ER effect strongly depends on the dispersing medium if the dielectric constant or conductivity of the dispersing medium is comparable to that of the dispersed phase [132] or if water exists in the system [133]. Experimentally, the zeolite/silicone oil system was found to show much stronger ER effect than the zeolite/mineral oil system, as shown in Figure 62. The same solid particulate material displays an ER effect in one medium, however, the ER effect disappears when the particle is dispersed into another medium [132].

The ER effect is greatly enhanced if the particulate material is mixed with a liquid that is also ER active.[134,135]. Polyhexyl isocyanate (PHIC) solution is a positive ER active material [136], and the viscosity of PHIC/p-xylene solution of concentration 23.3 vol% vs. the electric field is shown in Figure 63. When polymer resin particle dispersed in PHIC solution, the ER effect is greatly enhanced as shown in Figure 64. A similar phenomenon is also observed in the suspension of zeolite 3A particle dispersed in PHIC system. The enhancement may be related to the positive ER behavior of PHIC solution.

The ER effect is weakened if the particle material is dispersed in a liquid medium of the negative ER effect. *N*-4-methoxybenzylidine-4 butylaniline (MBBA) is an ER active material of negative ER effect. Zeolite particle is dispersed into silicone oil, mineral oil, and MBBA when 8 wt% water retained in zeolite particle, and the zeolite/MBBA suspension shows the yield stress in between zeolite/silicone and zeolite/mineral oil systems (see Figure 62). Once water was removed, the yield stress of zeolite/MBBA suspension was still less than that of zeolite/silicone oil suspension, implying that MBBA played a critical role in the system. However, the zeolite/MBBA system may show a unique temperature dependence, as the MBBA may change from nematic to isotropic phase with the increase of temperature. Without an electric field, the yield stress of the zeolite/MBBA suspension with particle volme fraction 31.5 vol% vs. temperature is shown in Figure 65. Around 37 ^{o}C there is an abrupt increase of the yield stress due to the microstructure change of MBBA. A same temperature dependence trend is expected when this suspension is reacted with an electric field, which is mostly favorable as ER fluids usually show a weakened ER effect as temperature increases.

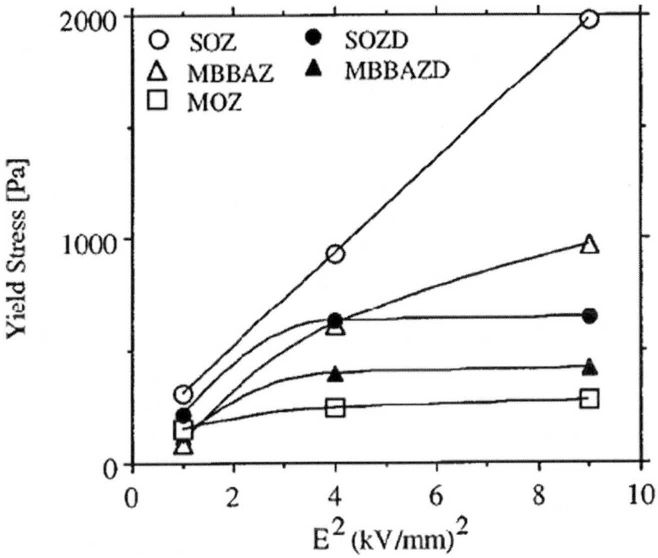

Figure 62 Dynamic yield stresses vs the square of electric field strength for the zeolite/silicone oil(SOZ), zeolite/mineral oil(MOZ), zeolite/MBBA (MBBAZ), dried zeolite/silicone oil (SOZD) and dried zeolite/MBBA (MBBAZD) systems. All suspensions have particle volume fraction 31.5 vol%. MBBA stands for *N*-4-methoxybenzylidine-4 butylaniline, Reproduced with permission from M. Jordan, A. Schwendt, D. Hill, S. Burton, N. Makris, J. Rheol. 41(1997)75. V. Sequeira, D. Hill, J. Rheol. 42(1998)203

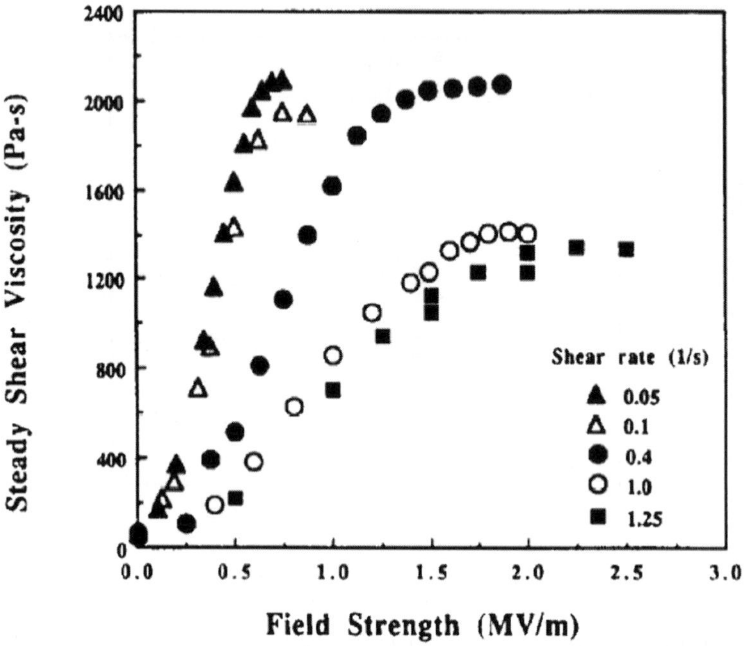

Figure 63 Viscosity of PHIC/p-xylene solution of concentration 23.3 vol% vs. electric field. PHIC stands for polyhexyl isocynate. Reproduced with permission from I-K Yang, and A.D. Shine, J. Rheol., 36(1992)1079

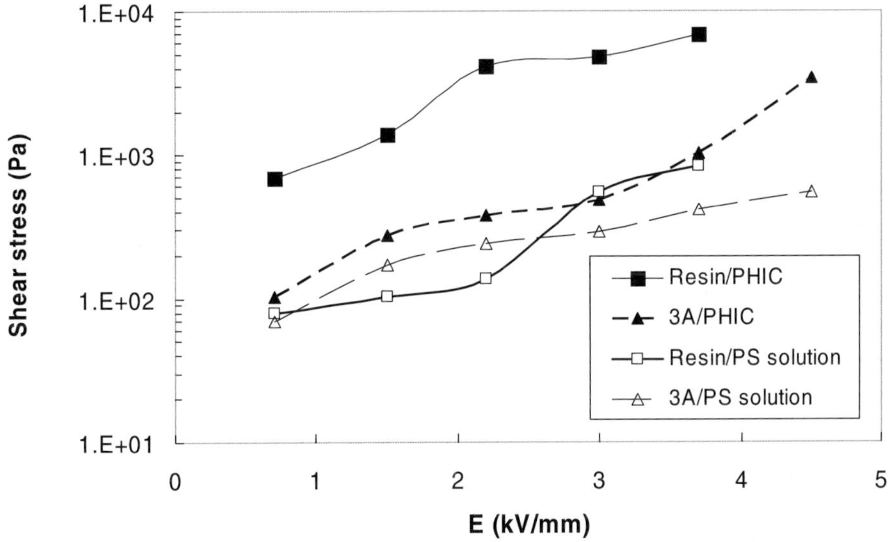

Figure 64 The shear stress of zeolite 3A/PHIC solution, 3A/PS solution, polymer resin/PHIC solution, and polymer resin/PS solution vs. the applied electric field. The PHIC solution refers to 10 wt% polyhexyl isocyanate/xylene solution, and PS solution refers to 10 wt% polystyrene/xylene solution. Redrawn from the data of G. Guist, F. Filisko, in Proc. Int. Conf. ER Fluids (Eds: M. Nakano, K. Koyama), World Scientific, Singapore 1998, p.5.

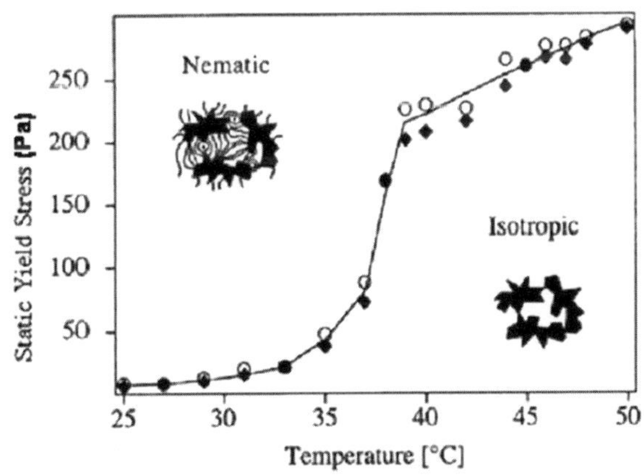

Figure 65 Static yield stress of the zeolite/MBBA suspension against temperature. The particle volume fraction is 31.5 vol%. Reproduced with permission from V. Sequeira, D. Hill, J. Rheol. 42(1998)203

The ER effect and particle stability could be improved if the particle is dispersed in a liquid mixture composed of two different liquids. Example is the mixture composed of 99-75 wt % silicone oil and 1-25 wt% of dodecyl benzene [137]. The dielectric constant, conductivity, and viscosity of the dispersing phase are dominant parameters for determining whether the dispersing phase has a strong impact on the ER effect of the whole suspension. The dielectric mismatch of the solid-to-liquid dielectric constant was believed to be a main reason for such differences. Example is the silica particle dispersed into different dispersing liquids [138]. The physical properties of liquid media are shown in Table 2, and the apparent viscosities of those silica suspensions obtained at 1.0 kV/mm is plotted against shear rate in Figure 66. The silica/trioctyltrimellitate suspension shows the strongest ER effect, which does not result from the initial high viscosity of trioctyltrimellitate. It looks like that there should be an optimal dielectric constant ratio of the dispersed particle to the medium. The dispersing medium of large dielectric constant weakens the ER effect.

Table 2 The physical properties of dispersing media*

Dispersing medium	Viscosity(mPa.s)	Dielectric constant
Silicone oil	105.0	2.416
Dioctylpharalate	70.0	4.629
Trioctyltrimellitate	230.0	4.314
Tricrecylphosphate	70.0	25.811

*Reproduced with permission from L.Rejon, M.A. Ponce, C. De La Luz, R.Nava, J. Intelligent Material Systems and Structures, 6(1995)840

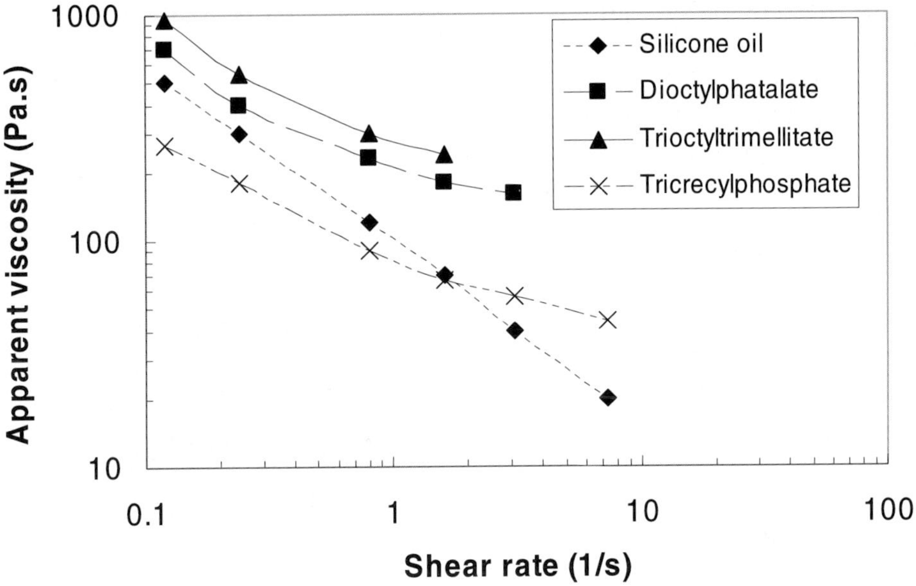

Figure 66 Apparent viscosity vs. shear rate for same silica particle dispersed in silicone oil, dioctylpharalate, trioctyltrimellitate, and tricrecylphosphate liquids. The particle concentration is 20 wt%, and the electric field is 1.0 kV/mm. Redrawn from the L.Rejon, M.A. Ponce, C.De La Luz, R.Nava, J. Intelligent Material Systems Structures, 6(1995)840

10 ELECTRODE PATTERN

The electrode pattern or configuration was found to have an impact on the ER effect, depending on what type of ER fluids sandwiched in between [139-143]. Instead of smooth surface electrodes, various patterned electrodes of a honeycomb-shaped metallic mesh structure, a concentric circle configuration, and a radial shape usually can increase the ER effect up to 2.3 times. Figure 67 shows several electrode surface patterns. Figure 68 shows the yield stress of composite particle with 1,3-butylene glycol dimethacrylate/butyl acrylate copolymer core and titanium hydroxide and phthalocyanine blue pigment shell dispersed in silicone oil (15 vol%) vs. the applied electric field measured using the plain electrode and honeycomb electrodes, respectively. The honeycomb electrode generates almost two times stronger ER effect at 4 kV/mm than the electrode of a smooth surface. Further study shows that the ER effect is dependent on the size of the hole in the honeycomb pattern [142]. Figure 69 shows the shear stress and the yield stress of a composite particle/silicone oil suspension vs. the metallic mesh size covered on the electrode surface under different electric fields. It seems that a maximum shear stress is obtained when the mesh size is about 100 μm under the electric fields 0.66 to 3.33 kV/mm. The shear stress measured with the metallic net electrode is approximated 1.8 to 2.3 times larger than those with the smooth electrode. The non-uniformity of the electric field on the patterned electrode and the shearing slip reduction due to the rough surface of the electode may be responsible for the enhanced ER effect.

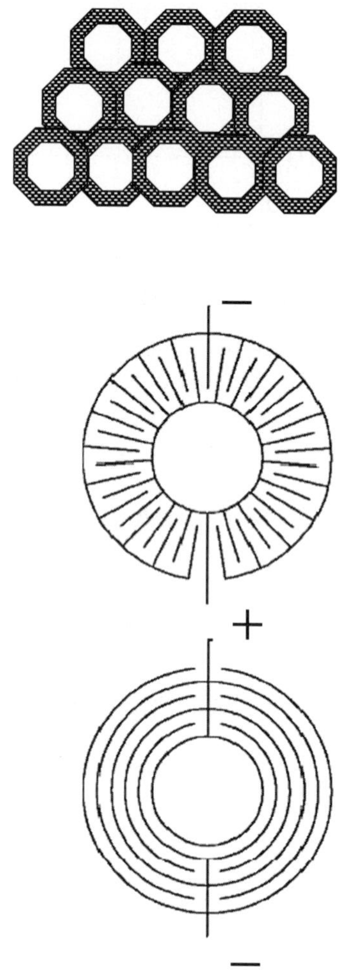

Figure 67 Patterned electrodes for improving ER effect.

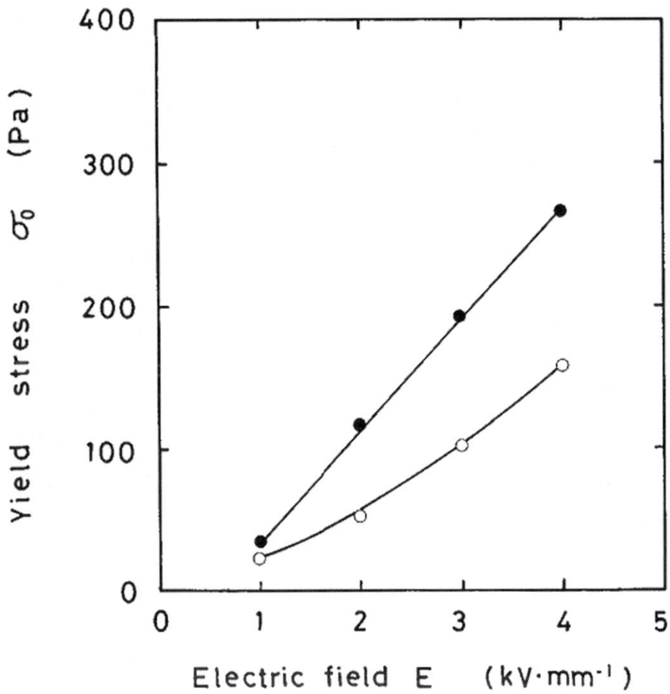

Figure 68 Yield stress of composite particle with 1,3-butylene glycol dimethacrylate/butyl acrylate copolymer core and titanium hydroxide and phthalocyanine blue pigment shell dispersed in silicone oil (15 vol%) vs. the applied electric field measured using ○ plain electrode and ● honeycomb electrode. Reproduced with permission from Y. Otsubo, J. Colloid and Interface Sci., 190(1997)466.

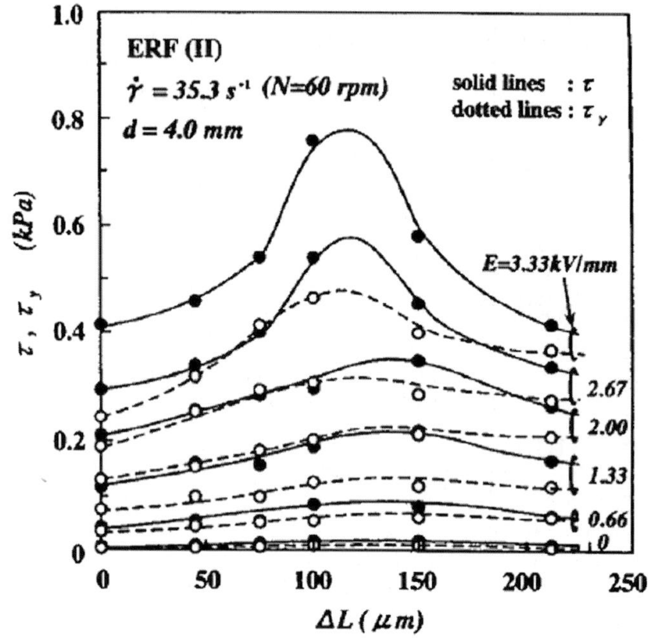

Figure 69 The shear stress and yield stress of a composite particle/siliceon oil vs. the metallic mesh size. The metallic mesh covers the electrode surface. The solid line shows the shear stress and the dotted line shows the yield stress. Reproduced with permission from R. Hanoka, M. Murakumo, A. Anzai, and K. Sakurai, IEEE Trans. Dielectric and Electrical Insulation, 9(2002)10

REFERENCES

[1] J. E. Stangroom, GB 2119392, 1983
[2] N.K. Jaggi, and J. Woestman, Bull. Am. Phys. Soc., 34(1989)1019
[3] R.Tao, J.T.Woestman, and N.K. Jaggi, Appl.Phys.Lett.55(1989)1844
[4] N. Felici, J-N. Foulc and P. Atten, Electrorheological Fluids, edited by R. Tao, and G. D .Roy, World Scientific, Singapore, 1994, p139
[5] X. Tang, C.Wu, and H. Conrad, J. Rheol., 39(1995)1059
[6] L. C. Davis, J. Appl.Phys. 81(1997)1985
[7] B. Khusid, and A. Acrivos, Phys. Rev. E, 52(1995)1669
[8] J. M. Ginder, Phys. Rev. E 47(1993)3418
[9] W.M. Winslow, U.S. Patent 2417850, 1947
[10] Y. Xu, R. Liang, J. Rheol. 35 (1991) 1355
[11] H. Block, J.P. Kelley, US Patent 1501635, 1987
[12] Y. Otsubo, J. Rheol. 36 (1992) 479

[13] A.V. Lykov, Z.P. Shulman, R.G. Gorodkin, and A.D. Matsepuro, J. Eng. Phys., 18(1970)979
[14] Yu. F. Deinega, and G.V. Vinogradov, Rheol. Acta, 23(1984)636
[15] T. Hao, Z. Xu, and Y. Xu, J. Colloid Interface Sci., 190(1997)334
[16] W. Wen, S. Men, and K. Lu, Phys. Rev. E. 55(1997)3015
[17] L.C. Davis, J. Appl. Phys. 81(1997)1985
[18] J. W.Pialet, and D.R.Ckark, Polym. Prepr. 35(1994)367
[19] H. J. Choi, M.S. Cho, and J.W.Kim, Appl.Phys.Lett.,78(2001)3806
[20] P.Gonon, and J-N Foulc, J. Appl.Phys. 87(2000)3563
[21] D.L. Klass, T.W. Martinek, J. Appl. Phys. 38 (1967) 67
[22] D.L. Klass, T.W. Martinek, J. Appl. Phys. 38 (1967)75.
[23] H. Block, K.M.W. Goodwin, E.M. Gregson, and S. M. Walker, Nature, 275(1978)632
[24] H. Block, J.P. Kelley, A. Qin, T. Watson, Langmuir 6 (1990)6.
[25] H. Block, W.D. Ions, G. Powell, R.P. Singh, M.S. Walker, Proc. R. SOC. London, Ser. A 352(1976)153;
[26] H.Block, E.M. Gregson, A. Ritchie, S.M. Walker, Polymer 24(1983)859;
[27] H. Block, W.D. Ions, S.M. Walker, J. Polym. Sci., Polym. Phys. Ed. 16(1978)989
[28] B.G. Barise, Macromoleculars, 7(1974)930;
[29] G.B. Jeffrey, Proc.R.Soc. A 102(1922)161
[30] H. Block, E. Kluk, J. McConnell, B.K.P. Scaife, J. Colloid Interface Sci. 101(1984)320
[31] K. Negita, and Y. Ohsawa, Phys. Rev. E, 52(1995)1934
[32] Y. Hu, and E. Lin, Int. J. Modern Phys. B, 16(2002)2562
[33] H. Uejima, Jpn. J. Appl. Phys., 11(1972)319
[34] D. Miller, C.A. Randall, A. Bhalla, R.E. Newnham, and J.H.Adair, Ferroelectric Lett., 15(1993)141
[35] K.Negita, and Y. Ohsawa, J. Phys. II France 5(1995)883;
[36] R. Hanaoka, S. Takata, Y. Nakazawa, T. Fukami, and K. Sakurai, Electric Eng. Jpan, 142(2)(2003)1
[37] K.W. Wagner, Arch. Electrotechnik 2 (1914) 371
[38] S.O. Morgan, Trans. Am. Electrochem. Soc. 65 (1934)109
[39] R.W. Sillars, JIEE 80 (1937) 378.
[40] T. Hao, J. Colloid Interface Sci. 206 (1998)240
[41] Z. Wang, R. Shen, X. Niu, K. Lu, and W.Wen, J. Appl.Phys., 94(2003)7832
[42] T. Hao, Akiko Kawai, and F. Ikazaki, Langmuir 14(1998)1256
[43] W.Wen, D. Zheng, and K.Tu, Rev. Sci.Instrum., 69(1998)3573
[44] C. S. Coughlin, and R. N. Capps, Proc. SPIE- Intn.Soc.Opt.Eng.2190(1994)19
[45] D.S. Keller, and D.V.Keller, J. Rheol., 35(8)(1991)1583
[46] Y.H. Shih, and H.Conrad, Intn.J.Mod.Phys.B, 8(20/21)(1994)2835
[47] C.W.Wu, and H. Conrad, J.Appl. Phys., 83(1998)3880
[48] M.Ota, and T. Miyamoto, J. Appl. Phys.76(9)(1994)5528
[49] Z. Tan, X. Zou, W. Zhang, Z. Jin, Phys. Rev. E 59(1999)3177
[50] A. Zaman, B. Moudgil, J. Rheol. 42(1998)21
[51] W. Wen, W. Tam, P. Sheng, J. Mater. Res. 13(1998)2783
[52] W. Tam, W. Wen, P. Sheng, Phys. Rev. B: Condens. Matter 279(2000)171

[53] J. Jun, S. Uhm, S. Cho, and K.Suh, Langmuir 20(2004)2429
[54] A. Lengálová, V. Pavlinek, P. Sáha, O. Quadrat, J. Stejskal, Colloids Surfaces A: Physicochem. Eng. Aspects 227 (2003)1
[55] D. J. Cumberland and R.J. Cawford, "The Packing of Particles", Elsevier, 1987
[56] S. Dukhin, Surface and Colloid Science, Ed: E. Matijevic, Wiley, New York 1971, p. 83
[57] R. Kanu, M. Shaw, J. Rheol. 42(1998) 657
[58] K. Yatsuzuka, K. Miura, N. Kuramoto, and K. Asano, IEEE Trans. IAS, 31(1995)457
[59] K. Asano, H. Suto, K. Yatsuzuka, J. Electrostatics 40&41 (1997)573
[60] A. Kawai, K. Uchida, and F. Ikazaki, Intn. J. Modern Phys. B, 16(2002)2548
[61] T.Hao, Z.Xu, Y.Xu, J.Colloid Interface Sci., 190(1997)334
[62] C.J.Gow, and C.F. Zukoski, J. Colloid Interface Sci., 136(1990)175
[63] K.D. Weiss, D.A. Nixon, J.D. Carlson, A.J. Margida, Polym.Preprints 35(1994)325
[64] T. Hao, J. Colloid Interface Sci. 206 (1998)240
[65] L.C. Davis, J. Appl. Phys. 72(1992)1334
[66] P. Atten, J-N. Foulc, N. Felici, Int. J. Mod. Phys. B 8(1994)2731
[67] J-N. Foulc, P. Atten, N. Felici, J. Electrostatic 33 (1994)103
[68] T. Hao, A. Kawai, and F. Ikazaki, Langmuir 14(1998)1256
[69] R. Sakurai, H. see, and T. Saito, J. Rheol., 40(3)(1996)395
[70] Yu. F. Deinega, K.K. Popko, N.Ya Kovganich, Heat-Transfer-Sov. Res. 10 (1978)50
[71] F.E. Filisko, D.R. Gamota, ASME 153 (1992) 5;
[72] F.E. Filisko, in: R. Tao Ed., Proc. Intern. Conf. On Electrorheological Fluids, World Sci, 1992,p. 116
[73] L.C.Davis, J. Appl. Phys. 73(1993)680;
[74] L.C.Davis, Appl. Phys. Lett. 60 (1992) 319
[75] C.F. Zukoski, ''Electrorheological Fluids, A Research Needs Assesment—Final Report,'' p. 5.3–1. DOE, Washington, DC, 1993
[76] K.Q. Lu, W.J. Wen, C.X. Li, and S.S. Xie, Phys. Rev. E, 52 (1995)6329
[77] T. Hao, Appl. Phys. Lett. 70 (1997)1956
[78] R. von Hippel, ''Dielectric Materials and Applications.'' Wiley, New York, 1954
[79] T. Hao, H. Yu, and Y.Z. Xu, J. Colloid Interface Sci. 184(1996)542
[80] T.Y. Chen, and P.F. Luchkham, Colloids Surf. A 78(1993)167;
[81] N. Sugimoto, Bull. JSME 20(1977)1476;
[82] G.G.P etrzhik, O.A. Chertkova, and A.A. Trapeznikov, Dokl. Akad. Nauk SSSR 253(1980)173
[83] D. Adolf, T. Garino, and B. Hance, *in* ''Proceedings of the International Conference on Electrorheological Fluids'' R. Tao, Ed., p. 167. World Scientific, Singapore, 1992
[84] J.E. Stangroom, Phys. Technol. 14 (1983) 290;
[85] Y. Chen, and H.Conrad, ''Developments in Non-Newtonian Flows,'' Vol. 175, pp. 199. Applied Mechanics Division, ASME, New York, 1993
[86] A. Kawai, K. Uchida, K. Kamiya, A. Gotoh, S. Yoda, K. Urabe et al., Int. J. Mod. Phys. B 10(1996)2849
[87] F. Ikazaki, A. Kawai, T. Kawakami, K. Edamura, K. Sakuri, H. Anzai et al., J. Phys. D: Appl..Phys. 31 (1998)336

[88] A. Lengalova, V. Pavlinek, P.Saha, J. Stejskal, T. Kitano, O.Quadrat, Physica A, 321(2003)411
[89] L.R ejon, O. Manero, and C. Lira-Galeana, Fuel, 83(2004)471
[90] M.S. Cho, Y.H. Cho, H.J. Choi, and M.S. Jhon, Langmuir, 19(2003)5875
[91] M.S. Cho, H.J. Choi, and W.S. Ahn, Langmuir, 20(2004)202
[92] M. Konishi, T. Nagashima, and Y. Asako, Proc. 6th intern.conf. on ER and MR suspensions and their applications, M. Nakano and K. Koyama ed., World Scientific, Singapore, 1998, p12
[93] C. Gehin, and J.Persello, Intern.J. Modern Phys. B, 16(2002)2494
[94] T. Saito, H. Anzai, S, Kuroda, and Z. Osawa, Proc. 6th intern.conf. on ER and MR suspensions and their applications, M. Nakano and K. Koyama, Ed., World Scientific, Singapore, 1998, p19
[95] T. Hao, and Y. Xu, Appl. Phys. Lett., 69(1996)2668
[96] S. Chwastcak, J. Colloid Interface Sci., 42(1973)295
[97] S. Wu, Polymer Interface and Adhesion, Marcel Dekker, New York, 1982
[98] H. G. Casimir and D. Polder, Phys. Rev. 73 (1948)360
[99] W. H. Keensom, Phys. Z. 22 (1922)643; *ibid.* 23 (1923)235
[100] R. J. Good, in Treatise on Adhesion and Adhesives, edited by R. L. Partick, Marcel Dekker, New York, 1967
[101] G.I.Skanavi, Dielectric Physics, translated by Yihong Chen, High Education Press, China, 1958
[102] T. Hao, Y Xu, Y. Chen, and M. Xu, Chinese Phys. Lett., 9(1995)573
[103] D.J. Klingenberg, C.F. Zukoski, Langmuir 6 (1990)15
[104] L. Marshall, C.F. Zukoski, and J. Goodwin, J. J. Chem. Soc., Faraday Trans. 1, 85(1989)2785
[105] A.M. Kraynik, R.T. Bonnecaze, J.F. Brady, in: R. Tao Ed., Proc. Intern. Conf. On ER fluids, World Scientific, 1992, p. 59
[106] H.J.H. Clercx, and G. Bossis, J.Chem.Phys., 103(1995)9426
[107] G. Bossis, E. Lemaire, O. Volkova, H. Clercx, J. Rheol. 41(1997)687
[108] H. J. H. Clercx and G. Bossis, Phys. Rev. E 48(1993)2721
[109] T. Hao, A. Kawai, and F. Ikazai, Langmuir, 16(2000)3058
[110] T. Hao, Y. Chen, Z. Xu, Y. Xu, Y. Huang, Chin. J. Polym. Sci. 12(1994)97
[111] S.A. Mezzasalma, and G.J.M. Koper, Colloid Polym. Sci, 280(2002)160
[112] X.Pan, and G.H. McKinley, J.Colloid and Interface Sci., 195(1997)101
[113] R.T. Bonnecaze, and J.F. Brandy, J.Rheol., 36(1992)73
[114] U.Y. Treasurer, F.E. Filisko, L.H. Radzilowski, J. Rheol. 35 (1991)1051
[115] H. Conrad, A.F. Sprecher, Y.Choi, Y. Chen, J. Rheol. 35 (1991)1393
[116] H. Conrad, Y. Li, and Y. Chen, J. Rheol. 39(1995) 1041
[117] J.W. Pialet, and D.R. Clark, Polymer Prepr. 35(1994)367
[118] T. Hao, and Y. Xu, J. Colloid Interface Sci. 181(1996)581
[119] Y. Lan, X. Xu, S. Men, and K. Lu, Appl.phys.Lett., 73(1998)2908
[120] K.Lu, Y. Lan, S. Men, X., Xu, X.,Zhao, S. Xu, Intern. J.Modern Phys. B, 15(2001)938
[121] Y. Asao, I. Yoshida, R. Ando, and S. Sawada, J. Phys. Soc. Jpn., 17(1962) 442
[125] T. Hanai, Kolloid Z. 171(1960) 23
[126] T. Hanai, Kolloid Z. 175(1961) 61

[127] T. Hanai, Bull. Inst. Chem. Res. (Kyoto Univ.), 39(1961)341
[128] D.A.G. Bruggeman, Ann. Phys. 24 (1935) 636
[129] T. Hanai, N. Koizumi, and R. Gotoh, ''Proceedings, Symposium on Rheological Emulsions'', P. Sherman, Ed., p. 91. Pergamon, Oxford,1963
[130] Y.C. Lan, S. Q. Men, X. Y. Xu, and K. Q. Lu, J. Phys. D: Appl. Phys. 33(2000)1239
[131] Y. Lan, S.Men, X.Xu, and K. Lu, Phys. Rev. E, 60(1999)4336
[132] T. Gario, D. Adolf, B. Hance, in Proc. Int. Conf. ER Fluids, R. Tao, Ed., World Scientific, Singapore 1992, p. 167.
[133] M. Jordan, A. Schwendt, D. Hill, S. Burton, N. Makris, J. Rheol. 41(1997)75
[134] G. Guist, F. Filisko, in Proc. Int. Conf. ER Fluids, M. Nakano, K. Koyama, Ed., World Scientific, Singapore 1998, p.5.;
[135] V. Sequeira, D. Hill, J. Rheol. 42(1998)203
[136] I-K Yang, and A.D. Shine, J. Rheol., 36(1992)1079
[137] S. Ono, R. Aizawa, Y. Asako, US Patent 5910269, 1999
[138] L.R ejon, M.A. Ponce, C.De La Luz, R. Nava, J. Intell. Material Systems Structures, 6(1995)840
[139] Y. Otsubo, J. Colloid Interface Sci., 190(1997)466
[140] N. Takesue, J. Furusho, and A. Inoue, J. Appl. Phys. 91(2002)1618
[141] R. Tao, Y.C.Lan, and X. Xu, Int. J. Mod. Phys. B., 16(2002)2622
[142] R. Hanoka, M. Murakumo, A. Anzai, and K.Sakurai, IEEE Trans Dielectric and Electrical Insulation, 9(2002)10
[143] B. Abu-Jdayil, and P.O. Brunn, J. Intell.Mat. Systems Structures, 13(2002)3

Chapter 6

Physics of electrorheological fluids

1. FORCES RELEVANT TO THE ER EFFECT

For the forces in a colloidal system, a good summarization has been made by Russell [1]. Since ER fluids work under a high electric field, the particle should be polarized to such an extent that the electrostatic force between them becomes important. Besides this electrostatic force, there are many other forces in ER fluids, as already observed in conventional colloidal suspensions. They are the hydrodynamic force on the particle due to the viscous continuous phase, the Brownian force on the particle due to the thermal motion of the continuous phase, short-range repulsive forces arising from the Born repulsion or steric interaction, adhesive force due to water [2] or surfactant [3], colloidal interaction such as the van der Waals attraction and the Derjaguin-Landau-Verwey-Overbeek(DLVO)-electrostatic repulsion forces. The structure and rheological properties of ER suspensions depend on the competition among all of the forces mentioned above. The "short range" forces means that they interact over a very short distance around 0.2 nm, of the length scale of a chemical bond. They decrease very quickly as the distance increases. Since they are restricted between the atoms involved in chemical reactions or molecular formations, they are also called the chemical forces [4]. The Born repulsion is such a short range force, arising from the overlap of electron clouds of two molecules incapable of forming covalent bonding. The steric repulsion is another type of the short-range force, which results from the polymeric interpenetration of two polymer coated particles. The long-range forces are those that can act upon each other over distances considerably greater than the "short-range" distance of the chemical bond dimension. Those forces are also called physical forces, as there is no bonding formation involved and only physical processes such as polarization are present. There are two kinds of fundamental long-range physical forces: The coulombic or electrostatic interactions, and the van der Waals forces. The electrostatic interactions result from the electrical forces between charged species, the strongest physical interaction equaling or exceeding the magnitude of covalent bonds. The van der Waals forces are the general term actually including three different types of atomic and molecular interactions from the permanent dipoles, induced polar actions, and the quantum mechanical forces (also called the London dispersion forces, or simply the dispersion forces). The dispersion force is universal in nature, as it results from the interactions between rapidly fluctuating dipoles due to the movement of outer shell electrons of an atom or a molecule. It may be the weakest force among the three, however, it universally exists in any system. There is another kind of force due to the presence of free polymer unadsorbed or weakly adsorbed in the system, called the depletion force. The unadsorbed polymer tends to move out of the narrow area formed by two particles, and the hydrodynamic "suction" effect thus is generated, creating the attraction force between two particles. These forces and their origins are summarized in Table 1. There are many books and literature available on these forces, and thus only a brief description is presented in this section for the convenience of deriving dimensionless units useful to describing the ER fluids.

Table 1 The forces in a colloidal system

Force	Origin
Hydrodynamic force	Viscous flow of medium
Brownian motion	Thermal motion of medium. Strong for particle of size less than 1μm
Born repulsive	Electron overlap
Steric force	Polymer chain interaction
Coulombic/electrostatic	Charge
Van der Waals forces	Permanent dipoles, induced polar and dispersion forces
Depletion attractive force	Smaller particle or free polymer or weakly adsorbed polymer
Surfactant or water adhesive forces	Water or surfactant bridge (Surface tension)

1.1 Hydrodynamic force

Since the colloidal particle is dispersed in a viscous fluid, the relative motion between the particle and the viscous liquid medium plays an important role in the flow behavior of a whole colloidal suspension. For a single spherical particle of radius r in a state of relatively moving in a Newtonian liquid of viscosity η, the frictional force F exerted on the particle can be expressed by Stokes' law:

$$\vec{F} = 6\pi r \eta \vec{v} \qquad (1)$$

where $\vec{v}$ is the velocity of the particle. Note that the $\vec{F}$ and $\vec{v}$ are vectors and both the direction and magnitude should be specified. Eq.(1) is also called the Stokes equation. There is another equation called the Navier-Stokes equation [5]

$$\mathbf{T}=-p\mathbf{I}+2\eta\mathbf{D} \tag{2}$$

where **T** is the stress tensor, p is the pressure, **D** is the stretching or rate of deformation tensor. The physical meaning of Eq.(2) is very clear: The stress on a simple fluid is the hydrostatic pressure plus the stress from the viscosity of the material. Assuming incompressibility, the density of the material is constant, i.e., trace **D** =0. Eq.(1) can be derived from Eq.(2) for flow over a sphere [6].

1.2 Brownian motion

Brownian motion of a particle is a result of the thermal motion of the molecular agitation of the liquid medium. Much stronger random displacement of a particle is usually observed in a less viscous liquid, smaller particle size, and higher temperature. A particle of size larger than 1 μm doesn't show a remarkable Brownian motion. There is much literature available on Brownian motion [7-9], and the Brownian motion is regarded as a diffusion process. For an isolated particle, i.e., there is no interparticle action, the diffusion coefficient D_o, can be expressed as the Stokes-Einstein equation:

$$D_o = \frac{k_B T}{6\pi\eta r} \tag{3}$$

where k_B is the Boltzmann constant; η is the viscosity of the liquid medium. If the concentration is relatively high, the diffuse coefficient D is concentration dependent [10]

$$D = D_0\,(1+V\,c) \tag{4}$$

where V is the virial coefficient, and c is the concentration. The virial coefficient V is positive for repulsive particle interaction and negative for attractive interaction.

1.3 Electrostatic force

The interaction force between two point charges can be expressed by Coulomb's law:

$$F_{el} = \frac{q_1 q_2}{4\pi\varepsilon_o\varepsilon_m d^2} \tag{5}$$

where q_1 and q_2 are the charge quantity of two species, d is the separation distance between two charged entities. ε_m is the dielectric constant of the medium in which two entities are dispersed. For two charges of same sign, F_{el} will be positive, and a repulsive force is expected; for two charges of opposite sign, F_{el} will be negative, and an attractive force is expected between them. For colloidal particles, the situation is relatively complicated, as they usually have the charge of the same sign, and they have an electrical double-layer around the surface. A repulsive interaction between them is therefore expected. For a spherical particle of radius r and charge Q, the surface potential ψ_s that results from the particle surface charge itself can be easily derived as [1,11]:

$$\Psi_s = \frac{Q}{4\pi\varepsilon_o\varepsilon_m r} \tag{6}$$

The actual surface potential ψ_o of a sphere should be the potential generated from the particle surface charge itself minus the one resulting from the counterions in the double-layer:

$$\Psi_o = \frac{Q}{4\pi\varepsilon_o\varepsilon_m r} - \frac{Q}{4\pi\varepsilon_o\varepsilon_m (r + 1/\kappa)} \tag{7}$$

where $1/\kappa$ is the double-layer thickness. Note that Eq.(6) and (7) are only valid for the case where the surface charge density is low [12]. The repulsive force between two charged spheres can be expressed in the following equations under the linearized Derjaguin approximation, i.e. the separation distance is small compared with the radius of the sphere [1,13]

Constant potential: $$F_{el} \approx 2\pi\varepsilon_o\varepsilon_m \left(\frac{k_B T}{ze}\right)^2 \kappa r \Psi_s^2 \frac{\exp(-\kappa d)}{1+\exp(-\kappa d)} \tag{8}$$

$$\text{Constant charge: } F_{el} \approx 2\pi\varepsilon_o\varepsilon_m \left(\frac{k_B T}{ze}\right)^2 \kappa r Q^2 \frac{\exp(-\kappa d)}{1-\exp(-\kappa d)} \quad (9)$$

where d is the surface-to-surface distance between two spheres. Note that Eq.(8) and (9) are only applicable to the spheres of a thin double-layer. There should be no double layer overlap in this case. For a concentrated colloidal system where the double-layer overlap is present, the repulsive interaction force can be calculated through determining the potential at the middle distance between two spheres, $\psi_{m.}$ [11,14].

$$F_{el} = k_B T n^o (\cosh\frac{\kappa d}{2})^{-2} \left(\frac{e\zeta}{k_B T}\right)^2 \quad (10)$$

where ζ is the Zeta potential.

1.4 van der Waals forces

1.4.1 Molecular level

A good summary of the van der Waals forces was given in [4] and [15]. Here only a skeleton description is presented. As mentioned earlier, there are three different van der Waals forces—that is, the interactions between permanent dipoles, induced dipoles, and the dispersion force. For the permanent dipole-dipole interaction, the maximum interaction energy occurs when the two dipoles are aligned in a line:

$$\psi_{d-d} = \frac{C\mu_1\mu_2}{4\pi\varepsilon_o\varepsilon_m d^3} \quad (11)$$

where μ is the dipole moment, d is the distance between two dipoles, and C is a constant depending on the orientation of dipoles. For two free orientation permanent dipoles, the Boltzmann angle-averaged dipole-dipole interaction can be expressed as:

$$\psi_{d-d} = -\frac{2}{3k_B T} \frac{\mu_1^2 \mu_2^2}{(4\pi\varepsilon_o\varepsilon_m)^2 d^6} \quad (12)$$

Eq.(12) is also called the Keesom equation, Unlike Eq. (11), the angle-averaged dipole-dipole interaction is inversely proportional to the 6th power of the distance. For the interaction between permanent dipole and induced dipole, the net dipole-induced dipole interaction can be expressed as the Debye equation:

$$\psi_{d-i} = -\frac{(\alpha_2\mu_1^2 + \alpha_1\mu_2^2)}{(4\pi\varepsilon_o\varepsilon_m)^2 d^6} \tag{13}$$

where α_1 and α_2 are the polarizibility of dipole 1 and dipole 2. For the interaction between induced-dipoles, i.e., the dispersion force, it can be expressed as the London equation

$$\psi_{i-i} = -\frac{3h}{2}\frac{\nu_1\nu_2}{\nu_1+\nu_2}\frac{(\alpha_2\alpha_1)}{(4\pi\varepsilon_o\varepsilon_m)^2 d^6} \tag{14}$$

where ν is the characteristic vibration frequency identified with the first ionization potential of the atom, i.e. the vibration frequency of the electrons vibrating independently. When the distance is larger than the frequency of the interaction between two induced dipoles, the interaction should be modified as:

$$\psi_{i-i} = -\frac{23hc}{8\pi^2}\frac{(\alpha_2\alpha_1)}{(4\pi\varepsilon_o\varepsilon_m)^2 d^7} \tag{15}$$

where c is the speed of light. Eq. (15) is also called the Casimir and Polder equation. The total attraction energy ψ_{attr} is the sum of the forces between the permanent dipoles, the permanent dipole and the induced dipole, and the induced dipole and the induced dipoles:

$$\psi_{attr} = \psi_{d-d} + \psi_{d-i} + \psi_{i-i} = -\frac{B}{d^6} \tag{16}$$

where B can be expressed as:

$$B = -\frac{1}{(4\pi\varepsilon_o\varepsilon_m)^2}\left[\frac{2\mu_1^2\mu_2^2}{3k_BT} + (\alpha_2\mu_1^2 + \alpha_1\mu_2^2) + \frac{3h}{2}\frac{\nu_1\nu_2}{\nu_1+\nu_2}\alpha_1\alpha_2\right] \tag{17}$$

Eq. (16) gives the total attraction energy at the molecular level. It indicates that the attraction energy becomes more negative as the separation distance decreases. When the separation distance becomes so small, to the extent that the electron clouds of two units start to overlap, a repulsive force named as the Born repulsion energy is generated and can be expressed as:

$$\psi_{repul} = \frac{B^{'}}{d^{12}} \tag{18}$$

where B′ is a constant. The total potential energy between two units should be the sum of the attractive and repulsive energy:

$$\psi_{total} = \frac{B^{'}}{d^{12}} - \frac{B}{d^{6}} \tag{19}$$

Eq. (19) is commonly called the Lennard-Jones 6-12 potential. In the literature, the terms "hydrophilic" and "hydrophobic" forces are also used. They actually are the van der Waals forces on the molecular level.

1.4.2 Macroscopic level

The van der Waals forces on the molecular level can be easily extended to the macroscopic level applied for the colloidal particles case, through integrating the attractive energy shown above over the total number of molecules in the area considered. The attractive energy between two identical blocks (same thickness and surface area) can be expressed as [15]:

$$\psi_{a-block} = -\frac{A}{12\pi}\frac{1}{d^{2}} \tag{20}$$

where A is the Hamaker constant, can be related to the materials density ρ, the molecular weight M, Avogadro's constant, N_A, and the constant B given in Eq. (17)

$$A = \left(\frac{\rho\pi N_A}{M}\right)^2 B \tag{21}$$

For two spheres of radius r_1 and r_2, the, the attraction energy can be expressed as:

$$\psi_{a-sphere} = -\frac{A}{6}\left[\frac{2r_1r_2}{f_1} + \frac{2r_1r_2}{f_2} + \ln\left(\frac{f_1}{f_2}\right)\right] \quad (22)$$

where f_1 and f_2 are the functions of r_1, r_2 and the surface-to-surface separation distance d

$$f_1 = d^2 + 2r_1d + 2r_2d \quad (23)$$

and

$$f_2 = d^2 + 2r_1d + 2r_2d + 4r_1r_2 \quad (24)$$

For two identical spheres with narrow separation distance, r » d, Eq.(22) can be simplified as:

$$\psi_{a-sphere} = -\frac{Ar}{12d} \quad (25)$$

Note the difference between Eq.(20) and (25). The attraction van der Waals force between macroscopic bodies is clearly dependent on the geometries of two units.

1.5 Polymer induced forces

When polymer is mixed with colloidal particles, the polymer may generate additional forces between colloidal particles, depending on the concentration of the polymer, solubility of polymer in the liquid medium, and the adsorption between the polymer chain and particle surface. At low polymer concentrations, the polymer chain may form a bridge spanning between two particles, leading to the bridging flocculation. At high concentrations, the polymer may form a brushlike layer on the particle surface, shielding the van der Waals forces and making the particle become stable. At a relatively high concentration, free non-adsorbed polymer chains exist in the system, and a depletion attractive force will be generated once the free polymer chains move out of the area formed by the particles. At a very high concentration, the depletion repulsive force will be generated due to unfavorable dimixing polymer chains in the depleted region. For simplicity, only the two most common forces, the steric repulsive force and the depletion attractive force, are briefly introduced below.

1.5.1 Steric repulsive force

Polymer can either be end-grafted or adsorbed on the particle surface. For end-grafted polymer case, the interaction energy between two such surfaces was theoretically addressed [16]:

$$\psi \approx 36 n_s k_B T \exp\left(-\frac{d}{R_g}\right) \qquad 2 < (d / R_g) < 8 \tag{26}$$

where n_s is the number density of polymer chain ends on the particle surface, d is the distance between the two surfaces, and R_g is the gyration radius of polymer chain. Eq. (26) indicates that the higher surface density, longer polymer chains, and shorter surface distance will generate a larger repulsive force. More detailed description on this steric repulsive force can be found in ref. [17,18]. For adsorbed polymer case, the situation is rather complicated. Theoretical description on this case can be found in ref. [19-22]

1.5.2 Depletion attractive force

For the non-adsorbing polymer case, the depletion force will be generated. The depletion mechanism was first theoretically addressed in ref. [23] using the excluded volume concept. Other approaches such as the density functional theory [24] and the virial expansion [25] were developed for deriving the exact expression for the depletion force. Simply, the interaction potential due to the depletion force can be expressed as [15]

$$\psi_{depl} = -(2\delta - d)P \tag{27}$$

where δ is the depletion layer thickness, and P is the osmotic pressure. δ can be scaled as the square root of the mean-square end-to-end distance of the polymer chain, $\delta \approx \langle R^2 \rangle^{1/2}$. The osmotic pressure can be related to the polymer properties and expressed as:

$$\frac{P}{nk_BT} = \frac{1}{N} + \frac{nv}{2} + \frac{n^2 w}{3} + \ldots. \tag{28}$$

where n is the segment density, $n \approx N / \langle R^2 \rangle^{3/2}$, and the values of N, v, and w for specific polymer can be related to the polymer molecular weight M, the length l_0 and mass m_0 of the monomer,

$$C_\infty = \frac{b_n^2}{l_0^2} \tag{29}$$

$$N = \frac{M}{C_\infty m_0} \tag{30}$$

$$w^{1/2} = \frac{C_\infty v_0 m_0}{N_A} \tag{31}$$

$$v = w^{1/2}(1 - 2\chi) \tag{32}$$

where C_∞ is called the characteristic ratio, b_n is the length of the rigid link, v_0 is the volume of the monomer, v is the excluded volume per bond, χ is the Flory-Higgins interaction parameter [26]. According to Vrij [27], the minimum depletion interaction potential ψ_{min} can be expressed as:

$$\psi_{\min} = -\frac{4\pi\delta^3}{3}\left(1 + \frac{3r}{2\delta}\right)\frac{P}{k_B T} \tag{33}$$

which is relatively weak in comparison with the steric interaction.

1.6 Adhesion force due to water or surfactant

Water or surfactant may form the bridge between two particles. The adhesive force between two spheres with a water bridge can be expressed as [28]:

$$F_{bridge} = 2\pi k \gamma_{wc} \cos\theta_{wp} \tag{34}$$

where γ_{wc} is the surface tension between the water or surfactant and the continuous medium, θ_{wp} is the contact angle between the water or surfactant and the particle, and k is a coefficient weakly depending on the volume of the water or surfactant within the bridge, of the value between 0.75 and 1. Clearly Eq.(34) indicates that this force is insensitive to the amount of water or surfactant. It thus couldn't give a reasonable explanation of the ER effect dependence on water amount described in the preceding chapter. A modified model on the adhesive force between rough particle surfaces was therefore proposed for accounting for the water amount dependence of the ER effect phenomenon [29]

$$F_{bridge} = 4\pi N_{br} \gamma_{wc} r_{br} \tag{35}$$

where N_{br} is the number of the water or surfactant bridges in the contact area, and r_{br} is the radius of the formed bridge. N_{br} is the function of the applied electric field and the water content, leading to the following expression:

$$F_{bridge} \sim 4\pi E W^{3/2} r^{5/2} \gamma_{wc} \left(\varepsilon_m / A_{pwm}\right)^{1/2} \tag{36}$$

where E is the applied electric field, W is the water volume per particle volume, r is the particle radius, ε_m is the dielectric constant of medium, A_{pwm} is the Hamaker constant between the particle, water or surfactant and the medium. This force does depend on the particle size, varies linearly with the applied electric field, and goes through a maximum as water or surfactant content increases [29]. For porous particles, a further modified model was proposed [30], and the number of bridges N_{br} is a function of the gap area and the applied electric field.

$$N_{br} \propto \frac{\varepsilon_m S}{\left(\varepsilon_w - \varepsilon_m\right)} \left(\frac{E_{local}}{E_{local}^c} - 1 \right) \tag{37}$$

where ε_w is the dielectric constant of water, S is the gap area formed by the bridges, E_{local} is the electric field in the gap, and E_{local}^c is the critical field that needs to be exceeded for forming the water or surfactant bridge.

A surfactant bridge model which combines the electrostatic force and the surface tension force was also proposed to explain the surfactant-activated ER effect phenomena [31, 32]:

$$F^{total} = F^{elec} + F^{surf}$$
$$= \frac{12\beta^2}{(r/d)-1}\left[\begin{array}{l} \frac{\pi}{4}\varepsilon_p\left(\frac{rEr_{br}(2r+d)}{\left(r_{br}^2+rd\right)}\right)^2 + \\ \frac{\pi}{4}\varepsilon_m\left(\frac{rE(2r+d)}{\left(r_{br}^2+rd\right)^{1/2}}\right)^2 \\ -\frac{\pi}{4}\varepsilon_m\left(\frac{r^{1/2}E(2r+d)}{(r+d)^{1/2}}\right)^2 \end{array}\right] + 2k\pi r\gamma_{wm}\cos\theta_{wp} \qquad (38)$$

where d is the gap size, the surface-to-surface distance between two particles, $\beta = (\varepsilon_p - \varepsilon_m)/(\varepsilon_p + 2\varepsilon_m)$, and ε_p is the dielectric constant of particle. The critical electric field E_c can be related to the surface tension of the surfactant as [31]:

$$E_c = \sqrt{\frac{2\gamma_{wm}}{\varepsilon_p d}} \qquad (39)$$

Since this model is based on the point-dipole approximation [33], the limitation is obvious, which will be addressed further in a future chapter.

1.7 Electric field induced polarization force

Since the dielectric constants of particle and the dispersing medium are different, the excess amount of charge appears on the particle surfaces under an electric field. The induced dipole moment μ can be expressed as [34]:

$$\mu = 4\pi\varepsilon_0\varepsilon_m\beta r^3 E \qquad (40)$$

When those two particles are far away from each other, they can be regarded as point-dipole, and the interaction energy between two point dipoles U_{pol} [34]

$$U_{pol} = -2\pi\varepsilon_0\varepsilon_m r^3(\beta E)^2\left[\frac{3\cos^2\theta - 1}{(d/2r)^3}\right] \tag{41}$$

where θ is the angle between the direction of applied electric field and the center-to-center connection of two particles.

1.8 Relative magnitude of interparticle interaction

Various dimensionless groups have been used to describe the relative importance of the forces described above. According to Gast and Zukoski [34], those forces can be simply scaled as shown in Table 2 and thus the dimensionless groups can be obtained by simply taking the ratio between them.

Table 2 The dimensionless groups

Energy	Scale	/thermal energy	/polarization
Thermal	k_BT		$\frac{k_BT}{\pi\varepsilon_0\varepsilon_m r^3(\beta E)^2}$
Van der Waals	$\frac{Ar}{12d}$	$\frac{Ar}{12dk_BT}$	
Electrostatic	$4\pi\varepsilon_o\varepsilon_m r\psi^2$	$\frac{4\pi\varepsilon_o\varepsilon_m r\psi^2}{k_BT}$	
Polarization	$\pi\varepsilon_0\varepsilon_m r^3(\beta E)^2$	$\lambda = \frac{\pi\varepsilon_0\varepsilon_m r^3(\beta E)^2}{k_BT}$	
Viscous	$6\pi\eta r^3\dot{\gamma}$	$P_e = \frac{6\pi\eta r^3\dot{\gamma}}{k_BT}$	$M_n = \frac{6\eta\dot{\gamma}}{\varepsilon_0\varepsilon_m(\beta E)^2}$

Here $\dot{\gamma}$ is the shear rate. The relative importance between the viscous and thermal forces is described by the Peclect number, Pe. The relative importance of the polarization to the thermal energy is described by the parameter λ. The Mason number, $M_n = P_e / \lambda$, is the ratio of the viscous to the polarization forces, designated after the work due to S. Mason and coworkers on the suspension structure in a combination of shear and an electric field [35-43]. In an ER suspension, the viscous force hinders the formation of the fibrillated structure, while the polarization force is responsible for the particle chaining.

1.9 Scaling analysis using the Mason number for ER fluids

The Mason number was claimed to be useful for scaling the ER effect under an electric field. The ER performance of the hydrated polymethacrylate with a broad size distribution around 9 μm dispersed into a chlorinated hydrocarbon was found to obey the following scaling equation [44]:

$$\frac{\eta}{\eta_\infty} = \frac{M_n^*}{M_n} + 1 \tag{42}$$

where η is the apparent viscosity, η_∞ is the high shear rate viscosity under zero electric field, M_n^* is a material constant independent of the electric field strength and shear rate but dependent on the particle volume fraction, dielectric properties, and viscosity. The experimental data for a particle volume fraction range 0.07-0.35, the electric field strength of 50-400 V/mm, shear rate 10^{-6}-10 s^{-1}, and temperature of 25-35°C, were found to collapse together, in the case where M_n^* was assumed to increase linearly with the particle volume fraction [44]. So $M_n^* = k\phi$, $M_n / M_n^* = kM_n / \phi$. This scaling is plotted in Figure 1 as η/η_∞ vs. M_n/ϕ at various experimental conditions.

A similar scaling behavior was also found in three kinds of silica (PPG Hi-Sil 132, 135 an 233) /silicone oil suspensions at particle weight fraction 9 wt%, electric field strength range 0-0.5 kV/mm, and shear rate range 0.028-0.28 s^{-1} [45]. η/η_∞ vs. M_n/M_n^* for those three suspensions at M_n^*=1.5 is shown in Figure 2. These results seem to indicate there is a transition happening in ER systems from the viscous controlled state to the polarization controlled state, which is the phase transition indeed and will be addressed in the next section

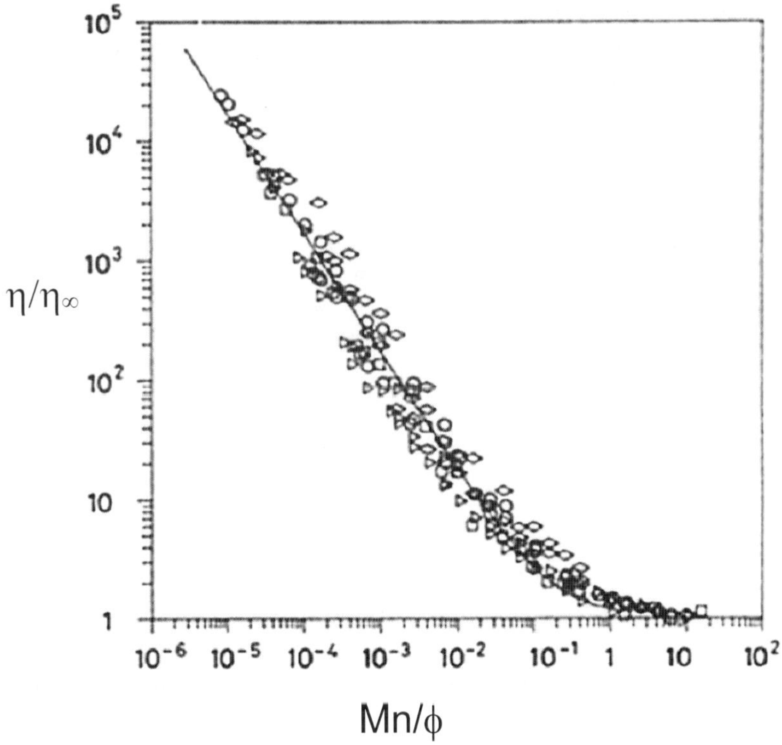

Figure 1 η/η_∞ vs. M_n/ϕ for the hydrated polymethacrylate/chlorinated hydrocarbon suspensions at various conditions: The particle volume fraction ranges 0.07-0.35, the electric field strength is of 50-400 V/mm, shear rate is 10^{-6}-10 s^{-1}, and temperature is between 25-35°C. Reproduced with permission from L. Marshall, C.F. Zukoski, and J.W. Goodwin, J. Chem. Soc. Faraday Trans. 85(1989)2785.

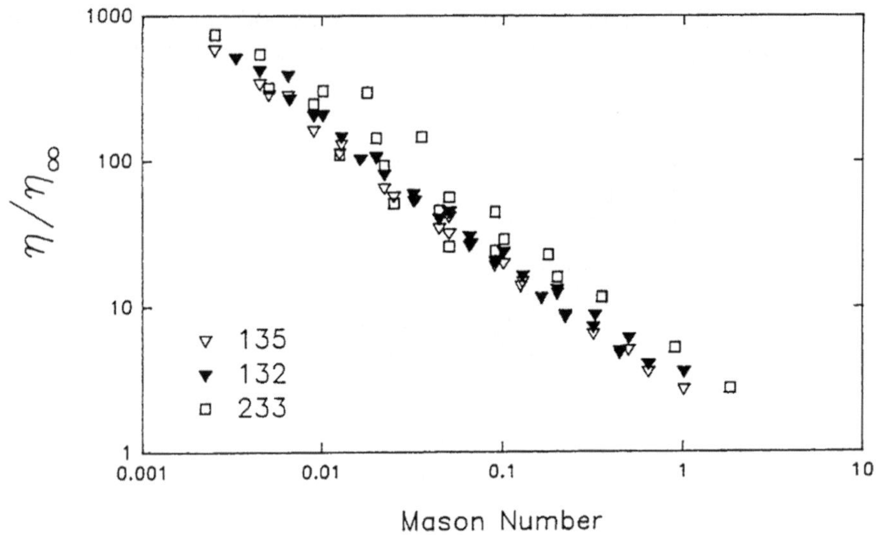

Figure 2 η/η_∞ vs.M_n/M_n^* for three silica/silicone oil suspensions (PPG Hi-Sil 132, 135 an 233) at various conditions: The particle weight fraction is 9 wt%. The electric field strength is in the range 0-0.5 kV/mm, and shear rate is 0.028-0.28 s^{-1}. Reproduced with permission from C.S. Coughlin, and R.N.Capps, SPIE, 2190(1994)19].

2. PHASE TRANSITION

2.1 Phase transition in colloidal suspensions

Phase transition is observed in colloidal suspensions even without the aid of an electric field. As the particle volume fraction increases, the equilibrium phase changes from a disordered state, to coexistence with a crystalline phase (close-packed particle arrangement), then to a glass state, and finally to a fully crystalline state [46,47]. A computer simulated phase diagram [46,48,49] is shown in Figure 3. Below the particle volume fraction 0.494, the colloidal suspension remains as a liquid, and becomes a colloidal crystal above the particle volume fraction 0.545. Metastable suspension with the coexistence of liquid and crystal is formed with the particle volume fraction in the range between 0.494 and 0.545. Once the particle volume fraction exceeds 0.545, particles will form a crystal structure. 0.7404 is the maximum packing fraction of the particle. Note that there is a glass transition state when the particle volume fraction is in the range of 0.58-0.63. A typical colloidal crystal is shown in Figure 4, where a close packed structure formed by monodispersed poly(styrene/sodium styrenesulfonate) latex on a glass surface, which was created by simply evaporating water.

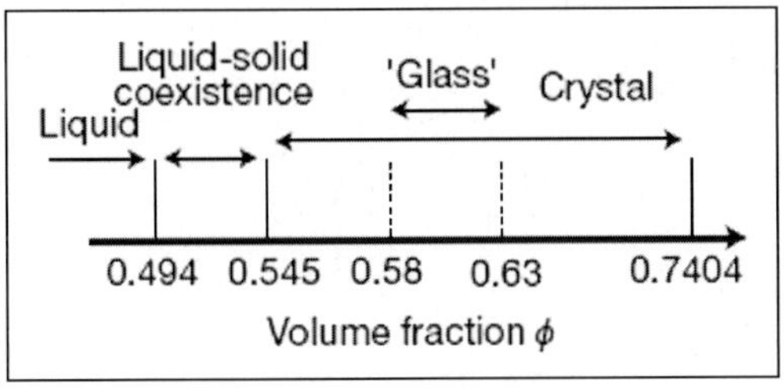

Figure 3 The computer simulated phase diagram of colloidal suspensions. Reproduced with permission from Z. Cheng, W.B. Russel, P.M. Chalkin, Nature 401(1999) 893.

The microstructure of a colloidal suspension is dependent on the interparticle forces discussed in the previous section. The relationship between the interparticle forces and the microstructure is shown in Figure 5. In the repulsive force dominant region, a loosely packed solid will be formed with the increase of the particle volume fraction, while in the attractive force dominant case, the fractal aggregate or colloidal gel will be formed. For the "hard sphere" case, i.e., there are no interparticle forces, the phase diagram is completely controlled by entropy. For this particular case, the densely packed colloidal crystal will be formed once the particle volume fraction exceeds 0.64. In the intermediate region where the repulsive and attractive forces are comparable, the microstructure is much more sensitive to the detail and magnitude of those two forces, leading to the liquid-crystal or aggregation coexistence situation.

Among the colloidal crystal structures, the face-centered cubic (fcc) lattice was found to be the most stable structure both theoretically [50] and experimentally [47, 51, 52], compared with other crystal structures such as, body-centered tetragonal (bct) and hexagonal closed-packed (hcp) structures. The Gibbs free-energy of fcc structure is more stable by around 0.005RT(where R is the gas constant, and T is temperature) relative to that of hcp structure, which has an identical close-packed volume [50].

Figure 4 Colloidal crystal of poly(styrene/sodium styrenesulfonate) latex formed on a glass surface after evaporation of water under infrared light irradiation. The scale bar in the top right is 1 μm. Reproduced with permission from Fang Zeng, Zaiwu Sun, Chaoyang Wang, Biye Ren, Xinxing Liu, and Zhen Tong, Langmuir, 18(2002)9116.

2.2 Phase transition in ER suspensions

The electric-field-induced phase transition in an ER suspension was found to be different from that in general colloidal suspensions. Tao and Martin [55, 56] predicted theoretically that the bct structure has an energy lower than that of the fcc (face-centered cubic) and other structures, based on dipolar interaction energy calculations. The dipolar interaction energy per particle for various crystal structures is shown in Table 3. The bct crystal structure is shown in Figure 6.

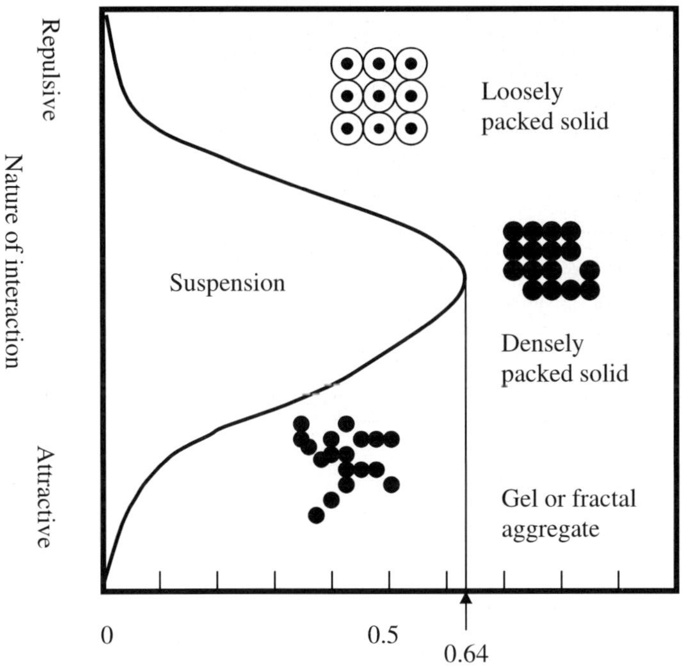

Figure 5 Schematic illustration of the relation between the interparticle forces and the microstructure observed in colloidal suspension. Redrawn from D.R.Ulrich, Chem. & Eng. News, January 1, 1990, page 28-35

Table 3 Dipolar interaction energy per particle for various structures

Structure	Energy per particle (unit $p^2/r^3\varepsilon_m$)*
bct lattice	-0.381268
fcc lattice	-0.3702402
hcp	-0.3700289
Separated chains	-0.300514
Cubic lattice	-0.261799

* Reproduced with permission from R. Tao, J.M. Sun, Phys. Rev. Lett. 67(1991)398. $p = \beta\varepsilon_m r^3 E_{loc}, \beta = (\varepsilon_p - \varepsilon_m)/(\varepsilon_p + 2\varepsilon_m)$

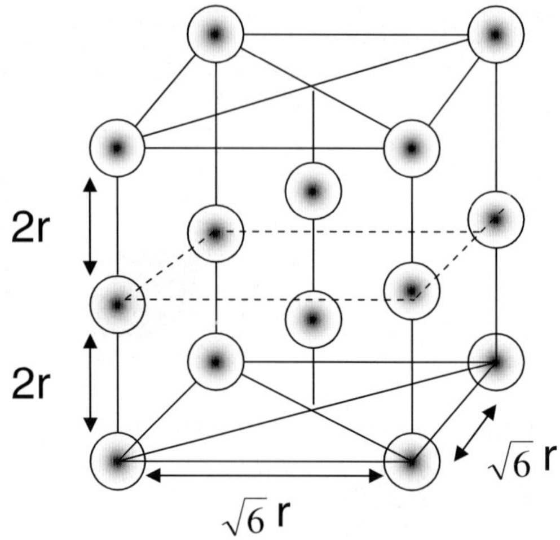

Figure 6 Three dimensional body-centered tetragonal (bct) structure. The particles have radius r and are not shown to scale.

The laser diffraction method [57] was employed to experimentally determine the crystal structure within the fibrillated columns by using a uniform glass microsphere/silicone oil system, and a bct structure was observed as predicted. The diffraction pattern is shown in Figure 7 for monodispersed glass beads of various sizes. The structure constants determined from the laser diffraction experiment were found to agree very well with the theoretical calculation based on dipolar interaction energy. Table 4 lists the experimentally determined and theoretically calculated structure constants for the bct structure formed by the silica spheres. The experimental data are consistent with the proposed bct structure.

Table 4 The bct structure constants of silica spheres *

Lattice plane	Sphere diam. (μm)	Experiment (μm)	Theory (μm)
(110)	20.0	a=34.1 b=21.1	34.6
(110)	40.7	a=69.1 b=38.9	70.540.7
(100)	40.7	a=54.8 b=43.8	49.840.7

* Reproduced with permission from T. Chen, R.N. Zitter, R. Tao, Phys. Rev. Lett. 68(1992)2555.

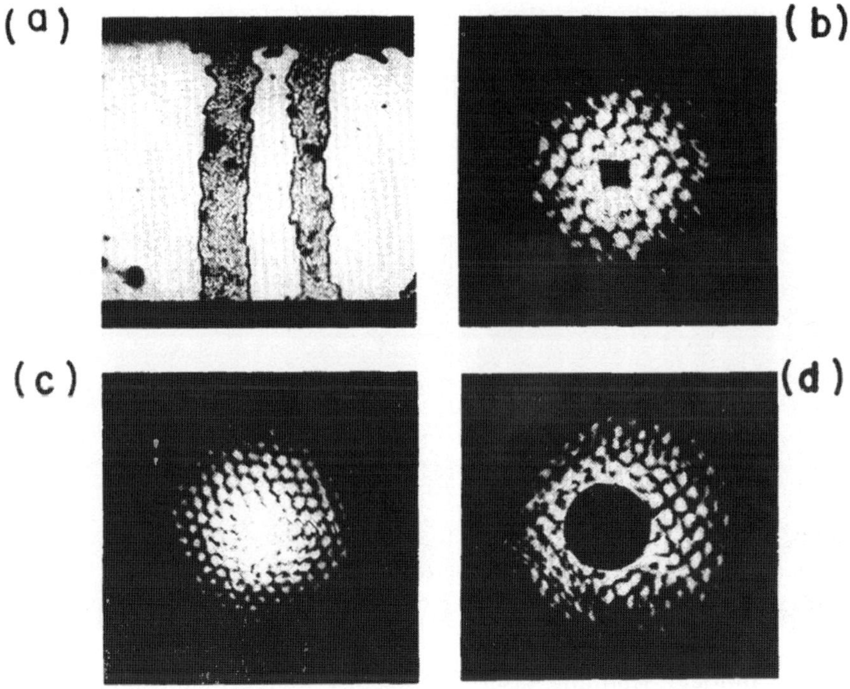

Figure 7 The laser diffraction pattern of two monodispersed glass microspheres of highly uniform diameter, either 20.0±1.8 or 40.7±1.7 mm, dispersed in a low viscosity silicone oil. a) Fibrillated chains formed between two electrodes of gap size 3 mm. b) Diffraction pattern of a (110) plane for 20.0 mm spheres; c) Pattern of a (110) plane for 40.7 mm spheres;d) Pattern of a (100) plane for 40.7 mm spheres. In b) and d), the centers are masked to suppress overexposure. Reproduced with permission from T. Chen, R.N. Zitter, R. Tao, Phys. Rev. Lett. 68(1992)2555.

The bct structure was also observed by using confocal scanning laser microscopy [58]. The bct crystal structure of the monodispersed silica sphere of radius 0.525 μm dispersed into an index matched mixture of 16 wt.% water and 84 wt.% glycerol under an electric field 1 kV/mm is shown in Figure 8 for the suspension of the particle volume fraction about 10 vol%. When the particle volume fraction reaches 45 vol%, without an applied electric field the particles are arranged in fcc structure as shown in Figure 9a. Under the reaction of the an electric field of 1 kV/mm and 500 kHz, the fcc structure is transformed into the bct structure as shown in Figure 9b. These results point out for the first time that an external electric field can induce a solid-solid phase transition in ER fluids, changing from a meta-stable liquid to a crystallized solid of fcc and finally bct structures.

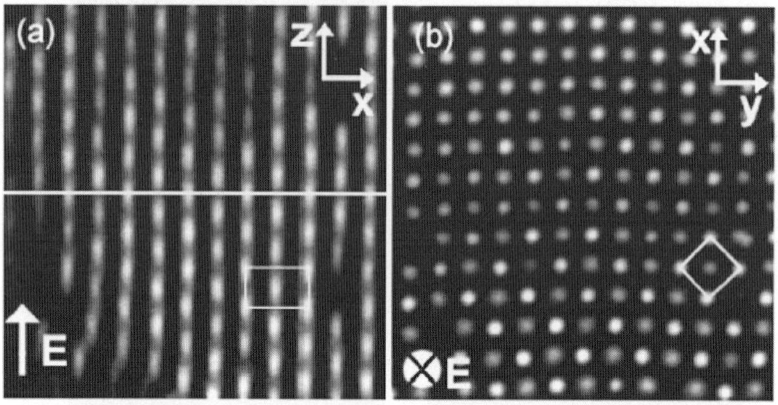

Figure 8. Confocal scanning laser microscopy image of the body-centered tetragonal crystal formed from the monodispersed silica spheres of radius 0.525 μm dispersed into an index match mixture of 16 wt.% water and 84 wt.% glycerol. The particle volume fraction is about 10 vol%, and the applied electric field is 1 kV/mm. a) A view along a plane parallel to the *E*-field; b) A view looking down the *E*-field showing the square. Reproduced with permission from Dassanayake, S. Fraden, A. van Blaaderen, J. Chem. Phys. 112(2000)3851

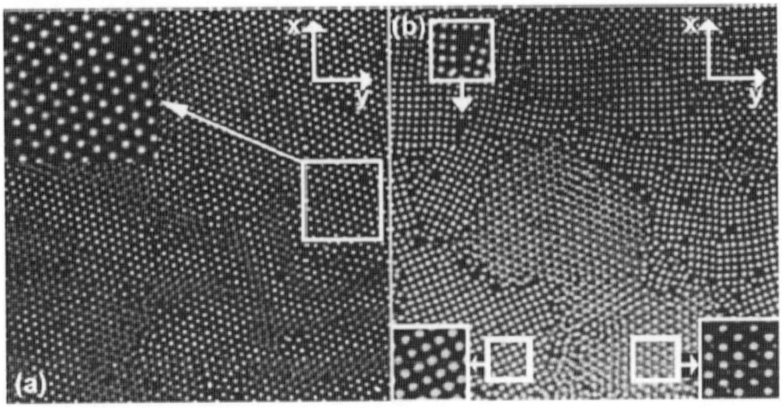

Figure 9. Electric field induced solid–solid transition from fcc structure under zero electric field to the bct structure under an electric field. The image shows raw confocal microscope data of a sample of volume fraction 45 vol%. a) A plane parallel to the electrodes before the *E*-field was turned on. b) The same area about 6 h after an *E*-field (about 1 kV/mm and 500 kHz) is applied perpendicularly to the image plane. Large areas of the crystal have transformed into bct order, identified by the square configurations. Reproduced with permission from Dassanayake, S. Fraden, A. van Blaaderen, J. Chem. Phys. 112(2000)3851

3. PERCOLATION TRANSITION

3.1 Percolation theory

Percolation theory represents one of the simplest models of disordered systems. It was developed to mathematically deal with disordered media, in which the disorder is defined by a random variation in the degree of connectivity [59]. The main concept of percolation theory is the existence of a percolation threshold, above which the physical property of whole system dramatically changes. A typical example of a percolation problem is that of the site percolation on a simple two-dimensional square lattice, as shown in Figure 10. The relevant entities could be either the squares determined by the gridlines or the points where these lines intersect. If the squares are chosen to be considered, this problem is called the site percolation, while if the points are chosen to be considered, it is called the bond percolation. For the example shown in Figure 10, the squares are chosen to be the relevant entities. The lattice is assumed to be empty originally with all sites unoccupied, and gradually the sites of the lattice are randomly occupied. If two occupied sites are nearest neighbors, a connection is supposed to be made between them. The probability with which each site is occupied could be used to define the average degree of connectivity, p. For p =0, there is no connectivity and every site is isolated. For p=1, all sites are connected.

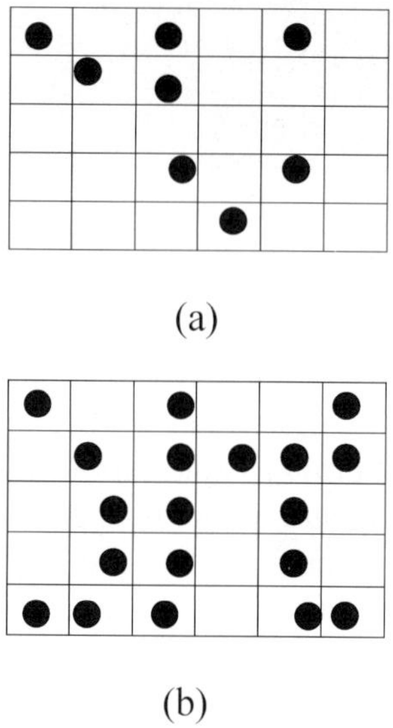

Figure 10 Two percolation probability, a) low probability; b) high probability

For a small probability a small number of the squares will be occupied, and for a large probability a large number of squares will be occupied. Once p is big enough, a cluster path that connects the top and bottom, left and right sides of the lattice will appear. Such a cluster is called a percolating path, and the critical probability, p_c, is called the percolation threshold. The presence of the percolated path represents a dramatic structural change of the lattice from a disconnected state to a connected one. Such a transition is called the percolation transition. For p is less than p_c, only isolated clusters exist in the system. For p is larger than p_c, there is always a percolated cluster, though some isolated clusters can still be present. The critical percolation point at various lattices is shown in Table 5.

Table 5 The percolation threshold of various lattices. Reproduced with permission from ref. [60]

Lattice	P_c (site percolation)	P_c (bond percolation)
3D-cubic (body-centered)	0.246	0.1803
3D-cubic (face-centered)	0.198	0.119
3D-cubic (simple)	0.3116	0.2488
2D-diamond	0.43	0.388
2D-honeycomb	0.6962	0.65271*
4-hypercubic	0.197	0.1601
5-hypercubic	0.141	0.1182
6-hypercubic	0.107	0.0942
7-hypercubic	0.089	0.0787
2D-square	0.592746	0.50000*
2D-triangular	0.50000*	0.34729*

* determined exactly.

For many percolation problems the critical thresholds are approximately the same for a 2-D or 3-D lattice when they are expressed as area or volume fractions [61].

Percolation transition is one kind of phase transitions (or critical phenomena). Unlike the melting or evaporation phase transition phenomena, which are second-order phase transitions, the percolation transition is a first-order phase transition without involving the temperature and volume changes in the system. It can be universally expressed as a power law or scaling law as shown below:

$$P \sim (p - p_c)^b \qquad p > p_c \tag{43}$$

where P is the fraction of occupied sites belonging to the percolation cluster, b is a critical exponent, which is independent of the lattice structure and the percolation type (site or bond percolation). b is only dependent on the dimensionality of space, 5/36 for 2-D and 0.41 for 3-D space. Note that in above equation p_c is not universal, dependent on the lattice structure and the percolation type (see Table 5).

Percolation theory could be applied in wide areas ranging from natural to social sciences, of both theoretical and practical interests. Application examples of the percolation theory in social sciences include elections, and the dissemination of new ideas and beliefs. Forest fires could be predicted using the percolation theory if the forest is assumed as a lattice, whose sites are occupied by trees. Making certain sites as burning, the percolation theory could be used to predict which other trees or sites were going to be ignited. In the physics and chemistry fields, the percolation theory can be widely used in the insulating-conducting transition phenomenon, gelation, polymerization, and colloidal crystallization process.

3.2 Percolation transition in ER suspensions

As mentioned above, a disordered or amorphous system can be understood with the aid of percolation theory, which deals with how the short-range finite connectivity would finally change to a long-range infinite connectivity, above a threshold probability [59]. The phase transition in the ER system could be the analog of a disorder-order transition, which may also be understood using the percolation theory. An investigation of the percolation transition in an ER fluid was carried out by Hao [62]. The oxidized polyacrylonitrile particles of conductivity 10^{-7} S/m and particle size 10 μm were dispersed in a low-viscosity silicone oil, forming several suspensions of various particle volume fractions between 0.1 and 0.5. The viscoelastic properties of such suspensions were measured in oscillation mode (strain amplitude =200%, frequency= 2 Hz) under the electric field between 0 and 2.5 kV/mm. It was found that the complex viscosity η^*, the real modulus G', and the imaginary modulus G'', increase sharply once the particle volume fraction exceeds a critical value ϕ_c, which is a constant and does not change with the applied electric field strength. The real modulus and complex viscosity obtained under the electric field 1.5 kV/mm are plotted against the particle volume fraction and shown in Figure 11. A similar relationship between the rheological properties and the particle volume fraction is observed in another four different electric fields, and even in a zero electric field. The general trend is that the rheological properties slightly increase when the particle volume faction is below 0.37, and abruptly increase once the volume fraction is larger than 0.37. The transition volume fraction point, 0.37, is called the critical or threshold volume fraction, and can be determined accurately with the double tangent lines crossing method. Table 6 shows the threshold volume fraction values obtained from the curves of the real modulus against the volume fraction in different electric fields. The threshold volume fraction value obtained from the curve of the complex viscosity against the particle volume fraction is slightly different from the one obtained from the curve of the real modulus against the particle volume fraction. The corresponding real modulus and the complex viscosity at this transition point are also shown in Table 6, named as the critical real modulus, $G_c^{'}$, and the critical complex viscosity, η_c^{*}, respectively. Note that the threshold volume fraction values are independent of the external electric field, and are a characteristic parameter related to the suspension itself. The threshold value in zero electric field is same as in an electric field, indicating that this transition is a sort of percolation transition related to how particles connect to each other.

Further evidence on the particle volume fraction induced percolation transition in those ER fluids were shown in the conductivity data. Figure 12 shows the dependence of the conductivity vs. the particle volume fraction. When the particle volume fraction is about 0.354, the conductivity dramatically increases, indicating that a long range connection between particles suddenly appears at this particle volume fraction. When the particle volume fraction is small, the particles are isolated or the contact number between particles is small, leading to low conductivity. When the network percolated structure is formed, the two electrodes are connected by those percolated paths, so the conductivity of the whole suspension increases. The increase of the rheological properties is also due to the microstructure change from the small particle clusters to the percolated network structure. Although the threshold values of the particle volume fraction determined by the curves of the real modulus and the conductivity against the particle volume fraction are not exactly same, both the real modulus and conductivity data support such a fact that there is a transition appearing at a critical particle volume fraction.

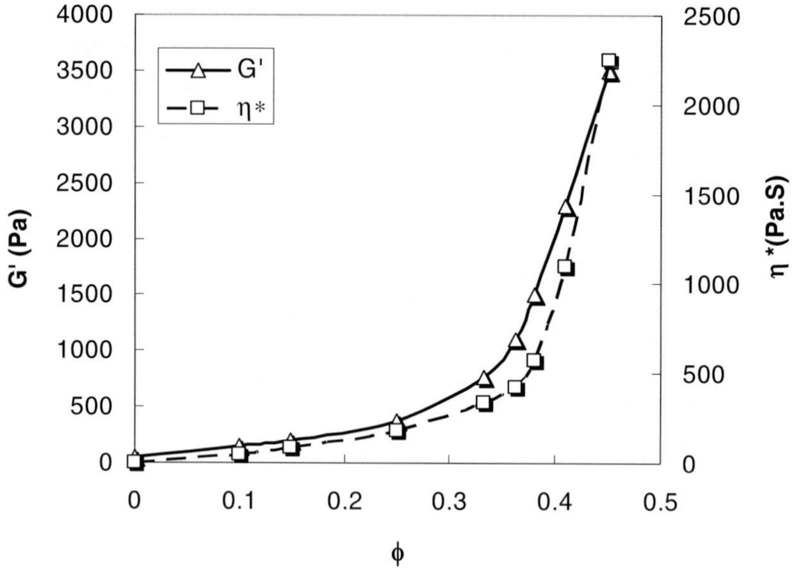

Figure 11 The real modulus and complex viscosity vs. the particle volume fraction for oxidized polyacrylonitrile/silicone oil suspensions. The applied electric field = 1.5 kV/mm. The strain amplitude = 200%, frequency = 2 Hz. Redrawn with permission from T. Hao, Y. Chen, Z. Xu, Y. Xu and Y. Huang, Chin. J. Polym. Sci., 12(1994)97

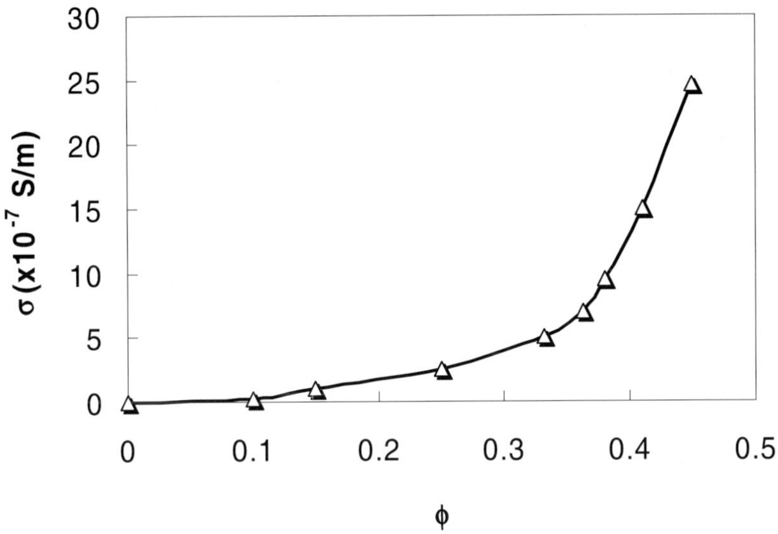

Figure 12 The conductivity (subtracted from the one of silicone oil) vs. the particle volume fraction for oxidized polyacrylonitrile/silicone oil suspensions. Redrawn with permission from T. Hao, Y. Chen, Z. Xu, Y. Xu and Y. Huang, Chin. J. Polym. Sci., 12(1994)97

As shown in Table 6, the critical real modulus and complex viscosity are dependent on the electric field. Figure 13 shows the critical real modulus $G_c^{'}$ and the complex viscosity η_c^{*} vs. the electric field. A linear increase trend with the electric field is found for both the critical real modulus and complex viscosity. Interestingly, the slope for the critical real modulus is exactly 1.3 times that for the complex viscosity, which may also be related to the universal scaling law. The dimensionless real modulus (G'/G'_c) against the dimensionless particle volume fraction (Φ/Φ_c) obtained at different electric fields is shown in Figure 14. All data collected at different electric fields overlap together regardless of the applied electric field strength, clearly indicating that the universal scaling law holds for this suspension and the transition is controlled by the percolation path.

Table 6 The threshold values in different electric fields. Reproduced with permission from ref. T. Hao, Y. Chen, Z. Xu, Y. Xu and Y. Huang, Chin. J. Polym. Sci., 12(1994)97

Electric field (kV/mm)	0	0.5	1	1.5	2	2.5
Threshold value ϕ_c	0.371	0.370	0.377	0.375	0.374	0.374
$G_c^{'}$ (Pa)	3.64	73.64	300.00	600.00	966.67	1350.0
η_c^*(Pa.s)	5.25	120	390	610	805	1100

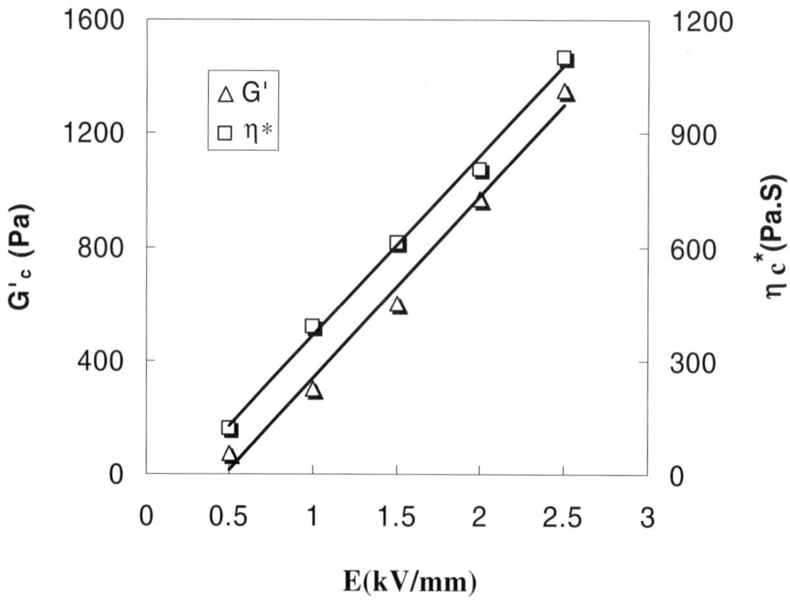

Figure 13 The critical modulus $G_c^{'}$ and critical complex viscosity η_c^* vs. the applied electric field. Redrawn with permission from T. Hao, Y. Chen, Z. Xu, Y. Xu and Y. Huang, Chin. J. Polym. Sci., 12(1994)97

According to the percolation theory, a network microstructure rather than the generally accepted fibrillated chain structure should be formed in an ER fluid when the particle volume fraction is above the critical point, 37 vol%. This was experimentally verified [62] and shown in Figure 15. The fibrillation phenomenon and small particle clusters are only observed in a dilute ER fluid. As the volume fraction increases, the small clusters can bind together and form big clusters, eventually forming a network structure, which is strongly anisotropic. The mechanical strength of the network structure in the direction of electric field should be larger than that in the other direction. Once the network structure has formed, the external electric field can not change its shape and can only make the network structure become much stronger. These two images are almost identical to each other, though the real modulus of the network structure shown in Figure 15b is much larger than that shown in Figure 15a. It is clear that the external electric field can not change the transition point when the network structure builds up. The fibrillated chain can only be formed when the particle volume fraction is less than 37 vol%. Above this critical value, a network structure was formed even under a high electric field, 2 kV/mm. Three kinds of particle-particle clusters or paths were presumed to be formed in an ER fluid: a) continuous paths, which start at one electrode and connect with another electrode; b) branched paths, which start from one electrode and end between the two electrodes; c) isolated paths, whose two ends exist in ER fluid but do not connect with any electrodes. The weight fraction of various paths can be calculated with the method shown below [63] and the calculated results are shown in Figure 16. The percolation path was found to start to appear at the particle volume fraction 24%, and to become dominant at the volume fraction of approximately 40%, where the whole ER suspension is occupied by a percolation network. The calculated critical volume fraction value agrees well with the experimentally measured one for oxidized polyacrylonitrile/silicone oil suspension. The calculation was carried out by applying the mathematical treatment developed by Flory [26] for molecular distribution and gelation in nonlinear polymers. When the extent of reaction α exceeds a critical value, monomers of functional groups f may convert to a gel of unlimited molecular weight.

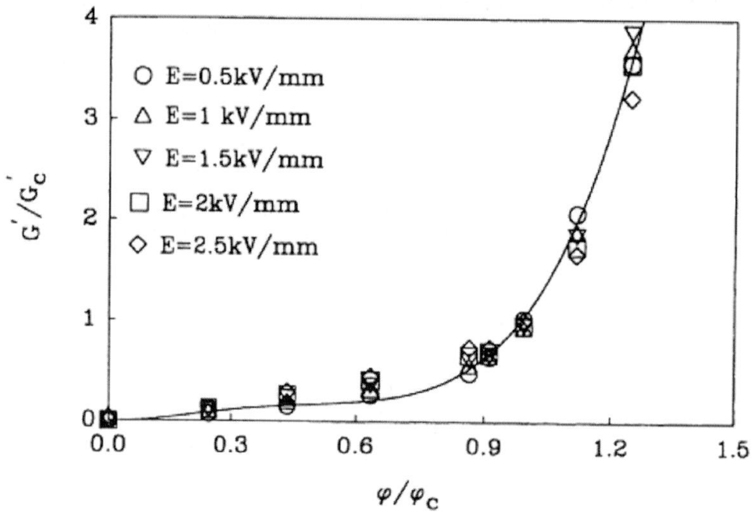

Figure 14 Dimensionless real modulus $G'/G_c^{'}$ against the dimensionless particle volume fraction ϕ/ϕ_c. Reproduced with permission from T. Hao, Y. Chen, Z. Xu, Y. Xu and Y. Huang, Chin. J. Polym. Sci., 12(1994)97

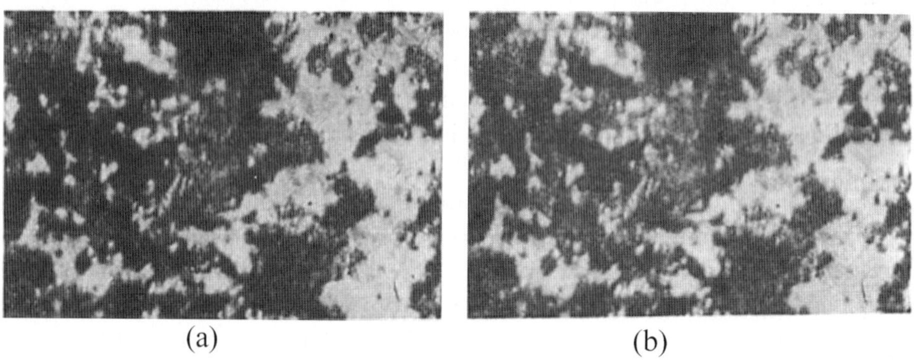

Figure 15 The photographs of network structure in the concentrated oxidized polyacrylonitrile/silicone oil of particle volume fraction 50 vol%. (a) E=0; and (b) E= 2kV/mm. The black area is occupied by the particles. Reproduced with permission from T. Hao, Y. Chen, Z. Xu, Y. Xu and Y. Huang, Chin .J. Polym. Sci., 12(1994)97.

The weight fraction of the sol or oligomers is given by:

$$W_s = \frac{\alpha'(1-\alpha)^2}{\alpha(1-\alpha')^2} \tag{44}$$

where α' is the smallest value of α obeying the following equation:

$$y = \alpha(1-\alpha)^{f-2} \tag{45}$$

thus the weight fraction of gel, Wg,

$$W_g = 1 - W_s \tag{46}$$

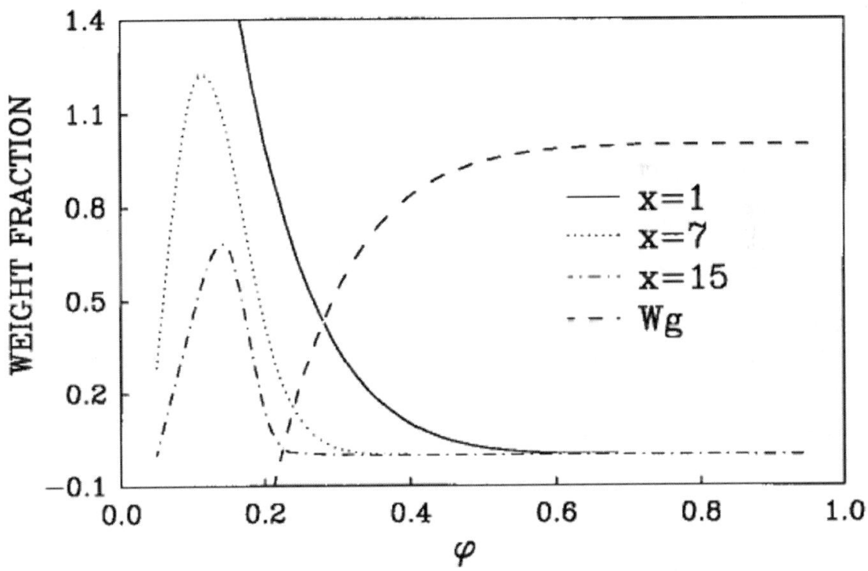

Figure 16 Weight fraction of various paths in an ER suspension calculated with Eqs. [46] and [47] under the assumption that functionality *f* is equal to 8 and the extent of reaction α is equal to the particle volume fraction. Reproduced with permission from T. Hao and Y. Xu, J. Colloid Interf. Sci., 181(1996)581

The weight fraction of a polymer chain of x amount of monomers, W_x, can be expressed as:

$$W_x = \frac{y^x\left(1-\alpha'\right)^2}{\alpha'}\left[\frac{f(fx-x)!}{(x-1)!(fx-2x+2)!}\right] \tag{47}$$

Several assumptions have to be made for applying Flory's theory to ER fluids. The extent of reaction α in Flory's theory could be replaced by the particle volume fraction. The sol or oligomer and gel can be considered as isolated particle aggregates and continuous percolation chains in ER fluids, respectively. The monomer functional groups f corresponds to the "contacting number" of one particle with the others around it, i.e., the coordination number of solid particles. Since a bct lattice structure is formed in ER fluids of monodispersed particles[57], f could be approximated as 8 even in an ER fluid of polydispersed particle size. In this way, the weight fraction of various particle aggregates can be estimated from Eq. (46) and (47) and depicted in Figure 16. The percolation path (or gel) starts to appear at the particle volume fraction 24 vol%, and reaches 100% at the particle volume fraction 40 vol%, where the entire ER system is occupied with the percolated network structure. This continuous network structure mainly contributes to the overall dc conductivity, and the weight fraction of the continuous paths, branched paths, and isolated paths determine the dielectric properties of ER suspensions.

In summary, like an ordinary colloidal suspension, ER fluids experience a percolation transition as the particle volume fraction increases. The critical particle volume fraction is determined by the intrinsic properties of an ER fluid, independent of the external electric field. Once the particle volume fraction exceeds a critical value, three-dimensional network structure build up. The generally accepted fibrillation phenomenon only appears in dilute suspensions and it is over-simplified for concentrated suspensions. Unlike the particle volume faction, the increase of electric field strength doesn't induce a sudden change of the rheological property, but it does induce a solid-solid phase transition as shown in Figure 9. The external electric field enhances the interaction between the particles and strengthens the network structure. However it can not change the shape of the formed network structure. The critical particle volume fraction is an important parameter for an ER fluid. Preparing an ER fluid of the particle volume fraction around the critical value would be better for a low viscosity at zero electric field and a strong ER effect under an electric field. Since the ER effect greatly depends on the particle volume fraction, it may be difficult to compare the ER effects of two ER fluids even in the same particle volume faction. However, it can be distinguished in the same electric field when the particle volume fractions of those two ER fluids are around their critical values.

4. RHEOLOGICAL PROPERTIES

The rheological properties of ER suspensions should be quite different from conventional non-ER fluids due to their substantially high mechanical strength under an electric field. In this section the rheological properties of ER suspensions are phenomenologically described. Derivation of the specific equations governing the rheological properties can be found in the chapters dealing with the ER mechanisms.

4.1 Steady shear behavior

The experimental [44, 64, 65], and computer simulation [66] results suggest that the shear stress τ of an ER fluid can be well expressed with the Bingham equation:

$$\tau = \eta_p \dot{\gamma} + \tau_d \tag{48}$$

where η_p is the plastic viscosity, $\dot{\gamma}$ is the shear rate, and τ_d is the Bingham or dynamic yield stress. The dynamic yield stress τ_d scales with the electric field as E^m, where E is the applied electric field, and m is two in low fields strength and $1 < m < 2$ in high fields. As pointed out by Kraynik [67], there are three yield stress, the elastic-limit yield stress, τ_y, the static yield stress τ_s, and the dynamic yield stress, τ_d. The elastic-limit yield stress is the one that materials cannot fully recover once the applied stress exceeds this value. The static yield stress is the minimum stress required to cause the fluid to flow. The dynamic yield stress is the one that can maintain the flow continuously once the stress exceeds the static yield stress (see Figure 17). Note that it is possible that the static and dynamic yield stress may coincide, or there is only static yield stress and no dynamic one. Barnes [68] contended that the dynamic yield stress is just an empirical value that depends on the experimental conditions. Although the existence of the dynamic yield stress is still controversial [68], the yield stress is widely accepted as a valuable parameter for characterizing the viscoelastic materials.

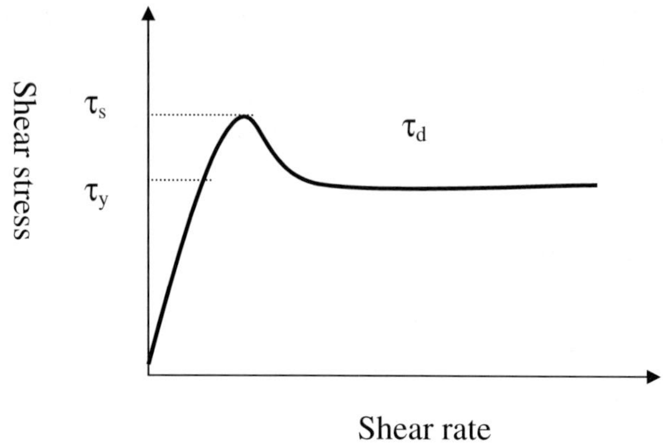

Figure 17 Illustration curve of shear stress vs. shear rate for defining the elastic-limit yield stress, τ_y, static yield stress, τ_s, and the dynamic yield stress, τ_d. Reproduced with permission from R.T. Bonnecaze and J.F. Brady, J. Rheol., 36(1)(1992)73

Many attempts have been made to determine the yield stress of an ER fluid, as this is the direct parameter evaluating how strong the ER suspension is under an electric field. In most cases, the yield stress is determined via extrapolating the shear stress to zero shear rate. The ER particle typically fibrillate under an electric field, bridging two electrodes together. The yield stress is the force that can break the particle bridges. Using microscopy, Klingenberg [65] directly observed how the ER particle behaves under both an electric field and a constant shear field. Figure 18 shows the videotaped images of hollow silica spheres of 57 μm in diameter dispersed in corn oil under an electric field 750 V/mm at various shear fields. Without a shear field the particle instantly formed straight and vertical bridges spanning between two electrodes (Figure 18a). Under a shear field but the shear rate is less than a critical value, the chains incline to the shearing direction but without rupture (Figure 18b). The middle point in between two electrodes is thought to be the weakest in terms of the particle interaction. Once the shear rate exceeds the critical value, all chains rupture in the middle point and move with the upper electrode in the upper half region, while the lower half parts of chains remain intact (Figure 18c). The ruptured "half" chains will re-form "single" chains at the next steady position via simply connecting with other half chains in the lower region.

Gamota [69] suggested that the rheological properties of ER fluids can be qualitatively understood in terms of three regions: a pre-yield, a yield, and a post-yield region. The pre-yield region can be experimentally characterized with a stress-control rheometer at the stress below the yield stress, or with a strain-control rheometer at very low shear rates, and or with a dynamic rheometer at low amplitudes. The post-yield region can be experimentally characterized even with a viscometer at a constant shear rate that is high enough to cause the ER material to flow. The yield region is complicated because of the difficulty of accurately determining the yield point. Unlike the extrapolation of the shear stress to zero shear rate for indirectly determining the yield stress, the air pressure driven viscometer is thought to be a direct means for obtaining the yield stress value. Figure 19 shows the yield stress of the oxidized polyacrylonitrile/silicone oil suspension measured with the air pressure driven viscometer against the square of the electric field strength. The perfect linear relationship between the yield stress and the square of the applied electric field strength is obtained. There is no yield stress saturation observed in the range of the applied electric field up to 3 kV/mm.

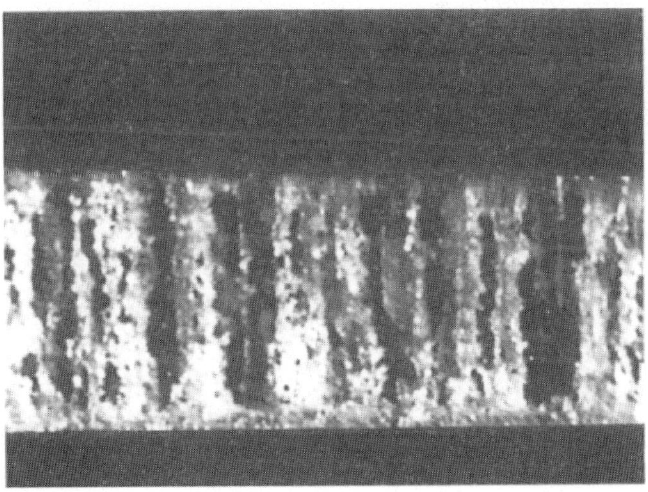

(a)

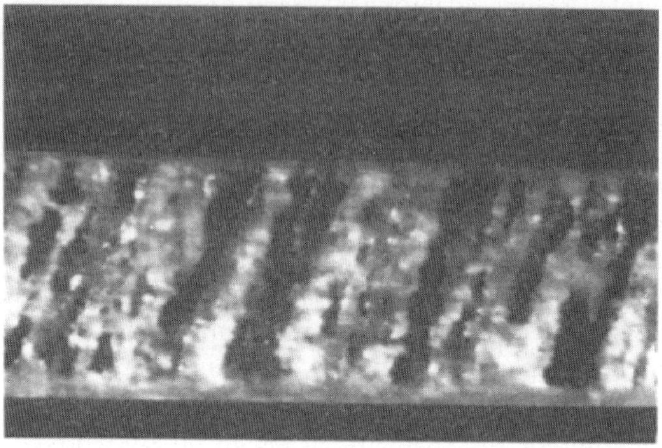

(b)

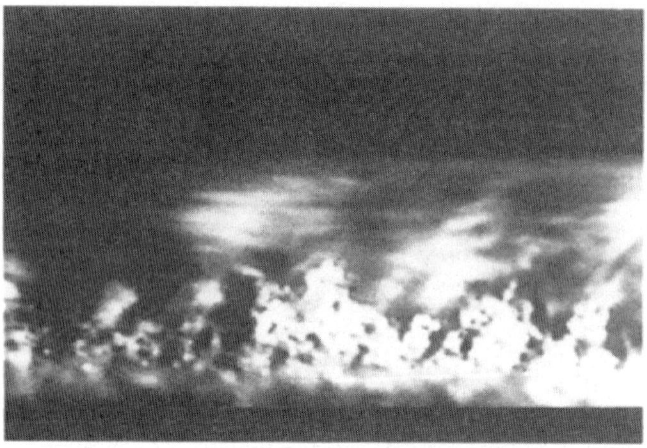

(c)

Figure 18 The microscopic picture of hollow silica spheres of 57 μm in diameter dispersed in corn oil under an electric field 750 V/mm. (a) The static structure under the electric field without shearing; (b) The structure under a strain less than the critical value. The top electrode is moving to the right side ; (c) The structure is under a shear rate 5 s^{-1}. The top electrode is moving to the right side. Reproduced with permission from D. J. Klingenberg and C. F. Zukoski IV, Langmuir, 6(1990)15.

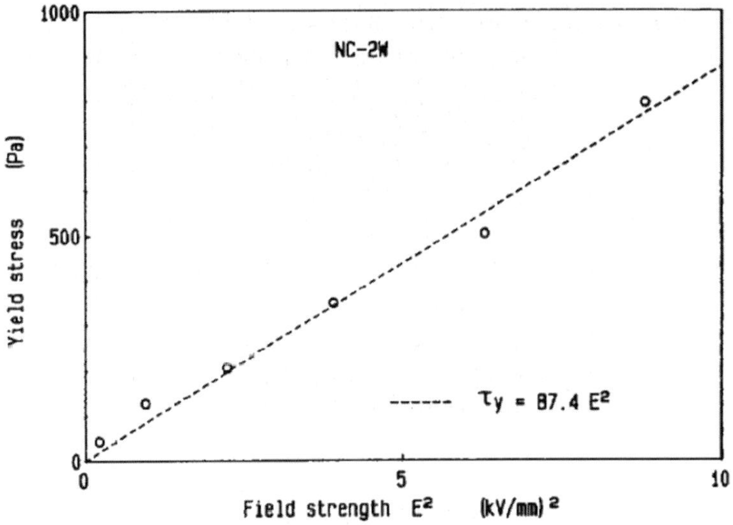

Figure 19 Yield stress vs. the square of electric field strength for the oxidized polyacrylonitrile/silicone oil suspension of the particle weight fraction 19.2 wt%. Reproduced with permission from Y. Xu and R. Liang, J. Rheol., 35(1991)1355.

The shear stress of an ER suspension is clearly a function of the applied electric field and the particle volume fraction. The flow curves of silica/silicone oil suspension with the particle volume fractions 0.2 and 0.4 at different electric fields are shown in Figures 20a and 20b, respectively. At zero electric field the flow is Newtonian for the suspension of 20 vol% particle and slightly pseudoplastic at low shear rates for the suspension of the particle volume fraction 40 vol%. Once the shear rate exceeds 100 s^{-1}, the flow for the suspension of 40 vol% also becomes Newtonian. The application of an electric field increases the shear stress and the flows become pseudoplastic over the entire range of shear rates. When the applied electric field strength is larger than 1.5 kV/mm, the flow curves show a plateau at low shear rates, and the plateau corresponds to the yield stress that increases with the applied electric field strength. The Bingham model clearly can be used to describe the flow behavior for 20 vol% suspension at shear rates below 300 s^{-1} and 40 vol% suspensions over the entire shear rate range at the electric fields below 3.0 kV/mm. Over 3.0 kV/mm the flow curves cannot be expressed by the Bingham model due to the decreased plastic viscosity from the pronounced shear thinning. The Herschel-Bulkley model shown in Eq. (49) may better fit the data. In comparison with the flow curve for 40 vol% suspension, the shear stress doesn't monotonically increase with shear rate for 20 vol% suspension. At high shear rates the shear stress decreases with the shear rate for 20 vol% suspension, which is obviously a result of the shear thinning. The flow behavior difference between 20 vol% and 40 vol% suspensions may be well explained with the assumption that this suspension has a critical volume fraction for percolation transition in between 20 vol% and 40 vol%. Under an electric field the suspension of the particle volume fraction 20 vol% may have a fibrillated chain structure, while the suspension of the particle volume fraction 40 vol% may have a percolated network structure. As shown in Figure 18 at a sufficiently high shear field the fibrillated chains will be ruptured and the reformation of chain structure may be unlikely due to the relatively low particle volume fraction, which leads to the shear thinning phenomenon. In a percolated network structure the rupture and reformation processes can occur simultaneously as there are enough particles surrounding each other. This assumption is verified in Figure 21, where the shear stress is plotted as the shear rate at various particle volume fractions from 10 vol% to 50 vol% at the electric field 2.0 kV/mm. When the particle volume fraction is below 30 vol%, the shear thinning occurs at high shear rates. Once the particle volume fraction is larger than 30 vol%, there is no shear thinning phenomenon over the entire shear rate range, indicating that there is a dramatic change of the microstructure at the particle volume fraction around 30 vol%.

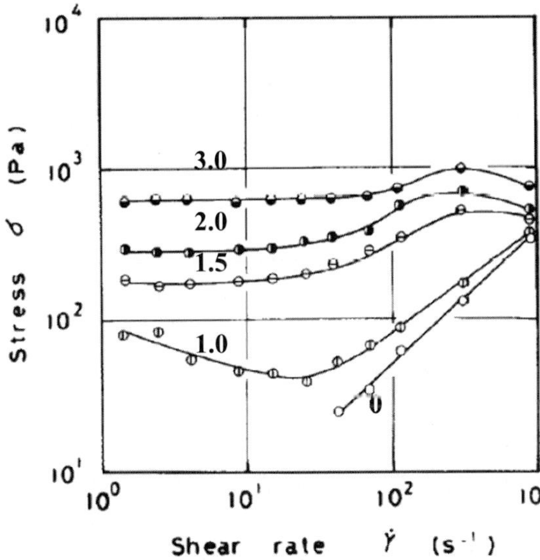

10⁴
10³
10²
10¹
Stress σ (Pa)
3.0
2.0
1.5
1.0
0
10⁰
10¹
10²
10³
Shear rate γ̇ (s⁻¹)

(a)

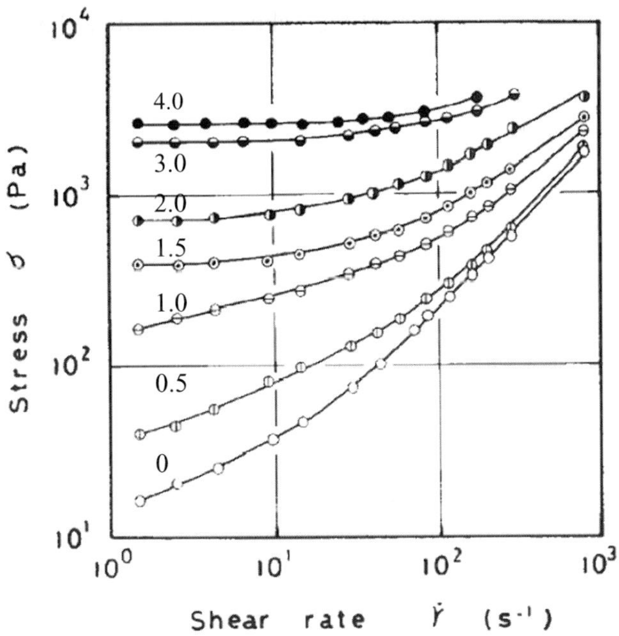

10⁴
10³
10²
10¹
Stress σ (Pa)
4.0
3.0
2.0
1.5
1.0
0.5
0
10⁰
10¹
10²
10³
Shear rate γ̇ (s⁻¹)

(b)

Figure 20 The flow curves of silica/silicone oil suspension with particle volume fraction 20 vol% (a) and 40 vol% (b) at different electric fields. The number on each curve is the applied electric field, kV/mm. Reproduced with permission from Y. Otsubo, J. Rheol., 36(1992)479.

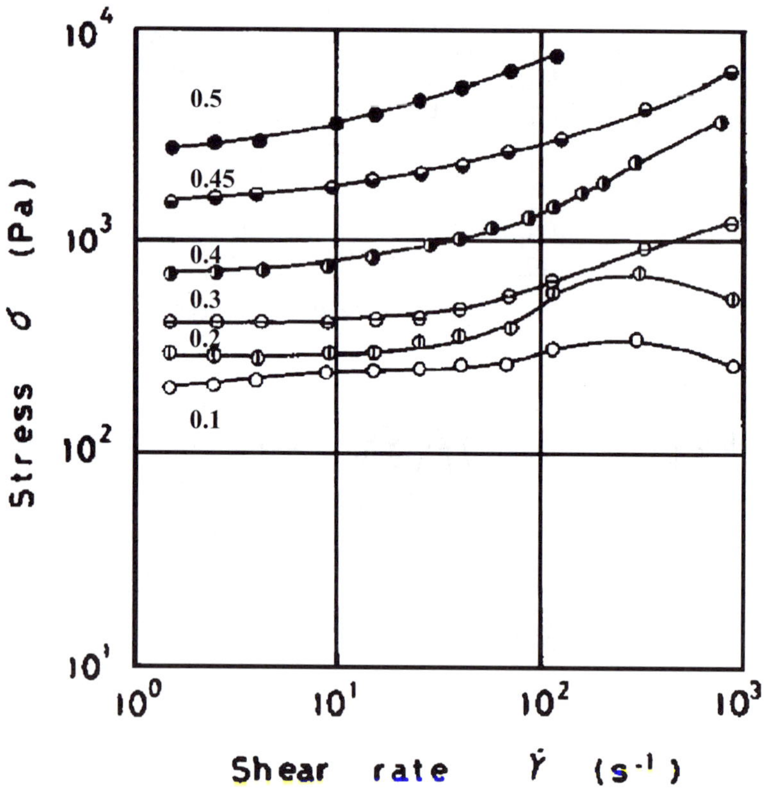

Figure 21 The flow curves of silica/silicone oil suspension with particle volume fractions from 10 vol% to 50 vol% at the electric field 2.0 kV/mm. The number on each curve is the particle volume fraction. Reproduced with permission from Y. Otsubo, J. Rheol., 36(1992)479.

The microstructure of an ER system definitely determines how this system behaves under an electric field. Figure 22 shows the shear stress of octylcynaobiphenyl vs. the electric field at shear rate 329.5 s^{-1} and various temperatures. As indicated in the literature [70-73], Octylcynaobiphenyl is a liquid crystal material, and has a phase transition from the smectic to the nematic phase at 306.72 K and from the nematic to the isotropic phase at 313.95 K. With the increase of temperature from 306.6 K to 312.8 K, octylcynaobiphenyl may have the different structures marked as a_c to b′ [73]. The ER property of octylcynaobiphenyl should depend on how the director is orientated in the fields. The shear stress passes through a maximum value when the liquid crystal material is in the smectic phase state. Once the material is in the nematic phase state, the ER effect becomes weak and saturates at the electric field strength above 0.7 kV/mm.

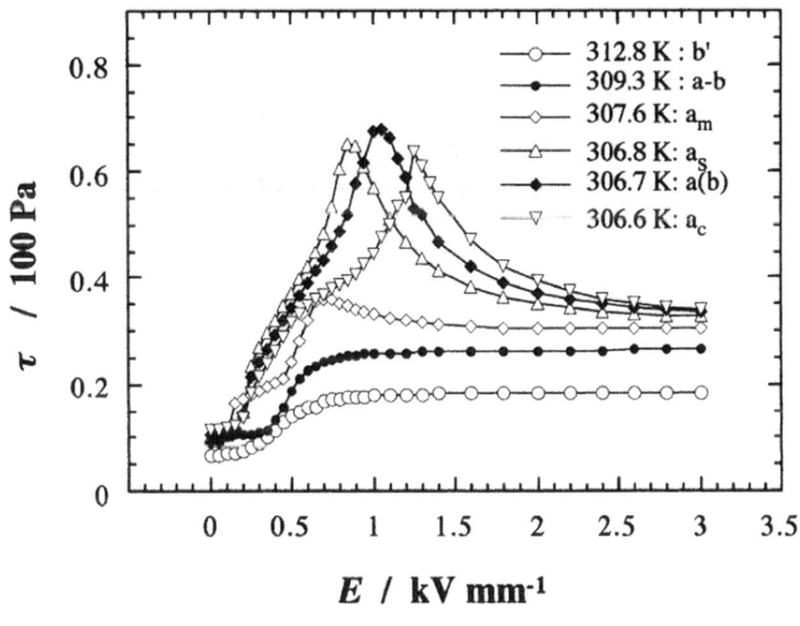

Figure 22 The shear stress of octylcynaobiphenyl vs. the electric field at shear rate 329.5 s^{-1} and various temperatures. Reproduced with permission from K.Negita, and S. Uchino, Mol. Cryst. Liq. Cryst., 378(2002)103

Similar to conventional colloidal suspensions, ER suspensions also show the shear thickening and shear thinning behaviors. Figure 23 shows the viscosity of the 3-(methacryloxy propyl)-trimethoxysilane coated monodispersed silica of size 242 nm dispersed in 4-methylcyclohexanol against shear rate at various electric fields between 0 and 600 V/mm. Without an electric field this suspension demonstrates a strong discontinuous shear thickening behavior. Under an external electric field, the shear thickening behavior is suppressed. Interestingly the shear thickening phenomenon is further weakened with the increase of the electric field strength.

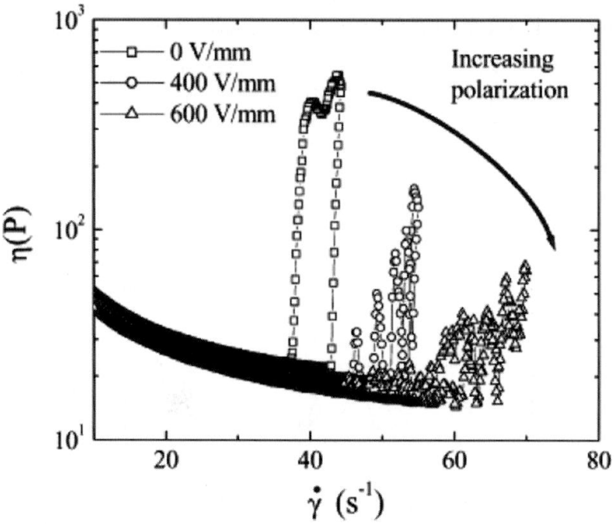

Figure 23 Viscosity of the 3-(methacryloxy propyl)-trimethoxysilane coated monodispersed silica of size 242 nm dispersed in 4-methylcyclohexanol vs. shear rate at various electric fields of fixed frequency 200Hz. The applied field is a zero mean square wave a.c. voltages. The particle volume fraction is 53 vol%. The shear thickening behavior is dramatically suppressed with the increases of the electric field Reproduced with permission from S.S. Shenoy N J. Wagner, and J.W. Bender, Rheol. Acta, 42(2003)287.

This phenomenon is not only related to the applied electric field strength but also the frequency of the orthogonal electric field. Figure 24 shows the apparent viscosity of the same suspension vs. the shear rate under a fixed electric field 600 V/mm of various frequencies. With the decrease of the field frequency the shear thickening phenomenon becomes weak and almost disappears at frequency around 40 Hz. Since the shear thickening behavior of this suspension results from the hydrocluster formation of silica particles, this phenomenon is easily understood in terms of the electric-field-strengthened particle interaction force under high electric fields and low frequencies. As one may already know, for many ER systems the ER effect is much stronger in a high electric field of a low frequency relative to a low electric field of a high frequency. The strong ER effect may compete with the hydrocluster formation of silica particles, thus weakening the shear thickening phenomena. Without the surface coating material, 3-(methacryloxy propyl)-trimethoxysilane, the silica particle/4-methylcyclohexanol suspension shows the shear thinning phenomenon [75].

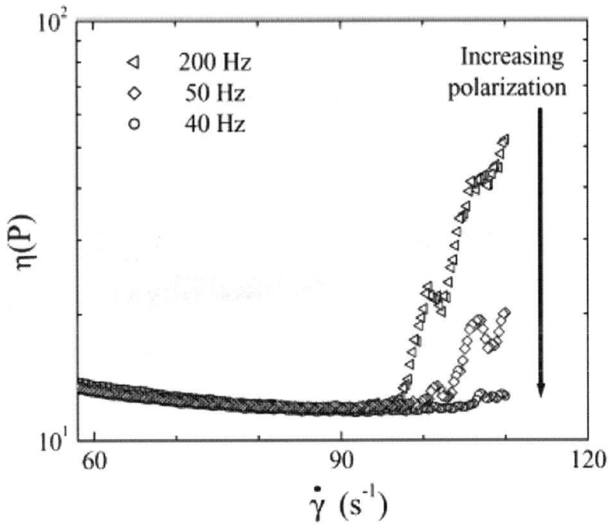

Figure 24 Viscosity of the 3-(methacryloxy propyl)-trimethoxysilane coated monodispersed silica of size 242 nm dispersed in 4-methylcyclohexanol vs. shear rate at various frequencies of fixed field 600 V/mm. The applied field is a zero mean square wave a.c. voltages. The particle volume fraction is 50 vol%. The shear thickening behavior is dramatically suppressed with the decrease of the field frequency. Reproduced with permission from S.S. Shenoy N.J. Wagner, and J.W. Bender, Rheol. Acta, 42(2003)287.

Figure 25 presents the reduced viscosity of monodispersed 0.75-μm-diam silica/4 methylcyclohexanol suspension of particle volume fraction 10 vol% vs. Peclet number ($Pe = 6\pi\eta_s a^3 \dot{\gamma} / k_B T$) at various electric fields from 400 V/mm to 1000 V/mm. The suspension shows a power-law dependence

$\eta \propto \dot{\gamma}^{-\Delta}$ of the apparent viscosity η on the strain rate $\dot{\gamma}$. Δ increases from 0.68 at 400 V/mm to 0.93 at 1000 V/mm [75]. This shear thinning phenomenon can be explained qualitatively with the model that the roughly prolate spheroidal droplets are assumed to form in the condensed ER phase [76]. The shear flow will rotate the ellipsoidal droplets and thus the long axis will deviate from the direction of the electric field, leading to the weak

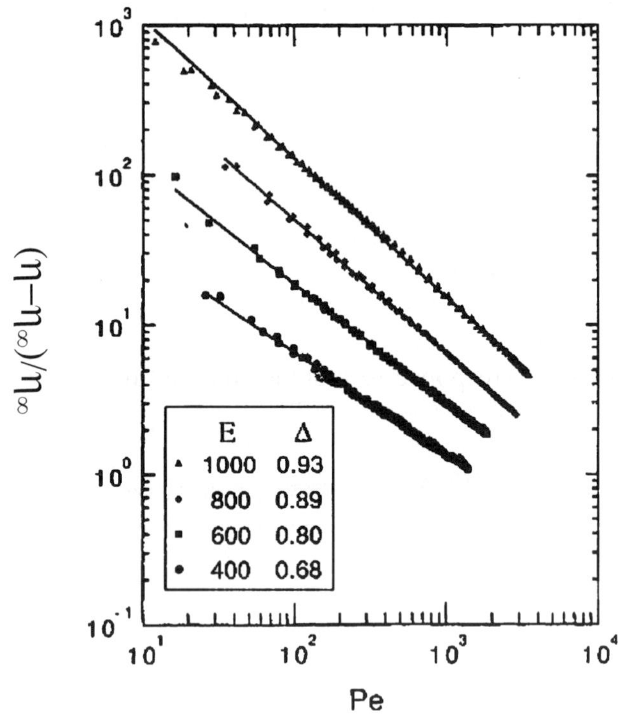

Figure 25 The reduced viscosity of monodispersed 0.75-μm-diam silica/4-methylcyclohexanol of particle volume fraction 10 vol% vs. Peclet number ($Pe = 6\pi\eta_s a^3 \dot{\gamma}/k_B T$) at various electric field from 400 V/mm to 1000 V/mm. The infinite shear-rate viscosity, η_∞, was measured as 0.45 Pa. Reproduced with permission from T. C. Halsey, J.E. Martin, and D. Adolf, Phys. Rev. Lett., 68(1992)1519.

interaction force between the droplets. As one may imagine, the larger the droplet size and shear rate are, the greater the rotation in the shear flow and thus the lower the apparent viscosity of the suspension will be. Of course under a continuous shearing process, the chain rupture and continuous reformation as evidenced with the video-image in Figure 18 may also account for the shear thinning phenomenon. The Herschel-Bulkley shear model is believed to better express this phenomenon, using parameter m and power-law index n [77, 78]

$$\tau = \tau_y + m\left|\dot{\gamma}\right|^{n-1}\dot{\gamma} \tag{49}$$

4.2 Dynamic rheological property

4.2.1 Strain dependence

Interests on the microstructure of ER fluids under an electric field stimulate the studies of the dynamic behaviors of ER fluids. The qualitative behavior of stress response to a sinusoidal strain may give the information on how the ER suspension responds to a mechanical stimulation. Figure 26 shows the oscilloscope traces of imposed strain and response stress of the hydrated particulate of the lithium salt of poly(methacrylate) dispersed in a chlorinated paraffin oil at various strains from 0.008 to 0.05 and a fixed electric field 2400 V/mm. The dashed lines represent the imposed strain and the solid lines represent the measured stress. At very low strains the stress remains in a perfect sinusoidal shape as the inputed strain, indicating that at such a small strain the ER suspension behaves elastically and it is in the linear response region. When the strain increases from 0.008 to 0.015, the stress response is no longer linear and the shape of stress curve deviates from the sinusoidal to the orthogonal-like shape. This may represent a transition region where the yield occurs. Further increase of the strain the stress response becomes nonsinusoidal with higher order harmonics. Such a stress-strain dependence is also found in anhydrolated zeolite, Linde 3A (molecular structure, $K_9Na_3[(AlO_2)_{12}(SiO_2)_{12}]$, dispersed in paraffin oil system, as shown in Figure 27. With the increase of the applied strain amplitude from 3.0 to 27.0 the stress response curve of this ER suspension deviates from a sinusoidal shape to a skewed saw-toothed one.

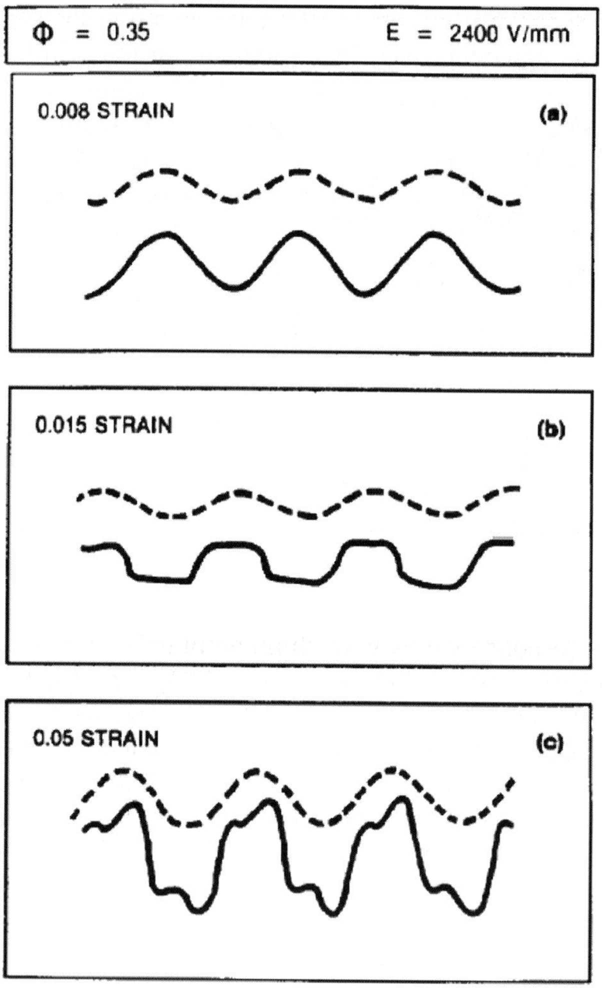

Figure 26 Oscilloscope traces of imposed strain (--) and response stress (-) at 0.008, 0.015, and 0.05 fractional strain for the hydrated particulate of the lithium salt of poly(methacrylate) dispersed in a chlorinated paraffin oil. The particle volume fraction is 0.35, the frequency is 10 rad/s and the applied electric field is 2.4 kV/mm. Reproduced with permission from W. S. Yen and P. J. Achorn, J. Rheol., 35(1991)1375.

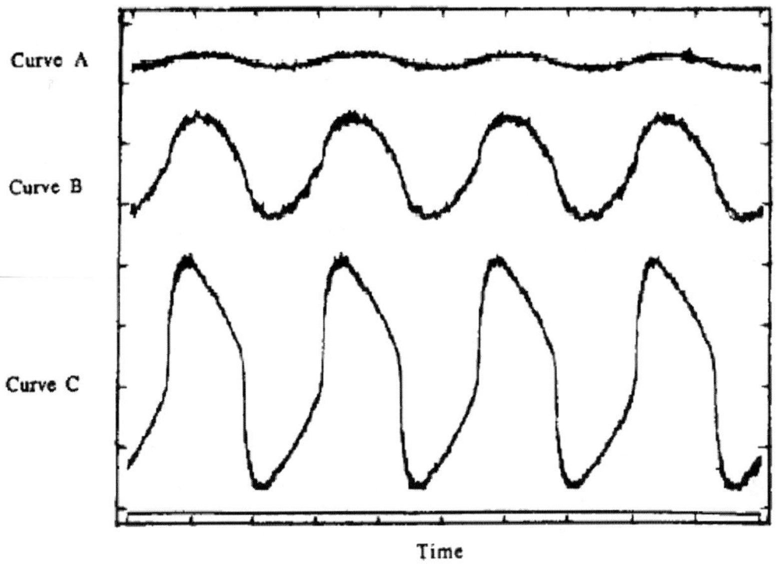

Figure 27 Stress response varies with strain amplitude for the anhydrolated zeolite, Linda 3A (molecular structure, $K_9Na_3[(AlO_2)_{12}(SiO_2)_{12}]$, dispersed in paraffin oil. A) thc strain amplitude is 3.0; B) the strain amplitude is 9.0; C) the strain amplitude is 27.0. For all three curves the applied electric field is 1.0 kV/mm. Reproduced with permission from D.R. Gamota and F.E. Filisko, J. Rheol., 35(1991)399.

Besides showing a strong dependence on the strain amplitude, the stress response also shows a strong dependence on the applied electric field. Figure 28 shows stress response curves of the same anhydrolated zeolite/paraffin oil at various electric field strengths ranging from 0 to 2.5 kV/mm. The curve A in Figure 28 is the imposed strain and the curve B is the stress response under zero electric field. In this case the stress is sinusoidal but lags the strain by a phase angle of 90°, indicating the material response is viscous. The curve C corresponds to the stress response under an electric field 1.0 kV/mm. The amplitude of the curve C substantially increases in comparison with the curve B and the phase angle shifts from 90°. However, the shape of the curve C remains essentially sinusoidal, indicative of this ER suspension deforming in a linear viscoelastic mode under such a condition. The phase angle and the amplitude of the stress are dependent on the applied electric field strength. The curve D is the stress response recorded under an electric field 2.5 kV/mm. The shape of the curve D deviates from the sinusoidal to the truncated, however, the frequency of this truncated curve is the same as the imposed strain, indicative of a transition from the linear mode to the nonlinear mode at such a high electric field. For the purpose of gaining further insight into the dynamic behaviors of ER suspensions under an oscillatory shear, the mechanical analog consisting of Voigt element in series with an elastic element, a Couloumb frictional element and a viscous element was proposed [69], and a nonlinear phenomenological model including an irreversible yielding term, a nonlinear softening elasticity, and viscous losses was further developed [80].

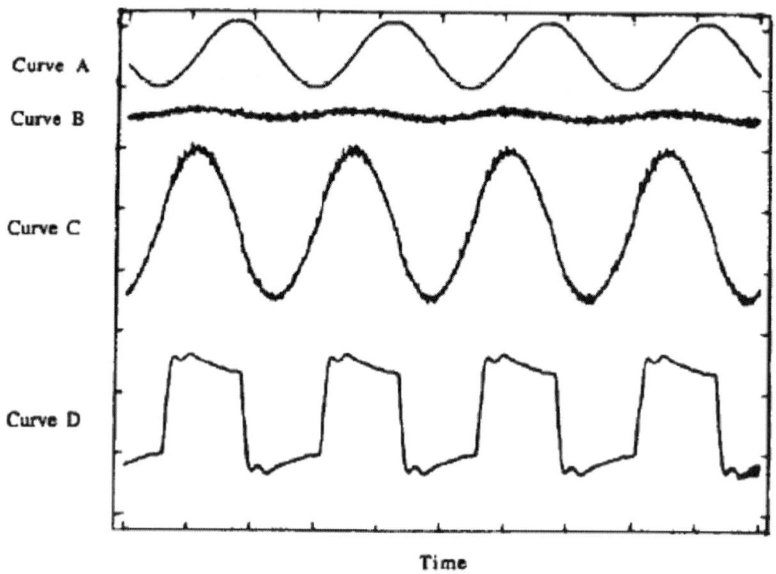

Figure 28 Stress response varies the applied electric field at a constant strain amplitude for the anhydrolated zeolite, Linda 3A (molecular structure, $K_9Na_3[(AlO_2)_{12}(SiO_2)_{12}]$, dispersed in paraffin oil. A) the applied strain; B) E=0.0 kV/mm; C) E=1.0 kV/mm; D) E=2.5 kV/mm. The strain amplitude is 1.0 and frequency is 10 Hz. Reproduced with permission from D.R. Gamota and F.E. Filisko, J. Rheol., 35(1991)399.

The flow mechanism of an ER suspension can also be obtained from the energy dissipated by the ER material. The area within the hysteresis loop of the stress-strain curve represents the energy dissipated per unit volume of the material per cycle. According to Gamota [69], there are two common energy dissipating mechanisms associated with ER suspensions: Viscous damping and the frictional or Coulomb damping. The energy dissipated in viscous damping per volume per cycle, U_v, is given by:

$$U_v = \oint \tau(t) d\gamma \tag{50}$$

where τ is the shear stress and γ is the shear strain. In the oscillatory flow, the relationship among the shear stress, strain, and strain rate is given as:

$$\tau(t) = \eta' \dot{\gamma} \tag{51}$$

$$\gamma = \gamma_0 \sin(\omega t) \tag{52}$$

$$\dot{\gamma} = \omega \gamma_0 \cos(\omega t) \tag{53}$$

where η' represents the dynamic viscosity, $\eta' = G'' / \omega$, where G'' is the loss modulus. Eq. (50) thus can be simplified as:

$$U_v = \oint \tau(t) \left(\frac{d\gamma}{dt} \right) dt = \pi \eta' \omega \gamma_0^2 \tag{54}$$

Eq. (54) indicates that the energy dissipated by the viscous damping is proportional to the square of shear strain amplitude, the dynamic viscosity and the frequency of the dynamic field. Substituting Eq. (53) into Eq.(52) one may obtain:

$$\tau = \pm \eta' \omega \sqrt{\gamma_0^2 - \gamma^2} \tag{55}$$

Rearranging Eq. (55) one may obtain:

$$\left(\frac{\tau}{\eta' \omega \gamma_0}\right)^2 + \left(\frac{\gamma}{\gamma_0}\right)^2 = 1 \qquad (56)$$

Eq. (56) indicates that an ellipse will be obtained if τ is plotted against γ. In other words, in viscous damping an elliptic hysteresis loop will be obtained. The energy dissipated per volume per cycle due to Coulomb damping in an oscillatory motion is given:

$$U_c = 4C_u N\gamma_0 \qquad (57)$$

where C_u is the Coulomb friction element parameter [81], and N is the normal stress. Eq. (57) indicates that the energy dissipated by the Coulomb damping is proportional to the shear strain amplitude. From the shape of the hysteresis loop one may tell the material is in the elasticity-dominant status or viscosity-dominant status. Figure 29 shows the hysteresis loops of anhydrolated zeolite, Linde 3A (molecular structure, $K_9Na_3[(AlO_2)_{12}(SiO_2)_{12}]$, dispersed in paraffin oil at various electric fields recorded during sinusoidal straining at a fixed amplitude 1.0 and frequency of 10 Hz. Without a stimulation of an external electric field, the area of the hysteresis loop is small and the shape of the hysteresis loop is exactly ellipsoidal, as expected with Eq. (56). With the increase of the applied field strength, the area of the hysteresis loop increases and the shape of the hysteresis loop changes from the ellipsoidal to rhombus-like figure. Note that at 1.0 kV/mm the shape of the hysteresis loop is still elliptical but the area increases substantially. The area of the hysteresis loop represents the energy dissipated per volume per cycle. The deviation from the elliptical shape indicates that the ER fluid changes from the viscous damping status to the Coulomb damping status.

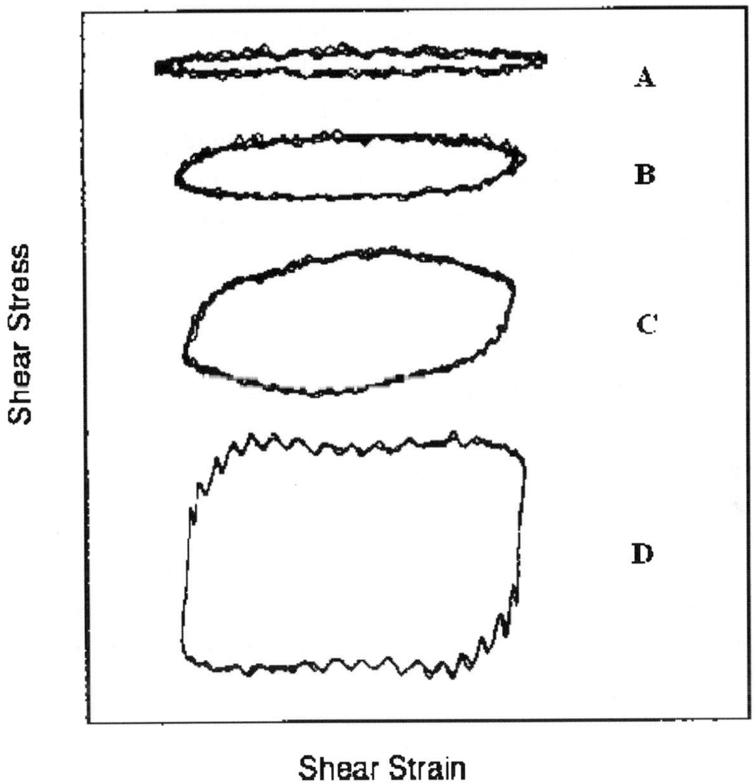

Figure 29 Hysteresis loop of anhydrolated zeolite, Linda 3A (molecular structure, $K_9Na_3[(AlO_2)_{12}(SiO_2)_{12}]$, dispersed in paraffin oil at various electric fields. A) E=0; B) E=1.0 kV/mm; C) E=2.0 kV/mm; and D) E=3.0 kV/mm. The strain amplitude is 1.0 and the frequency is 10 Hz. Redrawn from D.R. Gamota and F.E. Filisko, J. Rheol., 35(1991)399.

In addition to the shear stress, the storage modulus (G′) and loss modulus (G″) are two important parameters for characterizing ER suspensions. Linear viscoelastic behavior of ER suspensions is always addressed before the frequency sweep experiment. Typically at small strain amplitude both G′ and G″ show an independence of strain amplitude, and then decrease with the increase of strain amplitude. Figure 30 shows strain dependence of both the storage modulus (G′) and loss modulus (G″) for 1 μm in diameter silica/PDMS suspension with the particle volume fraction 17.1 vol% at various electric fields. Without an electric field, the loss modulus is almost one order of magnitude higher than the storage modulus and keeps constant up to a strain amplitude 10 %, which denotes the linear viscoelasticity region. With the increase of electric field the linear viscoelasticity region shifts to a low strain amplitude side. In the meanwhile the storage modulus becomes higher than the loss modulus once the applied electric field is larger than 250 V/mm.

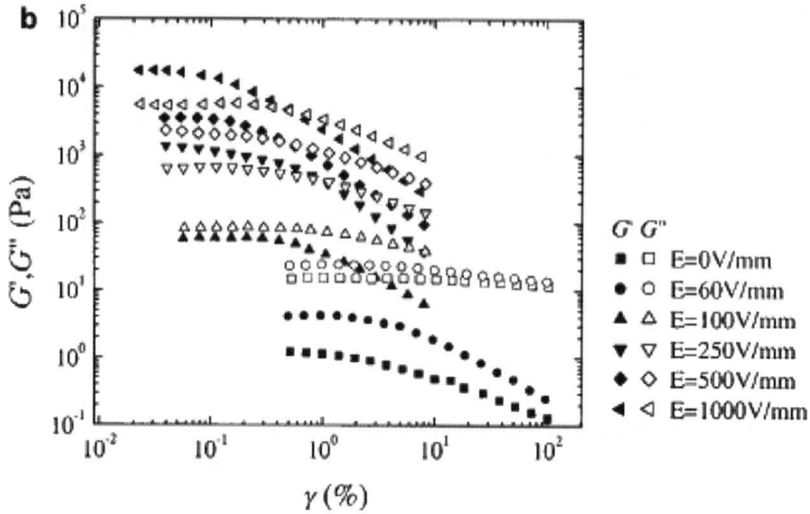

Figure 30 Strain dependence of both the storage modulus (G′) and loss modulus (G″) for 1 μm silica/PDMS suspension with the particle volume fraction 17.1 vol%. The frequency is fixed at 10 rad/s. Reproduced with permission from B.D. Chin and H.H. Winter, Rheol. Acta, 41(2002)265.

Not all ER suspensions have a linear viscoelasticity region and G′ and G″ don't always decrease with the strain amplitude. Such examples can be found in two similar aluminosilicate particulate materials dispersed in silicone oil systems [83]. Those two particulates have a common molecular formula that can be expressed as $(MO)_n \cdot (Al_2O_3)_x \cdot (SiO_2)_y \cdot (H_2O)_z$, where n, x, y, z all are integral numbers, and M represents the metallic atom. One is called permutite, and the other is called molecular sieve. After dispersed into silicone oil, they are named as PS and MS suspensions, respectively. Figure 31 shows the storage modulus G′, loss modulus G″, and absolute value of the complex viscosity $|\eta^*|$ vs. strain at the mechanical oscillatory frequency 5 Hz and zero electric field. For the PS suspension, G′, G″, and $|\eta^*|$ initially show a slightly decrease with the increase of strain amplitude, and then increase with it; Whereas for the MS suspension, G′, G″, and $|\eta^*|$ decrease with the increase of strain amplitude in a small strain range, and then gradually tend to level off . However, under an electric field of 0.5 kV/mm, the strain dependences of G′, G″, and $|\eta^*|$ of PS suspension becomes quite different than that under zero electric field: Instead of increasing with strain amplitude under zero electric field, those rheological parameters decrease slightly with strain amplitude, behaving similarly to the MS suspension. Figure 32 shows the strain dependences of G′, G″, and $|\eta^*|$ of those two suspensions under 0.5 kV/mm. For the MS suspension, the strain dependence of G′, G″, and $|\eta^*|$ at on-state and off-state electric field are hard to distinguish, except that the values of G′, G″, and $|\eta^*|$ obtained under 0.5 kV/mm are two orders of magnitude higher than that under zero electric field. Those phenomena were attributed to the microstructure differences between PS and MS suspensions.

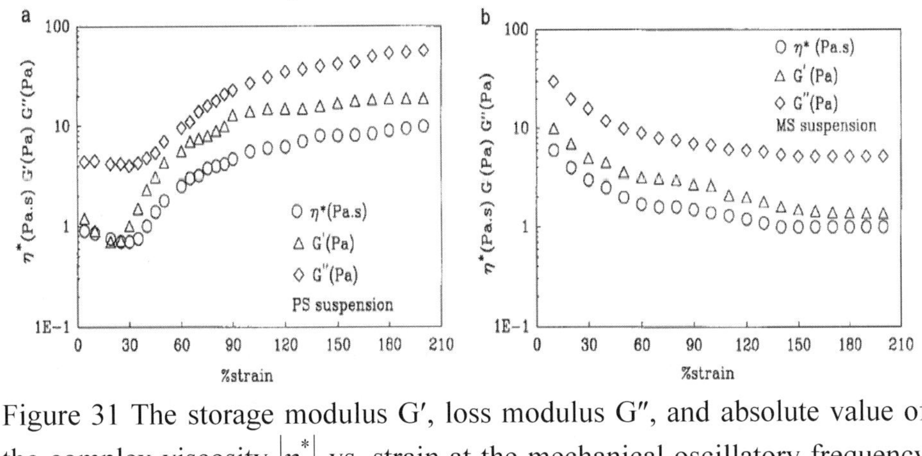

Figure 31 The storage modulus G′, loss modulus G″, and absolute value of the complex viscosity $|\eta^*|$ vs. strain at the mechanical oscillatory frequency 5 Hz and zero electric field. The particle volume fraction of those two suspensions are 35 vol%. (a) PS suspension, (b) MS suspension. Reproduced with permission from T. Hao and Y. Xu, J. Colloid Interf. Sci., 185(1997)324.

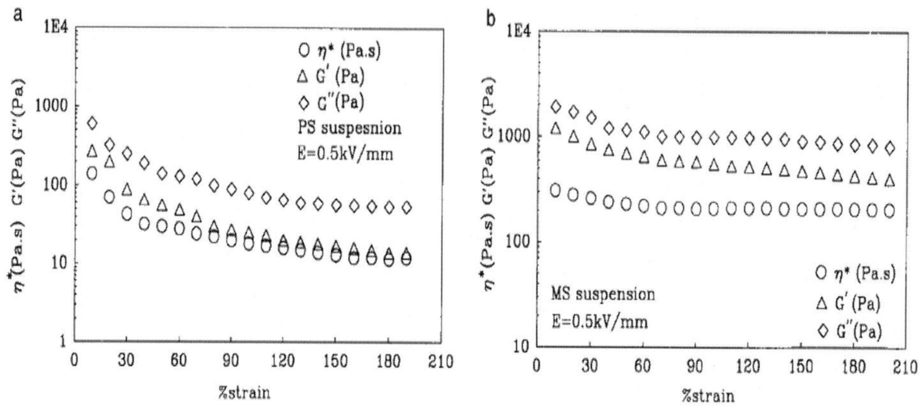

Figure 32 The storage modulus G′, loss modulus G″, and absolute value of the complex viscosity $|\eta^*|$ vs. strain at the mechanical field frequency 5 Hz and an electric field 0.5 kV/mm. The particle volume fraction of those two suspensions are 35 vol%. (a) PS suspension, (b) MS suspension. Reproduced with permission from T. Hao and Y. Xu, J. Colloid Interf. Sci., 185(1997)324.

Since the average diameter of PS and MS particles are 2.55 μm and 1.65 μm, respectively, and the PS has a relatively wide size distribution, those two suspensions may have a different percolation thresholds: The critical volume fraction of the MS suspension may be less than 35 vol%, while that of the PS suspension may be higher than 35 vol%. So a network structure would likely build up in the MS suspension at the particle volume fraction 35 vol%, whereas the PS suspension is still ahead of percolation transition at this concentration. Under zero electric field, the dispersed particles of the PS suspension would probably be randomly distributed in the medium, and there is almost no strong flocculation structures or comparatively large clusters involved in it. Once an external mechanical field is applied, the suspended particles would have a potential for rearrangement, and very likely form many large clusters even ordered structures. The mechanical-field-induced structure would be further ordered in larger strain amplitudes. As a result, G′, G″, and $|\eta^*|$ of the PS suspension increases with strain amplitude. Because of the slight wall sliding effect, G′, G″, and $|\eta^*|$ slightly decrease during small strain range. For the MS suspension, since a network structure or a strong flocculation structure has already formed under zero electric field and zero mechanical field, the applied mechanical field would destroy the existing structure instead of strengthening, resulting in a decrease in G′, G″, and $\left|\eta^{*}\right|$ with strain amplitude. Under an external electric field, a fibrillated structure will be formed in the PS suspension as the PS suspension has a percolation threshold higher than 35 vol%. While a network structure will be formed in the MS suspension. Both structures will be gradually destroyed with the increase of strain amplitude, leading to the decrease of G′, G″, and $\left|\eta^{*}\right|$ with strain under an electric field.

The loss modulus G″ may have a different strain dependence than the storage modulus G′. Figure 33 shows strain dependence of the storage and loss moduli scaled by the squared electric field strength for 20 wt% acidic alumina/PDMS suspension. At different combination of the electric field strength and oscillation frequency, G'/E_{rms}^2 shows a plateau value at small strain amplitudes and then decreases rapidly and becomes independent of ω/E_{rms}^2 at large strain amplitudes, where E_{rms} is the dc equivalent or root mean square (rms) of ac electric field. G''/E_{rms}^2 first increases with the strain amplitude and then passes through a maximum before decreasing with the increase of strain amplitude. The strain amplitude corresponding to the maximum G''/E_{rms}^2 may denote a critical strain value where a transition from linear to nonlinear deformation occurs.

The strain dependence of $\left|\eta^{*}\right|$ may be approximated with the Cox-Merz empiricism for suspensions [84]. The Cox-Merz rule indicates that $\left|\eta^{*}\right|$ as a function of $\omega\gamma_0$ is identical to the steady shear viscosity, η, as a function of the shear rate, $\dot{\gamma}$. On the basis of the Bingham model shown in Eq. (48), the steady shear viscosity can be expressed as:

$$\eta = \frac{\tau_y}{\dot{\gamma}} + \eta_p \tag{58}$$

For large shear rate the plastic viscosity can be approximated as the monolayer equivalent of the Einstein viscosity or the infinite shear rate suspension viscosity [85]:

$$\eta_p = \frac{1}{3\pi}\left(1 + \frac{5}{3}\phi\right) \tag{59}$$

where ϕ is the particle volume fraction. Replacing η with $\left|\eta^*\right|$, and $\dot{\gamma}$ with $\omega\gamma_0$, one may obtain:

$$\left|\eta^*\right| = \frac{\tau_y}{\omega\gamma_0} + \frac{1}{3\pi}\left(1 + \frac{5}{3}\phi\right) \tag{60}$$

Eq. (60) indicates that for large strain amplitude oscillation the complex viscosity may decrease linearly with the increase of the strain amplitude. As seen in Figure 32, the complex viscosities of PS and MS suspensions under 0.5 kV/mm approximately decrease linearly with the increase of strain amplitude, once the strain amplitude is larger than 30%. However, if the strain amplitude is small enough, a plateau region for the complex viscosity can still be observed. Figure 34 show the complex viscosity $\left|\eta^*\right|$ as a function of strain amplitude, $\omega\gamma_0$ in Figure 34a and as a function of strain amplitude scaled by the electric field strength squared, $\omega\gamma_0 / E_{rms}^2$, in Figure 34b at different combinations of electric field strength and oscillation frequency for a 20 wt.% acidic alumina/PDMS suspension. At very small value of either $\omega\gamma_0$ or $\omega\gamma_0 / E_{rms}^2$, the complex viscosity is independent of $\omega\gamma_0$ or $\omega\gamma_0 / E_{rms}^2$. Once $\omega\gamma_0$ or $\omega\gamma_0 / E_{rms}^2$ is large enough, the complex viscosity linearly decreases. Note that the complex viscosity data for different combinations of electric field strength and oscillatory frequency collapses onto a single curve at large $\omega\gamma_0 / E_{rms}^2$.

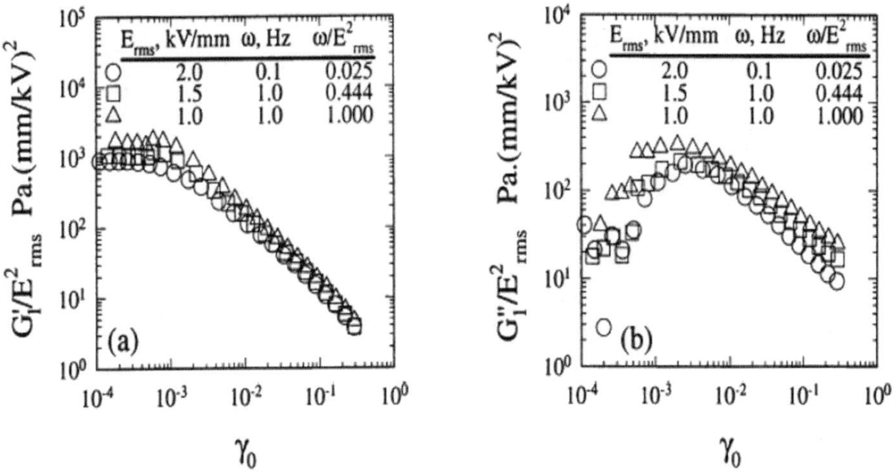

Figure 33. Strain sweep results for different combinations of electric field strength and oscillation frequency for a 20 wt.% acidic alumina/PDMS suspension: (a) storage modulus scaled by electric field strength squared, G'/E_{rms}^2, as a function of strain amplitude, γ_0, and (b) loss modulus scaled by electric field strength squared, G''/E_{rms}^2, as a function of strain amplitude, γ_0. E_{rms} is the dc equivalent or root mean square (rms) of ac electric field. Reproduced with permission from M. Parthasarathy and D. J. Klingenberg, J. Non-Newtonian Fluid Mech., 81 (1999) 83.

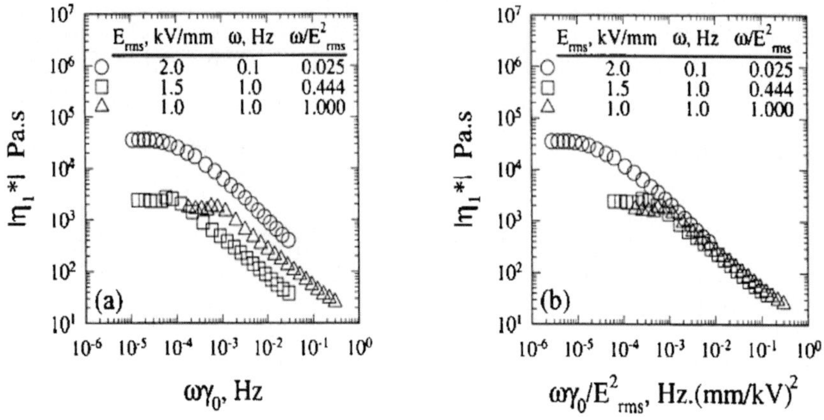

Figure 34 . (a) The magnitude of the complex dynamic viscosity, $|\eta^*|$ as a function of strain amplitude, $\omega\gamma_0$, and (b) the magnitude of the complex dynamic viscosity, $|\eta^*|$, as a function of strain amplitude scaled by the electric field strength squared, $\omega\gamma_0 / E^2_{rms}$, at different combinations of electric field strength and oscillation frequency for a 20 wt.% acidic alumina/PDMS suspension. Reproduced with permission from M. Parthasarathy and D.J. Klingenberg, J. Non-Newtonian Fluid Mech., 81 (1999) 83.

4.2.2 Frequency dependence

In a low dynamic frequency range (0.01-0.1 Hz), Vinogradov [87] observed that the storage modulus (G′) and loss modulus (G″) of diatomite/transformer oil were of similar magnitude and both increased with the increase of the electric field and the particle volume fraction. Brooks [88] found that G′ and G″ of lithium polymethacrylate/chlorinated hydrocarbon oil peaked at certain electric field. At low electric fields G′ is several orders of magnitude higher than G″, however at high electric fields both are of similar magnitude. Figure 35 shows G′ and G″ of oxidized polyacrylonitrile/silicone oil suspension of particle volume fraction 14.4 wt% at various electric field strengths vs. the mechanical field frequency. Over the frequency range from 0.1 to 200 rad/s, both G′ and G″ are independent of the frequency, indicating that the particle chains fully align with the electric field and the suspension behaves like a solid. A similar trend for both G′ and G″ was found in perchloric acid-doped polythiophene/silicone oil suspension [89] and even in an inorganic particle type ER suspension [83]. Figure 36 shows the storage modulus G′, loss modulus G″, and the absolute value of the complex viscosity $|\eta^*|$ of the molecular sieve particle/silicon oil vs. the mechanical field frequency. Both G′ and G″ don't change with the frequency, however, the complex viscosity linearly decreases with the frequency. According to the definition and the Maxwell model, the complex viscosity can be expressed as:

$$|\eta^*| = \frac{\sqrt{(G')^2 + (G'')^2}}{\omega} = \frac{G_m \tau}{\sqrt{1 + \omega^2 \tau^2}} \tag{61}$$

where G_m is the modulus of an ideal spring, ω is the angular frequency of an oscillatory mechanical field, and $\tau = \eta_m / G_m$, is the relaxation time, and the η_m is the viscosity of an ideal dashpot. When $1/\tau << \omega$, Eq. (61) can be simplified as:

$$|\eta^*| = G_m / \omega \tag{62}$$

In this case $|\eta^*|$ is inversely proportional to ω; the material would exhibit the elastic property as an ideal spring. When $1/\tau >> \omega$, Eq. (61) can be simplified as:

$$|\eta^*| = G_m \tau = \eta_m \tag{63}$$

In this case $|\eta^*|$ doesn't change with the angular frequency of an external mechanical field, and the material would exhibit the viscous property as an ideal dashpot. As shown in Figure 36, $|\eta^*|$ of the molecular sieve /silicon oil suspension decreases linearly with the frequency, which further indicates that at 0.5 kV/mm this suspension behaves like a elastic material.

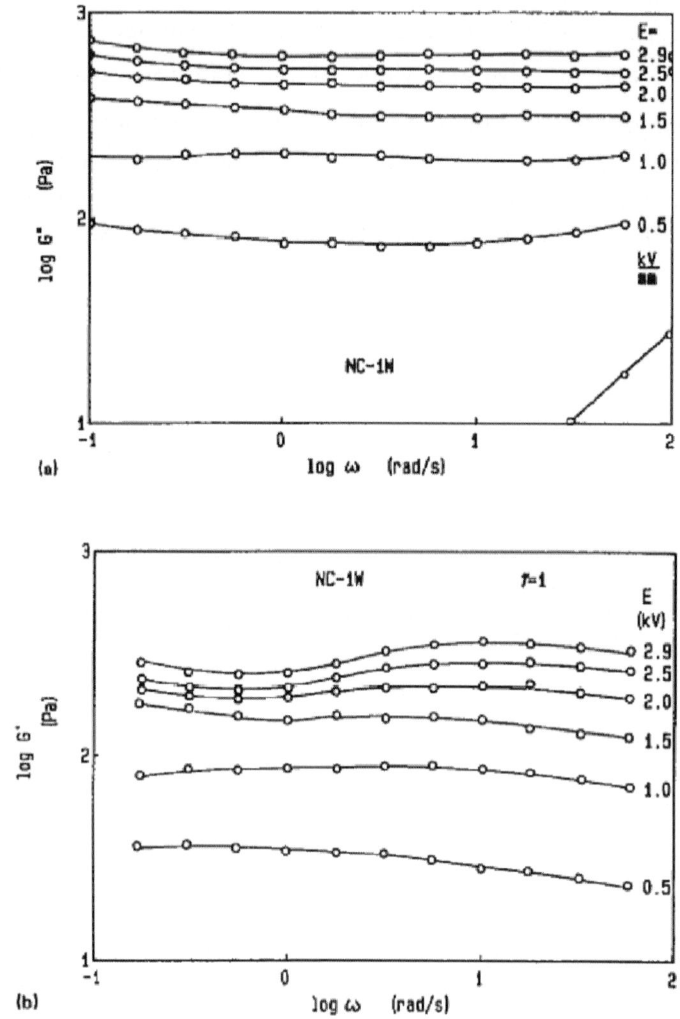

Figure 35 Frequency dependence of loss modulus G″ (a) and storage modulus G′(b) for oxidized polyacrylonitrile/silicone oil suspension of particle volume fraction 14.4 wt% at various electric field strengths. The strain is 1.0. Reproduced with permission from Y. Xu and R. Liang, J. Rheol., 35(1991)1355.

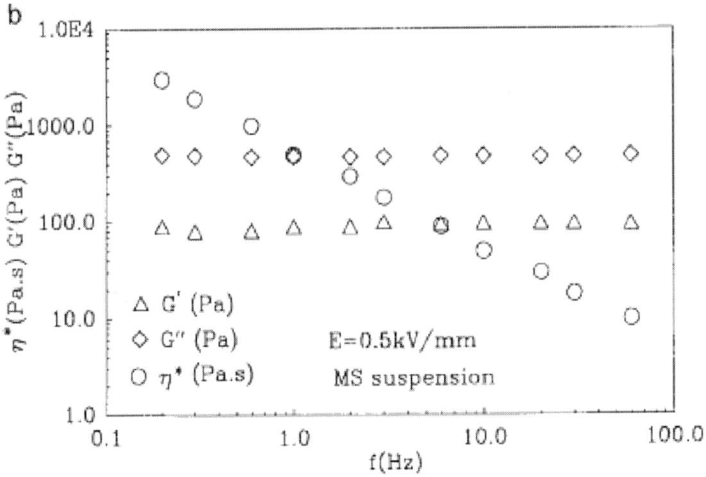

Figure 36 The storage modulus G′, loss modulus G″, and the absolute value of the complex viscosity $|\eta^*|$ of the molecular sieve/silicon oil vs. the mechanical field frequency at the strain amplitude 1 and the electric field 0.5 kV/mm. The particle volume fraction is 35 vol%. Reproduced with permission from T. Hao and Y. Xu, J. Colloid Interf. Sci., 185(1997)324.

Since the ER fluid usually changes from a liquid state (without an electric field) to a solid state under the reaction of an electric field, Chin [90] is interested in finding the gelation transition point using the rheological means. The mechanical loss tangent of monodispersed spherical silica dispersed in un-crosslinked PDMS, vinyldimethylterminated polydimethylsiloxane, is plotted again frequency in Figure 37. The measurement is performed at linear region at various electric fields and thus the strain amplitude is adjusted for this purpose. The slope of tgδ is negative under zero electric field, becomes flat at the electric field E=30 V/mm, and finally becomes positive at E=50 V/mm, indicating that there is a gelation transition around 30 V/mm. Such a low electric field can induce a structure change from the liquid state to the solid sate. With the increase of the electric field, the slope of tgδ becomes more positive and the absolute value of tgδ becomes much less than 1, suggesting that the field-induced solid becomes much stronger with the increase of the electric field.

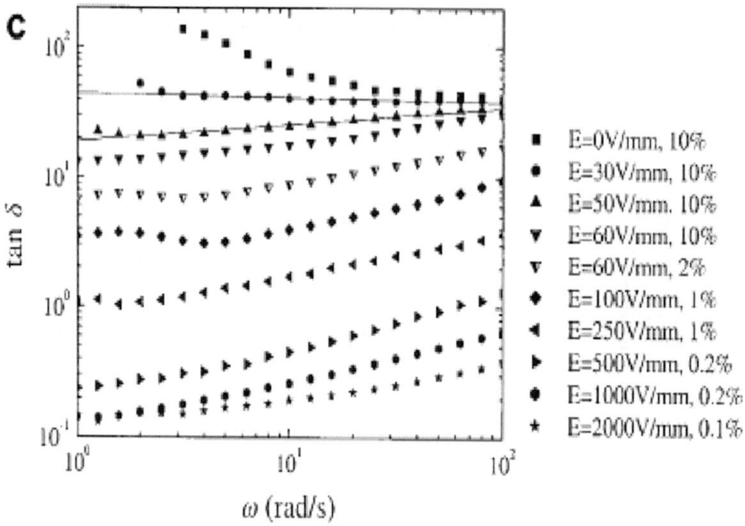

Figure 37 Frequency dependence of the mechanical loss tangent for 1 μm silica/PDMS suspension with the particle volume fraction 10.7 vol%. The strain amplitudes at each field strength are 10% for E≤ 60 V/mm, 1% for E=100 and 250 V/mm, 0.2% for E=500, and 1000 V/mm, and 0.1% for E=2000 V/mm. The change of the strain amplitude is for keeping the measurement in the linear region. Reproduced with permission from B. D. Chin and H. H. Winter, Rheol. Acta, 41(2002)265.

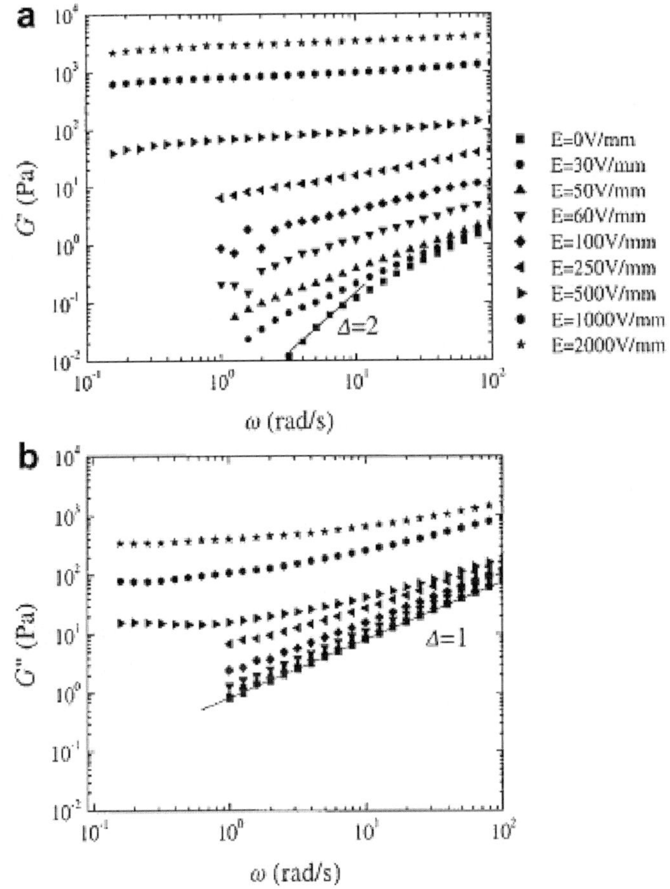

Figure 38 Frequency dependence of storage modulus G′ and loss modulus G″ for 1 μm silica/PDMS suspension with the particle volume fraction 10.7 vol%. The strain amplitudes at each field strength are 10% for E≤ 60 V/mm, 1% for E=100 and 250 V/mm, 0.2% for E=500, and 1000 V/mm, and 0.1% for E=2000 V/mm. The change of the strain amplitude is for keeping the measurement in the linear region. Reproduced with permission from B. D. Chin and H. H. Winter, Rheol. Acta, 41(2002)265.

The frequency dependence of storage modulus G′ and loss modulus G″ for 1 μm silica/PDMS suspension with the particle volume fraction 10.7 vol% is shown in Figure 38. Without an electric field the loss modulus shows the typical power law behavior with $G'' \propto \omega$, and the storage modulus approaches $G' \propto \omega^2$ in the terminal zone. Deviation of the power-law dependence for both G′ and G″ under an electric field may be attributed to the increase of the interparticle force with the applied electric field. Note that the power-law region is restricted to the frequency ranging from 10^0 to 10^2 rad/s for this system. Power-law dependence of G′ and G″ on frequency is an indication that the relaxation times associated with the self-similar structure in the system reside within this frequency range. Power-law relaxation in a limited frequency range may suggest non-uniform structural fluctuations [91, 92]. Since the network (or fibrillated) structure of an ER suspension under an electric field is non-ideal self-similar structure and the degree of the order of this structure is progressively increasing with the applied electric field, both the power and the frequency range where the power-law is valid are changing with the applied electric field, indicating there is a structure evolution process occurring in ER suspensions.

A possible mechanism of the structure evolution of an ER suspension under an electric field was proposed with the aid of percolation theory [62, 63]. Without an electric field, the particles are randomly distributed in the medium; With low electric fields the isolated particle clusters that stay in between two electrodes and without any contact with two electrodes will form. With the increase of the electric field, the branched clusters of one end attached to the electrode will form, and finally the continuous percolation clusters of both ends attached to the electrodes will form. The structure evolution under an electric field will determine how the rheological properties change with the electric field. Figure 39 schematically shows the structure evolution with the increase of electric field and the possible rheological property. The zero shear viscosity, yield stress, and the real and imaginary moduli are included in the illustration. The zero-shear viscosity is almost flat when the applied electric field is low. With the increase of the electric field, the zero-shear viscosity gradually increases. At the vicinity of the liquid-solid gellation point, the zero-shear viscosity dramatically increases to a value that is several orders of magnitude higher than the original one at low electric fields. For the storage and loss moduli they increase with the applied electric field, crossover each other around yielding point, and continuously increase with the increase of the electric field. Note that at low electric fields the loss modulus is higher than the storage modulus, and above certain electric field, the storage modulus becomes higher than the loss modulus, implying that the ER behaves a solid-like material once the applied electric field is strong enough. Similarly the yield point of the ER suspension occurs around the crossover point of the storage and loss moduli, and the yield stress is the shear stress at the electric field which the crossover occurs. The yield stress increases with the electric field, as the interaction force between the chained-particles will be strengthened before it reaches the saturation region.

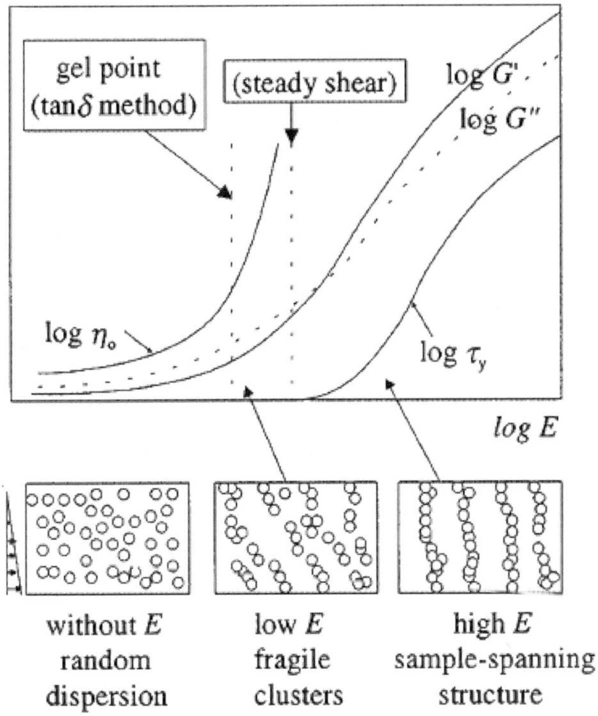

Figure 39 Schematic of structure evolution during the field-induced liquid-to-solid transition for an ER suspension. The gel point of a static sample occurs at low field strengths, as determined with tgδ method. Under a steady shear, the transition shifts to high field strengths, as determined with the yield stress method. Reproduced with permission from B. D. Chin and H. H. Winter, Rheol. Acta, 41(2002)265.

In qualitatively summarizing the flow behaviors of ER fluids, Parthasarathy [86] mapped out the frequency and strain dependence of the rheological properties of ER suspensions in the form of a Pipkin diagram [93], which is shown in Figure 40. The flow regimes are plotted as a function of the strain amplitude γ_0, and the dimensionless frequency,

$\varpi\left(\propto \omega / E^2\right)$. At very small strain amplitudes the ER suspension shows linear viscoelasticity, which may be independent of the moderate dimensionless frequency. The transition from the linear to nonlinear viscoelasticity occurs at the first critical strain amplitude, γ_1^{crit} , the curve a in Figure 40. Since the structural rearrangements are postponed to larger strain amplitudes as the frequency increases [94], the curve a should be the sigmoidal shape instead of a straight line. With the further increase of strain amplitude a transition from nonlinear viscoelasticity to the viscoplasticity occurs at the second critical strain amplitude, γ_2^{crit}, the curve b in Figure 40, where the elasticity disappears in the dynamic response. For sufficient small frequencies the deformation is quasi-static [95], and it depends only on the strain amplitude. At relatively high frequencies the hydrodynamic forces also degrade the structure, resulting in the plastic deformation at a smaller strain amplitude. Thus γ_2^{crit} decreases slightly with the increase of frequency. In the viscoplastic regime, the rheological response can be represented with models of a dynamic yield stress, such as the Bingham model. At a very high strain amplitudes and frequencies condition, the viscous contribution weighs up and the yield stress contribution becomes negligible. The ER suspensions behave like a Newtonian fluid. The curve c in Figure 40 represents the transition from the viscoplasticity to the Newtonian behavior. The viscoelasticity is essentially Newtonian when G″>>G′ at very high frequencies, which corresponds to the curve d in Figure 40. Note that the boundaries between the flow regimes shown in Figure 40 are qualitative and approximate, and the Figure 40 is only schematically illustrative.

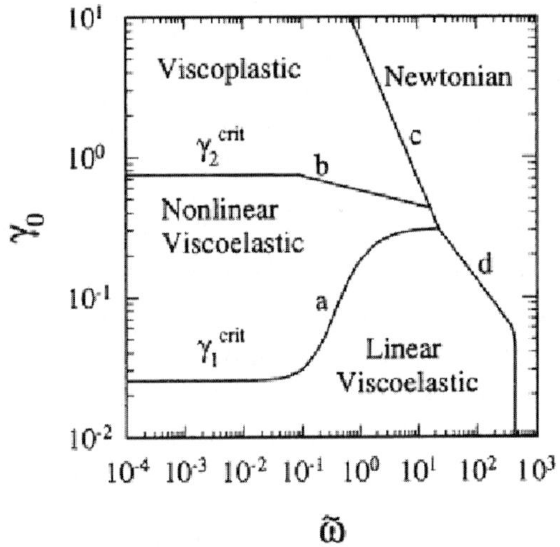

Figure 40 The Pipkin diagram of the dynamic rheological behavior of ER fluids. Reproduced with permission from M. Parthasarathy and D.J. Klingenberg, J. Non-Newtonian Fluid Mech., 81 (1999) 83.

4.2.3 Simulation results

There are huge efforts spent on directly calculating or simulating the rheological behaviors of ER suspensions on the basis of phenomenological and microstructure models. The phenomenological models are used to best fit the experimental data by employing the viscous, friction, and elastic elements in different combinations [96-98], as mentioned earlier. They are useful for engineering design purposes, however, they do not provide a fundamental understanding on why ER suspensions behave in such a way. The microstructure models take into account the particle polarization, aggregation, and fibrillation in an electric field, and calculate the interparticle forces based on dipole-dipole interaction and then extend the interaction forces to the rheological property [86, 94, 95, 99-104]. The microstructure models used for rheological calculation will be described in a future chapter under the category of the polarization models that were proposed for explaining the ER effect. In this section. only the rheological properties derived from those models are briefly addressed.

For a small amplitude oscillatory shear in which the ER suspensions are in the linear response region, the rheological behavior was simulated on the basis of the point-dipole approximation [100, 101]. With the increase of frequency the loss modulus increases first with the frequency and then decreases with the further increase of frequency after passing through a maximum. Figure 41 shows the three dimensionally simulated storage and loss moduli vs. the dimensionless frequency $\omega^*(\propto \omega / E^2)$ for 25 spheres, based on the point-dipole approximation for the electrostatic force and Stokes' drag for the hydrodynamic force. The dimensionless loss modulus $G^{''*}(\propto G''/E^2)$ scales with ω^* for small ω^* and $(\omega^*)^{-1}$ for large ω^*, passing through a maximum. The dimensionless modulus $G^{'*}(\propto G'/E^2)$ increases from a small ω^* plateau to a large ω^* plateau, implying that a transition from non-affine to affine deformation of percolating clusters occurs with the increase of frequency. Clearly the relaxation results from the frequency dependence of the microstructure of ER fluids.

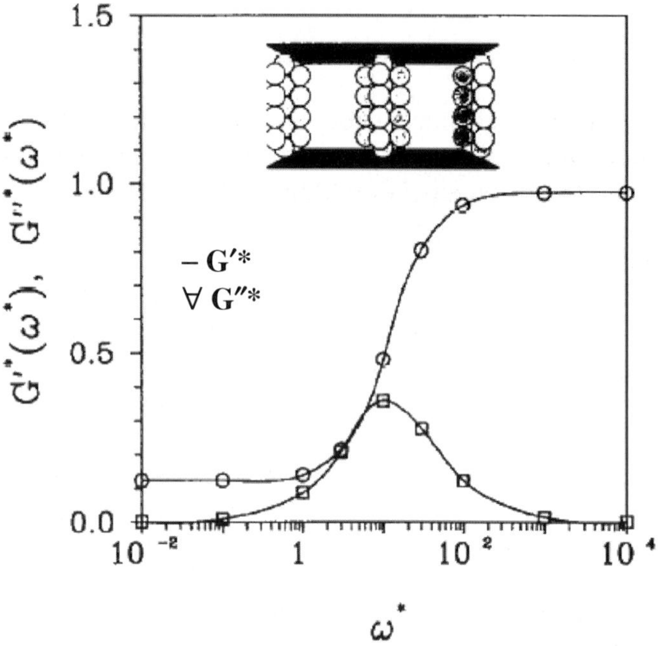

Figure 41 The three dimensionally simulated storage and loss moduli vs. the dimensionless frequency $\omega^* \propto \omega / E^2$ for 25 spheres. The three dimensional illustration contains periodic images of some spheres. The solid lines are cubic spline interpolations between the simulation data points. Reproduced with permission from D.J. Klingenberg, J. Rheol. 37 (1993)199.

For a large amplitude oscillatory shear (LAOS), the rheological property of the ER suspension is investigated by using the idealized polarization model in the particle level dynamic simulation [86,104], which is quite similar to the method used for the small amplitude oscillatory shear simulation. The LAOS behavior of ER suspensions are important as the ER devices usually operate in a dynamic mode with large deformation. Generally speaking, the LAOS behavior of complex fluids is very complicated due to their microstructures strongly depending on deformation history. According to Hyun [105,106]. the LAOS behavior of complex fluids can be classified at least into four types, which are schematically shown in Figure 42. The type I and II are strain thinning and strain hardening, respectively, and the type III and IV are weak strain overshoot (G′ decreasing and G″ overshooting) and strong strain overshoot (both G′ and G″ overshooting), respectively. The strain shinning is similar to the shear thinning, which is attributed to the chain orientation or alignment to the flow direction. The strain hardening and strain overshoot have a different physical origin from the strain thinning. The formation of network structure junctions among the chains is believed to contribute to the strain hardening and strain overshoot phenomena [106]. The overshoot of G″ in the type III may result from the balance between the formation and the destruction of the network junctions; while the type IV may result from the interaction between chain segments or strong junction formation in the network structure.

As indicated earlier the ER particle may form a network structure instead of the fibrillated chain structure once the particle volume fraction exceeds the critical volume fraction [62,83], due to the percolation transition. Under a large amplitude shear field the formation and destruction of network junctions may happen one after the other, and thus the type III LAOS behavior may best describe ER suspensions. Figure 43 shows the simulated storage and loss moduli vs. strain amplitude at different frequencies, using an idealized electrostatic polarization model of ER fluids that was implemented in the particle-level dynamics simulation. The storage modulus G′ and loss modulus G″ remain constant up to a certain strain amplitude ($\gamma_0 \approx 0.4$), which defines the linear response region. With further increase of the strain amplitude above 0.4, G′ decreases with γ_o^{-2} irrespective of frequency at large amplitudes. The loss modulus shows an overshoot at low frequencies and large amplitudes, and the overshoot disappears at large frequencies and amplitudes. This may indicate that the destruction of network junctions becomes dominant at large frequencies and amplitudes condition and the ER suspension simply shows the strain shinning behavior. Clearly, the LAOS behavior of ER suspensions falls into the type III in deed, which is also in consistent with the experimental and simulation results presented in ref. [86]

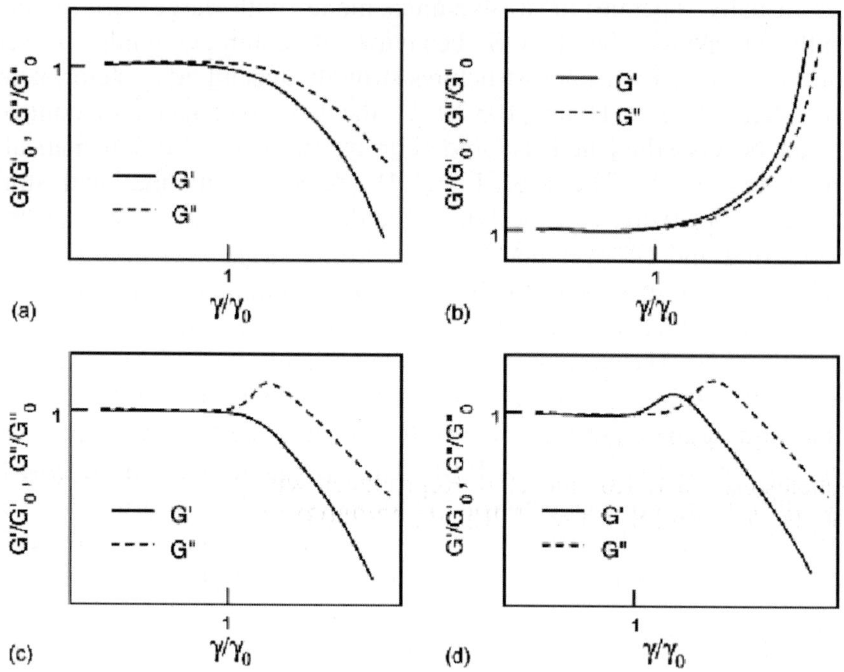

Figure 42 Schematic diagram of four types of LAOS behavior: (a) type I, strain thinning; (b) type II, strain hardening; (c) type III, weak strain overshoot; and (d) type IV, strong strain overshoot. Reproduced from Kyu Hyun, S. Kim, K.H. Ahn, S.J. Lee, J. Non-Newtonian Fluid Mech. 107(2002)51

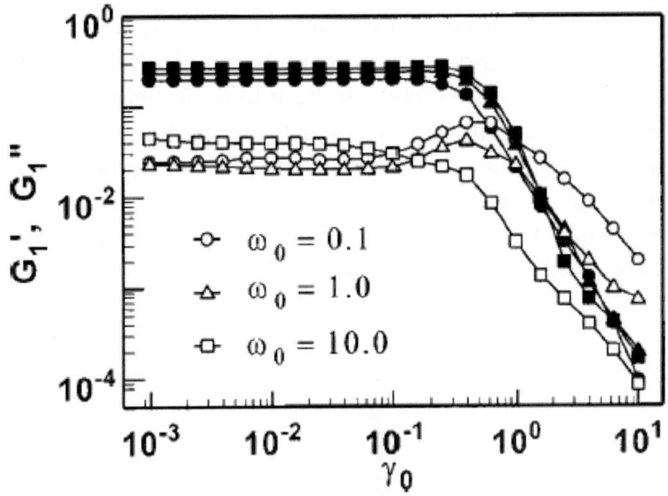

Figure 43 Fundamental storage modulus G_1' (filled symbols) and loss modulus G_1'' (open symbols) as a function of strain amplitude γ_0 at frequencies $\omega_0 = 0.1$, 1.0, and 10.0. Reproduced with permission from H. G. Sim, K. H. Ahn, and S. J. Lee, J. Rheol., 47(2003)879.

4.3 Transient shear

Transient shear test may shed light on how an ER suspension responds to a suddenly applied shear field. The investigation on the transient shear stress of the oxidized polyacrylonitrile/silicone oil suspension shows that at shear rate 1 s^{-1} the shear stress builds up rapidly to a constant value, and the steady shear stress value increases with the applied electric field strength. There is no stress overshoot observed in such an ER system [107]. However, this is really dependent on how strong the interparticle force in the ER suspension is. Figure 44 shows the transient shear stress behavior at the shear rate of 0.1 s^{-1} for 10 wt.% polypyrrole (PPy) coated polyethylene (PE) particle doped with 1.5 g and 0.75g $FeCl_3 \cdot 6H_2O$, respectively, in mineral oil. There is no overshoot observed in the PPy-PE/mineral oil suspension doped with 0.75 g $FeCl_3 \cdot 6H_2O$, while there is a strong shear stress overshoot in the PPy-PE/mineral oil suspension doped with 1.5 g $FeCl_3 \cdot 6H_2O$. The shear stress of the PPy-PE/mineral oil suspension doped with 1.5 g $FeCl_3 \cdot 6H_2O$ is almost ten times higher than that of the one doped with 0.75 g $FeCl_3 \cdot 6H_2O$, indicating that the interparticle force in the former suspension is much stronger than that in the latter. As one may imagine, the particles form the chain structure (or the network structure in high particle concentrations) under an electric field (Figure 44a). The shear field will break the chains at the middle in between two electrodes, resulting in the chain rupture and reformation simultaneously (Figure 44b). The steady shear stress may correspond to the chain rupture and reformation microstructure. The shear stress overshoot may imply that the shear stress corresponding to the chain structure without rupture is higher than the steady shear stress. A yielding point may exist for this suspension.

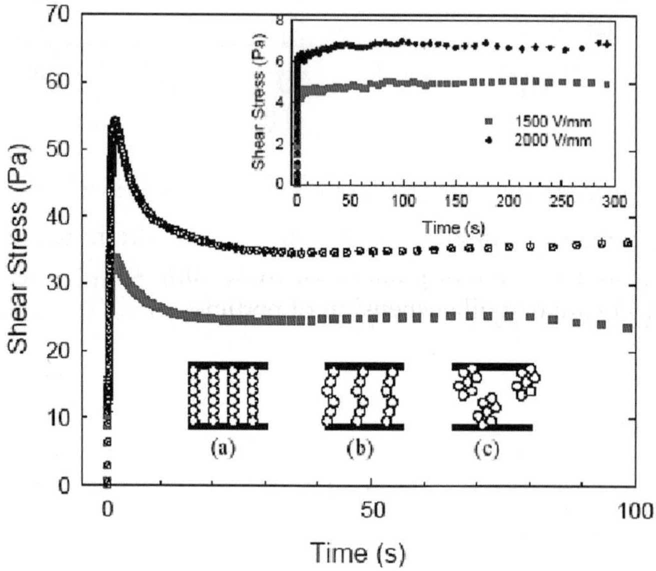

Figure 44. Transient shear stress behavior at the shear rate of 0.1 s^{-1} for 10 wt.% polypyrrole (PPy) coated polyethylene (PE) particle doped with 1.5 g $FeCl_3 \cdot 6H_2O$ dispersed in mineral oil. The inset is the same material doped with 0.75 g $FeCl_3 \cdot 6H_2O$. !, E=1.5 kV/mm; E=2.0 kV/mm. Reproduced with permission from Y. D. Kim, D. H. Park, Synthetic Metals 142 (2004) 147.

In contrast to the sudden flow, the sudden stop of the flow may provide the information on how soon the chain structure will rebuild up without a mechanical disturbation. Figure 45 shows the shear stress recovery phenomenon after the shear stops; the shear rate changes from 1 to 0 s^{-1}

under an electric field 2.5 kV/mm within 5 seconds, and 20 seconds, respectively, observed in the oxidized polyacrylonitrile/silicone oil suspension. When the shear rate suddenly changes from 1 to 0 s^{-1}, the shear stress quickly decays to a very low value within a second; and then gradually recovers to an even higher value than the steady shear stress. At the constant shear condition as shown in Figure 18, a ruptured half chain moves with the upper electrode to next equilibrium position and merges with another half chain that stayed with the bottom electrode, forming a single chain to span between two electrodes. The sudden stop of shearing prevents the re-formation of such a single chain, this leading to the sudden drop of the shear stress. Once the shearing fully stops, the single chains immediately build up between two electrodes without a mechanical disturbation, generating a high shear stress. The shear stress recovering to a higher value after the stop of shearing again indicates that the shear stress corresponding to the chain structure without rupture is greater than the shear stress corresponding to the chain rupture and reformation microstructure.

A shear rate sweep test may provide additional information on the microstructure evolution under a shear field. Figure 46 shows the shear rate sweep from 0 to 20 s^{-1} initially and then from 20 to 0 s^{-1} for the oxidized polyacrylonitrile/silicone oil suspension of particle weight fraction 19.2 wt% under zero electric field and 2.5 kV/mm. Under E=2.5 kV/mm, a hysteresis loop is observed and the loop area at low shear rates is larger than that at high shear rates. Without an electric field a small hysteresis loop is only observed at low shear rates and there is no hysteresis loop at high shear rates. When the shear rate changes from high to low, the ER suspension is always in the status of the chain rupture and reformation, so a steady shear stress is obtained. When the shear rate changes from low to high, the microstructure of the suspension changes from the fibrillated chain structure to the chain rupture and reformation structure, thus a slowly decreased shear stress is observed with the increase of the shear rate. Without an electric field, there is no pronounced structural difference between the suspension under a shear from the low to high shear rate or from the high to low shear rate. The hysteresis loop can only be observed in the low shear rates.

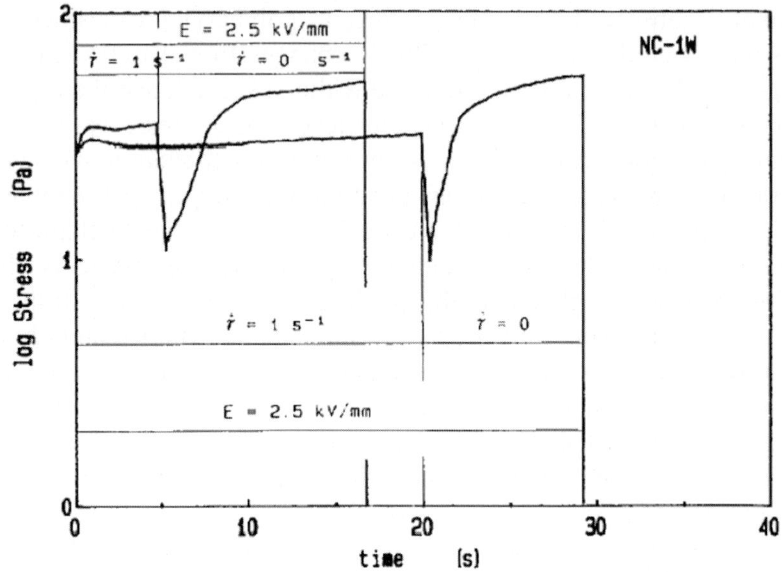

Figure 45 Shear stress recovery phenomenon after the shear stops, the shear rate changes from 1 to 0 s^{-1} under an electric field 2.5 kV/mm within 5 seconds, and 20 seconds, respectively. The ER suspension is the oxidized polyacrylonitrile/silicone oil suspension of particle volume fraction 19.2 wt%. Reproduced with permission from Y. Xu, and R. Liang, J. Rheol., 35(1991)1355.

The transient rheological property of ER suspensions suggest that due to the unique microstructure evolution of the ER suspensions under an electric field, the rheological parameters are more sensitive to the measuring condition in comparison with non-ER suspensions. For the purpose of comparing the ER effect of two ER suspensions, the transient and strain sweep experiments are necessary for determining the stead shear stress condition and the linear response region. The response time is another important parameter for evaluating how quickly the ER suspension may respond to an electric or mechanical stimulation.

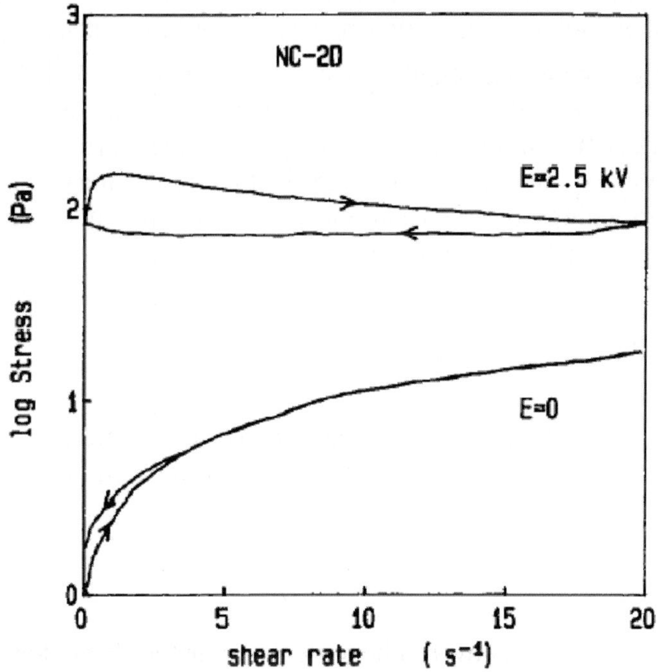

Figure 46 Hysteresis loop of the oxidized polyacrylonitrile/silicone oil suspension of particle weight fraction 19.2 wt% under zero electric field and 2.5 kV/mm. The cycle time is 40 s. Reproduced with permission from Y. Xu, and R. Liang, J. Rheol., 35(1991)1355.

4.4 Structure determination using scattering technology

The microstructure of ER fluids under the quiescent, steady and oscillatory shear fields was determined using the two-dimensional light scattering [109] and the small-angle neutron scattering [110] techniques. For a dispersion of monodispersed spherical particles of the radius a and containing n particles per unit volume, the normalized scattering intensity can be expressed as [110-112]:

$$I(q) = k(\rho_p - \rho_m)^2 \left(\frac{4}{3}\pi a^3\right)^2 nP(q)S(q) \tag{64}$$

where k is an instrumental constant related to the intensity of incident beam and the distance between the scattering sample and the detector. ρ_p and ρ_m are the average scattering length density of the particle and the medium, respectively. P(q) is the form factor (or intraparticle structure factor) describing the distribution of scattering within the particle; S(q) is the structure factor (or interparticle structure factor) describing the interference effects of correlations between particle positions. The q is the scattering wave vector and can be expressed as:

$$q = \left|\vec{Q}\right| = \frac{4\pi}{\lambda}\sin(\theta/2) \tag{65}$$

where λ is the light wavelength, and θ is the scattering angle. The characteristic length scale, L(t), of the structure domain can be obtained from the curve of I(q) vs. q:

$$L(t) = 2\pi / q_{\max} \tag{66}$$

where the q_{max} is the corresponded value of the intensity peak, I_{max}. Figure 47 shows time dependence of the characteristic length L(t) calculated from Eq. (66) for silica/4-methylcyclohexanol suspension with particle weight fraction 11 wt% at various electric fields. The kinetics of the structure coarsening can be well expressed by using the power law:

$$L(t) = L(0)\left[1 + (t/\tau)^{2/5}\right] \tag{67}$$

At earliest times the original characteristic length is about 1.9 mm. After that the length scale increase as $L(t) \sim t^{2/5}$ at all electric fields, but the growth rate increases with the applied field. Note that the original characteristic length scale L(0) also increases with the applied electric field. The electric field may induce different lengthening chains and the chains aggregate into columns at a different rate. In a shear field chains are expected to orientated with the shear field.

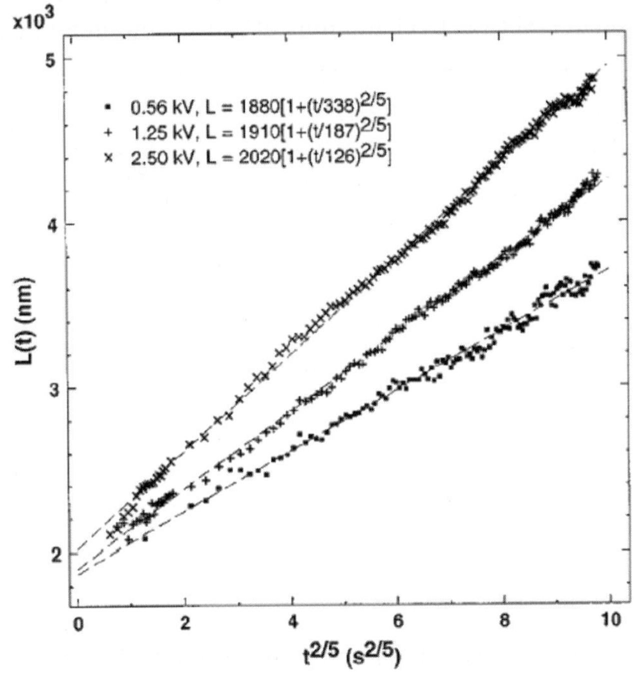

Figure 47 Time dependence of the characteristic length L(t), $L(t) = 2\pi / q_{\max}$, for silica/4-methylcyclohexanol suspension with particle weight fraction 11 wt%. The applied electric field is 0.56, 1.25, and 2.50 kV across 0.72 mm electrode gap. Reproduced with permission from J.E. Martin, J. Odinek, T.C. Halsey, and R. Kamien, Phys. Rev. E, 57(1998)756.

The orientation angle can be obtained from the Gaussian equation by fitting the scattering data:

$$I(\theta) = I(0)e^{-(\theta-\theta_{\max})^2 / 2d_{1/2}^2} \qquad (68)$$

where $d_{1/2}$ is the scattering half-width. $\theta_{\max}$ is the chain orientation angle, which is plotted against the cube root of shear rate in Figure 48. A good linearity is obtained between the orientation angle of chains and the cube root of the shear rate. Since the shear field is usually perpendicular to the electric field, Figure 48 may tell us that with the increase of shear rate the chains may orientate further far away from the direction of the electric field , leading to the weakened ER effect and thus the shear thinning phenomenon.

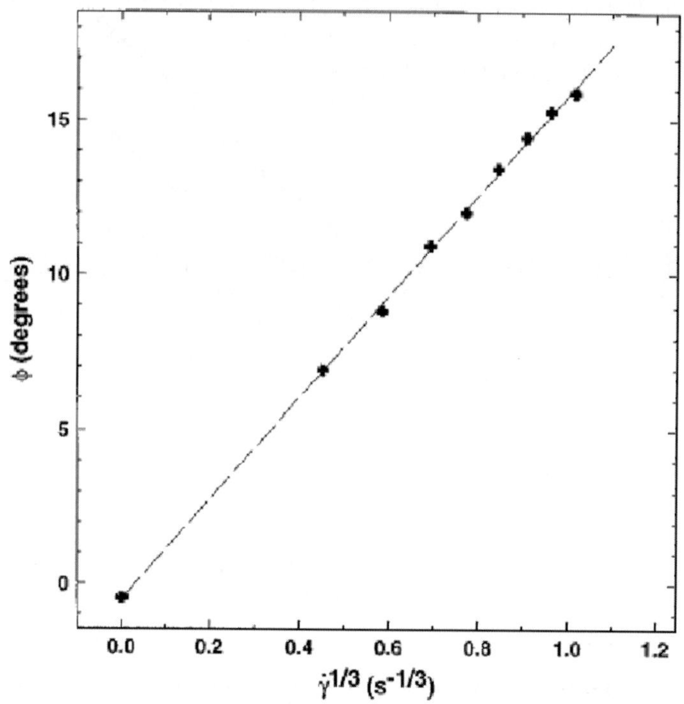

Figure 48 The orientation angle of chain or droplet vs. the cube root of the shear rate for silica/4-methylcyclohexanol suspension with particle weight fraction 7.5 wt%. The applied voltage is 1.2 kV at 400 Hz across the 1.0 mm gap. Reproduced with permission from J.E. Martin, J. Odinek, T.C. Halsey, and R. Kamien, Phys. Rev. E, 57(1998)756.

The small angle neutron scattering (SANS) studies on the silica/silicone oil at particle volume fraction ϕ=0.055 [110] revealed that this ER suspension can be well modeled using either the sticky hard sphere (SHS) model [113, 114] with a sticky parameter 0.4 or a fractal model with fractal dimension d_f=1.6 [110, 115]. Figure 49 shows the small angle neutron scattering (SANS) scattering intensity vs. the scattering wave vector q for the silica/silicone oil at particle volume fraction ϕ=0.055, under an electric field strength of 1kV/mm. At low scattering wave vectors both two models agree with data very well. The discrepancies at high scattering wave vector may arise from the intrinsic polydispersity of the silica particles. Using the fractal dimension obtained from the SANS measurement, the shear stress of this silica/silicone oil suspension with different water contents at 1.0 kV/mm can be well described with Eq. (69):

$$\tau = \eta_{\upsilon}^{-\upsilon} \eta_m \dot{\gamma} \tag{69}$$

where the parameter υ is described in term of fractal dimension as [116]:

$$\upsilon = \frac{6 - 2d_f + 2d_f/3}{3 - d_f} \tag{70}$$

η_{υ} is a function of the parameter υ

$$\eta_{\upsilon} = \eta_r^{-1/\upsilon} \tag{71}$$

η_r is the relative viscosity of the suspension to the liquid medium. For aggregated dispersions η_r can be correlated with the dimensionless shear rate G as [117]:

$$\eta_{\upsilon}^{\nu} + 2K\eta_{\upsilon} - 1 = 0 \tag{72}$$

where

$$\nu = \frac{\upsilon}{2.5\phi_m}$$

$$K = \frac{\phi}{2\phi_m}\left(\frac{2}{5G}\right)^{1/\upsilon} \tag{73}$$

ϕ_m is the maximum packing fraction and ϕ_m=0.64 for random dense packing. The dimensionless shear rate G can be expressed as:

$$G = \frac{\pi a^2 \eta_m \dot{\gamma}}{F_a} \tag{74}$$

F_a is the attractive interparticle forces. Eq. (72) may be solved numerically in terms of η_υ, and thus the shear stress can be plotted against G, which is shown in Figure 50. The calculations seem to well describe the data, though at low shear rates the theoretical values are slightly lower.

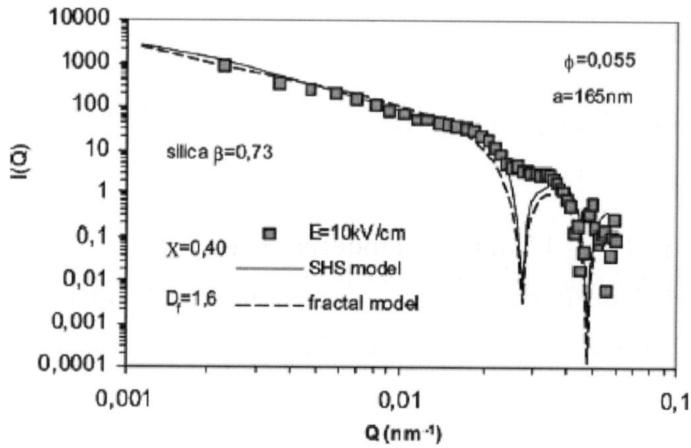

Figure 49 The small angle neutron scattering (SANS) scattering intensity vs. the scattering wave vector q for the silica/silicone oil at particle volume fraction ϕ=0.055, under an electric field strength of 1kV/mm. Two model curves are presented: sticky hard-sphere (SHS) and fractal model (a = 165 nm, χ = 0.40, d_f = 1.6). Reproduced with permission from C. Gehin, J. Persello, D. Charraut, and B. Cabane, J. Colloid Interf. Sci., 273 (2004) 658

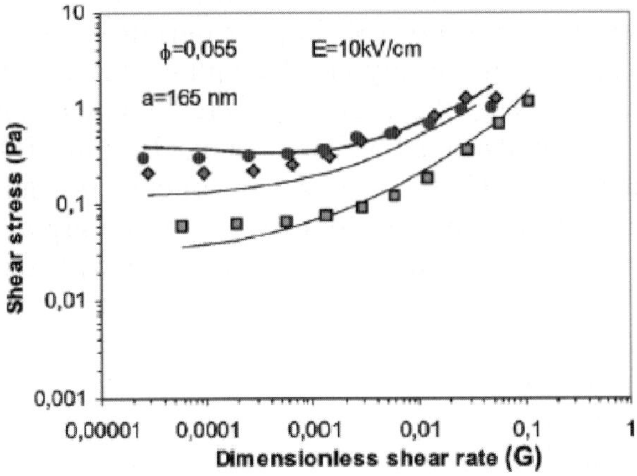

Figure 50 The shear stress vs. dimensionless shear rate for silica/silicone oil suspension with different water contents, • water volume fraction in silica is 1.3 vol%; ♦, water volume fraction in silica is 21.4 vol%; ■ water volume fraction in silica is 28 vol%, at particle volume fraction 5.5 vol% and electric field E=1 kV/mm. Reproduced with permission from C. Gehin, J. Persello, D. Charraut, and B. Cabane, J. Colloid Interf. Sci., 273 (2004) 658

5. CONDUCTIVITY MECHANISM

The dependence of the ER effect on the dispersed particle conductivity has been comprehensively investigated. As shown earlier, the strongest ER effect was found to occur at the conductivity about 10^{-7} S/m for the polyacenequinones/cereclor suspension [118], and there was no detectable ER activity observed once the conductivity shifts far away from this value. A similar result was obtained in oxidized polyacrylonitrile/silicone oil suspension and interpreted in view of Maxwell–Wagner polarization [119, 120]. The effect of the particle conductivity on the ER activity was also theoretically analyzed [121-123]. A conduction model was presented for understanding ER phenomena and ER mechanism [124]. The current density of an ER fluid is obviously a very important parameter that scales energy consumption in practical devices. Both the current density and ER activity are controlled by the conductivities of the dispersing and dispersed phases. The conductivity mechanism of ER fluids is thus of great significance both for controlling the current density and for optimizing the ER performance. Two conductivity models, the charging energy limited tunneling (CELT) and the quasi one-dimensional variable range hopping (Quasi-1d-VRH), were employed by Hao [125] to analyze the conductivity data for clarifying the conductivity mechanism. The conductivity data were also qualitatively interpreted with the percolation theory. Those results will be introduced later in this section.

5.1 Localization models

It is known that, in heterogeneous systems, electrons hop from one site to another mostly in two ways: crossing the energy barrier using the tunneling effect and hopping by heat excitation. The tunneling conductive mechanism dominates at low temperatures, whereas heat excited electron hopping occurs at comparatively high temperatures. Although ER fluids often operate at high temperatures where the tunneling mechanism does not work, this model is introduced as well since both models predict a very similar temperature dependence of conductivity.

5.1.1 Charging Energy Limited Tunneling (CELT)

The tunneling conductive mechanism model [126] assumed that an electron can be excited by an applied external electric field from the ground state to the excited state, and then crosses the energy barrier to the unoccupied neighboring molecular orbit; i.e., the minimum energy E_m must be much larger than the heat energy k_{BT}. For this reason, this conductive model is often called the charging energy limited tunneling model (CELT). The CELT model was originally proposed for application to systems of granular metal particles (diameters around 0.01 μm) randomly dispersed in an insulating oil [127], which is similar to ER suspensions. This model predicts that:

$$\sigma = \sigma_0 \exp\left[-(T_0 / T)^{1/2}\right] \tag{75}$$

for modest electric fields and

$$\sigma = \sigma_0^{'} \exp\left[-(E_0 / E)^{1/2}\right] \tag{76}$$

in the high field regime. Here T_0, E_0, σ_0, and σ_0' only weakly depend on temperature T and electric field strength E, and can be considered as constants, σ is the conductivity of system. The most probable separation between the center of two metallic grains d_p is given by [127]

$$d_p = k_B T_0 / 4eE_0 \tag{77}$$

or

$$d_p = k_B T / 4eE_c \tag{78}$$

where E_c is the critical (threshold) electric field above which σ increases rapidly according to Eq. (76) (high field regimes) and k_B is the Boltzmann constant, e is the electron charge. Eq. (76) could be rewritten as

$$\ln\sigma = \ln\sigma_0 - E_0 / E \propto -E_0 / E \tag{79}$$

Eq. (79) tells us that the natural logarithm conductivity of a system will decrease linearly with ($1/E$) under a given temperature.

5.1.2. Quasi-One-Dimensional Variable Range Hopping (Quasi-1d-VRH Model)

Once the energy barrier between two molecules is too wide (for particles, that is the distance between the two particle surfaces) , electrons can move by the hopping mechanism rather than by the tunneling mechanism. The electron hopping process from the localized state below the Fermi energy to the unoccupied state over the Fermi Energy is determined by the temperature and energy barrier between the beginning state and the terminal state [128]. The most probable hopping range is exponentially correlated to temperature, so this model is called the variable range hopping model. The Quasi-1d-VRH model [129] assumes that electrons hop only in a one dimensional chain and cannot hop in another direction; thus a temperature dependence of conductivity similar to that CELT predicted [128, 129] is given by

$$\sigma = \frac{\sigma_0}{T}\exp\left[-\left(\frac{T_0}{T}\right)^{1/2}\right] \tag{80}$$

Based on the ref. [128,129]

$$T_0 = \frac{8^3 \alpha^3}{9\pi k_B N(E_F)} \tag{81}$$

where $N(E_F)$ is the density of states at Fermi energy E_F, k_B is the Boltzmann constant, and α^{-1} is the decay length of the localized state. For the case of strong localization α can be expressed as [128]

$$\alpha \approx \left(\frac{8\pi^2 m E_b}{h^2}\right)^{1/2} \tag{82}$$

where E_b is the effective barrier height between localized sites, m is the effective mass of charge carrier, and h is the Planck constant. Consider a model in which electrons are trapped and localized in an effective potential $U(d)$ centered at d=0. With the application of electric field E, U is changed as [130]

$$U(d) = -\left(\frac{e^2}{4\pi\varepsilon_0 \varepsilon d}\right) - edE \tag{83}$$

where e is the electron charge, and ε_0 and ε are the dielectric constants of vacuum and material, respectively. The maximum barrier height $E_b = U_{max}$ between adjacent sites will be lowered by

$$\delta E_b = \delta U_{max} = \left(\frac{e^3}{\pi\varepsilon_0 \varepsilon}\right)^{1/2} E^{1/2} \tag{84}$$

Expanding E_b, substituting Eq. (82), and using Eq. (80) leads to the following expression based on the ref. [130]

$$\ln\sigma \propto -\left[\frac{T_0(E)}{T}\right]^{1/2} \propto \left(\frac{1}{4E_b}\right)\left(\frac{T_0}{T}\right)^{1/2}\left(\frac{e^3}{\pi\varepsilon_0\varepsilon}\right)^{1/2} E^{1/2} = KE^{1/2} \tag{85}$$

where

$$K = T_0^{1/2}\left[\left(\frac{e^3}{\pi\varepsilon_0\varepsilon}\right)^{1/2}\frac{T^{-1/2}}{4E_b}\right] \qquad (86)$$

$$= K'T^{-1/2}$$

and

$$K' = \frac{T_0^{1/2}}{4E_b}\left(\frac{e^3}{\pi\varepsilon_0\varepsilon}\right)^{1/2} \qquad (87)$$

K' is a constant for a given system.

5.3 Conductivity under a zero mechanical field

The conductivity behaviors of a water-free ER suspension containing oxidized polyacrylonitrile (OP) particles of the conductivity about 10^{-7} S/m were examined by Hao [125] with and without an oscillatory mechanical field at a relatively high electric field. The dispersing medium was silicone oil of the conductivity and the viscosity at room temperature 10^{-14} S/m and 50 cP, respectively. The conductivity of this suspension was converted from the dc current passing through the suspension. The dc current was found to decay with time and become stable within several-minutes, which is usually called dielectric absorption phenomenon. The time dependence of dc current under the electric field 2.0 kV/mm is shown in Figure 51. The dielectric absorption phenomenon results from the heterogeneity in the system, where the slow polarization such as the Maxwell-Wagner polarization is generated in an electric field [119]. For this particular ER suspension the current decay could not be observed until the applied electric field was above 0.6 kV/mm.

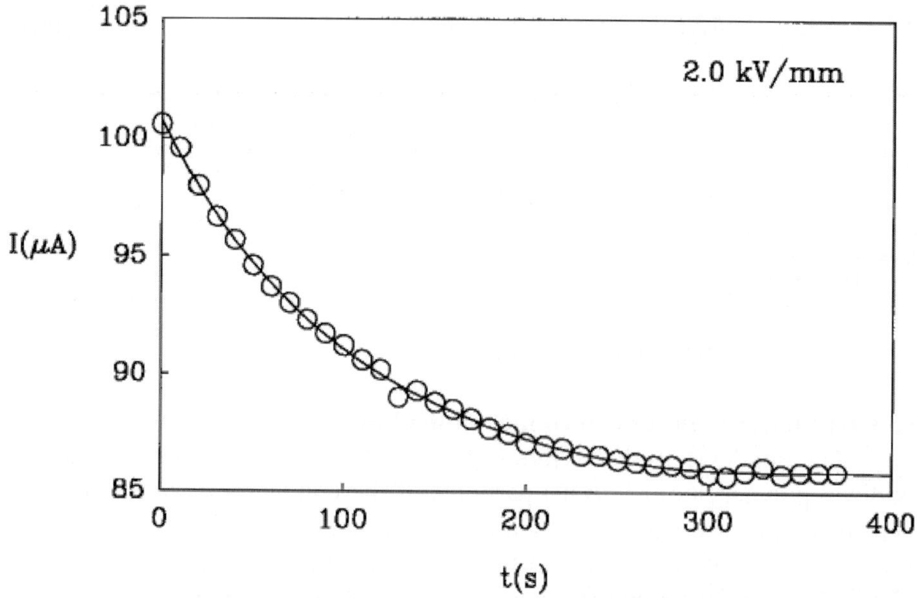

Figure 51 The dc current against time obtained at 2.0 kV/mm for oxidized polyacrylonitrile/silicone oil suspension. Particle volume fraction is 27 vol%. Reproduced with permission from T. Hao, and Y. Xu, J. Colloid Interf. Sci., 181(1996)581

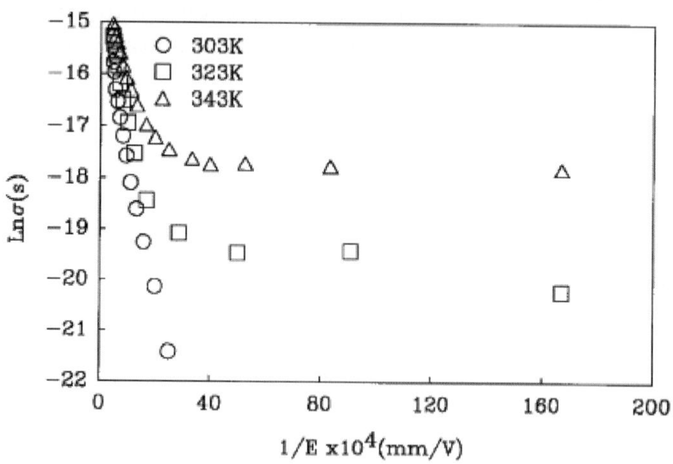

Figure 52 The conductivity σ(E) (S^{-1}) of the oxidized polyacrylonitrile/silicone oil suspension vs 1/E at three different temperatures. The particle volume fraction is 27 vol%. The maximum

electric field is 3 kV/mm. Reproduced with permission from T. Hao, and Y. Xu, J. Colloid and Interf. Sci., 181(1996)581

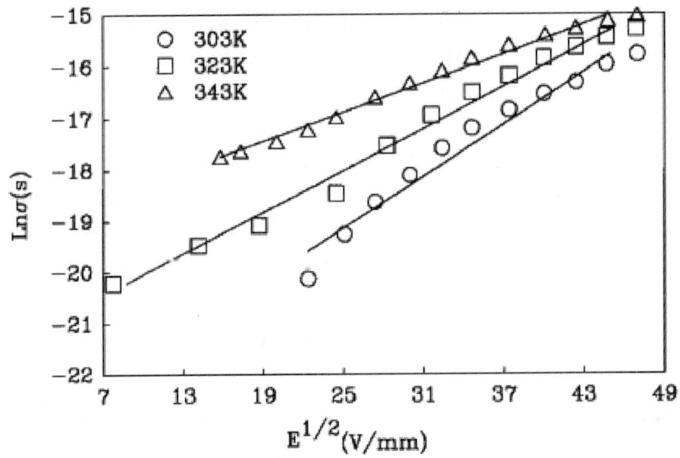

Figure 53 The conductivity σ(E) (S^{-1}) of the oxidized polyacrylonitrile/silicone oil suspension vs $E^{1/2}$ at three different temperatures. The data are same as Figure 52. The particle volume fraction is 27 vol%. The maximum electric field is 3 kV/mm. Reproduced with permission from T. Hao, and Y.Xu, J. Colloid Interf. Sci., 181(1996)581

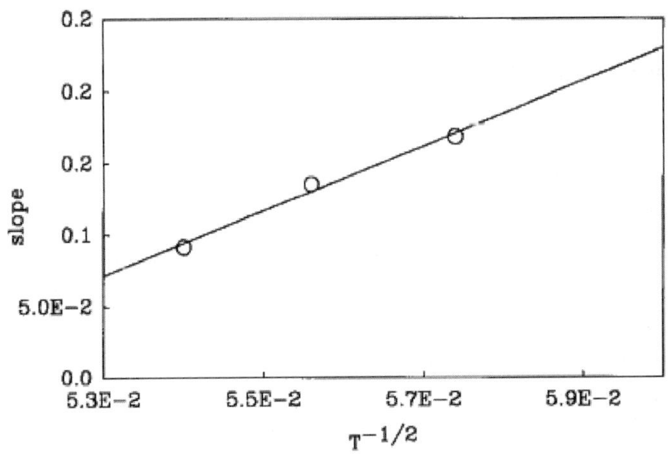

Figure 54 The slope for each of the lines in Figure 53 vs. temperature, $T^{-1/2}$. The particle volume fraction is 27 vol%. The maximum electric field is 3 kV/mm. Reproduced with permission from T. Hao, and Y.Xu, J. Colloid Interf. Sci., 181(1996)581

For the purpose of localizing the nonohmic conductive mechanism, the conductivity of polyarylonitrile/silicone oil suspension was measured at different temperatures in the absence of a mechanical field [125]. The leakage conductivity (the value of stabilized conductivity) was used for localization when the absorption current phenomenon took place. The experimental data were first analyzed within the framework of the CELT model. The oxidized polyacrylonitrile particles might be associated with highly conducting regions, and the silicone oil is associated with the low conducting regions. At T=300 K, using Eq.(78) E_c is estimated to be 129 V/mm. The divergence predicted by Eq. (76) could be observed once the experiments were carried out above the estimated threshold field. The experimental results are shown in Figure 52, plotted as lnσ(E) vs 1/E; σ is the measured sample conductivity (current/applied voltage), and the applied maximum electric field E is 3 kV/mm. A linear relation was not find between ln σ and 1/E. The main assumption of the CELT model is that the charging energy dominates the conduction when an electron hops from one particle to another [126]. This charging energy was $E_t \sim e^2 / \varepsilon_p r$, where r was the particle size. For oxidized polyacrylonitrile material, the dielectric constant ε_p is around 4 and the diameter is about 1 μm; E_t was negligibly small compared with $k_B T$ (T~300 K), and hence it cannot dominate the conduction. The same data were plotted as ln σ vs $E^{1/2}$ in Figure 53. A linear relationship between them was found, indicating that the data fit Eq. (85) reasonably well. Calculating the slope of each curve and assuming that they varied as $T^{-1/2}$, a straight line was also obtained, see Figure 54. Fundamentally, it could be concluded that the conductivity of the oxidized polyacrylonitrile-based ER suspension could be described by the Quasi-1d-VRH model, charge carriers of which localized along a one-dimensional chain, with an effective Coulombic barrier between adjacent sites. When the applied external electric field is not very strong (less than 100 kV/mm), the electron migration is mainly determined by heat excitation, and the tunneling effect can be neglected [126]. In Hao's experiment, the highest applied electric field is 3 kV/mm, far less than 100 kV/mm; thus the Quasi-1d-VRH model seemingly is able to cover the conductive behavior.

In the view of the morphology of ER suspensions, the Quasi-1d-VRH model also seems reasonable, as the filament chains may form one-dimensional paths for charge carriers. Although the anisotropic network structure could build up in concentrated ER suspensions [62], the hidden chains, spanned between two electrodes, still exist in the system [131]. Essentially, the Quasi-1d-VRH model is the specific consequence of strong disorder whose behavior can be predicted by the percolation theory [59]. As shown in last section, the rheological properties of the oxidized polyacrylonitrile-based ER suspension could be explained by means of the percolation model [62]. A critical particle volume fraction is found to exist in ER systems, and a random resistant network structure rather than the fibrillated chains would build up in a concentrated ER suspension under an electric field. The conductivity process can be analogously analyzed as below. Each node of the network could be considered as a localized electron site of the solid particle in this suspension, and each pair of nodes connected by an assumed wire whose conductance represents the tunneling transition or heat excitation rate between the corresponding sites. The conductances of the different wires could vary over many orders of magnitude, reflecting the enormous variation of hopping probabilities presented in the suspension. Assuming that all wires are removed from the network and then are put back one after another, the network would have the highest conductance at the beginning of this process, and then the conductivity would decrease. At first, the isolated wires would randomly distribute throughout the network, and then some wires would connect and form some larger clusters; eventually—at some critical conductance value—the percolation path would appear. At this moment the microconductance (hopping rate) controls the macroconductivity of the whole system. The wires whose conductivity is larger than the critical conductivity do not contribute to the macroconductance because they could form only isolated clusters or chains that could not span the whole system. Although the wires of smaller conductance could form the macropath, they would contribute little to the conductivity of the suspension because they are shorted out by the path of higher conductance. The current could pass only through the wires with the critical conductivity value. The isolated wires or chains may start from one electrode but come to an end at somewhere between the two electrodes (branched paths) . Although these chains can not contribute to the resultant conductivity, they can be polarized under an electric field and contribute to the overall dielectric polarization of the ER suspension.

5.2 Conductivity under an oscillatory mechanical field

Due to the unstable dc current value, the current measurement under an oscillatory mechanical field was carried out in a low electric field (in most cases, less than 0.6 kV/mm) by Hao [125]. Figure 55 shows the dc current curve of the OP suspension recorded by the pen recorder under dc electric field 0.5 kV/mm, a mechanical angular frequency $\omega=1$, and a strain amplitude of 50%. The stress and strain curves recorded by the rheometer in the same time period are also presented. Note that the dc current sinusoidally oscillates when a sinusoidal mechanical strain is applied, and the oscillatory frequency is equivalent to that of the stress and two times higher than that of the applied strain. Moreover, the maximum of the oscillated dc current corresponds to the minimum of the mechanical strain. Figure 56 shows dc current curves obtained under an angular frequency sweeping field at a strain amplitude of 50% and a strain sweeping field at an angular frequency ω=1, when the applied electric field is still 0.5 kV/mm. The stress and strain curves recorded in the same time frame are also presented.

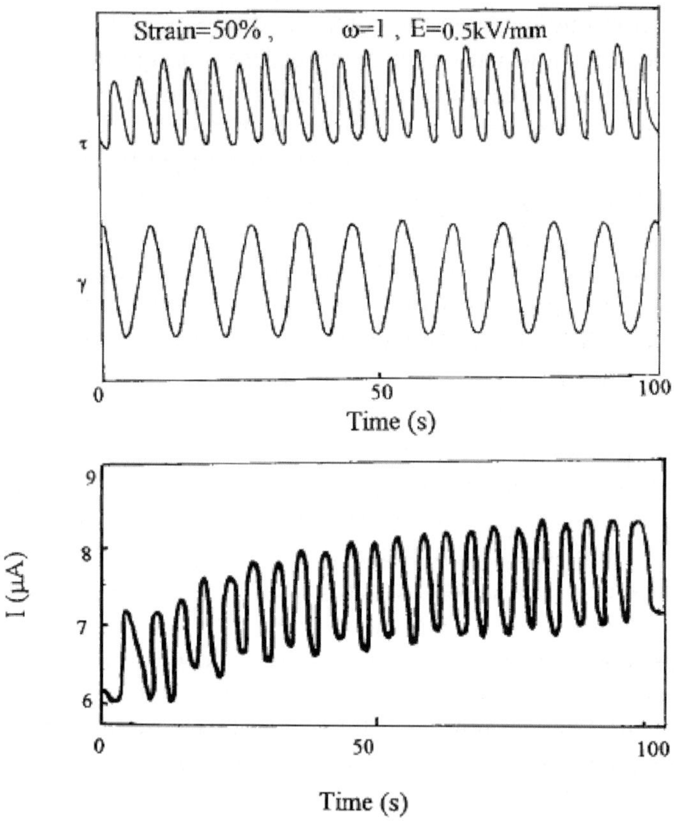

Figure 55 The dc current against time obtained at 0.5 kV/mm for oxidized polyacrylonitrile/silicone oil suspension. Particle volume fraction is 27 vol%. The mechanical angular frequency ω=1, and strain amplitude 50 %. The stress τ and the strain γ against time in the same time frame are also presented. Reproduced with permission from T. Hao, and Y.Xu, J. Colloid Interf. Sci., 181(1996)581.

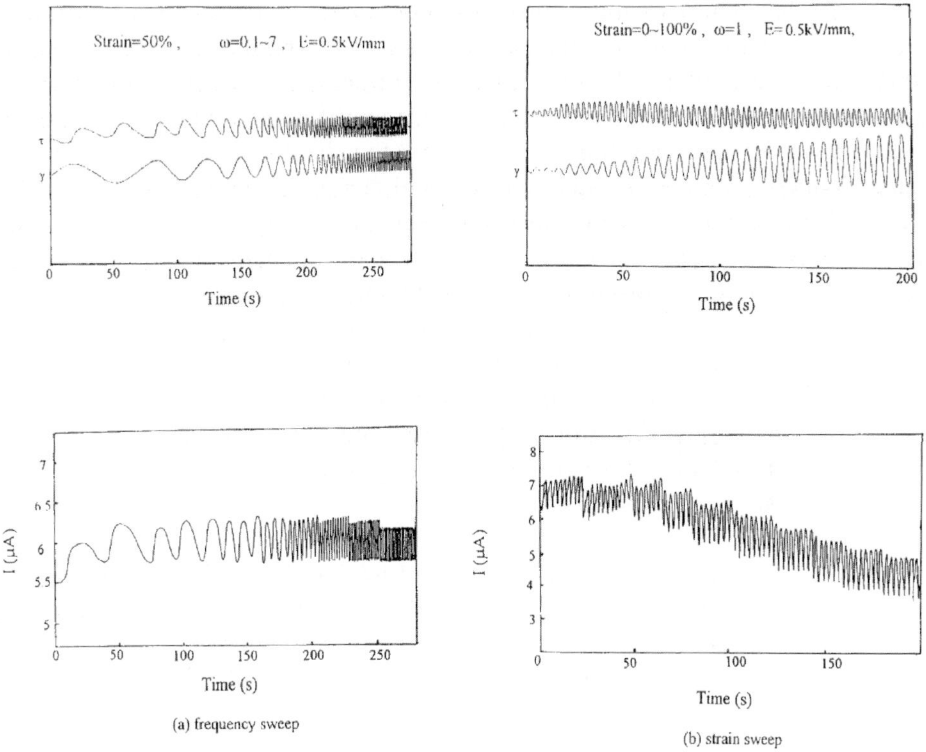

Figure 56 The dc current against time obtained at 0.5 kV/mm for oxidized polyacrylonitrile/silicone oil suspension (OP). a) Under a mechanical angular frequency sweep field at strain amplitude 50 %; b) under a strain sweep field at angular frequency 1. Particle volume fraction is 27 vol%. The stress τ and the strain γ against time in the same time frame are also presented. Reproduced with permission from T. Hao, and Y.Xu, J. Colloid Interf. Sci., 181(1996)581.

Under a given strain amplitude, the current average value was found to decrease only at high frequencies, whereas it obviously decreased over the entire strain sweeping range. However, the peak-to-peak value of current was not found to change obviously in either the applied mechanical frequency sweep field or the strain sweep field. Once the applied mechanical angular frequency was larger than 10, the measured dc current did not fluctuate, because it oscillated too fast to be recorded by the pen recorder.

The frequency of the oscillatory shear stress is not always two times higher than that of applied strain. It depends on the response time of the ER fluid [83]. Since the oscillation of dc current actually reflects the distortion information of fibrillated chains that in turn contributes to the shear stress, it is easy to understand that the dc current and the shear stress should have the same frequency, if the ER fluid can respond fast enough with the oscillatory strain. If the response time of an ER fluid is slower than the applied strain, the fibrillated chains may oscillate with the mechanical field, but the polarization-induced interparticle force is unable to be swiftly produced. The shear stress therefore oscillates at the same frequency as the applied strain, except there is a phase difference between them. The permutite/silicone oil suspension of the response time 1.1 second was found to show a same oscillatory frequency for both shear stress and shear strain, as shown in Figure 57. In such a slow suspension, the frequency of dc current is still two times than that of the applied strain.

Those special conductivity behaviors of the OP suspensions under the oscillatory mechanical field may be related to the fibrillated microstructure induced by the external electric field. Since the dispersing phase, silicone oil, is an insulating material, the magnitude of the dc current may be mainly determined by the conductivity of the dispersed particle, the number of particles in the fibrillated chains or columns, and the contact resistance between particles. Any chain distortions such as elongation, twist, entanglement, and breakage could modify the current value significantly. In a quiescent state, the chains are parallel to the direction of the external electric field and are inclined and elongated under a shear field [65]. The elongation magnitude should be determined by the applied mechanical strain amplitude, while contacting frequency between neighboring particles might be governed by the applied external mechanical frequency. In a frequency sweep field, the contacting probability of the particles should change with varied mechanical frequency, resulting in a fluctuating dc current. However, in a strain sweep field, the distance or the contact resistance between particles would increase with applied strain amplitude, resulting in the decay of the average current value. Under a steady oscillatory mechanical field the dc current would prefer a stable oscillatory behavior. When the oscillatory strain reaches the maximum, the chains should elongate to the greatest degree, and thus the dc current value will reach its minimum. The minimum values of dc current will appear twice in one period of strain oscillation, and thus the oscillatory frequency of dc current doubles that of the applied oscillatory strain.

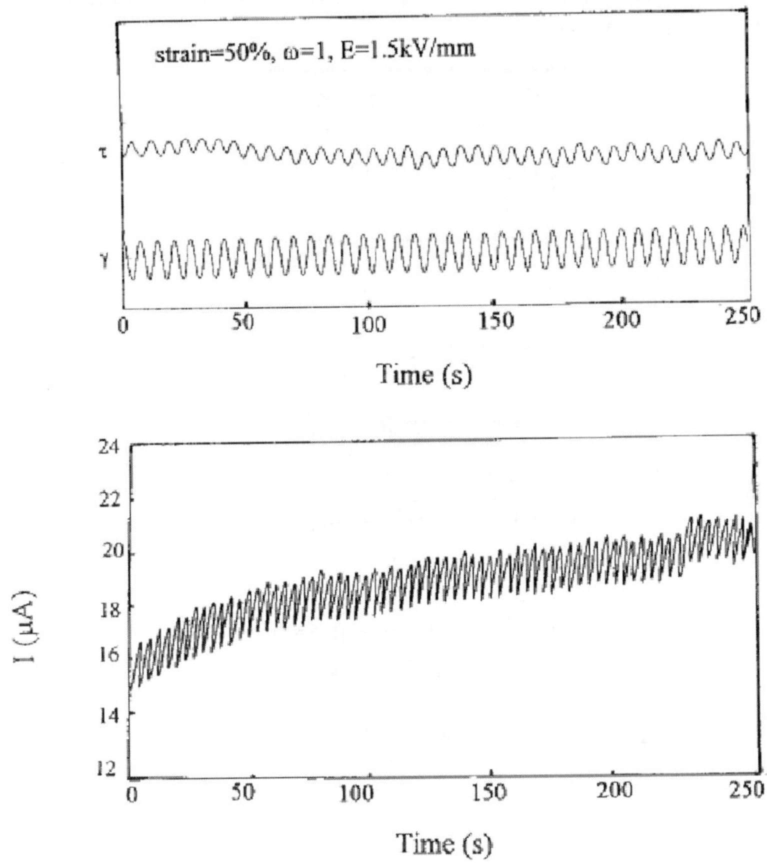

Figure 57. Oscillatory dc current of the permutite/silicone oil (PS) suspension recorded under a strain amplitude 50%, and the angular frequency 1, and an electric field of 1.5 kV/mm. The particle volume fraction is 35 vol%. The oscillatory shear stress τ and the applied strain γ recorded in the same time frame are also presented. Reproduced with permission from T. Hao, Y. Xu, J. Colloid Interf. Sci., 185(1997)324

If the shear stress of an ER fluid is generated from fibrillated chains, the oscillatory frequency of shear stress obviously should be equivalent to that of the dc current. This is the exact case for OP suspension, and the angular frequency of the shear stress doubles that of the applied oscillatory strain. However, for the PS suspension the frequency of shear stress is almost identical to that of imposed strain but is only half of the frequency of dc current. This may suggest that the chain rupture and re-formation process during shear is really slow.

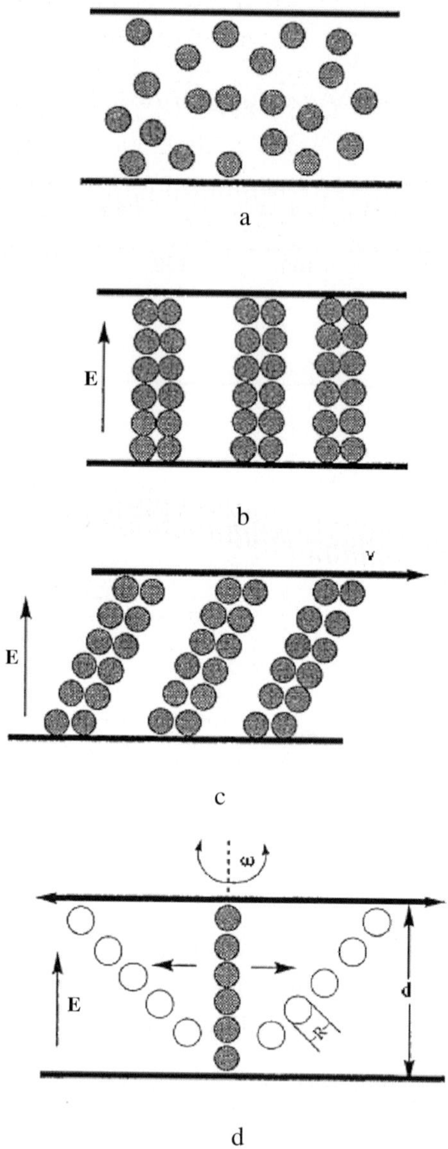

Figure 58 Schematic diagrams of an ER suspension under (*a*) off-state electric field; (*b*) on-state electric field; (*c*), both an electric and a shear fields; (*d*) both an electric and a dynamic shear field. Two bold lines stand for electrodes and parallel-plate of the rheometer. Reproduced with permission from T. Hao, J. Phys. Chem., 102(1997)1

The oscillatory dc current phenomena could be quantitatively described using the Quasi-One-Dimensional Variable Range Hopping (Quasi-1d-VRH Model) [132]. This model fits well for ER fluids for two reasons, 1) the fibrillated chains may provide a quasi-one-dimensional path for electron transfer; 2) the polydispersed particle system is facilitated for the variable range hopping process. The elongation and recovery of fibrillated chains would alternatively proceed under a dynamic mechanical field. These processes are schematically depicted in Figure 58. According to the Quasi-one-dimensional-Variable Range Hopping (Q-1d-VRH) model shown in last section in Eq. (85) and (86), the conductivity can be rewritten as:

$$\ln \sigma \propto (K_0/E_b)\, T^{-1/2}\, E^{1/2} \qquad (87)$$

Where

$$K_0 = T_0^{1/2}(e^3/\pi\varepsilon_0\varepsilon_p)^{1/2}/4 \qquad (88)$$

With

$$T_0 = 8^3\alpha^3/[9\pi k_B N(E_F)] \qquad (89)$$

where σ is conductivity of the system, σ_0 and K_0 are constant, α^{-1} is the decay length of the localized states, k_B is the Boltzmann constant, $N(E_F)$ is the density of states at Fermi energy E_F, E_b is the energy barrier between particles, T is temperature, E is the applied electric field strength (F= V/d, d is the gap between two electrodes, V is the field voltage), e is the electron charge, ε_0 is the vacuum dielectric constant. Assuming that under a dynamic field the Q-1d-VRH model still works, and E_b is directly proportional to the distance between two particle surfaces, so the interparticle distance variation with an applied dynamic field would obviously lead to dc current oscillation. Assuming N is particle number in one single chain and R is diameter of particle, then under a horizontal oscillatory strain field (see Figure 58d), $\gamma_{//} = \gamma_0 \sin \omega t$, E_b could be expressed as:

$$E_b = K_1\,[d\,(1+\gamma_{//}^2)^{1/2} - NR]/N \qquad (90)$$

where K_1 is a constant. So the conductivity σ therefore can be re-written as:

$$\ln \sigma \propto NKT^{-1/2} E^{1/2} [d (1+\gamma_{//}^2)^{1/2} - NR]^{-1}$$

$$= NKT^{-1/2} E^{1/2} [d (1+\gamma_0^2 \sin^2 \omega t)^{1/2} - NR]^{-1} \quad (91)$$

where $K=K_0/K_1$. The $\ln \sigma$ against ωt, based on Eq. (91), is schematically depicted in Figure 59b, which clearly indicates the frequency of the oscillatory dc current is exactly two times than that of the applied strain, and in addition, there is a $\pi/2$ phase difference between them. This agrees well with the experimental results shown previously. Theoretically, $T_0 = 8.2 \times 10^6$ K if the typical values of $\alpha^{-1} = 8\times10^{-10}$m, and $N(E_F) = 5\times10^{19}$cm^{3}ev^{-1}=3.12×10^{44} J^{-1}m^{-3} are used [128,133], quite close to the experimental value 2.7×10^6 K. For the oxidized polyacryonitrile/silicone oil system [125], the mean diameter of the particle is 1×10^{-6}m, and the particle number in one single chain (length is 1mm) is 909 if the particle surface distance is assumed as 1×10^{-7} m, $K_0=3.89\times10^{-21}$ $K^{-1/2}C^{-3/2}$ ($\varepsilon_p\approx5$ for oxidized polyacrylonitrile material), $\sigma_0 = 1.1\times10^{-7}$ S/m [125]. At quiescent state, $E_b = \alpha^2h^2/(8\pi^2m_e) = 1.03\times10^{-20}$ J (h is the Planck constant, m_e the electron mass), thus the constant K_l is 1.03×10^{-13} J m^{-1}. Under an dynamic field $\gamma_{//} = 0.5$ sint, T=298 K, E=5×10^5 V/m, the predicted oscillatory amplitude of dc current, according to the Eq. (91), is 1.64 μA, agreeing very well with the experimental value, around 1.4 μA, as shown see Figure 55.

If a vertical oscillatory field is applied, assuming original gap between two electrodes is d and a vertical oscillation is $\gamma_\perp = \gamma_0 \sin \omega t$, then fibrillated chains would vertically elongated, and the conductivity could be expressed as:

$$\ln \sigma \propto K\, NT^{-1/2} E^{1/2} / (d + \gamma_0 \sin\omega t - NR)$$

$$= K\, NT^{-1/2} V^{1/2}(d + \gamma_0 \sin\omega t)^{-1/2} (d + \gamma_0 \sin\omega t - NR)^{-1} \quad (92)$$

The $\ln \sigma$ as a function of ωt is schematically depicted in Figure 59c, where dc current displays a quite different behavior from that in a horizontal dynamic field. The frequency is almost as same as that of the applied strain. The non-symmetric oscillatory behavior indicates an extraordinarily large dc current value could be observed if γ_0 is large enough. By analyzing recorded dc current, one may easily tell which kind of vibration takes place, and what the amplitude and frequency are. Also, using specially designed electrodes, one can easily tell the oscillation direction, which would be very useful in mechanical measurement.

The analysis above is based on a presumption that fibrillated ER suspension is just elongated and inclined, and can not be destroyed under an applied dynamic field. In other words, the particle fibrillation rate is presumably thought to be always larger than the particle drift rate. If particle response time to an electric field is rather longer, recorded dc current curve was found to be distorted as shown in Figure 60 [83]. The response time of the permutite/silicone oil suspension was determined to be 1.1 second, far below the typical value of millisecond scale for many ER fluids. The distortion results from the slow response even to the mechanical field. Fortunately, the response time can be adjusted by varying the conductivity of dispersed particle [134] to the range that the recorded dc current truly reflects the chain deformation. Those findings may provide a new approach for real-time monitoring of mechanical signals, and it is another perfect example that the ER fluids can be used as an electronic-mechanical signal transferring interface.

In summary, the dc current absorption is observed in the oxidized polyacrylonitrile based-ER fluids. The conductive behaviors of ER suspensions with or without an oscillatory mechanical field are confined by the microstructure—the fibrillated chain structure induced by an external electric field. The dc current oscillates with the mechanical frequency and strain amplitude, implying that ER suspensions could be used as a mechanical sensor transferring a mechanical signal to an electric one. The conductive mechanism of an oxidized polyacrylonitrile-based ER suspension can be well described by a Quasi-1d-VRH model, where the localized charges hop from one localized state to another along the chains. This conduction model can be used to quantitatively describe the dc current oscillation phenomena.

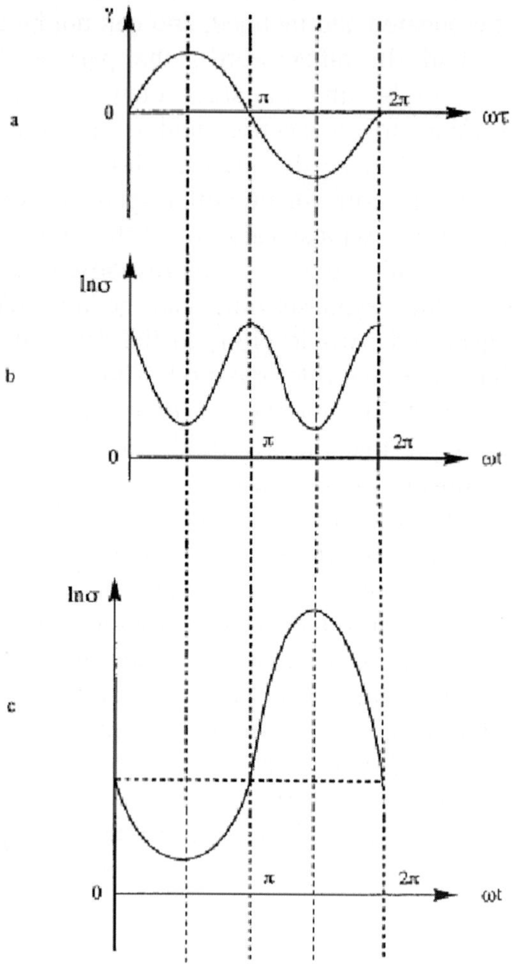

Figure 59 Schematic diagrams of (*a*) applied strain; (*b*) dc current ln σ against ωt under a horizontal dynamic field; (*c*) dc current ln σ against ωt under a vertical dynamic field. Reproduced with permission from T. Hao, J. Phys. Chem., 102(1997)1

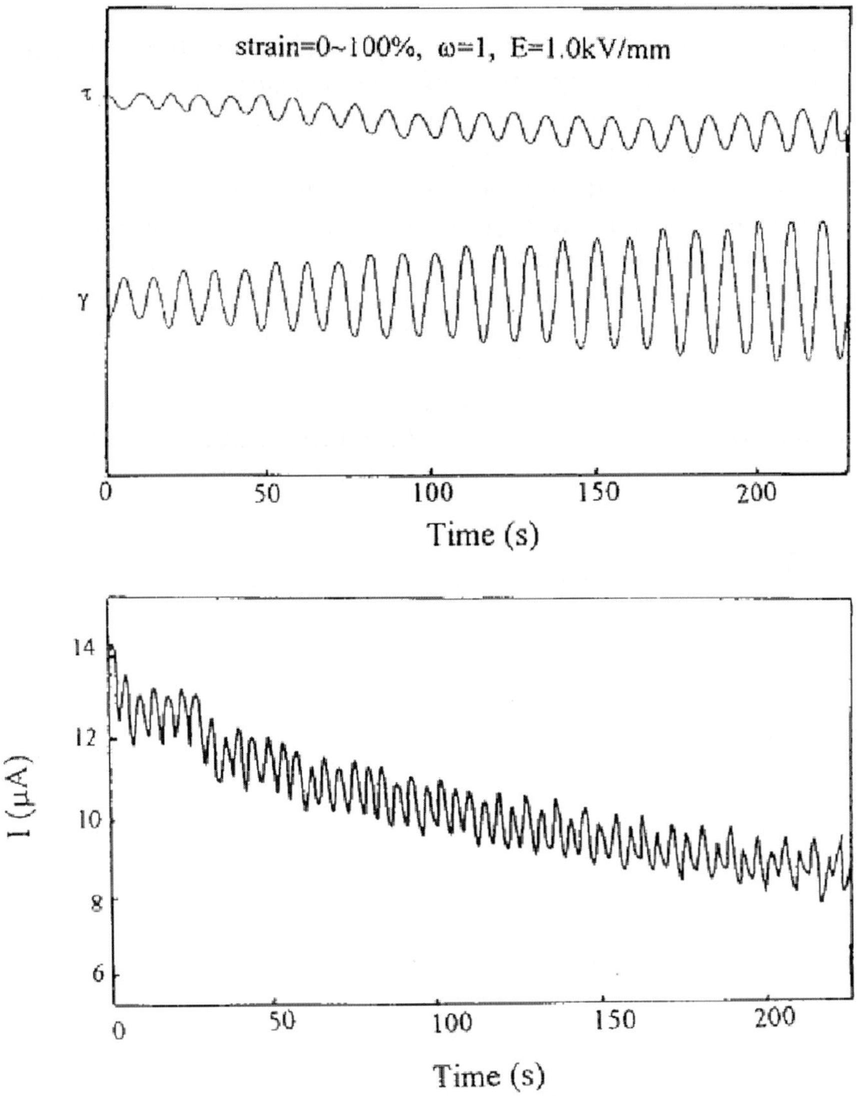

Figure 60. Oscillatory dc current of the permutite/silicone oil (PS) suspension recorded under a strain sweep field at the angular frequency 1 and an electric field of 1.0 kV/mm. The particle volume fraction is 35 vol%. The oscillatory shear stress τ and the applied strain γ recorded in the same time range are also presented. Reproduced with permission from T. Hao, Y. Xu, J. Colloid Interf. Sci., 185(1997)324

6. POLARIZATION PROCESS

Polarization processes are extremely important in ER fluids. Generally, there are four kinds of polarizations in a non-aqueous system containing no electrolytes or ions. They are electronic, atomic, Debye and the interfacial polarizations (the Wagner-Maxwell polarization). If the particulate material is an ionic solid, ionic displacement polarization should also be considered. The Debye and the interfacial polarizations are rather slow processes as compared with electronic and the atomic polarizations. Usually, the former two polarizations are called the slow polarizations, appearing at low frequency fields, whereas the last two are termed fast polarizations, appearing at high frequencies.

It would be very important to clarify which polarization process is responsible for the ER effect. Hao [135] investigated how the particle conductivity affects the response time of the ER suspension and how the particle surface properties affect the ER effect. It is concluded that the interfacial polarization contributes to the ER effect. This finding is consistent with his previous proposal that a large dielectric loss is required for a strong ER, effect, because only a material having a large dielectric loss could give a large interfacial polarization once it is dispersed into a liquid. This can be easily understood by considering where the dielectric loss comes from. In homogeneous two-component ER systems, the interfacial polarization is also found to be responsible for ER effect [136], implying that the interfaces even at microscopic scale also generates a strong enough Wagner-Maxwell polarization. Detailed polarization process and in turn the related dielectric properties associated with are addressed in next two chapters.

REFERENCES

[1] W. B. Russell, D.A. Saville, W.R.Schowalter, Colloidal Dispersion, Cambridge University Press, 1992
[2] J.E. Stangroom, Phys. Techn. 14(1983)290
[3] Y.D. Kim, D.J. Klingenberg, J. Colloid Interface Sci. 183(1996)568
[4] D. Myers, Surfaces, Interfaces, and Colloids, VCH publishers, 1991
[5] B.D. Coleman, H. Markovitz, and W. Noll, Viscometric flows of Non-newtonian Fluids, Springer-Verlag, 1966
[6] R.B. Bird, W.E. Stewart, and E.N. Lightfoot, Transport Phenomena, Wiley, New York, 1960
[7] E. Nelson, Dynamical theories of Brownian Motion, Princeton university Press, 1967
[8] Y. Pomeau, and P. Resibois, Phys. Rep. 19C(1975)64
[9] W.B. Russel, Ann. Rev. Fluid Mech., 13(1981)425

[10] T. Allen, Particle Size Measurement, Vol.1, Chapman & Hall New York, 1997, p 429
[11] R.J. Hunter, Zeta potential in colloid science, Academic press, 1981
[12] H. Ohshima, Journal of Colloid and Interface Science 247(2002)18
[13] B.V. Derjaguin, Kolloid Z. 69(1934)155
[14] E.J.W.Verway, and J.Th.G. Overbeek, The Theory of Stability of Lyophobic Colloids, Elsevier, 1948
[15] P.C. Hiemenz, and R.Rajagopalan, Principles of Colloid and Surface Chemistry, Marcel Dekker, 1997
[16] A.K. Dolan, and S.F. Edwards, Proc. Roy. Soc.Lond.A, 337(1974)509
[17] J.N. Israelachili, Intermolecular and surface forces, 2nd Ed., Academic Press, London, 1991;
[18] D.H. Napper, Polymeric Stabilization of Colloidal Dispersions, Academic Press, London, 1983
[19] A. Takahashi, and M. Kawaguchi, Adv. Polym.Sci., 46(1982)1
[20] P.-G deGennes, Adv. Colloid Interf Sci., 27(1987)189
[21] J.M. Scheutjens, and G.J. Fleer, Macromolecules, 18(1985)1882
[22] H.J. Ploehn, and W. B. Russel, Macromolecules, 22(1989)266
[23] S. Asakura, and F. Oosawa, J.Chem. Phys. 22(1954)1255
[24] B. Gotzelmann, R. Evans, and S. Dietrich, Phys.Rev.E, 57(1998)6785
[25] Y. Mao, M.E. Cates, and H.N.W. Lekkerkerker, Physica A, 222(1995)10
[26] J.P. Flory, Principles of Polymer Chemistry, Cornell Univ. Press, Ithaca, NY, 1953
[27] A. Vrij, Pure Appl. Chem.., 48(1976)471
[28] G. Mason, and W.C. Clark, Chem. Eng. Sci., 20(1965)859
[29] H. See, H. Tamura, and M. Doi, J. Phys. D: Appl.Phys. 26(1993)746
[30] H. Tamura, H.See, and M. Doi, J. Phys. D: Appl.Phys. 26(1993)1181
[31] Y.D. Kim, J. Colloid Interface Sci. 236(2001)225
[32] Y.D. Kim, and S.W. Nam, J. Colloid Interface Sci. 269 (2004) 205
[33] D.J. Klingenberg, and C.F. Zukoski, Langmuir, 6(1990)15
[34] A.P.Gast, and C.F.Zukoski, Adv. Colloid Interf Sci., 30(1989)153
[35] P.A. Arp, and S.G. Mason, Proc. Royal. Soc. A, 300(1970)421
[36] P.A. Arp, and S.G. Mason, Colloid Polym. Sci., 255(1977)566
[37] P.A. Arp, R.T. Foister, and S.G. Mason, Adv. Colloid Interf.Sci., 12(1980)256;
[38] R.S.Allan, and S.G.Mason, Proc. Roy.Soc., A267(1962)62
[39] C.E.Chaffey, and S.G.Mason, J.Colloid Interf. Sci., 19(1964)525
[40] C.E.Chaffey, and S.G.Mason, J.Colloid Interf. Sci., 27(1968)115
[41] A. Okagawa, and S.G.Mason, J.Colloid Interf. Sci., 47(1974)568
[42] A. Okagawa, R.G.Cox, and S.G.Mason, J.Colloid Interf. Sci., 47(1974)536
[43] P.A. Arp, and S.G. Mason, Colloid Polymer Sci., 255(1977)980
[44] L.Marshall, C.F.Zukoski, and J.W.Goodwin, J. Chem. Soc. Faraday Trans. 85(1989)2785
[45] C.S. Coughlin, and R.N.Capps, SPIE, 2190(1994)19
[46] Z. Cheng, W.B. Russel, P.M. Chalkin, Nature 401(1999) 893
[47] P.N. Pusey, W. van Megan, Nature 320 (1986) 340
[48] E.G. Hoover, and F.H. Ree, J. Chem. Phys., 49(1968)3609
[49] L.V. Woodcock, Ann. NY Acad. Sci., 371(1981)274

[50] L.V. Woodcok, Nature 385(1997)141
[51]; W.L. Vos, M. Megens, C.M. Kats, P. Bosecke, Langmuir 13(1997)6004
[52] H. Miguez, F. Meseguer, C. Lopez, A. Mifsud, J.S. Moya, L. Vazquez, Langmuir, 13(1997)6009
[53] Fang Zeng, Zaiwu Sun, Chaoyang Wang, Biye Ren, Xinxing Liu, and Zhen Tong, Langmuir, 18(2002)9116
[54] D. R. Ulrich, Chem. and Eng. News, January 1, 1990, page 28-35
[55] R. Tao, J.M. Sun, Phys. Rev. Lett. 67(1991)398;
[56] J. E. Martin, R. A. Anderson, and C. P. Tigges, J. Chem. Phys. 108(1998)3765
[57] T. Chen, R.N. Zitter, R. Tao, Phys. Rev. Lett. 68(1992)2555.
[58] U. Dassanayake, S. Fraden, A. van Blaaderen, J. Chem. Phys. 112(2000)3851
[59] R. Zallen, The Physics of Amorphous Solids, Wiley-Interscience, June 1, 1998
[60] D. Stauffer, and A. Aharony, Introduction to Percolation Theory, 2nd ed. London, Taylor & Francis, 1992
[61] H. Scher and R. Zallen, J. Chem. Phys. 53(1970)3759
[62] T. Hao, Y. Chen, Z.Xu, Y.Xu and Y. Huang, Chin. J. Polym.Sci., 12(1994)97
[63] T. Hao, and Y. Xu, J. Colloid Interf. Sci., 181(1996)581
[64] H. Uejima, Jpn.J.Appl.Phys. 11(1972)319
[65] D.J.Klingenberg, and C.F. Zukoski, Langmuir, 6(1990)15
[66] R.T. Bonnecaze, and J.F. Brady, J. Rheol., 36(1)(1992)73
[67] A.M.Kraynik, "ER fluids standards: Comments on ER fluid rheology", Proc. 2nd Intern.Conf. on ER fluids, J. D. Carlson, A.F. Sprecher, and H. Conrad, ed. Technomic, Lancaster, PA, 1990
[68] H.A. Barnes, and K.Walters, Rheol. Acta, 24(1985)323
[69] D.R. Gamota, and F.E. Filisko, J. Rheol., 35(1991)399
[70] D.F. Gu, and A. M. Jamiesen, J. Rheol, 38(1994)555
[71] C.R. Safina, L.B. Sirota, and R.J. Plano, Phys. Rev. Lett., 66(1991)1986
[72] K. Negita, Mol. Cryst. Liq. Cryst., 300(1997)163
[73] K.Negita, and S. Uchino, Mol. Cryst. Liq. Cryst., 378(2002)103
[74] S. S. Shenoy N. J. Wagner, and J. W. Bender, Rheol. Acta, 42(2003)287
[75] T. C. Halsey, J.E. Martin, and D. Adolf, Phys. Rev. Lett., 68(1992)1519
[76] T.C. Halsey, and W.Toor, Phys. Rev. Lett., 65(1990)2820
[77] D.Y.Lee, and N.M. Wereley,. Damping and isolation. Proceedings of SPIEConference on Smart Materials and Structures, T.T. Hyde, Ed., 3989(2000)244
[78] X.Wang , F. Gordaninejad, Damping and isolation. Proceedings of SPIE Conference on Smart Materials and Structures, T.T. Hyde, Ed., 3989(2000)232
[79] W. S. Yen and P. J. Achorn, J. Rheol., 35(1991)1375
[80] J.A. Powell, J.Rheol., 39(1995)1075
[81] W.N. Findley, J.S. Lai, and K.Onaran, Creep and elaxation of non-linear viscoelastic materials, North-Holland, New York, 1976
[82] B. D. Chin; and H. H. Winter, Rheol. Acta, 41(2002)265
[83] T. Hao, and Y. Xu, J. Colloid Interf. Sci., 185(1997)324
[84] D. Doraiswamy, A.N. Mujumdar, I. Tsao, A.N. Beris, S.C. Danforth, A.B. Metzner, J. Rheol., 35(1991)647
[85] J.F. Brady, and G. Bossis, J. Fluid. Mech., 155(1985)105

[86] M. Parthasarathy, and D. J. Klingenberg, J. Non-Newtonian Fluid Mech., 81 (1999) 83.
[87] G. V. Vinpgradov, Z.P. Shulman, Yu.G. Yanovskii, V.V. Barancheeva, E.V. Korobko, and I.V. Bukpvich, Inzh. Fiz. Zh., 50(1986)605
[88] D. Brooks, J. Goodwin, C. Hjelm, L. Marshall, and C.F. Zukoski, Colloids Surf., 18(1986)293
[89] D. Chotpattanont, A. Sirivat, and J.M. Jamieson, Colloid Polym Sci., 282(2004)357
[90] B. D. Chin; and H. H. Winter, Rheol. Acta, 41(2002)265
[91] D.J. Power, A.B. Rodd, L.Peterson, and D.V.Boger, J. Rheol. 42(1998)1021
[92] H.H. Winter, and F. Chambon, J. Rheol., 30(1986)367
[93] A.C. Pipkin, Lectures on viscoelasticity theory, Springer-Verlag, New York, 1972
[94] M. Parthasarathy, and D. J. Klingenberg, Rheol. Acta, 34(1995)430
[95] M. Parthasarathy, D. J. Klingenberg, Rheol. Acta, 34(1995)417
[96] D.R. Gamota, F.E. Filisko, J. Rheol. 35 (1991) 1411
[97] R.C. Erhgott, S.F. Masri, Smart Mater. Struct. 1 (1992) 275
[98] J.A. Powell, Smart.Mater. Struct. 2 (1993) 217; J.A. Powell, J. Rheol. 39 (1995) 1075
[99] M. Parthasarathy, K.H. Ahn, B.M. Belongia, D.J. Klingenberg, Int. J. Mod. Phy. B 8 (1994) 2789;
[100] D.J. Klingenberg, J. Rheol. 37 (1993) 199
[101] T.C.B. McLeish, T. Jordan, M.T. Shaw, J. Rheol. 35 (1991) 427
[102] T. Jordan, M.T. Shaw, T.C.B. McLeish, J. Rheol. 36 (1992) 441
[103] J.E. Martin, J. Odinek, J. Rheol. 39 (1995) 995
[104] H. G. Sim, K. H.Ahn, and S. J. Lee, J. Rheol., 47(2003)879
[105] Hyun, K., S. H. Kim, K. H. Ahn, and S. J. Lee, J. Non-Newtonian Fluid Mech. 107(2002)51
[106] H.G. Sim, K. H.Ahn, S. J. Lee, J. Non-Newtonian Fluid Mech. 112 (2003)237
[107] Y. Xu, and R. Liang, J. Rheol., 35(1991)1355
[108] Y. D. Kim, D. H.Park, Synthetic Metals 142 (2004) 147
[109] J.E. Martin, J. Odinek, T.C. Halsey, and R. Kamien, Phys. Rev. E, 57(1998)756
[110] C. Gehin, J.Persello, D. Charraut, and B.Cabane, J. Colloid Interf. Sci., 273 (2004) 658
[111] R.H. Ottewill, and A.R.Rennie, Modern aspects of colloidal dispersions, Kluwer Academic, 1998
[112] J.S. Pedersen, Adv. Colloid Interface Sci., 70(1997)171
[113] R.J. Baxter, J. Chem. Phys. 49 (1968) 2770
[114] A.T.J.M. Woutersen, R.P. May, C.G. de Kruif, J. Colloid Interface Sci. 151 (1992) 410
[115] J. Teixeira, J. Appl. Crystallogr. 21 (1988) 781
[116] R. Julien, R. Botet, Aggregation and Fractal Aggregates, World Scientific, Singapore, 1987
[117] A.A. Potanin, J. Colloid Interface Sci. 156 (1993) 143
[118] H. Block, and J.P.Kelly, Langmuir 6 (1990)6
[119] T.Hao, Y. Xu, Y. Chen, and Mao Xu, Chin. Phys. Lett. 12(1995)573
[120] T.Hao, Z.Xu, and Y.Xu, J. Colloid and Interf. Sci., 190(1997)334
[121] L.C.Davis, J. Appl. Phys. 73(2)(1993)680

[122] L.C.Davis, J. Appl. Phys. 72(4)(1992)1334
[123] R.A.Anderson, ''Proceedings, 3rd International Conference on ER Fluids'', R. Tao, Ed., p. 81. Wold Scientific, Singapore, 1992
[124] N. Felici, J.N. Foulc, and P Atten., ''Proceedings, 4th International Conference on ER Fluids'', R. Tao, Ed., p. 139. World Scientific, Singapore, 1995
[125] T.Hao, and Y. Xu, J.Colloid and Interf. Sci., 181(1996)581
[126] H. Meier, Organic Semiconductors, Dark- and Photo-Conductivity of Organic Solids, VCH, Weinheim/New York, 1974
[127] P. Sheng, and B. Abeles, Phys. Rev. Lett. 28(1972)34
[128] N.F.Mott, and E. Davis, Electron Process of Noncrystal Material, Clarendon, Oxford, 1979
[129] A.A.Gogolin, Phys. Rep.1 (1982)1; 5(1988) 269
[130] Z.H.Wang, E. Ehrenfreund, A.Ray, A. G. Macdiarmid, and A.J. Epstein, Mol. Cryst. Liq. Cryst. 189(1990) 263
[131] T. Hao, Z. Xu, Y. Li, Y. Chen, and Y.Xu, Advances in Rheology, Southern China University Sci. and Technol. Press, 1993, p554
[132] T. Hao, J. Phys. Chem., 102(1997)1
[133] E.A. Davis,.and N.F. Mott, Phil.Mag. 22(1970)903
[134] K.D. Weiss, D.A. Nixon, J.D. Carlson, A.J. Margida, Polymer Preprints, 35(1994) 325
[135] T. Hao, A. Kawai, and F. Ikazaki, Langmuir 14(1998)1256
[136] T. Hao, A. Kawai, and F. Ikazaki, J.Colloid Interf. Sci., 239(2001)106

Chapter 7

Dielectric properties of non-aqueous heterogeneous systems

Before the dielectric property of ER suspensions is specifically addressed, a general description of the dielectric property of non-aqueous systems is introduced first in this chapter. Comparing the dielectric properties of non-aqueous suspensions in general with that of ER suspensions in particular should be helpful for a better understanding of why the ER suspension can fibrillate under an external electric field.

1. BASIC DIELECTRIC PARAMETERS

When a dielectric material is sandwiched between two parallel electrodes as shown in Figure 1, this material will be polarized under an electric field, and the net charge is generated on the surfaces of two electrodes.

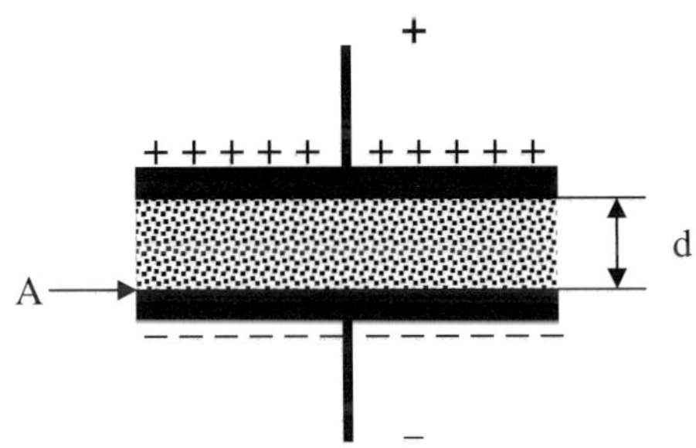

Figure 1 Schematic illustration of a dielectric in an electric field.

The relationship between the applied voltage V, capacitance C of the material, and the net charge q is:

$$V = \frac{q}{C} \tag{1}$$

If there is no dielectric inserted between two electrodes and the whole system is in the vacuum, Eq. (1) will be replaced by:

$$V_0 = \frac{q_0}{C_0} \tag{2}$$

In electromagnetism, the permittivity, ε_{pm}, is defined as the ratio of the electric displacement D to the applied electric field strength E:

$$\varepsilon_{pm} = \frac{D}{E} \tag{3}$$

where D is defined as

$$D = 4\pi \frac{q}{A} \tag{4}$$

A is the surface area of the electrode. E=V/d. Substituting Eq. (4) into Eq.(3)

One may obtain

$$C = \frac{\varepsilon_{pm} A}{4\pi d} \tag{5}$$

Eq. (5) can be used to calculate the capacitance of the parallel-plate capacitor. The absolute vacuum permittivity, $\varepsilon_0 = 8.85419\text{x}10^{-12}$ F/m, and the dimensionless relative permittivity, also called the dielectric constant, ε, is defined as:

$$\varepsilon = \frac{\varepsilon_{pm}}{\varepsilon_0} \tag{6}$$

or

$$\varepsilon = \frac{C}{C_0} \tag{7}$$

The permittivity ε_{pm} and magnetic permeability μ of a medium together determine the velocity v of electromagnetic radiation through that medium:

$$\varepsilon_{pm}\mu = \frac{1}{v^2} \tag{8}$$

In a vacuum,

$$\varepsilon_0 \mu_0 = \frac{1}{c^2} \tag{9}$$

where μ_0 is the permeability of vacuum, equal to $4\pi \times 10^{-7}$ N·A^{-2}, and c is the speed of light in vacuum, 299,792,458 m/s.

In an oscillatory electric field, the field-induced polarization will lag behind the applied electric field in some angle δ_E. The complex electric field E* and the complex displacement D* can be expressed as [1]

$$E^* = E_0 \exp(i\varpi t)$$
$$D^* = D_0 \exp i(\varpi t - \delta_E) \tag{10}$$

So the complex dielectric constant ε* will be

$$\varepsilon^* = \frac{D^*}{E^*} = \frac{D_0}{E_0}\exp(-i\delta_E) = |\varepsilon|(\cos\delta_E - i\sin\delta_E)$$
$$= \varepsilon' - i\varepsilon'' \tag{11}$$

where

$$\varepsilon' = |\varepsilon|\cos\delta_E \qquad \varepsilon'' = |\varepsilon|\sin\delta_E \tag{12}$$

thus

$$\tan\delta_E = \frac{\varepsilon''}{\varepsilon'} \tag{13}$$

ε' and ε'' are the real and imaginary components of the complex dielectric constant, ε' is usually called the dielectric constant and tan δ_E is termed the dielectric loss tangent.

2. KRAMERS-KRONIG RELATIONS

As shown above, the real and imaginary part of the complex dielectric constant are dependent on frequency, however, they are not independent of each other. ε' and ε'' can be correlated via the Kramers-Kronig relations [2,3], which describe the relation between the real and imaginary part of a certain class of complex-valued functions in physics and mathematics. The real part of the complex dielectric constant can be expressed as in the form of dielectric loss [4]

$$\varepsilon'(\omega_0) = \varepsilon_\infty + \frac{2}{\pi}\int_0^\infty \varepsilon''(\omega)\frac{\omega}{\omega^2 - \omega_0^2}d\omega \tag{14}$$

The dielectric loss can be separated into one part that results from dc conductivity and another part from relaxations. It can be expressed in the form of the dielectric constant ε' [4]

$$\varepsilon''(\omega_0) = \frac{\sigma_{dc}}{\varepsilon_0\omega_0} + \frac{2}{\pi}\int_0^\infty \varepsilon'(\omega)\frac{\omega_0}{\omega^2 - \omega_0^2}d\omega \tag{15}$$

in which ε_0 is the permittivity of vacuum, σ_{dc} is the dc conductivity, and ε_∞ is the high frequency dielectric constant. A major limitation of Kramers-Kronig relations in their practical use is that one needs to know the very broad frequency spectrum for converting the one parameter to the other. The broad-band dielectric analyzer can easily cover the frequency range between 10^{-6} to 10^8 Hz. Since the dc conductivity doesn't contribute to the dielectric constant, the dielectric loss converted from the dielectric constant only represents the second term in Eq. (15). Comparing the measured and converted function of dielectric loss, one may easily separate the dielectric loss that resulted from dc conductivity and from the polarization relaxations [4]. Note that in much of the literature the dc conductivity contribution term in Eq. (15) is omitted for Kramers-Kronig relations.

3. THE POLARIZATION TYPES AND THEIR RELATIVE RELAXATION TIME

Charged entities such as electrons, atoms, molecules and ions will be polarized under an electric field due to the separation of positive and negative charges, forming the dipoles. The polarizations resulting from the electrons, atoms and ions are commonly named electronic, atomic, and ionic polarizations. The polar molecule may re-orientate along the direction of the applied electric field, and this kind of polarization is called the Debye polarization. Besides those fundamental polarization types, there are other two important polarizations: The electrode polarization and the Wagner-Maxwell polarization. The electrode polarization is resulted from the electrolyte-formed electrical double layer at the surfaces of electrodes. The Wagner-Maxwell polarization is resulted from the heterogeneity of the system where the interfaces exist between two or more phases of different conductivities. Brief description on those polarizations is presented below.

3.1 Polarization type

3.1.1 The electronic polarization

The electronic polarization is the most common displacement polarization in atoms, ions, and molecules. The weakest bound electrons are the first to be displaced to the opposite direction of the applied electric field. This process is very quick, usually around 10^{-14}~10^{-15} second. Assuming that a charge entity q is elastically connected to the nuclear in the way f=kx, at the equilibrium point at x between the force with the nuclear and the force induced by the applied electric field E:

$$kx = qE \tag{16}$$

The dipole moment m

$$m = q \times x = \frac{q^2}{k} E \tag{17}$$

Eq.(17) indicates that the induced dipole moment is directly proportional to the applied electric field. The polarizability α is defined as the ratio between the induced dipole moment m and the applied electric field E

$$\alpha = \frac{m}{E} = \frac{q^2}{k} \tag{18}$$

The elastic constant k could be described as [5]:

$$k = k_1 \frac{q^2}{r^3} \tag{19}$$

k_1 is a constant and r is the radius of electron orbit. So Eq. (18) can be rewritten as:

$$\alpha = \frac{r^3}{k_1} \tag{20}$$

Eq. (20) shows that polarizability increases quickly with the increase of the orbit radius. α/r^3 is large, the polarizability is larger too. The ions of a large α/r^3 should have a large dielectric constant.

3.1.2 The atomic polarization

Atoms exist in a molecule in the form of ions no matter whether they are associated in ionic or co-valent bonds. Like the electrons moving under an electric field, ions can move along the direction of the electric field too. Ions of opposite charges are bound together through chemical bonds, forming a molecule. Under an electric field, those ions will move to the opposite direction, creating dipole moment. The polarizability of ions, α_i, can be expressed as [5]:

$$\alpha_i = \frac{(r_1 + r_2)^3}{n - 1} \tag{21}$$

where n is between 7 and 11, and r is the radius of ions. Eq.(21) has the similar form of Eq.(20), indicating the polarizability of ions may have a same magnitude of that of electrons. The atomic polarization is thus considered as one kind of ionic polarizations.

3.1.3 The ion polarizations

There is another kind of ion polarization that is related to the thermal motion of ions. The polarizability of such a thermal ion polarization, a_{iT}

$$\alpha_{iT} = \frac{q^2 d^2}{12 k_B T} \tag{22}$$

where d is the ion movement distance. When an electric field is applied, ions distribution in the dielectric will become asymmetric, even the ions are unable to move from the localized sites. The thermal energy will force ions to diffuse to the opposite direction of the applied electric field, till a new equilibrium ions distribution is established. This polarization is mainly controlled by the thermal motion of ions, and thus named as thermal ion polarization.

3.1.4 Debye polarization

Suppose that the polar molecules are randomly distributed in the system, under an electric field those molecules tend to orientate along the direction of this electric field. The polarization induced by the orientation of the dipoles is commonly named the Debye polarization. For a gaseous system, Debye derived an equation for calculation of the static dielectric constant, ε_0, the dielectric constant extrapolated at zero frequency [6]:

$$\frac{\varepsilon_0 - 1}{\varepsilon_0 + 2} = \frac{4\pi}{3} N\left(\alpha_e + \frac{\mu^2}{3k_B T}\right) \tag{23}$$

where N is the number of molecules per unit volume, α_e is the deformation polarizability due to the elastic displacement of the molecules, μ is the dipole moment of a molecule. α_e is given by the Mosotti-Clausius equation:

$$\frac{\varepsilon_\infty - 1}{\varepsilon_\infty + 2} = \frac{4\pi}{3} N\alpha_e \tag{24}$$

where ε_∞ is the dielectric constant at such high frequencies that the dipole polarization doesn't take place. Note that Eq. (23) is only applicable to the very dilute dipolar gases and the very dilute dipolar molecules in non-polar solvents [5,7] It is cannot be used for polar liquids and many solid dielectrics. More accurate theories for calculation of the static dielectric constant of condensed phase of polar molecules have been subsequently developed by Onsager [8], Kirkwood [9], and Frölich [10].

3.1.5 The electrode polarization

The electrode polarization is resulted from the accumulation of charge on electrode surfaces and the formation of the electrical double layers. When the applied electric field is a dc field, the ions are distributed in a double layer as described by Debye and Hückel [11,12]; when the applied electric field is an ac field, the ions in the suspension should respond to the charge on the electrodes, retarded by their drag force from the liquid medium. The double layer is thus sensitive to the frequency of the applied electric field, as the charges on the electrodes oscillate faster than the ions' movement. At high frequency, the ions are unable to move fast enough to form the double layer, so the electrode polarization disappears. The electrode polarization becomes serious at low frequency, leading to a very large dielectric constant. Figure 1 shows the dielectric constant of DI water and 0.02 M KCl solution. Water has dielectric constant 78. However, at low frequency below 200 Hz the dielectric constant increases to more than 10000 as frequency decreases to around 10 Hz, which is induced by the electrode polarization.

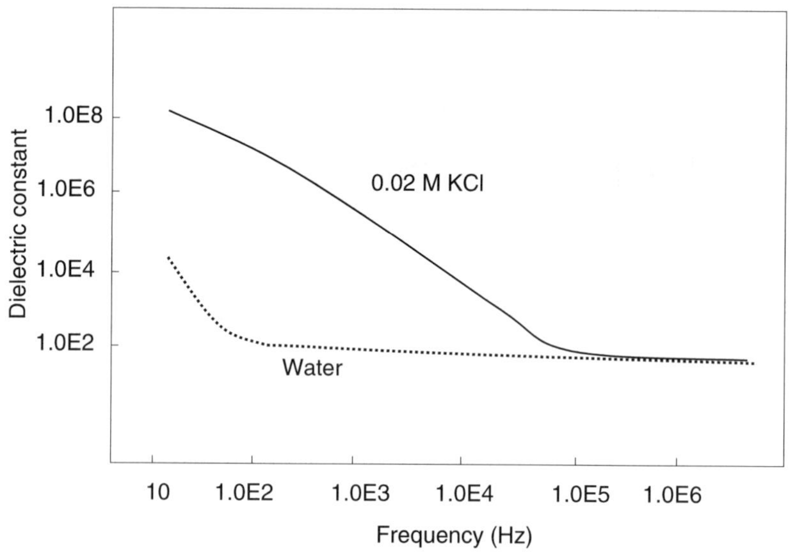

Figure 1 The measured dielectric constant of water and 0.02 M KCl solution as a function of frequency. Redrawn from the data of C. Gabriel, and S. Gabriel, the compilation of the dielectric properties of body tissues at RF and microwave frequency. King's College, London, UK, 1996.

For 0.02 M KCl aqueous solution, the electrode polarization becomes more serious. The dielectric constant reaches as high as 10^8 at around 10 Hz frequency. For monodispersed anionic polystyrene latex particle of a mean diameter 156 nm dispersed in 10^{-4} M NaCl aqueous solution, the strong electrode polarization was still observed in such low ion concentration at particle volume fraction 1.9 vol%, as shown in Figure 2 [13]. The electrode polarization always happens in a highly conductive aqueous system. However, even in a non-aqueous system the electrode polarization also can happen if there are electrolytes in even a very small concentration in the non-aqueous system. Note that forming an electrical double layer doesn't need a high concentration. Figure 3 shows the dielectric constant of the sodium bis (2-ethylhexyl)sulfosuccinate (AOT) water-in-oil microemulsion with iso-octane as the continuous phase at various water droplet volume fraction as the function of frequency [14].

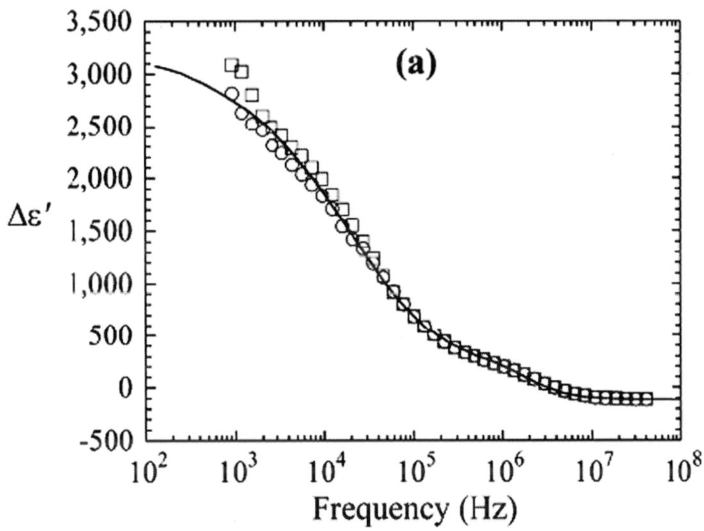

Figure 2 The dielectric increment Δε' of monodispersed anionic polystyrene latex particle of a mean diameter 156 nm dispersed in 10^{-4} M NaCl aqueous solution vs. frequency. The particle volume fraction is 1.9 vol%. Δε' was calculated using the equation $\varepsilon'_{appa} = \varepsilon_\infty\left[1+\left(\kappa^{3/2}D/\omega d^{1/2}\right)^2\right]+\phi\Delta\varepsilon'$, where κ is the reciprocal of the Debye length, D is the diffusion coefficient of ions, 2d is the spacer thickness between two electrodes, and ω is the frequency. Reproduced with permission from A. D. Hollingsworth, and D.A. Saville, J. Colloid Interf. Sci., 272(2004)235.

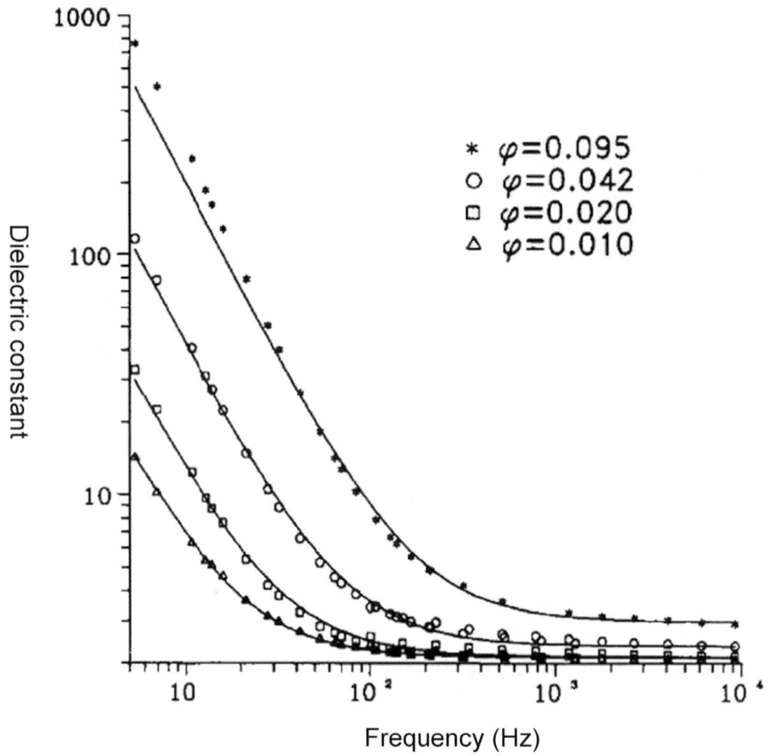

Figure 3 The dielectric constant of the sodium bis(2-ethylhexyl)sulfosuccinate (AOT) water-in-oil microemulsion with iso-octane as the continuous phase at various water droplet volume fraction. The molar water-to-surfactant ratio is 20. All spectra were taken at an electrode spacing of 1 mm. Reproduced with permission from P.A. Cirkel, J.M.P. van der Ploeg, and G.J.M. Koper, Physica A, 235(1997)269.

A strong electrode polarization was observed in low frequencies, and it became much stronger when the water droplet concentration increases from 1 vol% to 9.5 vol%. The electrode polarization disappeared at high frequency for all water-in-oil microemulsions of different water content. Since there is always such a strong electrode polarization in any system of electrolytes, dielectric measurement at low frequency becomes difficult. Correction methods for the electrode polarization in low frequency area have been developed [15-17]. Further description on the electrode polarization will be addressed in the next section.

3.1.6 The Wagner-Maxwell polarization

The Wagner-Maxwell polarization results from the interfaces between the dispersed phase and the dispersing medium of different conductivities. The charge carrier may be trapped at the interfaces, forming the space charge and generating the Wagner-Maxwell polarization. This polarization is related to the electric-field-induced charges that have nothing to do with the electrolytes in the system. Figure 4 shows the dielectric loss of two emulsions made of chlorinated paraffins dispersed in silicone oil vs. frequency. The chlorinated paraffin concentration of those two emulsions are 10 wt%. Both emulsions show a dielectric dispersion peak at low frequency, one at 1 Hz, and another one at 4 Hz, indicating that they have different relaxation time. Those two peaks are originated from the Wagner-Maxwell polarization at the liquid-liquid interfaces between the chlorinated paraffin and the silicone oil. Different peak positions are related to the conductivity of the dispersed chlorinated paraffin. For the solid particles dispersed into a liquid medium, like silica dispersed in silicone oil as shown in Figure 5, the solid-liquid interfaces will be generated, creating the Wagner-Maxwell type dielectric dispersion peak at low frequencies. If the silica particle surface is coated with more conductive polyaniline material, the dispersion peak shifts to high frequency range due to the higher conductivity of polyaniline, which is exactly following the trend as the Wagner-Maxwell theory predicts. The Wagner-Maxwell polarization is also related to the size of dispersed particles. Figure 6 shows the dielectric loss ε'' vs. frequency for polyaniline(PA) coated poly(methyl methacrylate)(PMMA) dispersed in silicone oil. The particle volume fraction is 10 vol%. The weight ratio of PA to PMMA in all suspensions is 0.2. The primary PMMA particle diameters of the PA coated samples, PAPMMA20-20, PAPMMA 20-45, PAPMMA 20-90, are 2, 4.5, and 9 μm, respectively. The large particle does move the dispersion peak to higher frequency area. How the particle conductivity and size affect the Wagner-Maxwell polarization will be discussed in more detail in the future.

In Wagner's original analysis the spherical and semiconductive particles are assumed to be sparsely distributed in an insulating medium of comparatively low dielectric loss [18,19]. The Wagner-Maxwell equations only hold for diluted suspensions of particle volume fraction less than 10 vol%. Various modifications on the Wagner-Maxwell equations have been made to fit for concentrated suspensions of the particle volume fraction larger than 10 vol% [20, 21]. There is no obvious difference between the Bruggeman and the Wagner equations if the particle volume fraction is less than 40 vol% [20, 21]. The Wagner-Maxwell polarization usually appears at low frequency less than 1000 Hz, depending on the conductivity of the dispersed phase. Detailed discussion on the Wagner-Maxwell polarization will follow further on.

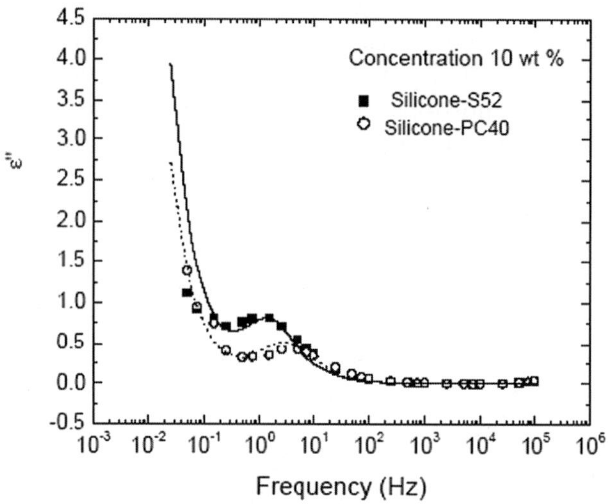

Figure 4 The dielectric loss of two emulsions made of chlorinated paraffins dispersed into silicone oil vs. frequency. The chlorinated paraffin concentration of those two emulsions are 10 wt%. Reproduced with permission from L. Rejon, B. Ortiz-Aguilar, H. de Alba, and O. Manero, Colloids Surf. A, 232(2004)87.

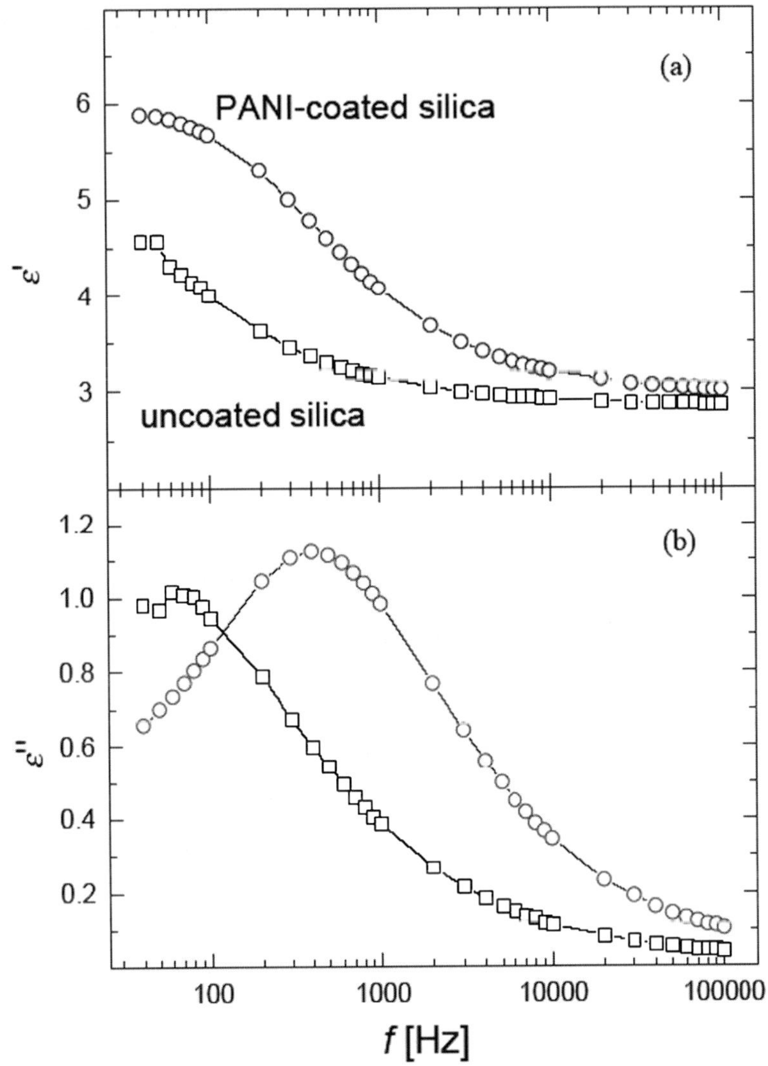

Figure 5 The dielectric constant and dielectric loss vs. frequency for the polyaniline-coated (O) or uncoated (□) silica dispersed in silicone oil. Reproduced with permission from A. Lengalova, V. Pavlnek, P. Saha, J. Stejskal, T. Kitano, O. Quadrat, Physica A, 321(2003)411.

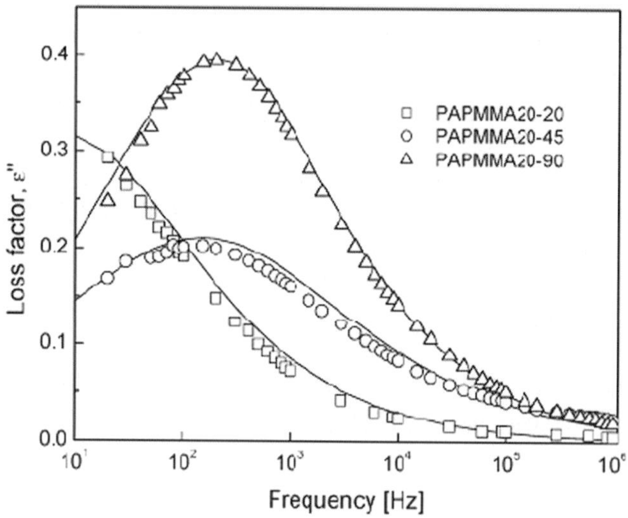

Figure 6 The dielectric loss ε" vs. frequency for polyaniline(PA) coated poly(methyl methacrylate)(PMMA) dispersed in silicone oil. The particle volume fraction is 10 vol%. The weight ratio of PA to PMMA is 0.2. The primary PMMA particle diameters of the PA coated PAPMMA20-20, PAPMMA 20-45, PAPMMA 20-90 are 2, 4.5, and 9 μm, respectively. Reproduced with permission from M.S. Cho, Y.H. Cho, H.J. Choi, and M.S. Jhon, Langmuir, 19(2003)5875.

3.2. Relative relaxation time of polarizations

The polarizations discussed above have different relaxation times, as they are governed by the different physical origins. Figure 7 schematically shows the wide frequency spectrum of the dielectric properties of a heterogeneous system. All polarizations are depicted in Figure 7 on the basis of their relative relaxation times. The dielectric dispersion of the electronic polarization appears at the highest frequency, more than 10^{15} Hz. With the polarization entity size increase, the dielectric dispersion peak gradually appears at low frequencies in the sequence of the atomic, Debye, interfacial polarization, and the electrode polarization. The Debye polarization usually appears at 10^6 Hz, the interfacial polarizations appears around 1000 Hz, and the electrode polarization appears below 100 Hz. The Debye, interfacial, and electrode polarizations are rather slow processes as compared with the electronic and the atomic polarizations. Usually, the former three polarizations are called the slow polarizations, whereas the last two are the fast polarizations. One may use the dielectric measurement technique to find out the polarization origin that should be responsible for a particular dispersion. The electronic, the atomic, and the Debye polarization may exist in any system no matter that it is homogeneous or heterogeneous, as long as there is dipole moment in the system. The Debye polarization doesn't typically happen in solid materials, as the dipole moment or molecular can be fixed in such a way that the dipole orientation becomes very unlikely. Only quite few solid materials show the Debye type polarization [10]. If there are electrolytes in the system, the electrode polarization definitely will present. The electronic, atomic, Debye, and the electrode polarization can appear either in a homogeneous system or a heterogeneous system. However, the Wagner-Maxwell polarization only appears in a heterogeneous system when there is an interface present.

The relaxation time of those polarizations are governed by different factors. For the electronic polarization the relaxation time is comparable to the speed of light. The relaxation time of atomics (ions) is comparable to the vibration frequency of each ion. The relaxation time of the Debye polarization is controlled by the potential barrier between two equilibrium positions where the dipole would orientate from one to other [10]. The relaxation time of the Debye polarization, τ_d, can be expressed as [10]:

$$\tau_d = (\pi / 2\omega_a) A e^{E_H / k_B T} \tag{25}$$

where $\pi/2\omega_a$ is the average time required by an excited molecule to turn from one equilibrium direction to the other, A is a factor that only varies slowly with temperature T, E_H is the energy barrier separating the two equilibrium positions, and k_B is the Boltzmann constant. For the interfacial polarization, the relaxation time τ_i of one system in which σ_p (the conductivity of the particle) » σ_m (conductivity of the medium) can be expressed as:

$$\tau_i = \frac{\varepsilon_0 (2\varepsilon_m + \varepsilon_p)}{\sigma_p} \tag{26}$$

where ε_0, ε_m, and ε_p are the dielectric constants of the vacuum, medium, and particle. Eq. (25) and (26) tell us that τ_d depends on the material state (liquid or solid), i.e., the molecule interaction and the environment, while τ_i depends on the dielectric properties of two components, especially on the particle conductivity. In liquid state E_H is smaller than that in solid state, thus τ_d is smaller too. A material of higher conductivity and lower dielectric constant will make τ_i become smaller.

For electrode polarization, the relaxation time , τ_e, should be controlled by the time of building up the electrical double layer. Suppose that the cations and anions have a same diffusion coefficient D of the dimension area/second, the ion velocity v_i in an electrical double layer of thickness κ^{-1} can be expressed as:

$$v_i = \frac{D}{\kappa^{-1}} \tag{27}$$

The time needed for building up an electrical double layer is:

$$\tau_i = \frac{\kappa^{-1}}{v_i} = \frac{\kappa^{-2}}{D} \tag{28}$$

Where κ^{-1} is the Debye screen length given by:

$$\kappa^2 = \frac{2ce^2}{\varepsilon_m \varepsilon_0 k_B T} \tag{29}$$

where c is the number concentration of electrolyte. Substituting Eq.(29) into Eq.(28):

$$\tau_e = \frac{\varepsilon_m \varepsilon_0 k_B T}{2ce^2 D} \tag{30}$$

The diffusion coefficient D of ion can be expressed by the Stokes-Einstein equation:

$$D = \frac{k_B T}{6\pi r_i \eta} \tag{31}$$

where r_i is the radius of ion and η is the viscosity of liquid medium. Substituting Eq.(31) into Eq.(30):

$$\tau_e = \frac{3\pi\varepsilon_m\varepsilon_0 r_i \eta}{ce^2} \tag{32}$$

Eq.(32) indicates that the relaxation time of the electrode polarization decreases with the number concentration of ions and increases with the size of ions, the dielectric constant and the viscosity of medium. Detailed discussion on the electrode polarization will be given further on.

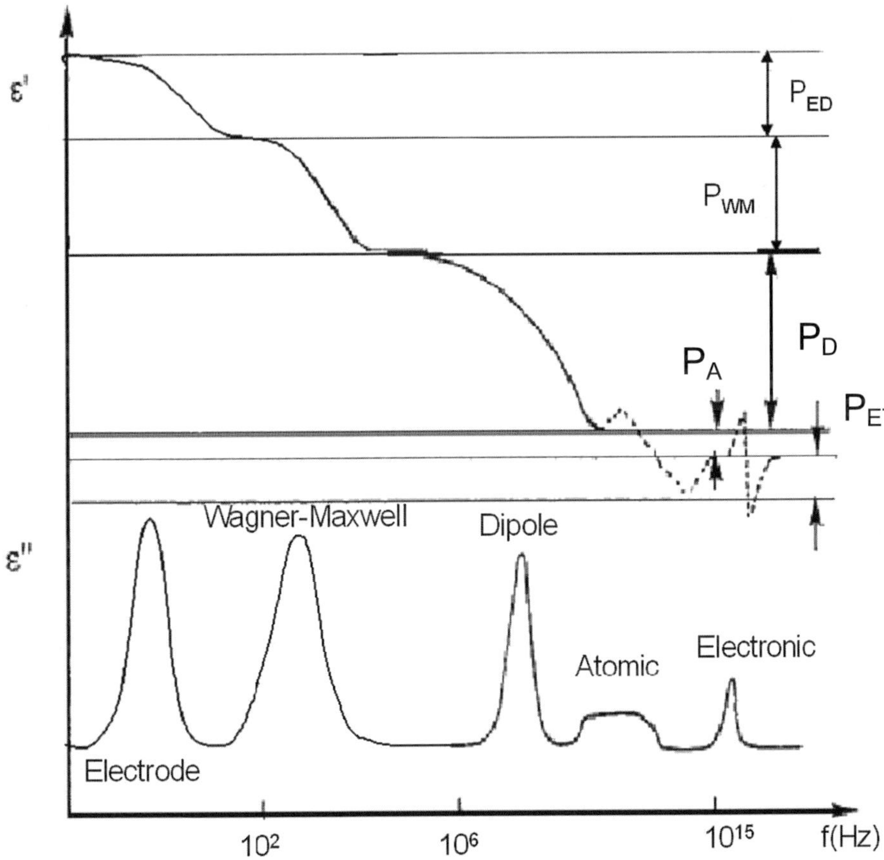

Figure 7. Schematic illustration of the dielectric constant and dielectric dispersion of a heterogeneous system vs. frequency.

3.3 Temperature dependence of the relaxation time

ER fluids can be generally classified into two categories based on the physical state of dispersed material. The heterogeneous ER fluid refers to the system containing microparticles (solid/liquid system) and the homogeneous ER fluid indicates that the dispersed phase is in a liquid state (liquid/liquid system). For heterogeneous ER fluid, one does not need to take the Debye polarization into consideration, as the solidification usually fixes the molecule with such rigidity in the lattice that there is little or no orientation of the dipoles even in an extremely strong electric field [10] (the solid phase also can prevent free rotation of the molecules even at temperatures near the melting-point [10]. However, the ion displacement polarization would probably occur if the particulate material is an ionic solid; if the dispersed phase is a liquid material, for example, the dispersed phase is a liquid crystalline polymer, the dipole orientation polarization, i.e., the Debye polarization, would take place under an electric field. The Maxwell-Wagner polarization will take place if the interface, either liquid-liquid or liquid-solid, exists in the suspension. If surfactant is added in the system, inverse micelles may form to create charged entities in the system and thus the electrode polarization should also be taken into account. Temperature dependence of the relaxation time of those polarization may provide a way for identifying which polarization exists in the ER fluids.

For thermal ion polarization, the relaxation time associated with ion movement can be expressed as: [5]

$$\tau_{ion} = \frac{1}{2\nu} e^{\frac{U}{k_B T}} \tag{33}$$

where U is the activation energy between two equilibrium positions, ν is the ion oscillatory frequency. Typically, $U=10^{-19}$ J, $\nu=10^{12}$ Hz. At T=300 K, τ_{ion} is in the scale of 10^{-2} second. With the increase of temperature, τ_{ion} becomes small.

According to Debye [22], since most Debye polarization happens in liquid medium, the relaxation time of Debye polarization can be expressed on the basis of Stokes' law

$$\tau_d = \frac{6\pi\eta r^3}{2k_B T} = \frac{3\pi\eta r^3}{k_B T} \tag{34}$$

where r is the radius of dipole. Eq. (34) assumes that dipole behaves like a sphere experiencing frictional force when it rotates in an electric field. Suppose that the viscosity of liquid medium following the simple Arrhenius viscosity equation, as shown in Chapter 2, Eq. (10)

$$\eta = Ae^{\frac{E_a}{k_B T}} \tag{35}$$

Substituting Eq.(35) into Eq. (34) leads to:

$$\tau_d == \frac{3\pi A r^3}{k_B T} e^{\frac{E_a}{k_B T}} \tag{36}$$

Note the difference between Eq. (25) and (36). Eq. (36) indicates that there is a relatively complicated relationship between the relaxation time of Debye polarization and temperature in comparison with thermal ion polarization. The relaxation time of Debye polarization is dependent on the size of the molecule or the dipole, which is exactly true when the size scale is large, such as polymer chains.

For the interfacial polarization, according to Eq. (26), the relaxation time is the function of dielectric constants of both the medium and particle and the conductivity of the particle. For most solid materials especially for ionic solids, the conductivity can be expressed as [5; 23]:

$$\sigma_p = \sigma_0 e^{-\frac{E_b}{k_B T}} \tag{37}$$

where E_b is the activation energy. Substituting Eq. (37) into Eq. (26) leads to

$$\tau_i = \frac{\varepsilon_0 \left(2\varepsilon_m + \varepsilon_p\right)}{\sigma_0} e^{\frac{E_b}{k_B T}} \tag{38}$$

Since both ε_m and ε_p are function of temperature, they may be approximately expressed as:

$$\varepsilon_m = \varepsilon_m^r \left(1 + (T - T_r)\frac{d\varepsilon_m}{dT}\right) \tag{39}$$

and

$$\varepsilon_p = \varepsilon_p^r \left(1 + (T - T_r) \frac{d\varepsilon_p}{dT} \right) \tag{40}$$

where ε_m^r and ε_p^r are dielectric constant of the medium and the particle at reference temperature, T_r. $\frac{d\varepsilon_p}{dT}$ and $\frac{d\varepsilon_m}{dT}$ will be given an expression in chapter 8. Substituting Eq. (39) and Eq. (40) one may obtain :

$$\tau_i = \frac{\varepsilon_0 \left(2\varepsilon_m^r + \varepsilon_p^r \right) + \varepsilon_0 (T - T_r) \left(2\varepsilon_m^r \frac{d\varepsilon_m}{dT} + \varepsilon_p^r \frac{d\varepsilon_p}{dT} \right)}{\sigma_0} e^{\frac{E_b}{k_B T}} \tag{41}$$

A more complicated relationship between the relaxation time of the interfacial polarization and temperature is obtained. Usually for ER fluids, $\frac{d\varepsilon_m}{dT} < 0$, and $\frac{d\varepsilon_p}{dT} > 0$.

As we discussed earlier, for the electrode polarization there is a relaxation time related to the double layer build up, which is expressed in Eq. (32). Again using Arrhenius viscosity equation to replace the viscosity term in Eq. (32) leads to

$$\tau_e = \frac{3\pi\varepsilon_m \varepsilon_0 r_i A}{ce^2} e^{\frac{E_a}{k_B T}} \tag{42}$$

There should be another relaxation time related to the ions transversing the measurement cell (see section 6 in this chapter)

$$\omega_1 = \frac{4D}{d^2} \tag{43}$$

where ω_1 is another characteristic angular frequency, D is the diffuse coefficient, d is the distance between two electrodes. Using the Stokes-

Einstein equation expressed in Eq.(31), Eq. (43) can thus be rewritten as:

$$\omega_1 = \frac{2k_B T}{3\pi d^2 r_i \eta} \tag{44}$$

Again using the Arrhenius viscosity equation to replace the viscosity term leads to

$$\omega_1 = \frac{2k_B T}{3\pi d^2 r_i A} e^{-\frac{E}{RT}} \tag{45}$$

Eq.(42) can be further written as:

$$\ln(\frac{1}{\tau_e}) = \ln(\frac{ce^2}{3\varepsilon_m \varepsilon_0 r_i A}) - \frac{E}{RT} \tag{46}$$

Eq. (46) indicates that the natural logrithium of the relaxation time of double layer build up should have a linear relationship with (1/T), which can be used to determine if the dielectric relaxation peak is resulted from the electrode polarization. Note that $1/\tau_e = 2\pi f_m$, where f_m is the frequency corresponding to the dielectric peak position. Figure 8 shows an example, from ref. [24] for zeolite/silicone oil suspension in which the ion transportation is dominant, and the electrode polarization contributes to the dielectric dispersion peak. Note that the Arrhenius viscosity equation is only an approximation for many materials, and the constant is actally temperature dependent. According to the derivation made in Chapter 2 on the basis of the rate process theory, the viscosity of a liquid can be exactly expressed as:

$$\eta = AT^{3/2} e^{E_0/k_B T} \tag{47}$$

Where A is a constant. Substituting Eq.(47) into Eq.(32) and (44) leads to:

$$\omega_0 = 1/\tau_e = \frac{ce^2}{3A\varepsilon_m \varepsilon_0 r_i} T^{-3/2} e^{-E_0/k_B T} \tag{48}$$

$$\omega_1 = \frac{2k_B T^{-1/2} e^{-E_0/k_B T}}{3\pi\ d^2 A r_i} \tag{49}$$

Eq.(48) and (49) indicate that ω_0 is much more sensitive to temperature than ω_1. As temperature decreases, ω_0 moves to a high frequency range, and ω_1 moves to a low frequency range. The relaxation time corresponding to those two characteristic frequencies are

$$\tau_e = \frac{3A\varepsilon_m \varepsilon_0 r_i}{ce^2} T^{3/2} e^{E_0/k_B T} \tag{50}$$

$$\tau_1 = \frac{3\pi d^2 A r_i}{2k_B} T^{1/2} e^{E_0/k_B T} \tag{51}$$

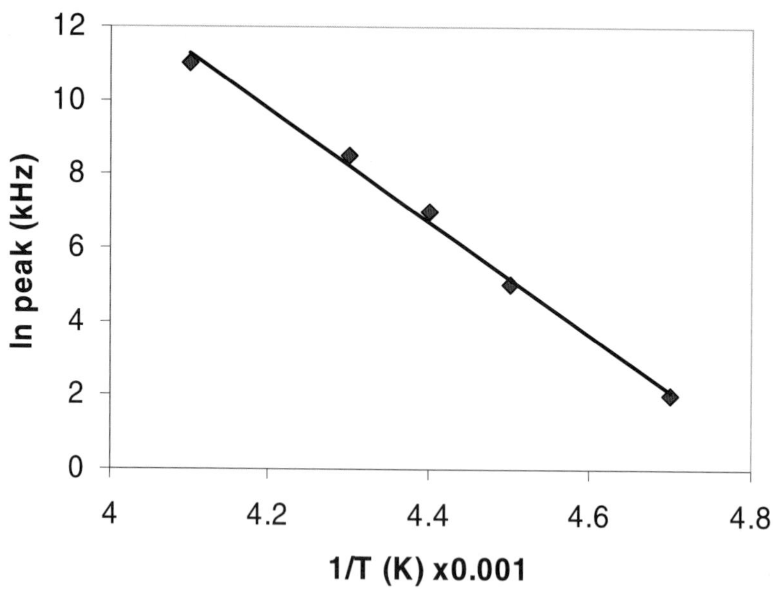

Figure 8 The peak position vs. the reciprocal of temperature for zeolite 4A/silicone oil suspension. The ions from zeolite are responsible for the dielectric dispersion peak. Redrawn from F.E. Filisko, and D.R. Gamota, Recent Adv. Non-newtonian Flows, ASME, 153(1992)75.

4. DIELECTRIC RELAXATION

Under an electric field, the entities are polarized, departing from one equilibrium state to a new equilibrium state. This process is time related one and can be characterized with a relaxation time either by a single value or a function.

4.1 Single relaxation time

Under a dc step field E, the electric displacement D takes time to reach an equilibrium state, as shown in Figure 9.

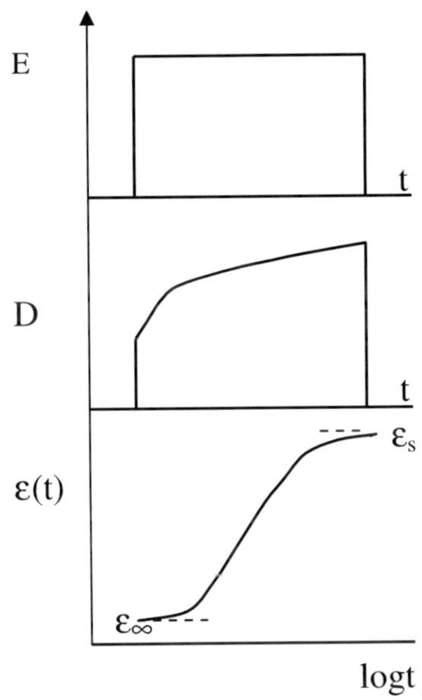

Figure 9 Schematic illustration of the electric displacement D and dielectric constant ε(t) under a step dc electric field.

Considering that D=εE, one may assume $[\varepsilon(t)-\varepsilon_\infty]E$ is proportional to an internal parameter p:

$$[\varepsilon(t)-\varepsilon_\infty]E = kp \tag{52}$$

where k is a constant. The parameter p is a measure of the degree to which the polarization departs from the equilibrium state. Suppose that the equilibrium value of p, denoted p_e, is proportional to E:

$$p_e = k_1 E \tag{53}$$

where k_1 is a constant. The rate of change of p from the equilibrium state p_e is:

$$\dot{p} = \frac{dp}{dt} = -\frac{p - p_e}{\tau} \tag{54}$$

where τ is the relaxation time. Integrating Eq.(54)

$$p = p_e\left(1 - e^{-t/\tau}\right) \tag{55}$$

At equilibrium state, $\varepsilon(t)=\varepsilon_s$, the static dielectric constant, Then one may have

$$kk_1 = \varepsilon_s - \varepsilon_\infty \tag{56}$$

When an ac electric field, $E = E_0 \exp(i\omega t)$, is applied to a material, the dielectric displacement, $D = D_0 \exp i(\omega t - \delta_E)$, and $D^* = \varepsilon^* E^*$. The solution to Eq. (54) is:

$$p^* = \frac{p_e}{1 + i\omega t} \tag{57}$$

Multiplying k leads to

$$\varepsilon^* = \varepsilon_\infty + \frac{(\varepsilon_s - \varepsilon_\infty)}{1 + i\omega\tau} \tag{58}$$

Separating the real and imaginary part of Eq. (58):

$$\varepsilon' = \varepsilon_\infty + \frac{(\varepsilon_s - \varepsilon_\infty)}{1+\omega^2\tau^2}$$
$$\varepsilon'' = (\varepsilon_s - \varepsilon_\infty)\frac{\omega\tau}{1+\omega^2\tau^2} \tag{59}$$

Eq. (59) is known as the Debye equation. ε″ shows a maximum at $\omega_{max} = 1/\tau$, which forms a basis for determining the relaxation time of the Debye polarization. A method for checking Eq.(59) was proposed by Cole-Cole [25] via plotting ε″ against ε′. From Eq. (59) one can obtain:

$$\left[\varepsilon' - \frac{\varepsilon_s + \varepsilon_\infty}{2}\right]^2 + (\varepsilon'')^2 = \left(\frac{\varepsilon_s - \varepsilon_\infty}{2}\right)^2 \tag{60}$$

Eq.(60) indicates a semicircle of radius $\frac{\varepsilon_s - \varepsilon_\infty}{2}$ will be obtained in the Cole-Cole plot, which is uninfluenced by the frequency range or the relaxation time.

4.2 Multiple relaxation times

Cole and Cole [25] modified single relaxation Debye equations (58) by replacing the term $(1+i\omega\tau)$ with $1+(i\omega\tau)^\alpha$, where α is a parameter, 0<α≤1. Then

$$\varepsilon' = \varepsilon_\infty + \frac{(\varepsilon_s - \varepsilon_\infty)\left[1+(\omega\tau)^\alpha \cos\alpha\pi/2\right]}{1+2(\omega\tau)^\alpha \cos\alpha\pi/2 + (\omega\tau)^{2\alpha}} \tag{61}$$

$$\varepsilon'' = \frac{(\varepsilon_s - \varepsilon_\infty)\left[(\omega\tau)^\alpha \sin\alpha\pi/2\right]}{1+2(\omega\tau)^\alpha \cos\alpha\pi/2 + (\omega\tau)^{2\alpha}} \tag{62}$$

The Cole-Cole curve of Eq. (61) and (62) gives a depressed semicircle of radius $\frac{\varepsilon_s - \varepsilon_\infty}{2}\operatorname{cosec}\frac{\alpha\pi}{2}$. However, the Cole-Cole curve is still symmetrical about ωτ=1. When α=1, Eq.(61) and (62) reduces to the Debye equation. Figure 10 schematically shows how the parameter α changes the

Cole-Cole plot. Note that log ε" vs. log ω has the slope α at low frequency and the slope –α at high frequency, which may be used for determining the parameter α.

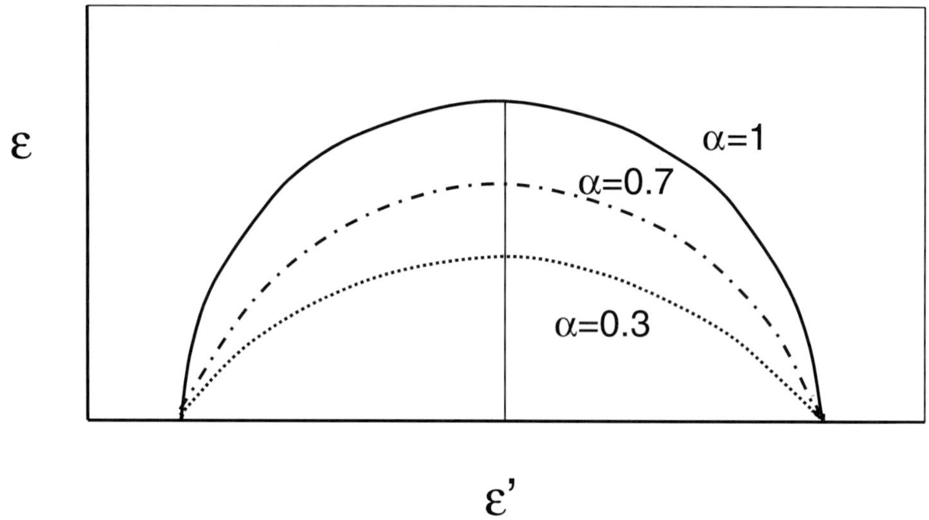

Figure 10 Schematic Cole-Cole plot of different parameter α.

It is easy to see that the Cole-Cole plot is not a symmetrical semicircle, but rather a skewed curve. Davidson [26] replaced the term $(1+i\omega\tau)$ with $(1+i\omega\tau)^{\alpha}$ in Eq. (58), resulting in the following equation:

$$\varepsilon' = \varepsilon_{\infty} + (\cos\theta)^{\alpha} \cos\alpha\theta(\varepsilon_s - \varepsilon_{\infty}) \quad (63)$$

$$\varepsilon'' = (\cos\theta)^{\alpha} \sin\alpha\theta(\varepsilon_s - \varepsilon_{\infty}) \quad (64)$$

where $\tan\theta = \omega\tau$. Eq.(63) and (64) can describe many systems adequately[1], indicating that many practical systems have multiple relaxation times. In most cases the Havilliak-Negami (HN) equation [27] can be used to describe the dielectric data of the heterogeneous system:

$$\varepsilon^* = \varepsilon_\infty + \frac{(\varepsilon_s - \varepsilon_\infty)}{[1+(i\omega\tau)^\alpha]^\beta} \qquad 0<\alpha,\ \beta<1 \tag{65}$$

When $\alpha=1$, $\beta=1$, Eq.(65) reduces to the Debye equation; $\alpha\neq1$, $\beta=1$, it reduces to the Cole-Cole equation; $\alpha=1$, $\beta\neq1$, it reduces to the Davidson-Cole equation.

5. DIELECTRIC PROPERTY OF MIXTURE

For a mixture of two components having the dielectric constant of small difference, the dielectric constant of the mixture, ε_{mix}, can be expressed as following equation [5]:

$$\varepsilon_{mix}^k = \phi_1\varepsilon_1^k + (1-\phi_1)\varepsilon_2^k \tag{66}$$

where ϕ_1 and ε_1 are the volume fraction and the dielectric constant of the first component, respectively. ε_2 is the dielectric constant of the second component. k is a parameter determined by those two components distributed in the system. When those two components are aligned to the direction perpendicular to the electric field, k= -1 (see Figure 11a)

$$\frac{1}{\varepsilon_{mix}} = \frac{\phi_1}{\varepsilon_1} + \frac{(1-\phi_1)}{\varepsilon_2} \tag{67}$$

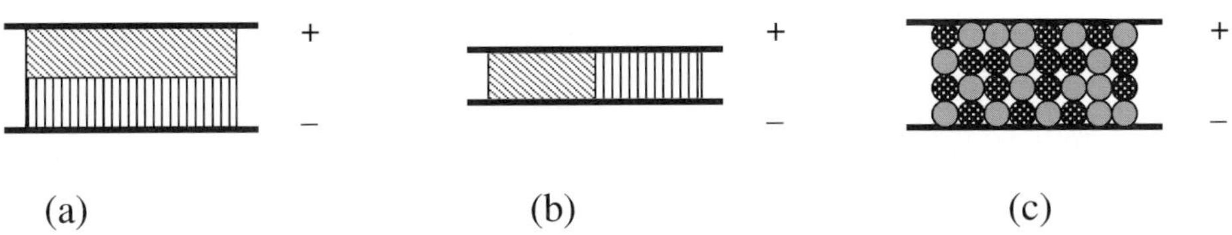

Figure 11 Schematic illustration of the mixture of two dielectric components. (a) two components stacked together; (b) two components aligned in parallel; (c) two components homogeneously mixed together.

When two components are aligned in parallel as shown in Figure 11b, k=1

$$\varepsilon_{mix} = \phi_1 \varepsilon_1 + (1 - \phi_1)\varepsilon_2 \tag{68}$$

In the case that two components are homogeneously mixed together, k approaches to zero. Differentiate Eq.(66) and divide k for both sides

$$\varepsilon_{mix}^{k-1} d\varepsilon_{mix} = \phi_1 \varepsilon_1^{k-1} d\varepsilon_1 + (1 - \phi_1)\varepsilon_2^{k-1} d\varepsilon_2 \tag{69}$$

when k«1, Eq.(69) can be approximated as:

$$\frac{d\varepsilon_{mix}}{\varepsilon_{mix}} \approx \phi_1 \frac{d\varepsilon_1}{\varepsilon_1} + (1 - \phi_1)\frac{d\varepsilon_2}{\varepsilon_2} \tag{70}$$

Integrate Eq.(70) and consider $\varepsilon_{mix} = \varepsilon_1$ at ϕ_1=1. one may obtain:

$$\ln \varepsilon_{mix} = \phi_1 \ln \varepsilon_1 + (1 - \phi_1)\ln \varepsilon_2 \tag{71}$$

Eq. (71) is the commonly used logarithm relationship for a mixed dielectric system. Note that this relationship is only valid in the condition that

$$\varepsilon_1^k \approx \varepsilon_2^k \quad \text{at } k \ll 1 \tag{72}$$

Similarly the conductivity and the dielectric loss tangent of the mixture should also follow the logarithm relationship:

$$\ln \sigma_{mix} = \phi_1 \ln \sigma_1 + (1 - \phi_1)\ln \sigma_2 \tag{73}$$

$$\ln tg\delta_{mix} = \phi_1 \ln tg\delta_1 + (1 - \phi_1)\ln tg\delta_2 \tag{74}$$

For a mixture where one particulate material dispersed into another continuous material, the dielectric constant becomes relatively complicated. It is dependent on how particles distribute in the continuous phase. There are three typical microstructures that particles may take, the ordered, random and fractal structures, as shown in Figure 12. For dilute systems of nonagglomerating particles (the particle volume fraction is less than 10 vol%), no matter that the particle takes the ordered or random distribution structure, the dielectric constant of such a system can be expressed by the Maxwell-Garnett-Sillars formula [28]

$$\varepsilon_{mix} = \varepsilon_m \left(1 + \frac{\phi K}{1 - \phi K} \right) \tag{75}$$

where

$$K = \frac{(\varepsilon_p - \varepsilon_m)}{\varepsilon_m + A(\varepsilon_p - \varepsilon_m)} \tag{76}$$

where ε_p and ε_m are the dielectric constant of particle and dispersing medium, respectively. ϕ is the particle volume fraction. A is dependent on the particle shape and the form. For spheres A=1/3. A is of the value between 0 and 1, Eq. (75) is only valid for the dilute system in which there is no interaction between particles.

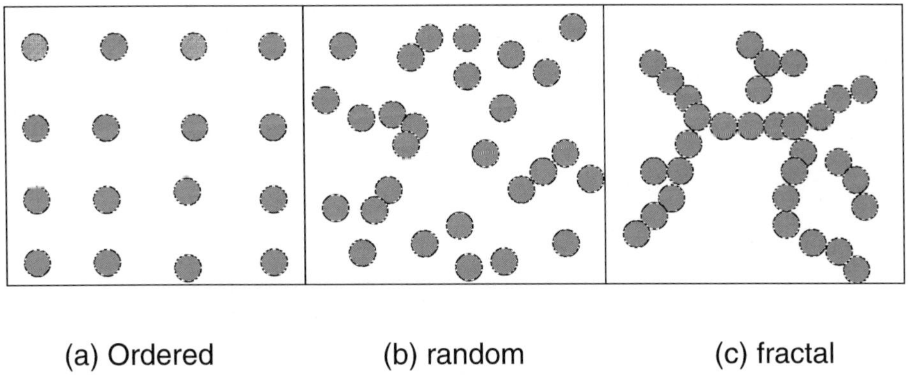

Figure 12 Particle distribution in a continuous phase, (a) ordered structure (particle crystal); (b) random distribution; (c) fractal structure.

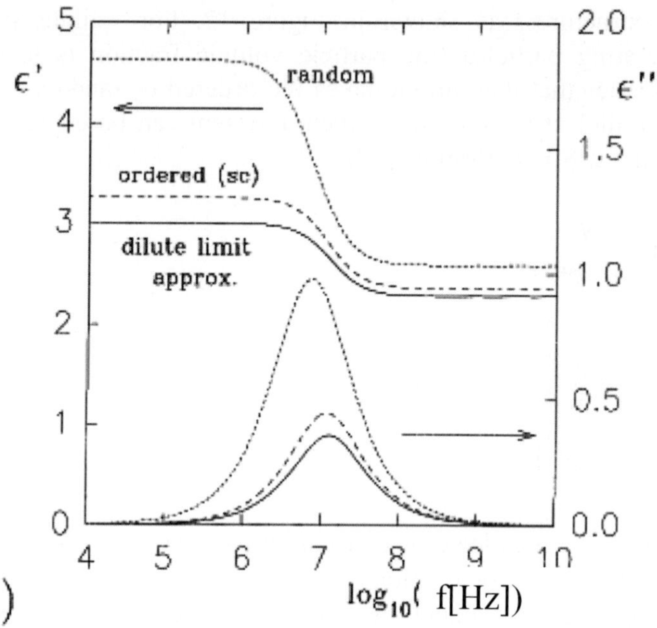
ε'
ε''
random
ordered (sc)
dilute limit
approx.
log10(f[Hz])

a)

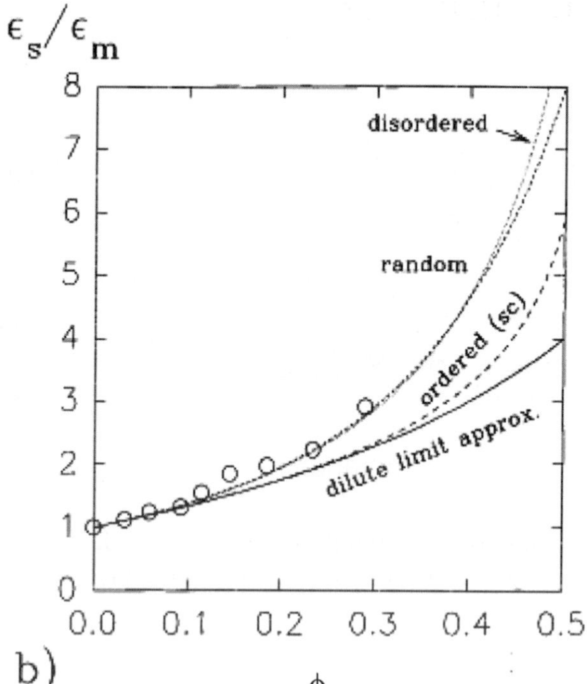
εs/εm
disordered
random
ordered (sc)
dilute limit approx.
ϕ

b)

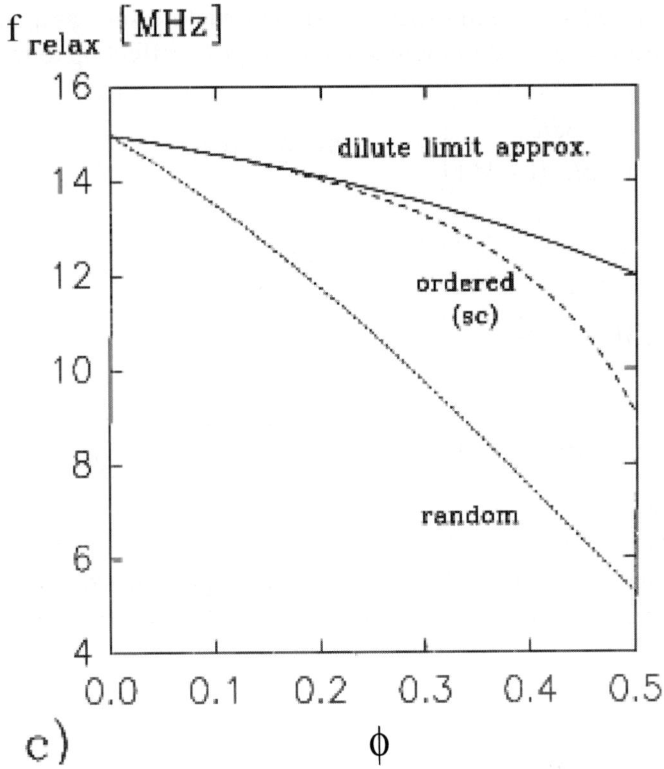

Figure 13 Dielectric property of well-separated spheres embedded in an insulating matrix. (a) the dielectric constant and dielectric loss vs. frequency; (b) The ratio of the static dielectric constant of the mixture to that of the dispersing medium vs. particle volume fraction; (c) the relaxation frequency $f_{relax}=1/\tau$ vs. particle volume fraction. Dilute limitation is calculated from Eq. (75). The ordered structure of simple cubic lattice (sc) is calculated from Eq. (77). The random distribution structure is calculated from Eq. (80). The disordered structure means the particle aggregates formed in the system and is calculated from Eq. (75) with K assumed as $K=1+(\phi/\phi_m)(1/\phi_m-1)$, where the random close packing is assumed, $\phi_m=0.63$, the maximum packing fraction. Reproduced with permission from R. Pelster, and U. Simon, Colloid Polym. Sci., 277(1999)2

For a concentrated system of ordered particle structure such as a cubic lattice, Eq.(75) has been extended to include the multipoles effect up to n=7 [29, 30]

$$\varepsilon_{mix} = \varepsilon_m \left(1 + \frac{3\phi K_n}{1 - \phi K_n (1 + M)} \right) \qquad \text{for } \phi \le 0.46 \tag{77}$$

where

$$K_n = \frac{(\varepsilon_p - \varepsilon_m)}{[\varepsilon_p + \varepsilon_m (1 + n)/n]} \tag{78}$$

$$M = b_1 K_5 \phi^{11/3} + c_1 K_7 \phi^5 - a\phi^{7/3} \times \frac{1 + c_2 K_5 \phi^{11/3} + c_3 K_5^2 \phi^{22/3}}{K_3^{-1} + b_2 \phi^{7/3} + c_4 K_5 \phi^6} \tag{79}$$

with a=1.3045, b_1=0.01479, b_2=0.4054, c_1=0.1259, c_2=0.5289. c_3=0.06993, c_4=6.1673. For a concentrated system of random particle spatial distribution, the Hanai-Bruggeman formula can be used to describe the dielectric property [31, 32]:

$$(1 - \phi)^3 \frac{\varepsilon_{mix}}{\varepsilon_m} = \left(\frac{\varepsilon_{mix} - \varepsilon_p}{\varepsilon_p - \varepsilon_m} \right)^3 \qquad \text{for } \phi < \phi_c \tag{80}$$

where ϕ_c is the percolation threshold particle volume fraction. The dielectric property of the particle dispersion system of different particle packing structure is compared in Figure 13, based on Eq. (75, 77, 80). The random particle distribution structure gives the highest dielectric constant. Clearly, the dielectric tool can be used for detecting the particle packing structure.

6. DIELECTRIC PROPERTY OF NON-AQUEOUS SYSTEMS WITH CHARGING AGENT

6.1 Charging agent or dispersant

The charging agent or dispersant used in non-aqueous system is a surfactant for making particles charged or well dispersed, forming a stable suspension. As mentioned in chapter one, the charging agent usually will form an inverse micelle structure, trapping all polar impurities or residues from non-aqueous oil phase into the center core. Commonly used charging agents include calcium diisopropylsalicylate [33], tetraisoamylammonium picrate [33], polyisobutylene succinimide, OLOA 1200 (from Chevron) [34-36], di-(2-ethylhexyl) sodium sulfosuccinate (Aerosol OT, AOT from American Cyanamid) [37-39], Octyl pheol ethoxylate (Triton X-100) [40], phosphatidylcholine (lecithin)[41], etc. The molecular structures of those charging agents are shown in Figure 14. As one may notice, all charging agents may contain both a positive and negative group, except Triton X-100, which is more commonly used as a dispersant.

6.2 Charging mechanism based on the conductivity data

As one may already knows, the conductivity of a pure organic liquid is about 10^{-13} S/m. Once a charging agent is added in and forms the inverse micelles, the conductivity of whole system can easily reach 10^{-7} S/m, which is a several order of magnitude increase. Such a dramatic increase makes one believe that the substantial amount of ions is created in the system. The next question is how the inverse micelles gain the charge, and where the charge comes from. The charging mechanism is directly related to the dielectric property, and needs to be addressed first.

A common phenomenon observed for the inverse micelle system is that the conductivity increases linearly with the concentration of charging agent. Figure 15 shows the conductivity of AOT/dodecane solution vs. AOT concentration ranging from 0.8 to 200 mM, from ref.[45]. A very good linear relationship is obtained. A similar linear relationship between the conductivity of the micelle solution and the charging agent concentration was observed in OLOA 1200/dodecane [42], AOT/benzene [43], and lecithin/hydrocarbon (Isopar H from Exxon) system [44]. This common phenomenon is obviously backed up by a common charging mechanism in the non-aqueous micelle system.

The conductivity of a liquid medium containing ions is well understood [5]. Usually the thermal motion is not strong enough to dissociate the molecule and the motion of molecule will contribute to the conductivity. Due to the collision from neighboring molecules, the molecule will travel in a distance in the scale of its size and then stay there. The relative number of the molecules staying at the ground state and the free state or activation state determine the conductivity. Suppose that the potential at ground state is U_1, and that at the activation state is U_2, and the number of molecules per unit volume is n_o, then:

$$n_1 = An_0 e^{-\frac{U_1}{k_B T}} V_1 \tag{81}$$

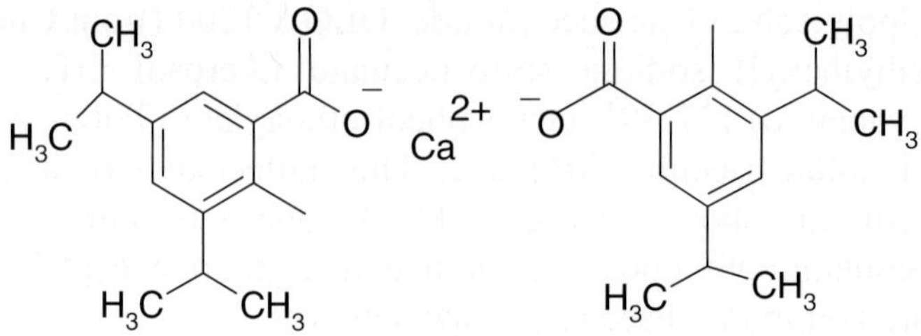

Calcium diisopropylsalilyte

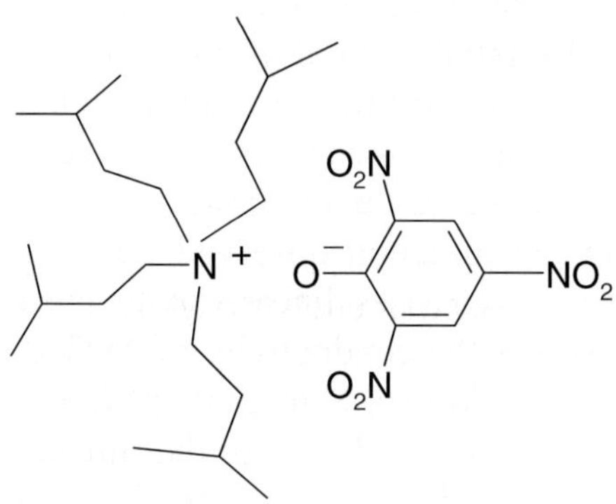

tetraisoamylammonium picrate

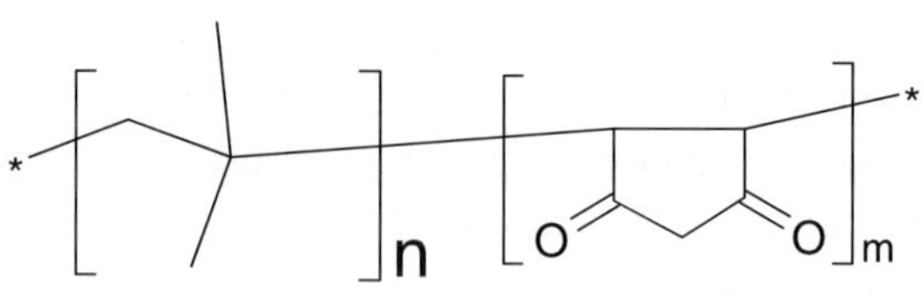

Polyisobutylene succinimide, OLOA 1200

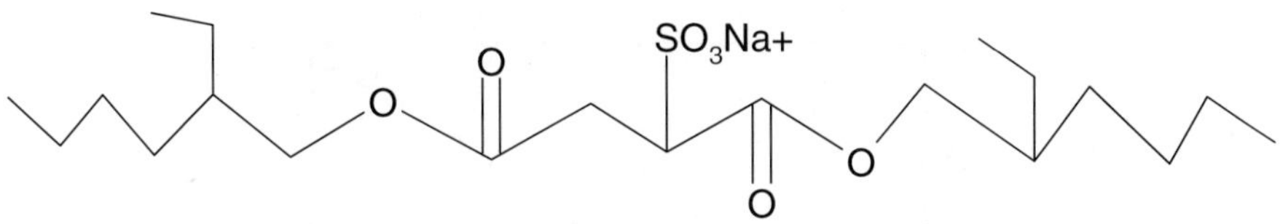

di-(2-ethylhexyl)sodium sulfosuccinate, AOT

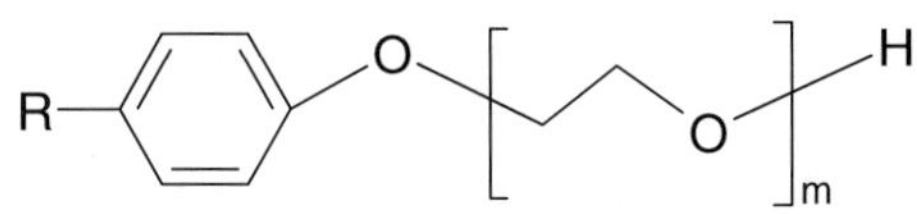

R=Octyl, m=9.5 (average)
Octylphenol ethoxylate, Triton X-100

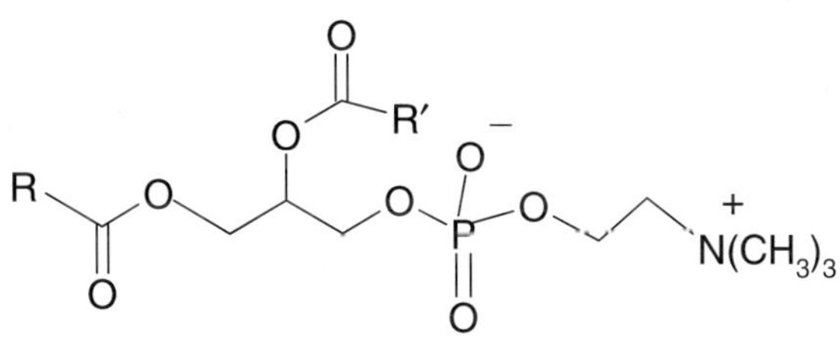

Phosphatidylcholine, Lecithin

Figure 14 Commonly used charging agents or dispersants in nonaqueous system.

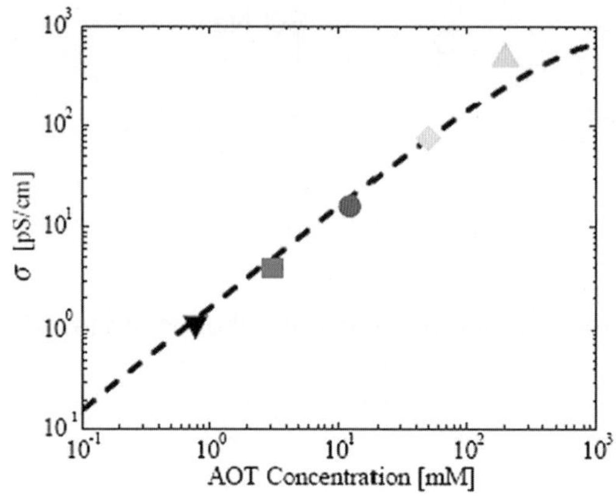

Figure 15. The conductivity of AOT/dodecane solution vs. AOT concentration, ranging from 0.8 to 200 mM. Redrawn with permission from M.F. Hsu, E.R. Dufresne, and D.A. Weitz, Langmuir, 21(2005)4881

$$n_2 = An_0 e^{-\frac{U_2}{k_B T}} V_2 \tag{82}$$

where V_1 and V_2 are the volumes occupied by each state of molecule. A is a constant. The dissociation constant α can be expressed as:

$$\alpha = \frac{n_2}{n_0} = \frac{n_2}{n_1 + n_2} = \frac{e^{-\frac{U_2 - U_1}{k_B T}}}{1 + e^{-\frac{U_2 - U_1}{k_B T}}} \tag{83}$$

$U_2 - U_1 = U_0$, the dissociation energy.

$$\alpha = \frac{e^{-\frac{U_0}{k_B T}}}{1 + e^{-\frac{U_0}{k_B T}}} = \frac{1}{1 + e^{\frac{U_0}{k_B T}}} \tag{84}$$

For a low dielectric constant liquid, U_0 is very large, $U_0 >> k_B T$, making $\alpha \approx 0$. Without an external electric field, the molecule motion is random. The number of molecule in one of three perpendicular axes in a space is $\frac{n_0}{3}$, and the number of molecule in the positive axis is $\frac{1}{2} \bullet \frac{n_0}{3}$. So in a unit time period the number of molecule overcoming the energy barrier, U_0, is:

$$n = \frac{n_0}{6} \nu e^{-\frac{U_0}{k_B T}} \tag{85}$$

where ν is the molecule vibration frequency at certain position, and it is equal to the hopping times of a molecule trying to overcome the energy barrier, U_0 per unit time. Once an electric field E is applied, the excess amount of ions moving along the direction of the electric field is:

$$\Delta n = \frac{n_0}{6} \nu \left(e^{-\frac{U_0 - \Delta U}{k_B T}} - e^{-\frac{U_0 + \Delta U}{k_B T}} \right) \tag{86}$$

where

$$\Delta U = \frac{qE\delta}{2} \tag{87}$$

where q is the charge of ions, δ is the ion free walking distance, in the scale of 10^{-9}m. Usually $\Delta U << k_B T$, which can be confirmed by a simple calculation as follows. Suppose $\Delta U = k_B T$, and at room temperature 300 K,

$$E = \frac{2k_B T}{q\delta} = \frac{2 \times 1.38 \times 10^{-23} \times 300}{1.6 \times 10^{-19} \times 10^{-9}} = 5.18 \times 10^7 \quad \text{V/m} \tag{88}$$

The applied electric field is far less than 5.18 $\times 10^7$ V/mm, making the assumption, $\Delta U << k_B T$, always valid. So Eq. (86) can be approximated by using the equation $e^x \approx 1 + x$ ($|x| < 1$):

$$\Delta n = \frac{n_0}{6} v e^{-\frac{U_0}{k_B T}} \frac{2\Delta U}{k_B T} = \frac{n_0 q E \delta v}{6 k_B T} e^{-\frac{U_0}{k_B T}} \tag{89}$$

The velocity of the excess amount of ions moving along the electric field is the relative number of the excess amount of ions multiplying the moving distance per second

$$v_E = \frac{\Delta n}{n_0} \delta \tag{90}$$

so the mobility of ion is:

$$\mu_E = \frac{v_E}{E} = \frac{q \delta^2 v}{6 k_B T} e^{-\frac{U_0}{k_B T}} \tag{91}$$

If the concentration of ion is known, then the conductivity of the ion contributed by ion motion along the direction of the electric field thus can be expressed as:

$$\sigma = n_0 q \mu_E = \frac{n_0 q^2 \delta^2 v}{6 k_B T} e^{-\frac{U_0}{k_B T}} \tag{92}$$

Eq. (92) indicates that the conductivity of ions is dependent on the ion free walking distance, the vibration frequency of the ion, temperature, and the energy barrier for the ion to move away from neighboring ions. It is independent of the viscosity of the liquid medium and the applied electric field. The important feature in Eq. (92) is that the conductivity of the ion linearly increases with the ion concentration. It is worth mentioning that there is another simple way for deriving the conductivity of an ion. Suppose that the ion is spherical and travels in an electric field as in electrophoresis, one may obtain the mobility of such an ion by balancing the friction force governed by Stokes' law ($f = 6\pi\eta a v$) and the electrical force ($f = qE$):

$$\mu_E = \frac{q}{6\pi a \eta} \tag{93}$$

Then the conductivity due to the ion motion is:

$$\sigma = n_0 q \mu_E = \frac{n_0 q^2}{6\pi a \eta} \tag{94}$$

Eq. (94) also predicts a linear relationship between the conductivity and the ion concentration. However, the conductivity depends on the ion size and the viscosity of the liquid. Note that the Stokes' law and the equations developed from it apply to spherical and rigid particle only. There is a nonslip stationary surface between the liquid and the solid particle for particle to experience a frictional force. The inverse micelle is a soft entity without a clear boundary between the micelle surface and the liquid medium. Its shape can be easily distorted under a frictional force on the surface. Caution should thus be taken for applying Eq. (94) to the inverse micelles. Eq. (92) is therefore preferred for describing the conductivity of the micelle system in the following discussion.

For inverse micelles, Eq. (92) indicates that the conductivity is proportional to the square of the free walking distance. The free walking distance is related to the size of the micelle, the distance between the micelles, and the interaction between the micelles. The size of inverse micelle is uniform and the percentage of charged micelles is believed to be very small, implying that the distance between the charged inverse micelles is large. The interaction between the charged micelles is thus negligible, and the free walking distance can be reasonably assumed to the size of the micelle. Suppose that the diameter of the micelle is d_m, then Eq. (92) can approximated as:

$$\sigma = \frac{n_0 q^2 d_m^2 \nu}{6 k_B T} e^{-\frac{U_0}{k_B T}} \tag{95}$$

Eq. (95) indicates that the conductivity of micelle may be directly proportional to the square of the diameter of the micelle, which is quite different from Eq. (94), where the conductivity is inversely proportional to the micelle diameter. If the dramatic conductivity increase in the presence of the inverse micelles is believed to be resulted from the charged micelle, then the large micelle size may be favorable for large conductivity, as the micelles are able to travel a large distance, according to Eq. (95). In contrast, Eq. (94) predicts that the conductivity increases as the decrease of the micelle size, which contradicts to the physical picture on how the charged micelle contribute to the conductivity and conflict with the experimental fact that the very small amount of charging agent can substantially increase the conductivity of whole system.

There are several charging mechanisms proposed for explaining how inverse micelles attain charges in nonaqueous medium [34, 46-50], as discussed in Chapter 1. Those mechanisms can be roughly classified as: a) Preferential ion adsorption mechanism. The ions from either impurities or a trace amount of water can be trapped by the inverse micelles and make the inverse micelle charged, In this mechanism, the charge is uncontrollable. b) Dissociation of weak electrolyte, The micelle behaves like a weak electrolyte, dissociated into a positively charged or negatively charged parts. However, this mechanism will lead to the conductivity variation with the square root of the concentration; c) Collision mechanism. The micelles become charged by exchange of charge between two uncharged micelles due to collision. This mechanism predicts that the conductivity linearly increases with the concentration. However, according to the rate theory, it is hard to believe that the probability of charge exchange between micelles is extremely low without an additional driving force. Those three mechanisms are schematically depicted in Figure 16. According to the first mechanism, the charged micelle concentration is dependent on the concentration of impurities and independent of the concentration of micelles, which is contradictory to the experimental observation shown in Figure 15. According to the second mechanism, suppose the original micelle is M_{AB}, and an equilibrium reaction takes place like:

$$M_{AB} \Leftrightarrow M_A^+ + M_B^- \tag{96}$$

The equilibrium constant of such a reaction can be expressed:

$$K_{eq} = \frac{[M_A^+][M_B^-]}{[M_{AB}]} \tag{97}$$

The charged micelle concentration is:

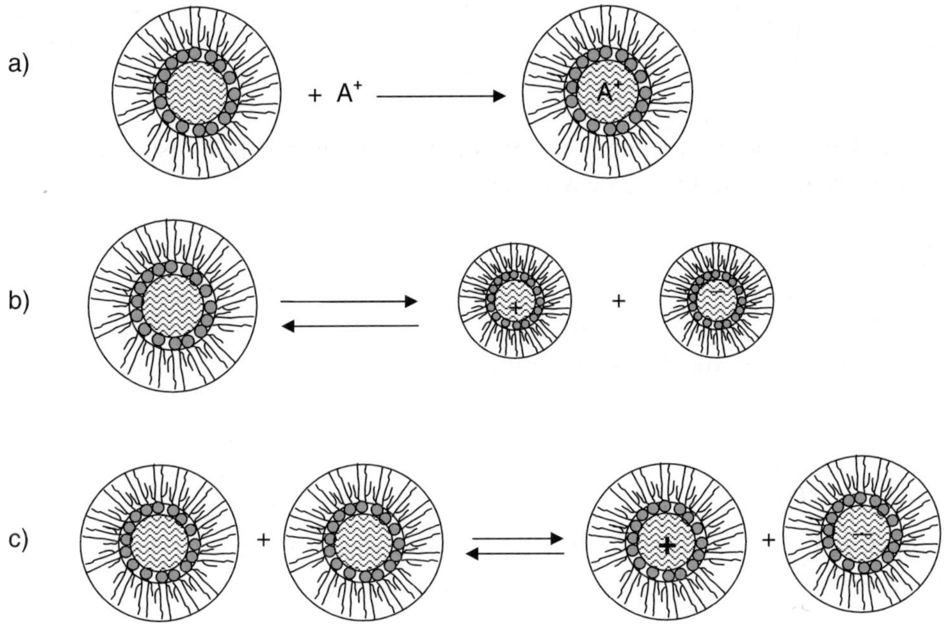

Figure 16 Schematic illustration of micelle charging mechanism. a) Preferential ion adsorption; b) weak-electrolyte-like ionization; c) collision between micelles.

$$n^{+} = n^{-} = K_{eq}^{1/2} n_{0}^{1/2} \tag{98}$$

Substituting Eq. (98) into Eq. (95) and considering both positive and negative micelles will contribute to the conductivity leads to:

$$\sigma = \frac{K_{eq}^{1/2} n_0^{1/2} q^2 d_m^2 \nu}{3k_B T} e^{-\frac{U_0}{k_B T}} \tag{99}$$

Eq. (99) indicates that the conductivity should vary with the square root of the initial concentration of the micelle. For the third mechanism, the equilibrium reaction is:

$$M_A + M_B \Leftrightarrow M_A^{+} + M_B^{-} \tag{100}$$

The equilibrium constant can be expressed as:

$$K_{eq} = \frac{[M_A^+][M_B^-]}{[M_A][M_B]} \quad (101)$$

Since there is no difference between the micelles A and B statistically, their concentration should be same and equal to :

$$[M_A] = [M_B] = \frac{n_0}{2} \quad (102)$$

Thus the charged micelle concentration is:

$$n^+ = n^- = \frac{1}{2} K_{eq}^{1/2} n_0 \quad (103)$$

Substituting Eq. (103) into Eq. (95) and considering both positive and negative micelles will contribute to the conductivity leads to:

$$\sigma = \frac{K_{eq}^{1/2} n_0 q^2 d_m^2 \nu}{6 k_B T} e^{-\frac{U_0}{k_B T}} \quad (104)$$

Eq. (104) predicts a linear relationship between the conductivity and the micelle concentration. Although the third mechanism can successfully explain the conductivity behavior of the inverse micelle, it is unlikely to happen as the Coulombic attraction is so strong in low dielectric constant medium, and the probability of such a reaction is so low based on the rate theory.

The charge in the micelle is most likely induced by the external electric field. If a species contains weakly bound charge pairs, the external electric field may induce the dissociation of such pairs and create more charged species. This phenomenon was first addressed by Onsager [51] who found that the equilibrium constant of such a process is directly proportional to the electric field strength and inversely proportional to the dielectric constant of the medium. In other word, in nonaqueous systems this effect is much more profound than that in aqueous systems [52]. Based on the conductivity data, Denat contended [53, 54] that the inverse micelles may contain a small fraction of completely ionized (strong electrolytes) component, which may be induced by an external electric field. Lane [55] found that the charging takes place by partitioning ionic species into the inverse micelles. A very likely physical picture for micelle charging may be that weakly bound charge pairs on the charging agent or in the inverse micelles dissociate under an electric field, and the dissociated ions are subsequently stabilized in the polar pool of another neighboring inverse micelle. This charging mechanism is schematically depicted in Figure 17, and can be simply expressed as:

$$M \overset{E}{\Leftrightarrow} M_A^+ + B^- \tag{105}$$

$$M + B^- \rightarrow (MB)^- \tag{106}$$

Eq. (105) shows that under an electric field an inverse micelle dissociates into a positively charged micelle and a negative ion. This process is an equilibrium one and the equilibrium constant is dependent on the applied electric field. The negative ion is immediately trapped into the polar pool of another neighboring micelle, making this micelle negatively charged. The overall process can be expressed as:

$$M + M \overset{E}{\Leftrightarrow} M_A^+ + (MB)^- \tag{107}$$

Comparing Eq. (107) with Eq. (100), one may easily reach the conclusion that Eq. (107) will lead to the same conductivity equation, Eq. (104). This means that the electric-field-induced charging mechanism predicts that the conductivity of an inverse micelle system should increase linearly with the concentration of the inverse micelle, which is consistent with experimental observation. The difference between Eq. (107) and Eq. (100), or the difference between the electric-field-induced charging mechanism and the collision mechanism, is that the former will show the conductivity dependence on the electric field, as the equilibrium constant of Eq. (105) is dependent on the electric field. However, the collision charging mechanism will predict the conductivity independent of the electric field. Randriamalala [56] found that the conductivity of AOT solutions indeed increases with applied electric field as predicted by Onsager's theory up to 5.6×10^6 V/m over a wide AOT concentration range, implying that the electric-field-induced charging mechanism is reasonable.

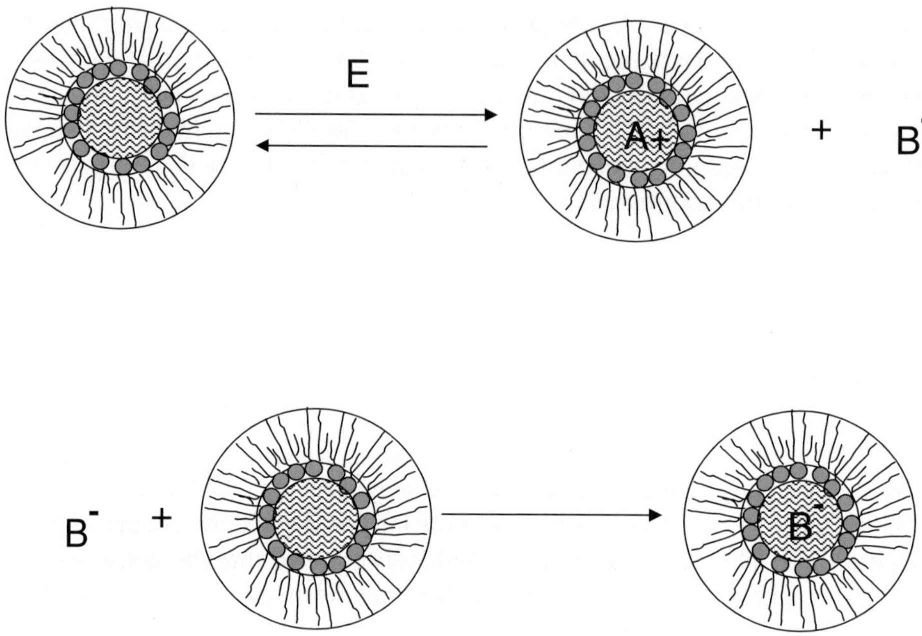

Figure 17 The electric-field-induced charging mechanism in inverse micelle systems.

In any case, the fraction of charged micelle is equal to $K_{eq}^{1/2}$, independent of the charging agent concentration, which is supported experimentally [45]. Again, however, for the electric-field-induced charging mechanism the fraction of charged micelle is dependent on the applied electric field, while for the collision mechanism it is independent of the applied electric field.

In a word, the conductivity of inverse micelle systems is essentially controlled by ionic transportation under an electric field, and the ions are the charged inverse micelles. The electric-field-induced charging mechanism may well describe how inverse micelles are charged and well explain the observed experimental phenomena to date. Since the charged micelles are major components, the dielectric property of the inverse micelle systems may simply controlled by the electrode polarization process, which will be described in next section.

6.3 The electrode polarization in non-aqueous systems

The electrode ploarization has been extensively addressed both in aqueous and non-aqueous systems [57-59]. Since the electrode polarization will substantially distort the dielectric property at low frequency, various theories [57, 58, 60, 61], and correction methods [15, 59, 62, 63] have been developed for modeling the electrode polarization behavior at low frequencies and for finally removing the artifically high dielectric constant resulting from the electrode polarization. Under the assumptions that the polarizability is only slightly modified by the electrolyte, and the electrolyte only carries unit charge [57, 58], the complex dielectric constant of a system can be expressed as:

$$\varepsilon^* = \varepsilon_m \left[1 - \frac{1}{1 + i\omega/\omega_0} \left(1 - \sqrt{\frac{\omega_1/\omega_0}{1 + i\omega/\omega_0}} \right) \right]^{-1} \tag{108}$$

where ε_m is the dielectric constant of the continuous phase, ω is the frequency, ω_0 and ω_1 are two characteristic frequencies related to the building up the double layer and the electrolytes traversing the measurement cell, respectively. ω_0 can be expressed as:

$$\omega_0 = \kappa^2 D \tag{109}$$

where D is the diffusion coeficient, and κ is the inverse Debye screening length given by:

$$\kappa^2 = \frac{2ce^2}{\varepsilon_m \varepsilon_0 k_B T} \tag{110}$$

where c is the number concentration of electrolyte. ω_1 can be expressed as:

$$\omega_1 = \frac{4D}{d^2} \tag{111}$$

where d is the gap size between two electrodes. The ratio between ω_0 and ω_1 is:

$$\frac{\omega_1}{\omega_0} = \left(\frac{2\kappa^{-1}}{d} \right)^2 \tag{112}$$

Note that κ^{-1} is usually in the nanometer scale and d is in millimeter scale, ω_1 is usually two or three orders of magnitude smaller than ω_0. The inverses of ω_0 and ω_1 are the times related to the building up the double layer and the electrolyte diffusively crossing the measurement cell. At sufficiently high frequency, the real part of the dielectric constant separated from Eq. (108) can be expressed as:

$$\varepsilon' = \varepsilon_m \left[1 + \sqrt{\frac{\omega_1}{2}} \omega_0 \omega^{-3/2} + \ldots \right] \tag{113}$$

Eq.(113) clearly indicates that at high frequency the dielectric constant decays with $\omega^{-3/2}$. For a dielectric material, the effective conductivity σ_{eff} is usually expressed as:

$$\sigma_{eff} = \omega \varepsilon'' = \omega_0 \varepsilon_m \left[1 - \sqrt{\frac{\omega_1}{2}} \omega^{-1/2} + \ldots \right] \tag{114}$$

Thus at very high frequency,

$$\sigma_{eff} \approx \omega_0 \varepsilon_m = \frac{2e^2 cD}{k_B T} \tag{115}$$

Eq.(115) shows that the effective conductivity linearly incerases with the number concentration of electrolytes. The direct implication of this derivation is that the electrode polarization effect will make the real part of the dielectric constant become subtantially high only when the frequency is below ω_1. In this frequency range, the charges accumulate at the surface of two electrodes, forming the double layer. Above ω_1, the real part of the dielectric constant decays with the frequency as $\omega^{-3/2}$, which has been observed both in the aqueous [62] and non-aqueous systems [15,64]. And also, it will vary with the gap size between two electrodes, acccording to Eq. (113). Such an example is given in Figure 18 for the sodium bis(2-ethylhexyl)sulfosuccinate (AOT) water-in-oil microemulsion with iso-octane as the continuous phase.

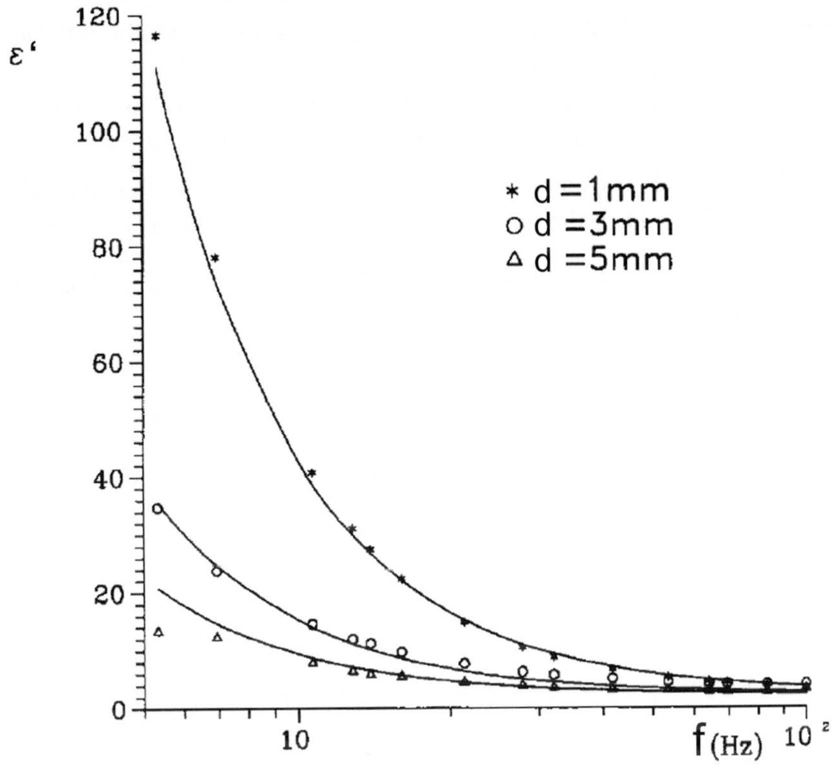

Figure 18 The dielectric constant of the sodium bis(2-ethylhexyl)sulfosuccinate (AOT) water-in-oil microemulsion with iso-octane as the continuous phase vs. frequency obtained at different gap size between two electrodes. The molar water-to-surfactant ratio is 20. Reproduced with permission from P.A. Cirkel, J.M.P. van der Ploeg, and G.J.M. Koper, Physica A, 235(1997)269.

6.4 Inverse micelle size calculated from the dielectric property

The size of the inverse micelle can be roughly estimated with the conductivity equation shown in Eq.(94). However, as mentioned earlier, this equation is only valid for solid particle and may give a wrong estimation for the size of inverse micelles. Since the dielectric spectroscopy can provide the information on the relaxation time of the micelle moving under an electric field, the size of micelle is thus able to be determined via the dielectric measurement.

Einstein [65] derived that the average of the square of the displacement from the initial position of a particle can be expressed as:

$$\overline{x^2} = 2D_{rel}\tau \tag{116}$$

where D_{rel} is the diffusion coefficient relative to another particle and $D_{rel} = 2D$, where D is the self-diffusion coefficient [66, 67]. τ is the time. Since the root mean square (rms) displacement $r_{rms} = (3\overline{x^2})^{1/2}$, Eq. (80) can be rewritten as:

$$r_{rms} = (12D\tau)^{1/2} \tag{117}$$

The relaxation time τ can be correlated with the dielectric dispersion peak frequency, f_{max}, at which the dielectric loss tangent shows a maximum value,

$$\tau = \frac{1}{2\pi f_{max}} \tag{118}$$

and D can be expressed using the Stokes-Einstein equation:

$$D = \frac{k_B T}{6\pi\eta a} \tag{119}$$

Substituting Eq. (118) and (119) into Eq. (117) leads to:

$$r_{rms} = \left(\frac{k_B T}{\pi^2 \eta a f_{max}}\right)^{1/2} \tag{120}$$

r_{rms} can be assumed as the interparticle spacing, which is derived in chapter 2, shown in Eq. (25)

$$r_{rms} = 2\left(\sqrt[3]{\phi_m / \phi} - 1\right)a \tag{121}$$

where ϕ_m is the maximum packing fraction of particle, depending on the packing structure. Combining Eq.(120) and (121) leads to:

$$a = \left[\frac{k_B T}{4\pi^2 \eta f_{\max} \left(\sqrt[3]{\phi_m / \phi} - 1 \right)^2} \right]^{1/3} \tag{122}$$

Eq. (122) can be used for estimating the micelle size. Since f_{max} is the function of the conductivity, and thus the function of concentration, a may not change with the micelle concentration.

7. DIELECTRIC PROPERTY OF NON-AQUEOUS SUSPENSIONS WITHOUT ELECTROLYTE

Without electrolyte, the dielectric property of non-aqueous suspension will be mainly governed by the Maxwell-Wagner polarization, which will be discussed in detail in this section.

7.1 The Wagner-Maxwell model for dilute suspensions

Maxwell [68] theoretically addressed the dielectric dispersion of a heterogeneous system by means of the static conductivity. His formula was late generalized by Wagner [18] to the frequency spectrum of the dielectric property. The dielectric dispersion resulted from the heterogeneity of the system can be well described with the Maxwell-Wagner model. The corresponded polarization that contributes to this type dielectric dispersion is thus called as the Wagner-Maxwell polarization, or the interfacial polarization. The Wagner-Maxwell model assumes that the dispersed particle is spherical and sparsely distributed throughout a continuous material of comparatively low dielectric loss. Since the interparticle distance is so big, the interaction between particles is negligible, which requires the particle volume fraction is usually less than 10 vol%. As mentioned earlier, in a charge-free region, i.e., there is no free charge at the interface between the particle and the dispersing medium, the Laplace equation holds:

$$\Delta\psi=0 \tag{123}$$

where Δ is the Laplace operator, $\Delta = \frac{\partial^2}{\partial x^2} + \frac{\partial^2}{\partial y^2} + \frac{\partial^2}{\partial z^2}$, ψ is the electrostatic field potential. The dielectric displacement D is defined as:

$$\mathbf{D} = \varepsilon_0 \varepsilon \mathbf{E} \tag{124}$$

Since there is no free charge at the interfaces S, the dielectric displacement D should not change through the interface boundary

$$D_m\big|_S = D_p\big|_S \tag{125}$$

Considering $E = -grad\Psi$, Eq. (124) and (125) leads to

$$\varepsilon_m \frac{\partial \Psi_m}{\partial n}\bigg|_S = \varepsilon_p \frac{\partial \Psi_p}{\partial n}\bigg|_S \tag{126}$$

where n is the normal to the surface. Also, the electrostatic potential should be equal at the interface on both sides:

$$\Psi_m\big|_S = \Psi_p\big|_S \tag{127}$$

The solution of Laplace's equation under the boundary conditions expressed in Eq.(126) and (127) is presented [69] as:

$$\Psi_m(r,\theta) = -Er\cos\theta + \frac{\varepsilon_p - \varepsilon_m}{2\varepsilon_m + \varepsilon_p} a^3 E \frac{\cos\theta}{r^2} \tag{128}$$

where r is the distance to the center of the particle, and θ is the angle between the radius **r** and the direction of **E**. a is the radius of the particle. The first term in Eq. (128) stands for the potential of the applied homogeneous field, and the second term gives the potential change due to the polarization of the particle [33]. It is know that [71] the potential produced by a dipole orientated along the axis θ=0 in a medium can be expressed as:

$$\Psi_m(r,\theta) = \frac{D_m}{4\pi\varepsilon_m\varepsilon_0} \frac{\cos\theta}{r^2} \tag{129}$$

where D_m is the dipole moment. Comparing the second term of Eq. (128) with Eq.(129), one may easily reach that the dipole moment induced at an interface is:

$$D_m = 4\pi\varepsilon_0\varepsilon_m \frac{\varepsilon_p - \varepsilon_m}{2\varepsilon_m + \varepsilon_p} a^3 E \tag{130}$$

Suppose that in a volume V the number of small particles dispersed in is N and there is no interaction between particles, then the total dipole moment in this volume V is:

$$D_{total} = D_m N = 4\pi\varepsilon_0\varepsilon_m \frac{\varepsilon_p - \varepsilon_m}{2\varepsilon_m + \varepsilon_p} a^3 EN \tag{131}$$

If the volume V is assumed as a big particle of radius b and of dielectric constant ε, then according to Eq.(130), the total dipole moment D_{total} can be expressed in a different way as:

$$D_{total} = 4\pi\varepsilon_0\varepsilon_m \frac{\varepsilon - \varepsilon_m}{2\varepsilon_m + \varepsilon} b^3 E \tag{132}$$

Eq.(131) should be identical to Eq.(132), resulting in

$$\frac{\varepsilon - \varepsilon_m}{2\varepsilon_m + \varepsilon} = \frac{\varepsilon_p - \varepsilon_m}{2\varepsilon_m + \varepsilon_p} \phi \tag{133}$$

Where ϕ is the particle volume fraction. Eq.(133) can be rewritten as in a more complex form:

$$\frac{\varepsilon^* - \varepsilon_m^*}{2\varepsilon_m^* + \varepsilon^*} = \frac{\varepsilon_p^* - \varepsilon_m^*}{2\varepsilon_m^* + \varepsilon_p^*} \phi \tag{134}$$

Where * denotes the complex dielectric constant. Analogously the complex conductivities can be expressed as:

$$\frac{\sigma^* - \sigma_m^*}{2\sigma_m^* + \sigma^*} = \frac{\sigma_p^* - \sigma_m^*}{2\sigma_m^* + \sigma_p^*} \phi \tag{135}$$

Rearranging Eq.(134) and (135) leads to:

$$\varepsilon^* = \varepsilon_m^* \frac{2\varepsilon_m^* + \varepsilon_p^* - 2\phi(\varepsilon_m^* - \varepsilon_p^*)}{2\varepsilon_m^* + \varepsilon_p^* + \phi(\varepsilon_m^* - \varepsilon_p^*)} = \varepsilon_h + \frac{\varepsilon_l - \varepsilon_h}{1 + i\omega\tau} + \frac{\sigma_l}{i\omega\varepsilon_0} \tag{136}$$

and

$$\sigma^* = \sigma_m^* \frac{2\sigma_m^* + \sigma_p^* - 2\phi(\sigma_m^* - \sigma_p^*)}{2\sigma_m^* + \sigma_p^* + \phi(\sigma_m^* - \sigma_p^*)} = \sigma_l + \frac{i\omega\tau(\sigma_h - \sigma_l)}{1 + i\omega\tau} + i\omega\varepsilon_0\varepsilon_h \tag{137}$$

Where

$$\varepsilon_h = \varepsilon_m \frac{2\varepsilon_m + \varepsilon_p - 2\phi(\varepsilon_m - \varepsilon_p)}{2\varepsilon_m + \varepsilon_p + \phi(\varepsilon_m - \varepsilon_p)} \tag{138}$$

$$\varepsilon_l = \varepsilon_m \frac{\sigma_l}{\sigma_m} + \frac{9(\varepsilon_p\sigma_m - \varepsilon_m\sigma_p)\sigma_m\phi}{[2\sigma_m + \sigma_p + \phi(\sigma_m - \sigma_p)]^2} \tag{139}$$

$$\sigma_l = \sigma_m \frac{2\sigma_m + \sigma_p - 2\phi(\sigma_m - \sigma_p)}{2\sigma_m + \sigma_p + \phi(\sigma_m - \sigma_p)} \tag{140}$$

$$\sigma_h = \sigma_m \frac{\varepsilon_h}{\varepsilon_m} + \frac{9(\sigma_p\varepsilon_m - \sigma_m\varepsilon_p)\varepsilon_m\phi}{[2\varepsilon_m + \varepsilon_p + \phi(\varepsilon_m - \varepsilon_p)]^2} \tag{141}$$

τ is the relaxation time of the Maxwell-Wagner polarization, and can be expressed as:

$$\tau = \frac{1}{2\pi f_0} = \frac{2\varepsilon_m + \varepsilon_p + \phi(\varepsilon_m - \varepsilon_p)}{2\sigma_m + \sigma_p + \phi(\sigma_m - \sigma_p)}\varepsilon_0 \tag{142}$$

and

$$(\sigma_h - \sigma_l)\tau = (\varepsilon_l - \varepsilon_h)\varepsilon_0 \qquad (143)$$

Eq.(136) to (143) are called Maxwell-Wagner equations. Those equations predict a single relaxation process taking place when $\varepsilon_m \sigma_p \neq \varepsilon_p \sigma_m$. For most non-aqueous systems, the dispersed particle always has a much higher conductivity than the dispersing medium, i.e., $\sigma_p >> \sigma_m$. Thus Eq.(142) can be simplified as:

$$\tau = \frac{\varepsilon_0 (2\varepsilon_m + \varepsilon_p)}{\sigma_p} \qquad (144)$$

Eq. (140) reduces to:

$$\sigma_l = \sigma_m \frac{1 | 2\phi}{1 - \phi} \qquad (145)$$

The real and imaginary parts of the complex dielectric constant can be easily separated from Eq.(136)

$$\varepsilon' = \varepsilon_\infty \left[1 + \frac{K}{(1 + \omega^2 \tau^2)} \right] \qquad (146)$$

$$\tan \delta = \frac{K \omega \tau}{1 + K + \omega^2 \tau^2} \qquad (147)$$

in those equations

$$\varepsilon_\infty = \varepsilon_m \left[1 + \frac{3\phi(\varepsilon_p - \varepsilon_m)}{2\varepsilon_m + \varepsilon_p} \right] \qquad (148)$$

$$K = \frac{9\phi\varepsilon_m}{2\varepsilon_m + \varepsilon_p} \qquad (149)$$

Eq.(146) to (149) are simplified Maxwell-Wagner equations, which only hold for the suspensions made from conductive particles dispersed in an insulating liquid medium with the low particle volume fractions less than 10 vol%.

7.2 Dilute suspensions of spherical particle with shell

Maxwell [68] first addressed the dielectric behavior of dispersions with spherical particles covered with shells. Pauly and Schwan [72] subsequently extended Maxwell's treatment for obtaining the frequency dependence of the dielectric constant of the systems. Maintaining the same procedure as used in Wagner's theory, the dielectric constant of those systems can be expressed as:

$$\frac{\varepsilon_m-\varepsilon}{2\varepsilon_m+\varepsilon}=\frac{(\varepsilon_m-\varepsilon_{shell})(2\varepsilon_{shell}+\varepsilon_p)+(\varepsilon_{shell}-\varepsilon_p)(2\varepsilon_{shell}+\varepsilon_m)V}{(2\varepsilon_m+\varepsilon_{shell})(2\varepsilon_{shell}+\varepsilon_p)+2(\varepsilon_m-\varepsilon_{shell})(\varepsilon_{shell}-\varepsilon_p)V}\phi \tag{150}$$

Where ε_{shell} is the dielectric constant of the shell, V is the volume ratio between the bare particle and the coated particle, and ϕ is the volume fraction of the particle with shell. The complex dielectric constant can be written as:

$$\varepsilon^*=\varepsilon_h+\frac{\varepsilon_l-\varepsilon_i}{1+i\omega\tau_1}+\frac{\varepsilon_i-\varepsilon_h}{1+i\omega\tau_2}+\frac{1}{i\omega\varepsilon_0}\sigma_l \tag{151}$$

Where ε_h, ε_i, ε_l are the dielectric constant of the suspension at high, intermediate, and low frequency, and can be expressed as follows:

$$\varepsilon_h=\varepsilon_m\frac{B}{D} \tag{152}$$

$$\varepsilon_i=\frac{1}{\varepsilon_0 C(\tau_1-\tau_2)}\left[(\varepsilon_0\varepsilon_m-\sigma_m\tau_2)(E-A\tau_2)-\left(\frac{\varepsilon_0\varepsilon_m}{\tau_1}-\sigma_m\right)B\right] \tag{153}$$

$$\varepsilon_l=\frac{1}{C^2}\left[\varepsilon_m AC-\frac{\sigma_m}{\varepsilon_0}(AF-CE)\right] \tag{154}$$

$$\sigma_l=\frac{\sigma_m A}{C} \tag{155}$$

$$\tau_1 = \frac{F + \sqrt{F^2 - 4CD}}{2C} \tag{156}$$

$$\tau_2 = \frac{2D}{F + \sqrt{F^2 - 4CD}} \tag{157}$$

with

$$A = (1 + 2\phi)\sigma_{shell} a + 2(1 - \phi)\sigma_m b \tag{158}$$

$$B = [(1 + 2\phi)\varepsilon_{shell} c + 2(1 - \phi)\varepsilon_m d]\varepsilon_0^2 \tag{159}$$

$$C = (1 - \phi)\sigma_{shell} a + (2 + \phi)\sigma_m b \tag{160}$$

$$D = [(1 - \phi)\varepsilon_{shell} c + (2 + \phi)\varepsilon_m d]\varepsilon_0^2 \tag{161}$$

$$E = [(1 + 2\phi)(\varepsilon_{shell} a + \sigma_{shell} c) + 2(1 - \phi)(\varepsilon_m b + \sigma_m d)]\varepsilon_0 \tag{162}$$

$$F = [(1 - \phi)(\varepsilon_{shell} a + \sigma_{shell} c) + (2 + \phi)(\varepsilon_m b + \sigma_m d)]\varepsilon_0 \tag{163}$$

$$a = (1 + 2V)\sigma_p + 2(1 - V)\sigma_{shell} \tag{164}$$

$$b = (1 - V)\sigma_p + (2 + V)\sigma_{shell} \tag{165}$$

$$c = (1 + 2V)\varepsilon_p + 2(1 - V)\varepsilon_{shell} \tag{166}$$

$$d = (1 - V)\varepsilon_p + (2 + V)\varepsilon_{shell} \tag{167}$$

Eq. (151) clearly indicates that there are two relaxation peaks in the suspension of the coated particles, as shown in Figure 19. Those two relaxation times overlap together when $F^2 = 4CD$. In this case, there is only one relaxation time.

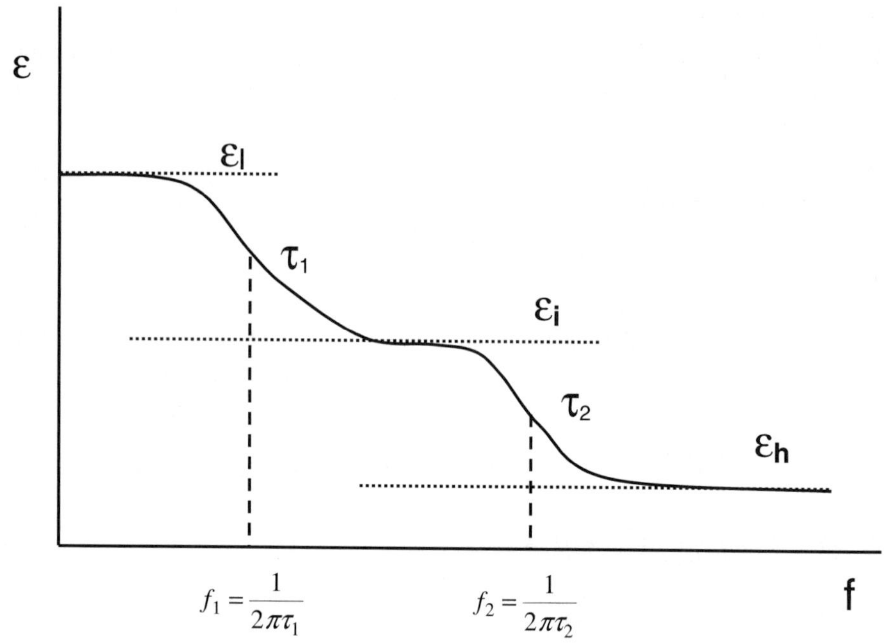

Figure 19 Schematic illustration of the frequency dependence of the dielectric constant of the suspension with particles of shells Redrawn from T. Hanai, Electrical Properties of Emulsions, in Emulsion Science, P. Sherman, ed., Academic Press, 1968, p394:

7.3 The Hanai model for concentrated suspensions

Since the Maxwell-Wagner model was derived for a dilute system, it was extended by Hanai [73-75] for describing concentrated dispersions by employing a similar mathematical procedure used by Bruggeman [76]. In Hanai's treatment, a concentrated suspension is obtained by successively adding infinitesimally small quantities of dispersed phase. The Maxwell-Wagner equation (134) still holds for such a process. When the concentration changes from ϕ' to $\phi'+\Delta\phi'$, the dielectric constant changes from ε^* to $\varepsilon^*+\Delta\varepsilon^*$. Finally Eq.(134) leads to

$$\frac{2\varepsilon^* + \varepsilon_p^*}{3\varepsilon^*\left(\varepsilon^* - \varepsilon_p^*\right)}\Delta\varepsilon^* = \frac{-\Delta\phi'}{1-\phi'} \tag{168}$$

Suppose that the system will finally reach the dielectric constant ε* when the particle volume fraction reaches ϕ via the successive addition of infinitesimal small quantities of the dispersed particles, then this process can be summed up as:

$$\int_{\varepsilon_m^*}^{\varepsilon^*} \frac{2\varepsilon^* + \varepsilon_p^*}{3\varepsilon^*\left(\varepsilon^* - \varepsilon_p^*\right)} d\varepsilon^* = \int_0^{\phi} \frac{-d\phi}{1-\phi} = \log(1-\phi) \tag{169}$$

Note that the left side of Eq.(169) is integrated along a certain contour from ε_m^* to ε*. It can be evaluated as [T. Hanai, Electrical Properties of Emulsions, in Emulsion Science, P. Sherman, ed., Academic Press, 1968, p394]:

$$\int_{\varepsilon_m^*}^{\varepsilon^*} \frac{2\varepsilon^* + \varepsilon_p^*}{3\varepsilon^*\left(\varepsilon^* - \varepsilon_p^*\right)} d\varepsilon^* = \frac{1}{3}\log\left(\frac{\varepsilon_m^*}{\varepsilon^*}\right) + \log\left(\frac{\varepsilon^* - \varepsilon_p^*}{\varepsilon_m^* - \varepsilon_p^*}\right) \tag{170}$$

Therefore Eq.(169) can be rewritten as:

$$\left(\frac{\varepsilon_m^*}{\varepsilon^*}\right)^{1/3}\left(\frac{\varepsilon^* - \varepsilon_p^*}{\varepsilon_m^* - \varepsilon_p^*}\right) = 1-\phi \tag{171}$$

Eq.(171) is called Hanai's equation, which can be considered as the complex form of Bruggeman's equation [76], shown in Eq. (171a).

$$\left(\frac{\varepsilon_m}{\varepsilon}\right)^{1/3}\left(\frac{\varepsilon - \varepsilon_p}{\varepsilon_m - \varepsilon_p}\right) = 1-\phi \tag{171a}$$

Separating the real and imaginary part of Eq.(171) leads to

$$\frac{\left[\left(\varepsilon' - \varepsilon_p'\right)^2 + \left(\varepsilon'' - \varepsilon_p''\right)^2\right]\left[\left(\varepsilon_m'\right)^2 + \left(\varepsilon_m''\right)^2\right]^{1/3}}{\left[\left(\varepsilon_m' - \varepsilon_p'\right)^2 + \left(\varepsilon_m'' - \varepsilon_p''\right)^2\right]\left[\left(\varepsilon'\right)^2 + \left(\varepsilon''\right)^2\right]^{1/3}} = (1-\phi)^2 \tag{172}$$

and

$$\tan^{-1}\frac{\varepsilon'\varepsilon''_m-\varepsilon'_m\varepsilon''}{\varepsilon'\varepsilon'_m+\varepsilon''_m\varepsilon''}=3\tan^{-1}\frac{\left[\left(\varepsilon'-\varepsilon'_p\right)\left(\varepsilon''_m-\varepsilon''_p\right)-\left(\varepsilon'_m-\varepsilon'_p\right)\left(\varepsilon''-\varepsilon''_p\right)\right]}{\left(\varepsilon'-\varepsilon'_p\right)\left(\varepsilon'_m-\varepsilon'_p\right)+\left(\varepsilon''_m-\varepsilon''_p\right)\left(\varepsilon''-\varepsilon''_p\right)} \quad (173)$$

For a special case where the conductivity of the dispersed particle is much higher than that of the dispersing medium, i.e., $\sigma_p >> \sigma_m$, the high frequency dielectric constant ε_{hf} $(\omega\to\infty)$ can be expressed as:

$$\left(\frac{\varepsilon_m}{\varepsilon_{hf}}\right)^{1/3}\left(\frac{\varepsilon_{hf}-\varepsilon_p}{\varepsilon_m-\varepsilon_p}\right)=1-\phi \quad (174)$$

The low frequency dielectric constant ε_{lf} $(\omega\to 0)$

$$\varepsilon_{lf}=\varepsilon'_m\frac{\sigma_l\left(\sigma_p-\sigma_l\right)}{\sigma_m\left(\sigma_p+2\sigma_l\right)} \quad (175)$$

where σ_l can be calculated from:

$$\frac{\left(\sigma_p-\sigma_l\right)}{\sigma_p}\left(\frac{\sigma_m}{\sigma_l}\right)^{1/3}=1-\phi \quad (176)$$

7.4 Particle shape effect on the dielectric property

The Maxwell-Wagner equation for diluted suspensions and the Hanai equation for concentrated suspensions were derived for the spherical particles dispersed in a continuous medium. The particle shape deviated from the spherical shape has been observed to change the dielectric properties of the whole suspension predicted either by the Maxwell-Wagner or the Hanai equations. Sillars [77] extended the Maxwell-Wagner equation for orientated non-spherical particles such as needle-shaped (prolate), flat-disk, and ellipsoidal particles. Here “orientated” means that the longest axis is aligned to the direction of the externally applied electric field. For randomly orientated ellipsoidal particles, Fricke [78] derived an equation on the basis of the Maxwell-Wagner equation. In the case where the ellipsoidal particles have a high eccentricity, Fricke’s equation turns out to be limited to the very low particle concentrations [79]. Various dielectric mixing theories or models have been developed for accounting for the particle shape effects for two-phase systems [80-84]. According to Bánhegyi [83], all those models can be classified into three groups: a) The matrix inclusion type formulae. Those formulae are derived under the assumption that the particles are dispersed in a continuous matrix phase in such a low concentration that there is no interaction between particles. Obviously, the Maxwell-Wagner and extended Maxwell-Wagner-Sillars or Bruggeman-Boyle equations fall in this category. Even Hanai equation should belong to this group, though it is valid for concentrated suspensions. Since most of those equations deal with low particle volume fractions, there is no percolation threshold issue involved; b) The statistic mixture type formulae. Those equations are derived on the basis of mean field or average polarizability arguments, and originally developed for dealing with the metal-insulator composites. Since metal is significantly different from ionic compounds like polymer or glass, the numerical results from those models cannot be used for emulsions or polymer based composites. The Böttcher-Hsu equation belongs to this category [85, 86], and it predicts a percolation threshold; c) The symmetrical integral formulae. Those equations are derived using the symmetric integration technique, and again there is no percolation threshold predicted. Looyenga equation [87] is one of those. All those three categories equations incorporates the particle shape influence on the dielectric properties, which can be expressed using a parameter called the depolarization factor along one of the axes of the ellipsoid, A. A is defined as [88]:

$$A^j = \int_0^\infty \frac{(abc)du}{2(u+j^2)\sqrt{(u+a^2)(u+b^2)(u+c^2)}}, j = a,b,c \tag{177}$$

where $A^a + A^b + A^c = 1$, and a, b and c are the dimensions of the ellipsoidal particle in x, y and z axes as shown in Figure 20. A^j is the depolarization factor for a particle whose rotation axis j is aligned parallel to the applied electrical field.

For a special case of spheroidal particles, b can be equalized as c, and then A becomes a function of the aspect ratio a:b. Eq.(177) can be simplified as [84]

$$A^a = \frac{1}{1 + 1.6(a:b) + 0.4(a:b)^2}$$

$$A^b = 0.5(1 - A^a) \tag{178}$$

Thus for a sphere, $A^{a.b.c.} = 1/3, 1/3, 1/3$, for a thin disk, $A^{a.b.c.} = 1,0,0$, and for a long needle, $A^{a.b.c.} = 0, 0.5, 0.5$. The relationship between the depolarization factor and the aspect ratio is depicted in Figure 21. It clearly shows that as the particle shape changes from a flat disk to a long needle, A^a decreases from 1 to almost zero, and in contrast A^b increases from zero to 0.5. The dielectric property of the two-components mixture strongly depends on how the particles align with the applied electric field.

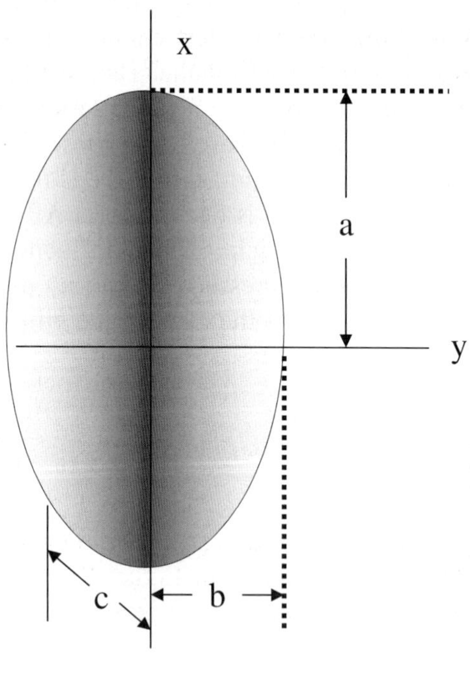

Figure 20 An ellipsoidal particle of three axes.

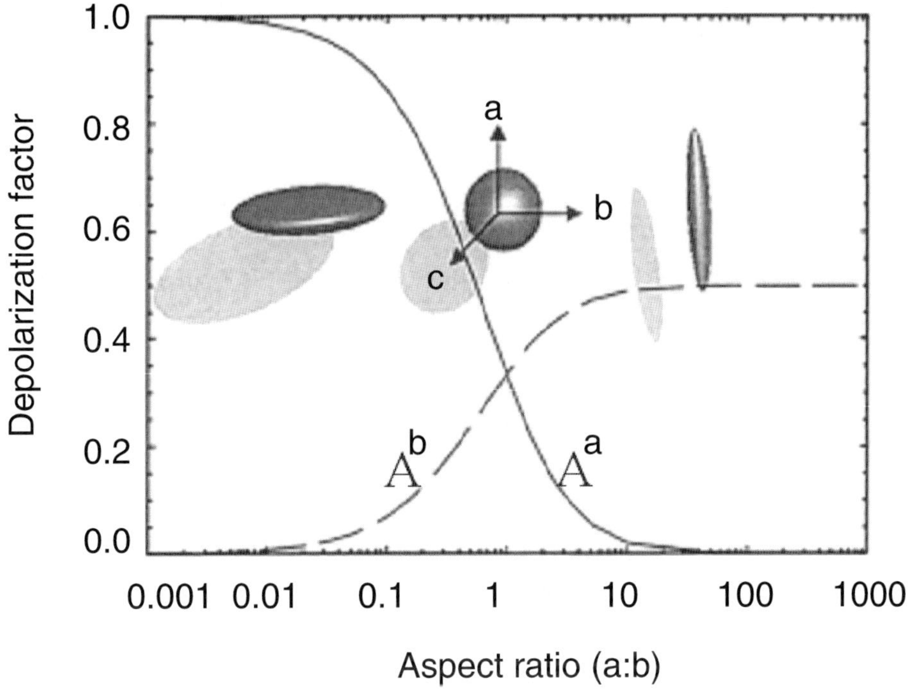

Figure 21 The depolarization factor A as a function of the aspect ratio a:b. Redrawn with permission from S.B. Jones, and S.P. Friedman, Water Resour. Res., 16(1980)574. A^a is the depolarization factor for a particle whose rotation axis a is aligned parallel to the electrical field, and A^b is the one corresponding to the rotation axis b.

7.4.1 The Maxwell-Wagner-Sillars equation and its extensions

The complex dielectric constant of a suspension ε^* of orientated ellipsoidal particles with the dielectric constant ε_p^* at the particle volume fraction ϕ dispersed in a continuous medium with a complex dielectric constant ε_m^*, can be calculated from the Maxwell-Wagner-Sillars equation [77]:

$$\varepsilon^* = \varepsilon_m^* \frac{\varepsilon_m^*(1-\phi)(1-A)+\varepsilon_p^*[\phi + A(1-\phi)]}{\varepsilon_m^* + A(1-\phi)(\varepsilon_p^* - \varepsilon_m^*)} \tag{179}$$

Separating the real and imaginary parts of the complex dielectric constant leads to the Debye-type equations [89]:

$$\varepsilon' = \varepsilon_\infty + \frac{\varepsilon_s - \varepsilon_\infty}{1 + (\omega \tau_{MWS})^2} \tag{180}$$

$$\varepsilon'' = (\varepsilon_s - \varepsilon_\infty)\frac{\omega \tau_{MWS}}{1 + (\omega \tau_{MWS})^2} \tag{181}$$

where ε_s and ε_∞ are the low frequency and high frequency limiting dielectric constants, and τ_{MWS} is the relaxation time of the interfacial polarization. They can be expressed as:

$$\tau_{MWS} = \frac{\varepsilon_m + A(1-\phi)(\varepsilon_p - \varepsilon_m)}{\sigma_m + A(1-\phi)(\sigma_p - \sigma_m)} \tag{182}$$

$$\varepsilon_s = \varepsilon_m \frac{\sigma_m + [A(1-\phi)+\phi](\sigma_p - \sigma_m)}{\sigma_m + [A(1-\phi)](\sigma_p - \sigma_m)} + \phi\sigma_m \frac{[\sigma_m + A(\sigma_p - \sigma_m)](\varepsilon_p - \varepsilon_m) - [\varepsilon_m + A(\varepsilon_p - \varepsilon_m)](\sigma_p - \sigma_m)}{[\sigma_m + A(1-\phi)(\sigma_p - \sigma_m)]^2} \tag{183}$$

$$\varepsilon_\infty = \varepsilon_m \frac{\varepsilon_m + [A(1-\phi)+\phi](\varepsilon_p - \varepsilon_m)}{\varepsilon_m + A(1-\phi)(\varepsilon_p - \varepsilon_m)} \tag{184}$$

For the case that ellipsoidal particles are randomly orientated in the matrix, Fricke [78, 90] develops the formulae on the basis of the Maxwell-Wagner-Sillars equation for describing the dielectric property of such a system. However, as pointed out by Grosse [79], Fricke's equations are not valid for the ellipsoidal particles with high eccentricity. In the relative high concentration (depending on the aspect ratio of the ellipsoidal particles), Fricke's equations reduce to the Maxwell-Wagner-Sillars equation. Since the two-phase randomly orientated system could be equalized to an isotropic mixture, Sihvola and Kong [91] developed a general form equation accounting for the summation over all three axial dimensions of the particle:

$$\varepsilon' = \varepsilon_m + \left\{ \left[\sum_{j=a,b,c} \frac{\phi(\varepsilon_p - \varepsilon_m)[\varepsilon_m + p(\varepsilon' - \varepsilon_m)]}{3[\varepsilon_m + p(\varepsilon' - \varepsilon_m) + A^j(\varepsilon_p - \varepsilon_m)]} \right] \times \left[1 - \sum_{j=a,b,c} \frac{\phi A^j(\varepsilon_p - \varepsilon_m)}{3[\varepsilon_m + p(\varepsilon' - \varepsilon_m) + A^j(\varepsilon_p - \varepsilon_m)]} \right]^{-1} \right\} \tag{185}$$

For the anisotropic case, the resulting effective dielectric constant is strongly dependent on how the particles align with respect to the applied electric field:

$$\varepsilon_j' = \varepsilon_m + \left\{ \left[\frac{\phi(\varepsilon_p - \varepsilon_m)[\varepsilon_m + p(\varepsilon_j' - \varepsilon_m)]}{[\varepsilon_m + p(\varepsilon_j' - \varepsilon_m) + A^j(\varepsilon_p - \varepsilon_m)]} \right] \times \left[1 - \frac{\phi A^j(\varepsilon_p - \varepsilon_m)}{[\varepsilon_m + p(\varepsilon_j' - \varepsilon_m) + A^j(\varepsilon_p - \varepsilon_m)]} \right]^{-1} \right\} \tag{186}$$

Where p is a parameter which reduces Eq.(185) and (186) to the different mixing rules. For example, p=0, they reduce to the Maxwell-Garnett rule [92]; $p = 1 - A^j$, they reduce to the Polder and van Santen or Bruggeman rules[21,80]; p=1, they reduce to the coherent potential mixing rule [93]. The Maxwell-Garnett rule represents a diluted system, and it generally predicts an effective dielectric constant close to the dielectric constant of the matrix. The Polder and van Santen or Bruggeman rule predicts a system of relative high particle concentration without the significant inter-particle force. The coherent potential rule predicts the largest contrast between the effective dielectric constant of the mixture and that of the matrix [94]. As an example, the dielectric constant of a two-phase solid/water mixture with the particle volume fraction 50 vol% and the particle dielectric constant ε_p=5 is computed against the particle aspect ratio from 10^{-3} to 10^{3} using Eq.(185) for the isotropic mixing case and using Eq. (186) for the anisotropic mixing case and depicted in Figure 22. The dielectric constants for the isotropic case, ε, and for the anisotropic case, ε_a and ε_b are calculated for showing the difference between the models. The Maxwell-Garnett shows the dielectric constant closer to that of the matrix. In comparison with the spherical particle, the needle-shaped particle leads to the high ε_a as the larger dipole is induced at the longer length of the particle. Accordingly the disk particle leads to a high ε_b due to the large disk diameter in b axis. The major change of the dielectric constants induced by the particle shape occurs at the aspect ratio between 0.1 and 10. Beyond this region, the particle shape effect becomes negligible. When the aspect ratio is 1, i.e., the spherical particle case, the dielectric constant of the isotropic mixing system is equal to that of the anisotropic mixing system.

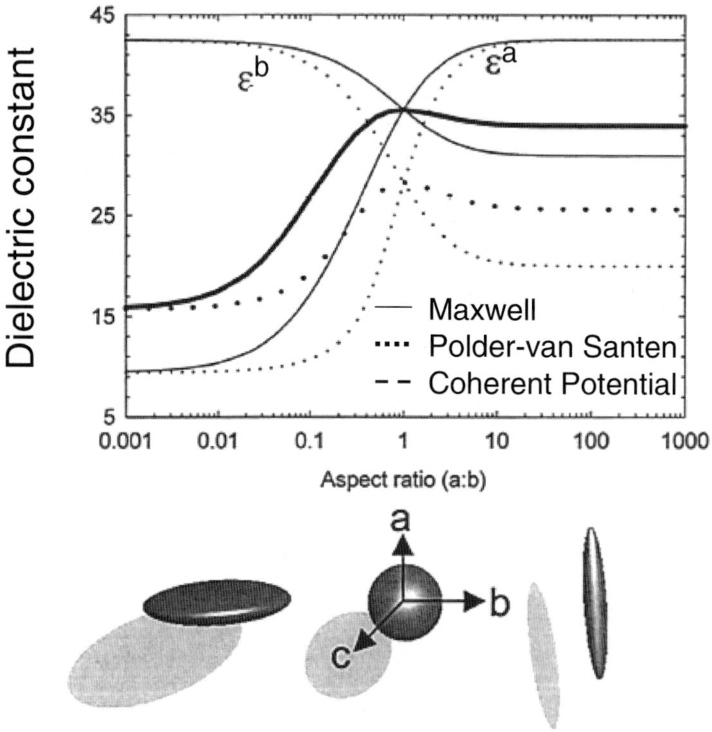

Figure 22 The dielectric constant as a function of particle aspect ratio for particles dispersed in water. ε_m=80, ε_p=5, ϕ=0.5. Reproduced with permission from S.B. Jones, and S.P. Friedman, Water Resour. Res. , 36(2000)2821

Boned and Peyrelasse [95] extended the asymmetric integration technique used by Bruggeman and Hanai to randomly orientated ellipsoidal mixture and obtained the following formula:

$$\left(\frac{\varepsilon^* - \varepsilon_p^*}{\varepsilon^* - \varepsilon_m^*}\right)\left(\frac{\varepsilon_m^*}{\varepsilon^*}\right)^A = 1 - \phi \qquad (187)$$

when A =1/3, i.e., the spherical particle case, Eq.(187) reduces to the Hanai equation, Eq. (171). Eq.(187) should work for the concentrated suspension as the Hanai equation does.

7.4.2 The Böttcher-Hsu equation

The mean field statistical mixture formulae have been mainly developed for dealing with the metal-insulator composites. The simplest statistical mixture formulae are developed by Böttcher [85] and extended to orientated ellipsoidal systems by Hsu [86]:

$$\frac{\left(\varepsilon_m^* - \varepsilon^*\right)(1-\phi)}{\varepsilon^* + \left(\varepsilon_m^* - \varepsilon^*\right)A} + \frac{\left(\varepsilon_p^* - \varepsilon^*\right)\phi}{\varepsilon^* + \left(\varepsilon_p^* - \varepsilon^*\right)A} = 0 \qquad (188)$$

This equation predicts that a percolation transition occurs at ϕ=A [86]. The static dielectric constant diverges at this concentration proportionally to $|\phi - A|^{-1}$ if the dispersed particle is a conductor.

7.4.3 Looyenga equation

The symmetrical integral formula was first introduced by Landau and Lifshitz [96] on the basis of general electrodynamic arguments for a two-phase isotropic mixture. The same formula was obtained by Looyenga [87] using the model of concentric spheres and symmetrical integration. This equation is further extended to orientated ellipsoidal systems by many researchers independently [97-99]:

$$\left(\varepsilon^*\right)^{1-2A} = (1-\phi)\left(\varepsilon_m^*\right)^{1-2A} + \phi\left(\varepsilon_p^*\right)^{1-2A} \qquad (189)$$

It would be interesting to compare Eq.(189) with previous mixture equations that are shown in Section 5. The special case of Eq. (189) for spherical particles can be easily obtained by assuming A =1/3.

7.4.4 Comparisons between the mixture equations

Numerical results from the above three type equations are compared by Bánhegyi [83]. The dielectric constant and loss of two-phase spherical particle mixture are calculated with the Maxwell-Wagner-Sillars equation, the Böttcher-Hsu equation, and the Looyenga equation using the parameters $\varepsilon_m = 2$, $\varepsilon_p = 8$, $\sigma_m = 10^{-16}$ S/m, $\sigma_p = 10^{-8}$S/m, and shown in Figure 23 against frequency for different concentration levels ranging from 0.1 to 0.9. In the case of the Maxwell-Wagner-Sillars equation (Figure 23a), there is a dielectric loss peak for the whole concentration range. The dielectric loss peak continuously shifts to a lower frequency with the increase of the particle volume fraction. The Böttcher-Hsu equation shows a percolation threshold at the particle volume fraction $\phi = A = 1/3$ for the spherical particle case (Figure 23b). Below this threshold particle volume fraction, the particles are "isolated" between each other and the dielectric loss peak, or the so-called interfacial polarization peak, appears. The Böttcher-Hsu equation predicts a similar behavior as the Maxwell-Wagner-Sillars equation does. Above this threshold particle volume fraction, the particles form a continuous network structure in the insulating matrix, and thus the interfacial loss peak diminishes. The charge accumulation at the interfaces couldn't be established due to the conductive paths formed by the particles. The divergence of the dielectric constant at $\phi = A$ is an unique feature of the Böttcher-Hsu equation. The dielectric properties predicted by the Looyenga equation are shown in Figure 23c. The low frequency behavior predicted by the Looyenga equation is quite different from that predicted by either the Maxwell-Wagner-Sillars equation or the Böttcher-Hsu equation. There is no any dielectric loss peak in the whole concentration range.

If the particle is non-spherical and orientated, the dielectric properties of the whole suspension depend on the orientation of particle axes relative to the direction of the electric field. Figure 24 shows the dielectric constant and dielectric loss vs. frequency calculated by using different models for prolate (needle-shape) particle along the axis perpendicular to the axis of rotation at different particle volume fraction marked in the graphs. The aspect ratio of the particle is 10, and the parameters are $\varepsilon_m = 2$, $\varepsilon_p = 8$, $\sigma_m = 10^{-16}$ S/m, $\sigma_p = 10^{-8}$ S/m. The situation is similar to the spherical system shown in Figure 23. The only difference is that the threshold particle volume fraction of the Böttcher-Hsu equation is close to 0.5, rather than 0.3. This is not a coincidence, and the threshold particle volume fraction $\phi = A$ still holds. Assuming that the prolate is an ellipsoid with two equal axes, the depolarization factor A can be spitted into two parts, one is parallel with the axis of the rotation (||), and another is perpendicular to it (⊥) [100]:

$$A_{\|} + 2A_{\perp} = 1 \tag{190}$$

For a prolate with aspect ratio 10, $A_{\|}$ =0.0203, and $A_{\perp}$ =0.4899. So the threshold particle volume fraction is 0.4899 for this case. If the particle is an oblate (disk-shape) with an aspect ratio 1/10, then $A_{\|}$ =0.8608, and $A_{\perp}$ =0.0696. Figure 23 shows the dielectric constant and loss vs. frequency calculated by using the three equations for oblate (disk-shape) particle along the axis of rotation at different particle volume fractions ranging from 0.1 to 0.9. In this case all three equations predict a dielectric loss peak of different magnitude. The threshold particle volume fraction predicted by the Böttcher-Hsu equation shifts to a very high value, around 0.9, which is again following the rule ϕ=A. With the increase of the particle volume fraction, the dielectric loss peak shifts to the low frequency field, and all peaks appear below 100 Hz.

A detailed comparison between those three equations was discussed in ref. [83]. The main conclusions are; a) The sharp differences between Maxwell-Wagner-Sillars equation, the Böttcher-Hsu equation, and the Looyenga equation will be gradually observed as the conductivity difference between the components increases; b) the phase heterogeneity influences the dielectric relaxation strength, the difference between the dielectric constant at low and high frequency where the dielectric loss peak appears; c) at low frequencies the difference among those three equations becomes significant.

It would be more helpful if the predictions obtained from those equations can be quantitatively compared with the experimental results. Figure 26 shows the calculated and experimental dielectric constant vs. frequency for nitrobenzene/water emulsion. The nitrobenzene volume fraction is 0.5. The parameter used for calculation is that the dielectric constant and conductivity of nitrobenzene and water is 35.15, 6.249 $\times 10^{-6}$ S/cm, 78.0, 7.286 $\times 10^{-5}$ S/cm, respectively. The experimental data were obtained at temperature 20°C. All calculated curves are similar, and the only difference is the relaxation strength.

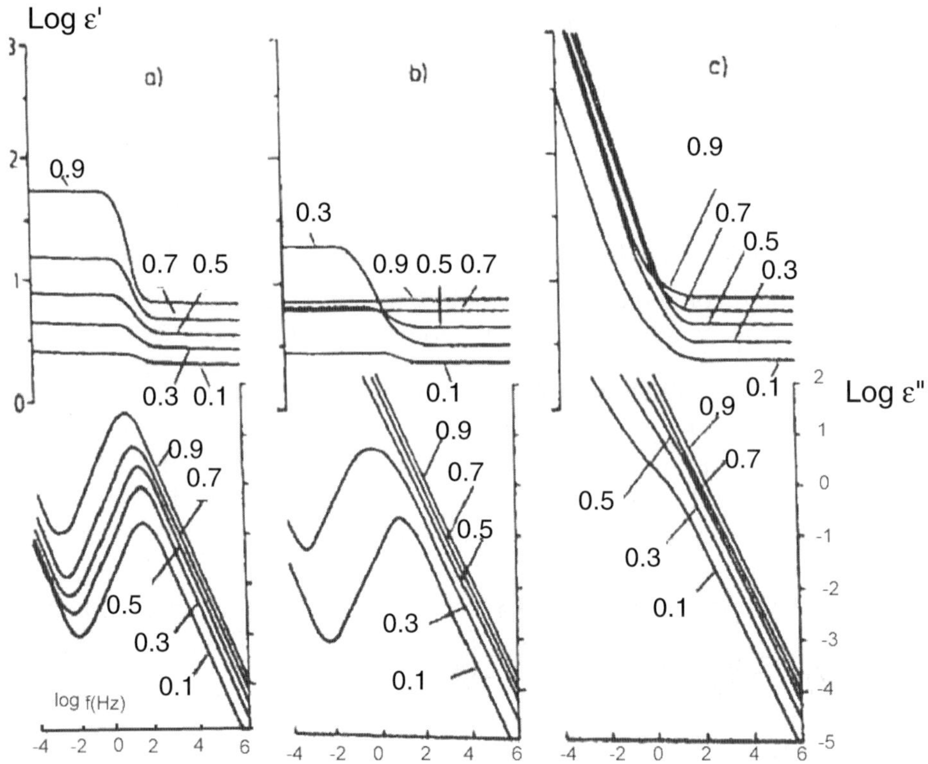

Figure 23 The dielectric constant and loss vs. frequency calculated by using different models for spherical particle case at different particle volume fraction marked in the graphs. a) Maxwell-Wagner-Sillars equation; b) Böttcher-Hsu equation; c) Looyenga equation. Parameters are $\varepsilon_m = 2$, $\varepsilon_p = 8$, $\sigma_m = 10^{-16}$ S/m, $\sigma_p = 10^{-8}$ S/m. The particle volume fraction changes from 0.1 to 0.9 with the interval 0.2. Reproduced with permission from G. Bánhegyi, Colloid Polym. Sci., 266(1988)11.

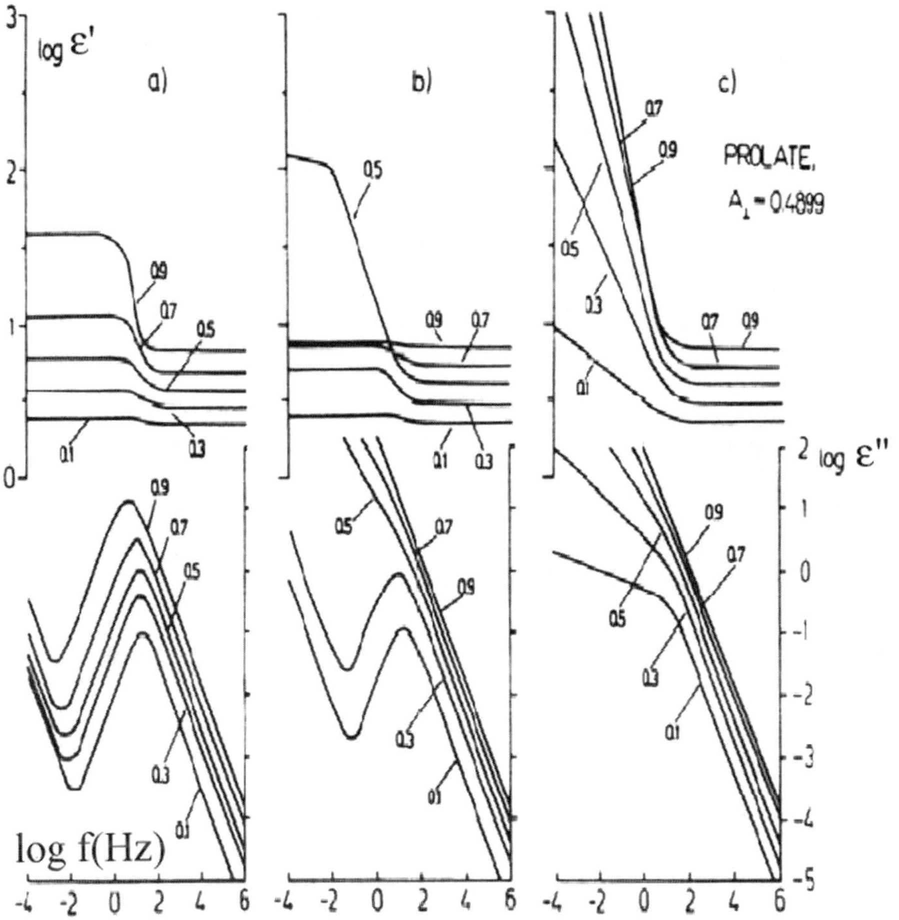

Figure 24 The dielectric constant and loss vs. frequency calculated by using different models for prolate (needle-shape) particle along the axis perpendicular to the axis of rotation at different particle volume fraction marked in the graphs. The aspect ratio of the particle is 10. a) Maxwell-Wagner-Sillars equation; b) Böttcher-Hsu equation; c) Looyenga equation. Parameters are $\varepsilon_m = 2$, $\varepsilon_p = 8$, $\sigma_m = 10^{-16}$ S/m, $\sigma_p = 10^{-8}$ S/m. The particle volume fraction changes from 0.1 to 0.9 with the interval 0.2. Reproduced with permission from G. Bánhegyi, Colloid Polym. Sci., 266(1988)11.

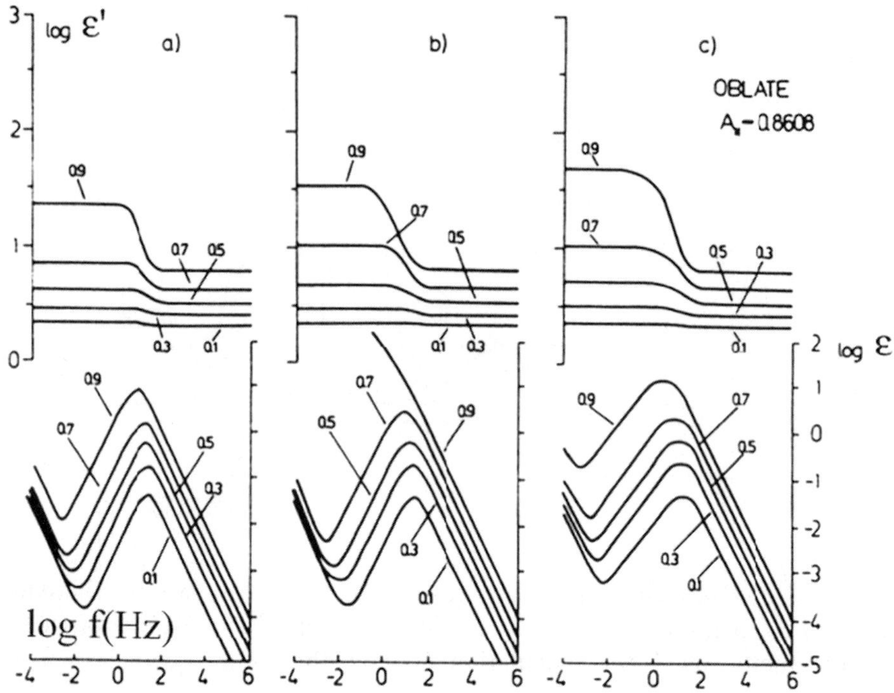

Figure 25 The dielectric constant and loss vs. frequency calculated by using different models for oblate (disk-shape) particle along the axis of rotation at different particle volume fraction marked in the graphs. a) Maxwell-Wagner-Sillars equation; b) Böttcher-Hsu equation; c) Looyenga equation. Parameters are $\varepsilon_m = 2$, $\varepsilon_p = 8$, $\sigma_m = 10^{-16}$ S/m, $\sigma_p = 10^{-8}$ S/m. The particle volume fraction changes from 0.1 to 0.9 with the interval 0.2. Reproduced with permission from G. Bánhegyi, Colloid Polym. Sci., 266(1988)11.

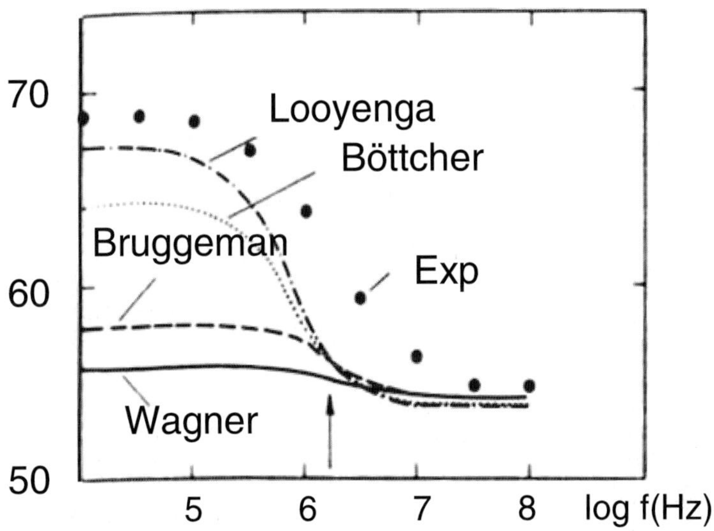

Figure 26 The calculated and experimental dielectric constant vs. frequency for nitrobenzene/water emulsion. The nitrobenzene volume fraction is 0.5. The parameter used for calculation is that the dielectric constant and conductivity of nitrobenzene and water is 35.15, 6.249 $\times 10^{-6}$ S/cm, 78.0, 7.286 $\times 10^{-5}$ S/cm, respectively. Temperature is 20°C. Reproduced with permission from G. Bánhegyi, Colloid Polym Sci., 266(1988)11.

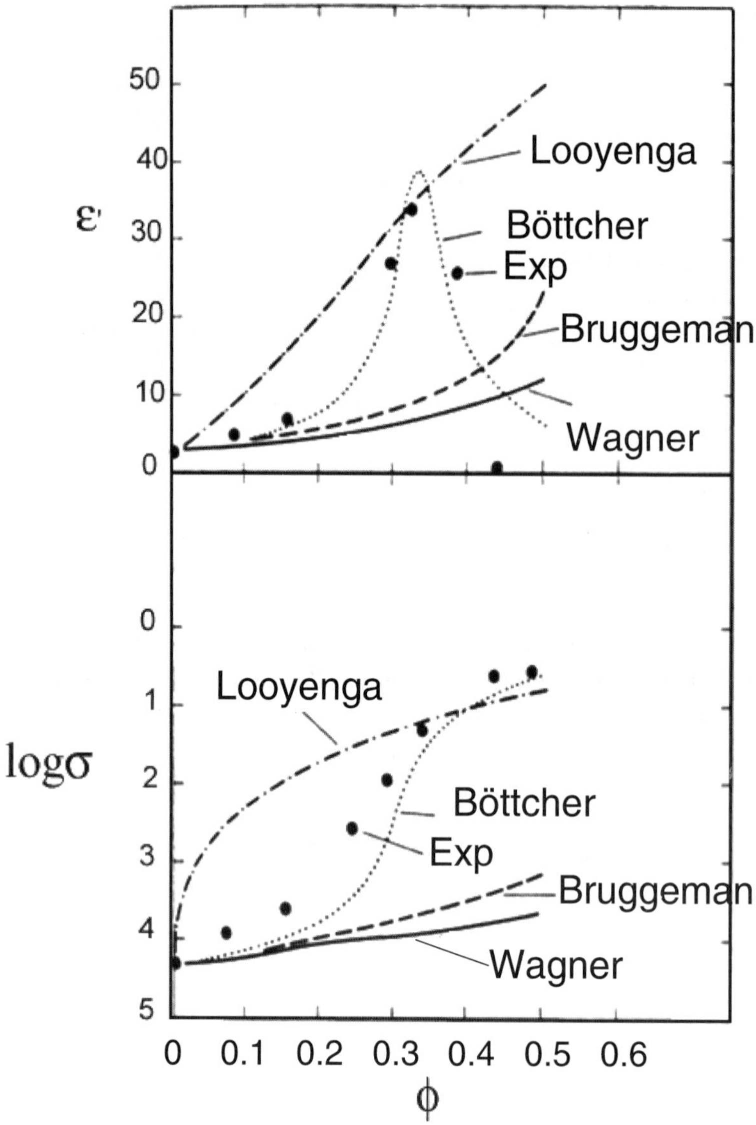

Figure 27 The calculated and experimental dielectric constant (top) and conductivity (bottom) vs. particle volume fraction for carbon black/PVC composite at 915 MHz. The parameter used for calculation is that the dielectric constant and conductivity of carbon black and PVC is 0, 10^0 S/cm, 3, 5×10^{-5} S/cm, respectively. Experimental data were obtained at room temperature. Reproduced with permission from G. Bánhegyi, Colloid Polym. Sci., 266(1988)11.

The Looyenga equation gives the best fit to the experimental data, and the Wagner-Maxwell-Sillar equation gives the worst prediction. This is an insulator/conductor case. For the conductor/insulator case, carbon black/PVC composite is taken as an example, and the calculated and experimental dielectric constant (top) and conductivity (bottom) vs. the particle volume fraction for carbon black/PVC composite at 915 MHz are shown in Figure 27. The parameter used for calculation is the dielectric constant and conductivity of carbon black and PVC are 0, 10^0 S/cm, 3, 5×10^{-5} S/cm, respectively. Experimental data were obtained at room temperature. A percolation transition at the particle volume fraction around 0.32 is well predicted by the Böttcher-Hsu equation, which also gives the best fit to both the dielectric constant and conductivity experimental data. Interestingly, the Looyenga equation gives a good fit in the vicinity of the threshold particle volume fraction, and the Wagner-Maxwell-Sillar equation gives a reasonable fit when the particle volume fraction is low, below 0.2.

In summary, the Wagner-Maxwell-Sillar equation can be used only if the dispersed phase doesn't have the "metallic" conductivity with respect to the dispersing continuous phase. In addition, there should be no direct contact between the dispersed particles, i.e., the Wagner-Maxwell-Sillar equation only holds for the dilute dispersions. The Böttcher-Hsu equation holds for the situation where the conductive and dielectric properties of the dispersed and dispersing phases are sharply different. When the two components are similar, the Böttcher-Hsu equation usually gives a quite similar result to the Wagner-Maxwell-Sillar equation. In addition, the Böttcher-Hsu equation predicts a well defined percolation threshold at ϕ=A. The Looyenga equation usually can not be used on theoretical grounds, and it is less reliable than the other two equations. Note those are general rules, and there are always exceptional cases.

8 DC TRANSIENT CURRENT

For a heterogeneous system, slow polarizations should always occur once an electric field is applied. Those slow polarizations include the Debye polarization, the interfacial polarization, and the electrode polarization if there is electrolyte in the system. Those slow polarizations will result in an unstable dc current: dc current decays with time and finally become stable, as shown in Figure 28. Once the applied electric field is turned off, the dc current goes down to zero quickly and continuously drops to a negative current and then gradually decays to zero. The charge and discharge curves are usually symmetrical in shape of opposite signs. This phenomenon is called the dc current adsorption phenomenon, which is similar to the charge and discharge phenomenon observed in a capacitor. Another source that can lead to this phenomenon is the electrochemical reaction nearby the electrode surface, forming an insulating layer that may hinder the electron transferring into the electrode. In both a heterogeneous and a homogeneous solid or liquid, the dc dispersion current will be observed. The so-called space charges are believed to be directly responsible for it. The space charge can be generated from the electrode polarization in which the ions form the electrical double layer and accumulate at the electrode surface; it can be also generated from the Maxwell-Wagner polarization in which the charges accumulate at the interfaces. The hindered dipole orientation (Debye polarization) associated with the rotation of dipole systems can also contribute to the space charge.

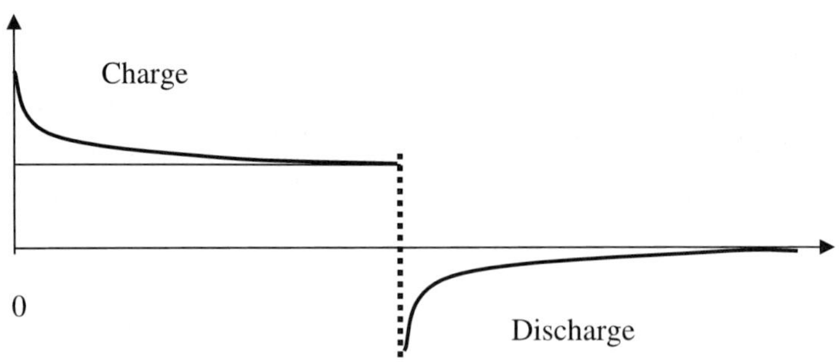

Figure 28. dc current vs. time. After an electric field is applied, dc current immediately jumps to a high value and then exponentially decay to a steady value; once the applied electric field is removed, dc current drops to a negative value, and then decays to zero.

The dc current decay function has been determined experimentally. At low temperature the dc current decay curve of a solid material can given as [5]:

$$I = kt^{-n} \tag{191}$$

where k and n are constant. This equation becomes invalid when t=0, as the current is unable to become infinite at very short time period. So Eq.(191) is only suitable when t>1 second. In case that t< 1, another equation should be used [5]:

$$I = k_1(t + t_0)^{-n} \tag{192}$$

where k_1 and n are constant. For mica material, $t_0=2.1\text{x}10^{-2}$ second, and n=0.87. Since dc current decay is governed by two different equations, this whole process is classified into the fast process governed by Eq. (192) and the slow process governed by the Eq.(191). The fast process is related to the fast polarizations, and the slow process is related to the slow polarizations. For a pure and symmetrical crystal material, there should be no dc current adsorption phenomenon. For a crystal of low symmetry, dc current should decay with time substantially. For an amorphous or ceramic material, dc current decay even happens at high temperatures. For a heterogeneous material, dc adsorption current always happens. After decaying for a long time, the dc current finally levels off, and this value is called the leakage current. By separating the leakage and adsorption currents from the current decay curve, the dielectric loss of the whole system can be calculated, which will be addressed in detail later.

8.1 Calculate the space charge amount from dc current decay curve

As described earlier, the space charge (or called surface charge) can be produced through the following mechanisms [7,101]: (i) the Debye polarization; (ii) the interfacial polarization (the Maxwell-Wagner polarization), associated with the displacement of the charge carriers over a microscopic distance; (iii) the leakage conduction, associated with the formation of space charges. In addition to those three, the electrode polarization associated with the double layer formation should also contribute to the space charge formation, though it is hard to separate from the leakage conduction contribution. So the apparent current, $I(t)$, can be separated into two components, the absorption current due to the space charge, dq/dt, and the leakage current, $i(t)$, due to the charge transfer between the dielectric and electrodes [101]:

$$I(t) = i(t) + \frac{dq}{dt} \tag{193}$$

If at t_1 the current levels off, then the total amount of space charge can be expressed as [102]:

$$q = \int_0^{t_1} [I(t) - i(t)] dt \tag{194}$$

From Eq.(194), one may estimate the particle surface charge amount, which was demonstrated by Hao [102]. The dc current passing through the aluminosilicate particle dispersed in silicone oil of particle volume fraction 35 vol% under an electric field 1.5 kV/mm is shown in Figure 29 against time. Since the absorption current usually exponentially decreases with time, ln $I(t)$ against t is plotted in Figure 30 using the same data of Figure 29. Two linear regions are observed: Line I stands for the current absorption part and line II stands for the leakage current part. From line I, one may get the expression for I(t) as:

$$I(t) = 131.6 e^{-0.0015t} \tag{195}$$

Assuming that the cross point between line I and line II is the starting point for the steady current, then t_1 =185.5 s. From line II, $i(t)$ can be estimated as 92.75 μA. Then from Eq.(194), q =4.92x10^{-2} C. This is the total charge ever bounded by all particles. Suppose that the surface area of the electrode is A and the gap size between two electrodes is d, then the volume of the sample to be tested is:

$$V=Ad \tag{196}$$

The particle volume Vp is:

$$Vp=V\phi \tag{197}$$

ϕ is the particle volume fraction. The single particle volume $V_{sp}=4\pi r^3/3$; r is the particle radius. So the particle number N in our geometry is:

$$N = V_p / V_{sp} = \frac{3Ad\phi}{4\pi r^3} \tag{198}$$

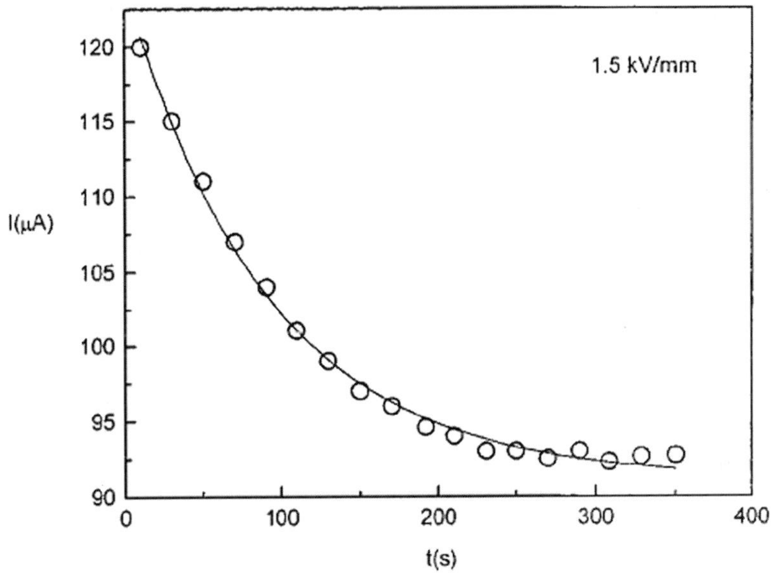

Figure 29 dc current passing through the aluminosilicate/silicone oil suspension of particle volume fraction 35 vol% under an electric field 1.5 kV/mm, plotted as current vs. time. Reproduced with permission from. T. Hao, A. Kawai, and F. Ikazaki, Langmuir, 14(1998)1256.

Thus the charge on each individual particle can be estimated. In this reference, the radius of the electrode is 0.025 m, d= 0.001 m, ϕ=0.35, and r=1.275 x 10^{-6} m; thus N =7.92 x10^{10} and one single particle bounds charge $q_s = q/N$=6.21x10^{-12} C.

The bound surface charge is obviously related to the applied electric field E. If a particle surface charge density is dq and the static dielectric constant is ε_{sp}, dq can be expressed as [10]:

$$d_q = \frac{\varepsilon_{sp} E}{4\pi} \tag{199}$$

If the particle surface area is S, then

$$q_s = d_q S = k_2 E \tag{200}$$

where $k_2 = \frac{S\varepsilon_{sp}}{4\pi}$, is a constant. From Eq.(200), the dielectric constant of a single particle can be estimated.

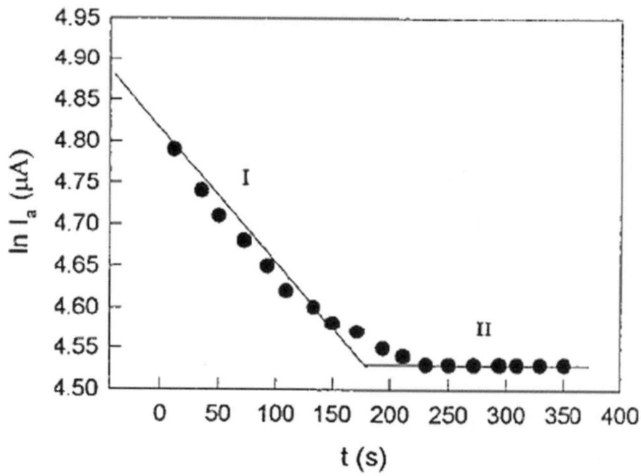

Figure 30 dc current passing through the aluminosilicate/silicone oil suspension of particle volume fraction 35 vol% under an electric field 1.5 kV/mm, plotted as ln I vs. time using the same data of Figure 29. Reproduced with permission from T. Hao, A. Kawai, and F. Ikazaki, Langmuir, 14(1998)1256.

8.2 Calculate the dielectric property of the material from the dc adsorption current

A general theory, assuming that the superposition principle holds, has been developed for converting the time dependent charge-discharge current to the frequency dependence of the complex dielectric constant [1,103]:

$$\begin{aligned} \varepsilon^*(\omega) &= \varepsilon'(\omega) - i\varepsilon''(\omega) \\ &= \varepsilon_\infty + \int_0^\infty \beta(t) e^{-i\omega t} dt - i\frac{\sigma_0}{\omega C_0} \end{aligned} \tag{201}$$

where ε_∞ is the dielectric constant at high frequency and is due to the electronic and atomic polarization. $\beta(t) = \frac{I(t)}{C_0 V}$, and I(t) is the charge-discharge dc current. C_0 is the capacitance of the electrodes when the sample is replaced by air, and V is the applied voltage, t is the time, σ_0 is the dc conductivity of the sample, corresponding to the leakage current. Note that the last right term in Eq. (201) $i\frac{\sigma_0}{\omega C_0}$ is the dielectric loss resulted from dc conductivity. The relaxation component of the complex dielectric constant from Eq. (201) is:

$$\varepsilon^* = \varepsilon_\infty + \int_0^\infty \beta(t)e^{-i\omega t}dt \tag{202}$$

Suppose that β(t) takes the form as shown in Eq. (195), $\beta(t)=ke^{-t/\tau}$, and then Eq. (202) becomes:

$$\varepsilon^*(\omega) = \varepsilon_\infty + \frac{\varepsilon_s - \varepsilon_\infty}{1 + i\omega\tau} \tag{203}$$

where $k = (\varepsilon_s - \varepsilon_\infty)/\tau$, ε_s is the static dielectric constant, τ is the relaxation time. Eq. (203) indicates if the dc adsorption transient current has the form, $\beta(t)=ke^{-t/\tau}$, then there is a single Debye relaxation process in the system. However, as discussed earlier, β(t) may be more complicated, making analytical results impossible. Assuming that β(t) has the form as shown in Eq. (191):

$$\beta(t) = kt^{-n} \tag{204}$$

then Eq.(202) becomes:

$$\begin{aligned} \varepsilon'(\omega) &= \varepsilon_\infty + \omega^{n-1}k\Gamma(1-n)\sin\frac{n\pi}{2} \\ \varepsilon''(\omega) &= \omega^{n-1}k\Gamma(1-n)\cos\frac{n\pi}{2} \end{aligned} \tag{205}$$

where Γ denotes the gamma function. Eq.(205) can be used to calculate the dielectric constant and loss at any frequency, which is dependent on how the charge-discharge current function is chosen. A simplification of Eq. (205) was made by Hamon [104]. He noticed that the dielectric loss of Eq. (205) can be rewritten as:

$$\varepsilon''(\omega)=\frac{\beta(t)}{\omega}\left[\frac{\Gamma(1-n)\cos n\pi/2}{(\omega t)^{-n}}\right] \tag{206}$$

If the frequency is chosen to satisfy the equation:

$$\omega t=\left[\Gamma(1-n)\cos n\pi/2\right]^{-1/n}\approx\frac{\pi}{5} \qquad \text{at } 0.1<n<1.2 \tag{207}$$

then

$$\varepsilon''(\omega)=\frac{\beta(t)}{\omega} \tag{208}$$

and

$$\varepsilon'(\omega)=\varepsilon_{\infty}+\varepsilon''\tan\frac{n\pi}{2} \tag{209}$$

The major advantage of Eq. (208) is that parameters k and n are not involved in the calculation, though n is needed for calculating the dielectric constant. It should be noted that this transient dc current method is only valid for the frequency 10^{-4} to 0.1 Hz [1]. The reason is obvious: The dc current decay function is only valid in a certain time period.

REFERENCES

[1] N.G. McCrum, B.E. Read, and G. Williams, Anelastic and Dielectric Effects in Polymeric Solids, Dover Publications, New York, 1991
[2] H.A. Kramers, Nature (London), 117(1926)775
[3] R. de L.Kronig, J. Opt. Soc. Am., 12(1926)547
[4] P.A.M. Steeman, and J. van Turnhout, Colloid Polym.Sci. 275(1997)106
[5] G.I. Skanavi, The Dielectric Physics, High Education Press, Beijing, China, 1958. Translated by Yihong Chen
[6] P. Debye, Polar Molecules, Chem. Catalog, New York, 1929
[7] C. P. Smyth, Dielectric behavior and structure, McGraw-Hill, New York, 1955

[8] L. Onsager, J. Am. Chem. Soc., 58(1936)1486
[9] J.G. Kirkwood, J. Chem. Phys., 7(1939)911
[10] H. Frölich, Theory of Dielectrics, Oxford University Press, Oxford, 1949
[11] P. Debye, and E. Hückel, Z.Phys. 24(1923)185
[12] R.J. Hunter, Zeta potential in colloid science, Academic Press, 1981
[13] A.D. Hollingsworth, and D.A. Saville, J. Colloid Interf. Sci., 272(2004)235.
[14] P.A. Cirkel, J.M.P. van der Ploeg, and G.J.M. Koper, Physica A, 235(1997)269
[15] H.P. Maruska, and J.G. Stevens, IEEE Trans. Electr. Instr., 23(1988)197
[16] Y. Feldman, E. Polygalov, I. Ermolina, Yu Polevaya, and B. Tsentsiper, Meas. Sci. Technol. 12(2001)1355
[17] T.J. Katsube, and N. Scromeda-Perez, Geological Survey of Canada, current research 2003-H2, 2003
[18] K.W. Wagner, Arch. Electrotechnik, 2(1914)371,
[19] S. O. Morgan, Trans. Am. Electrochem.Soc., 65(1934)109; R.W. Sillars, JIEE 80(1937)378
[20] T. Hanai, N. Koizumi, and R. Gotoh, Proc. Symp. Rheol. Emulsions, Ed. P. Sherman, Oxford, Pergamon, 1963, p91
[21] D.A.G. Bruggeman, Ann. Phys. 24(1935)636
[22] Debye, Chem. Rev.,19(1936)171
[23] J. Bisquert, G. Garcia-Belmonte, Russian J. Electrochem., Vol. 40(3)(2004)352
[24] F.E. Filisko, and D.R.Gamota, Recent Adv. Non-newtonian Flows, ASME, 153(1992)75
[25] R.H. Cole, and K.S. Cole, J. Chem.Phys. 9(1941)341
[26] D.W. Davidson, and R.H. Cole, J.Chem.Phys., 18(1950)1417
[27] S. Havrilliak, and S. Negami, J. Polym. Sci., C, 14(1966)99
[28] R. Pelster, and U. Simon, Colloid Polym Sci., 277(1999)2
[29] D. R. McKenzie, and R.C. McPhedran, Nature, 265(1977)128
[30] D. R. McKenzie, and R. C. McPhedran, Proc. R Soc Lond Ser A, 359(1978)45
[31] L. K. H. van Beek, Progress in Dielectrics, J.B. Birks, Ed., Heywood, London, 1967, pp 69-114;
[32] S. S. Dukhin, Dielectric Properties of Disperse Systems, in "Surface and Colloid Science", E.Matijevic, Ed., Wiley-Interscience, New York, 1971
[33] J.L.van der Minne, and P.H.J. Hermanie, J. Colloid Sci., 8(1953)38
[34] R.J. Pugh, T. Matsunaga, and F.M. Fowkes, Colloids Surfaces, 7(1983)183
[35] R.J. Pugh, and F. M. Fowkes, Colloids Surfaces, 9(1984)33
[36] R.J. Pugh, and F. M. Fowkes, Colloids Surfaces, 11(1984)423
[37] A. Kitahara, T. Kobayashi, and T. Tachibana, J. Phys. Chem., 66(1962)363
[38] H.F. Eicke, and H. Christen, J. Colloid Interface Sci., 48(1974)281
[39] W.R. Heffiner, and M.A. Marcus, J. Colloid Interface Sci., 124(1988)617
[40] D. Zhu, K. Feng, and Z.A. Schelley, J. Phys. Chem., 96(1992)2382
[41] P. A. Cirkel, M. Fontana, and G. J. M. Koper, J. Dispersion Sci. Technol. 22(2001)221
[42] I. D. Morrison, and S. Ross, Colloidal Dispersion, John Wiley and Sons, Inc., New York, 2002, p320
[43] A. Denat, B. Gosse, and J.P. Gosse, Rev. Phys. Appl., 16(1981)673

[44] T.G. Davis, G.A. Gibson, R.H. Luebbe, and K. Yu, Proc. 5th Int. Congress on advances in Non-impact Printing Technol., SPSE, Springfield, VA, 1989, pp404-416
[45] M.F. Hsu, E.R. Dufresne, and D.A. Weitz, Langmuir, 21(2005)488
[46] I.D. Morrison, Dispersion and Aggregation: Fundamentals and applications, B.M. Moudgil, and P. Somasundaran, Ed., The Eng. Foundation, New York, 1994, p55-73
[47] I.D. Morrison, Colloids and Surfaces A, 71(1993)1
[48] A. Kitahara, T. Satoh, S. Kawasaki, and K. Konno, J. Colloid Interface Sci., 86(1982)105
[49] F.M. Fowkes, Discuss. Faraday Soc., 46(1966)246;
[50] M. Kosmulski, Interfacial Dyanamics, N. Kallay Ed., Marcel Dekker, 2000, pp273-312
[51] L. Onsager, J. Chem Phys., 2(1934)599
[52] A. Alj, J. P. Gosse, B. Gosse, A. Denat, and M. Nemamcha, Rev. Phys. Appl., 22(1987)1043
[53] A. Denat, B. Gosse, and J. P. Gosse, Rev. Phys. Appl., 16(1981)673
[54] A. Denat, B. Gosse, and J. P. Gosse, J. Electrostatics, 12(1982)197
[55] G. A. Lane Proc. of SPIE, 1253(1990)29
[56] Z. Randriamalala, A. Denat, J. P. Gosse, and B. Gosse, IEEE Trans. Elect. Insul., EI-20 (1985)167
[57] P.A. Cirkel, J.P.M. van der Ploeg, and G.J.M. Koper, Physica A, 235(1997)269;
[58] C. Chassagne, D. Bedeaux, J.P.M. van der Ploeg, and G.J.M. Koper, Colloids Surfaces A, 210(2002)137
[59] Yu Feldman, E.Polygalov, I. Ermolina, Yu Polevaya, and B. Tsentsiper, Meas. Sci. Technol., 12(2001)1355
[60] C. Chassagne, D. Bedeaux,G.J.M. Koper, J. Phys. Chem. B 105(2001)11743
[61] M. Scott, R. Paul, K.I.S. Kaler, J.Colloid Interface Sci., 230(2000)388
[62] H.P.Schwann, in " Physical Techniques in Biological Research, Vol. 6, Part B, W.L. Nastuk, Academic Press, New York, 1963, p323
[63] C. Grosse, M. C.Tirado, J. Non-Crystalline Solids, 305(2002)386
[64] S.L. Srivastava, and R. Dhar, Indian J. Pure Appl. Phys., 29(1991)745
[65] A. Einstein, Investigation on the Theory of Brownian Movement, New York: Dover, 1956
[66] U.Balucani, M.Gori, and R. Vallauri, Chem. Phy. Lett., 152 (1985)119; [67] C. A. Bearchell, J. A. Edgar, D. M. Heyes, and S.E. Taylor, J. Colloid Interface Sci., 210(1999)231
[68] J.C. Maxwell, Electricity and Magnetism, Vol.1, Clarendon Press, Oxford, 1892
[69] L.D. Landau, E.M. Lifshitz, and L.P. Pitaevskil, Electrodynamics of Continuous Media: Volume 8 (Course of Theoretical Physics), Butterworth-Heinemann; 2 edition, 1984
[70] S. S. Dukhin, Dielectric Properties of Disperse Systems, in "Surface and Colloid Science", E. Matijevic, Ed., Wiley-Interscience, New York, 1971
[71] A. von Hippel, Dielectrics and Waves, Wiley, London, 1954
[72] H. Pauly, and H.P. Schwan, Z. Naturf., 14b(1959)125
[73] T. Hanai, Kolloidzeitschrift, 171(1960)23
[74] T. Hanai, Kolloidzeitschrift, 175(1961)61
[75] T. Hanai, Bull. Inst. Chem. Res. Kyoto Univ., 39(1961)341

[76] D.A.G. Bruggeman, Ann. Phys., 24(1935)636
[77] R. W. Sillars, JIEE, 80(1937)378
[78] H. Fricke, Phys. Rev. 24(1924)575
[79] C. Grosse, Ferroelectrics, 86(1988)181
[80] D. Polder, and J.H.V. van Santen, Physica, 12(1946)257
[81] G.P. de Loor, Appl. Sci. Res., 3(1953)479
[82] A.H. Sihola, and J.A.Kong, IEEE Trans.Geosci.Remote Sens., 26(1988)420;
[83] G. Bánhegyi, Colloid Polym. Sci., 266(1988)11
[84] S.B. Jones, and S.P. Friedman, Water Resour. Res., 16(1980)574
[85] C.J.F. Böttcher, Rec. Trav. Chim. 64(1945)47
[86] W.Y. Hsu, T.D. Gierke, J.C. Molnar, Macromol.,16(1983)1945
[87] H. Looyenga, Physica, 31(1965)401
[88] L.D. Landau, and E.M. Lifshitz, Electrodynamics of Continuous Media, Permagon, Tarrytown, N.Y., 1960
[89] B. Lestriez, A.Maazouz, J.F. Gerard, H. Sautereau, G. Boiteux, G. Seytre, and D.E. Kranbuehl, Polymer, 39(1998)6733
[90] H. Fricke, J.Phys. Chem., 57(1953)934
[91] A. H. Sihvola, and J.A. Kong, IEEE Trans. Geosci. Remote Sens., 26(1988)420
[92] J.C. Maxwell-Garnett, Philos.Trans. R. Soc.London, Ser.A, 203(1904)385
[93] L. Tsang, J.A. Kong, R.T. Shin, Theory of Microwave Remote Sensing, John Wiley, New York, 1985
[94] S.B. Jones and S.P. Friedman, Water Resour. Res., 36(2000)2821
[95] C. Boned and J. Peyrelasse, Colloid Polym. Sci., 261(1983)600
[96] L.D. Landau, and E.M. Lifshitz, Electrodynamics of Continuous Media, Pergamon Press, London, 1960, p46
[97] G. Bánhegyi, Colloid Polym Sci., 264(1986)1030
[98] M.H. Boyle, Colloid Polym Sci., 263(1985)51
[99] W.E.A. Davies, J. Phys. D 4(1971)318
[100] L.K.H. van Beek, Progress in Dielectrics, J.B. Birks, ed., Vol.7, Heywood, London, 1967, p69
[101] B. Gross, J. Chem. Phys., 17(1949)866
[102] T. Hao, A. Kawai, and F. Ikazaki, Langmuir, 14(1998)1256
[103] M.F. Manning, and M.E. Bell, Rev. Mod. Phys., 12(1940)215
[104] B.V. Hamon, Proc. Inst. Electr. Engr.,99, Pt.IV, Monograph 27, 1952

Chapter 8

Dielectric properties of ER suspensions

The dielectric property of an electrorheological (ER) fluid would be very important for understanding the ER phenomena, as the ER effect is induced by an external electric field and the polarizations should play an critical role. In Chapter 7, the dielectric properties of non-aqueous suspensions are described and the general trends mentioned earlier should of course be observed in the ER suspensions, though ER suspensions usually operate under a very high electric field in comparison with the electric field applied in dielectric measurement. In this chapter, experimental evidence on how the dielectric properties of the ER suspensions of spherical or quasi-spherical particles correlate with their ER performance is introduced first. The theoretical efforts for explaining those experimental observations are summarized in the next section, and an extracted dielectric mechanism for the ER effect is thus described. On the basis of the dielectric mechanism, a yield stress equation is developed for describing the rheological response of ER fluids. The particle shape effect on the dielectric property of ER suspensions and then on the ER effect is given, followed by a brief conclusion. In summary, an attempt is made in this chapter to provide a clear physical indication of the ER response mechanism through the dielectric understanding of the ER phenomena.

1 INTRODUCTION

As mentioned before, the ER phenomenon was first reported in an insulating oil containing starch or flour particulate material by Winslow in 1947 [1]. The first example of using the dielectric tool to investigate the ER phenomenon was carried out by Klass in 1967 [2]. It was found that the dielectric constant increases linearly with the particle volume fraction, and a double-layer mechanism was proposed to explain the observed experimental results. The interaction among the double layers, as well as the interaction between the double layer and the particle, would become much stronger once an electric field is applied. The double layer can be polarized and then be distorted, thus the electrostatic force between particles could dramatically increase, generating the remarkable ER effect. A more detailed dielectric investigation on how the dielectric constant of whole suspension changes with the water content in the suspension and the particle volume fraction was carried out by Uejima [3]. At that time, the "water" [4] and the electrical double layer [2] were believed to be crucial to the ER effect, since all ER fluids are only active if the dispersed particles adsorb a trace amount of water or surfactants. Since anhydrous ER fluids [5,6] were innovated in 1985, it has been realized that the "water" or the electrical double layer mechanism should be largely modified, as they are unable to explain the considerably strong ER effect in an ER fluid that does not contain any amount of water. Block [7] studied the dielectric property of a water-free ER system and proposed a FMP theory (flow modified polarization) to understand the ER effect. According to the FMP theory, there would be a resonance between the applied electric field and the shear field. The dielectric constant and loss of the ER suspension would have a peak when the shear rate is 4π times that of the frequency of the applied electric field. An extra surface charge can be generated once the particle is polarized in an electric field, which would drive the particle to re-arrange in the suspension. This is the possible reason that the ER suspension can be fibrillated to form a bridge between two electrodes. However, the physical picture of how the particle dielectric property can change the ER performance is not clear in Block's theory.

Anderson [8] and Gario [9] studied how the conductivity and dielectric constant of both the liquid medium and particulate material affect the ER effect. They found that the ER effect becomes stronger once the dielectric constant ratio of the particle-to-liquid increases. Davis [10-12] theoretically addressed this issue and found that the ER effect is controlled by the conductivity ratio of the particle-to-liquid at low electric frequency fields and by the dielectric constant ratio at high frequency fields, which is the well-known polarization model. The dielectric constant mismatch between the suspended particle and the medium is believed to be essential. However, the ER effect of the barium titanate suspension ($BaTiO_3$, its dielectric constant around 2000, depending on its crystallization state), inactive under a dc field [13] and active after adsorbing a small amount of water [14] or being stimulated by an ac field [9] suggests that the polarization model does not adequately reveal the physical origin of the ER effect. Khusid and Acrivos' theory [15] takes into account both the electric field induced particle aggregation process and the interfacial polarization process, and it can almost explain all the currently observed ER phenomena, especially the ones that cannot be appropriately understood with the polarization and the conduction models. However, some discrepancies with the experimental results still exist. This is because the assumptions of

Khusid and Acrivos' theory—the dispersed particle and the suspending medium have no intrinsic dielectric dispersion and that the variation of the applied electric field is very slow compared with the rate of polarization—are not always valid for the ER fluids.

The dielectric studies on anhydrous ER fluids provide a deep insight into the ER mechanism. Filisko [16-19] suggested that the ER effect of anhydrous ER fluid stems from the intrinsic property of a dispersed particle; the mobile charges (ions or electrons) on the surface of particles that are able to move freely on the surface but cannot move off of it would be essential. Thus the ER effect should somewhat correlate with the dielectric dispersion in a low frequency field. However, they also found the ER effect does not correlate directly with the charge concentration. Block [7] suggested that the polarization rate as well as its magnitude are very important for the ER response. Too high a rate of polarization and too low a rate of polarization were experimentally found to be unfavorable for a strong ER effect. Kawai [20] and Ikazaki [21,22] experimentally found that for a good ER fluid the dielectric relaxation frequency should be between 100 and 10^5 Hz and the difference of the dielectric constant below and above the relaxation frequency must be large. These experimental results, in my opinion, imply that the ER effect perhaps correlates with the particle dielectric loss, as charges moving within the particle under an electric field would be one of the causes of the dielectric loss. Qiu [23] theoretically addressed the influence of polarization rate on the ER effect and came to a good agreement with the experimental results. Hao's series of papers revealed why the polarization rate is important, and unambiguously pointed out that it is the interfacial polarization that contributes to the dielectric dispersion observed in the low frequency [24-30]. An empirical criterion was thus proposed for selecting the ER material, and was also theoretically verified to be a direct consequence of the entropy decrease associated with the ER effect under an electric field. On the basis of this physical insight, a yield stress equation was derived for describing the rheological behavior of ER fluids.

2 DIELECTRIC PROPERTY OF THE ER SUSPENSIONS OF SPHERICAL OR QUASI-SPHERICAL PARTICLES

The purpose of the dielectric investigation is to correlate the dielectric properties of the ER materials to the ER effect and to provide the predictive guidance for making high performance ER fluids. Before doing this, the following questions should be answered: 1) Which kind of polarization should be responsible for the ER effect; 2) Whether the Wagner-Maxwell equation can describe the dielectric property of the ER system, as the interfaces clearly exist in ER fluids; 3) How the dielectric constant or dielectric loss of the materials used for fabricating ER fluids correlates with the ER effect. In other words, is there a criterion that could be extracted from the dielectric experimental facts for designing the high-performance ER fluids? Those questions will be answered in this section.

As described in previous chapters, the ER Effect is most likely controlled by the slow polarizations, as the response time of the ER effect is on a millisecond scale, and the electronic and atomic polarizations are a much faster process. For many good ER fluids the dielectric dispersion is experimentally observed to appear in a low frequency field [18, 20-22, 26,31,32] further indicating that the ER effect is controlled by the slow polarizations. For example, the dielectric loss tangents of two kinds of ER suspensions consisting of aluminosilicate powder dispersing into polydimethylsiloxane (PDMS) oil (powder treated at 550°C and 750°C for 8 hours and named as AS1 and AS2, respectively) are plotted against frequency in Figure 1. These two ER suspensions give a very strong ER effect, and the dielectric dispersion appears in a low-frequency field less than 10^3 Hz, indicating that the slow polarization is the major contributor to the dielectric dispersion. Another example is shown in Figure 2 for the dielectric loss of sulphonated poly(styrene-co-divinylbenzene) particles (SSD) of different diameters (5,15, and 50 μm) dispersed in silicone oil vs. frequency. The dielectric loss peaks appear at the frequency around 1000 Hz, and the exact position varies with the size of the particle dispersed in the suspension. Experimental results obtained by Filisko [18,32] also suggest that the ER effect might be associated with the low-field dielectric dispersion. Besides the dielectric loss peak, another different piece of experimental evidence is that the dc current passing through the ER fluid is not stable, exponentially decreasing with time and then leveling off, which was found both in a polymeric ER fluid (oxidized polyacrylonitrile/PDMS) [33] and an inorganic ER fluid (aluminosilicate//PDMS), as shown in Figure 51, Chapter 6.

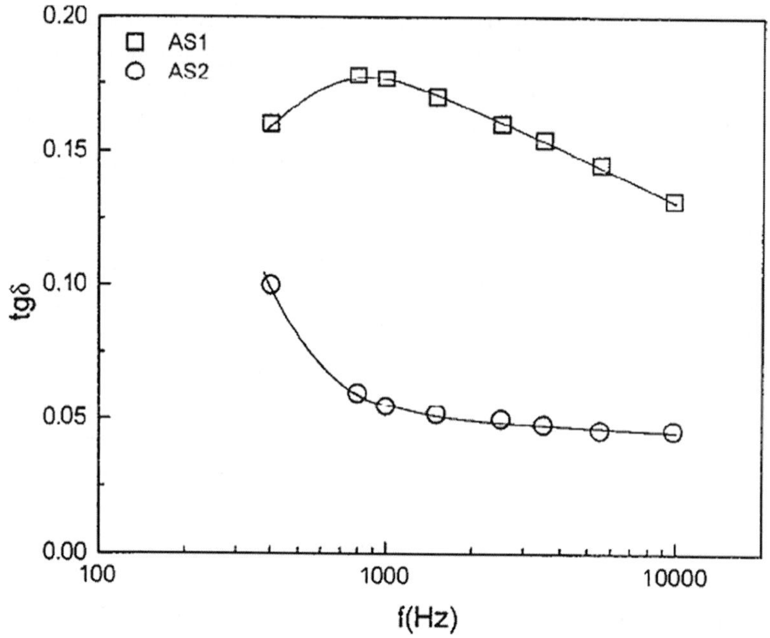

Figure 1 The dielectric loss tangents of the aluminosilicate powder dispersed into the polydimethylsiloxane (PDMS) oil (powder treated at 550°C and 750°C for 8 hours and named as AS1 and AS2, respectively) as a function of frequency. Reproduced with permission from T. Hao, A. Kawai, and F. Ikazaki, Langmuir, 14(1998)1256.

This is the so-called dielectric adsorption phenomenon physically resulting from the heterogeneity of the system and the slow polarizations generated under an external electric field. It further indicates that the slow polarizations rather than fast polarizations are dominant in ER systems. All these experimental facts suggest that in ER systems the slow polarizations would likely be the most important process.

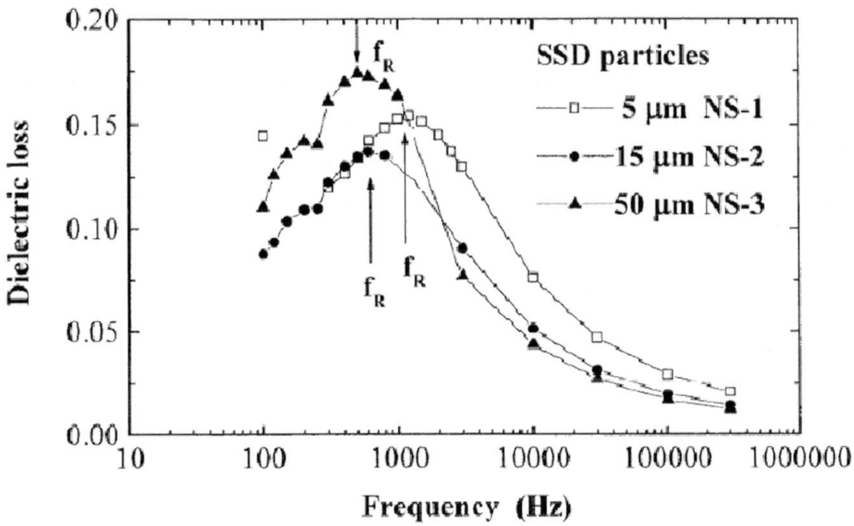

Figure 2 The dielectric loss of the sulphonated poly(styrene-co-divinylbenzene) particles (SSD) of different diameter diameters (5,15, and 50 μm) dispersed in silicone oil vs. frequency. The particle volume fraction is 6.8 vol% and water content is 2.8 wt%. Reproduced with permission from F. Ikazaki, A. Kawai, T. Kawakami, K. Edamura, K. Sakurai, H. Anzai, and Y. Asako, J. Appl. Phys. *D,* 31(1998)336

Since most ER suspensions don't contain the charging agent, the electrode polarization can be ruled out as contributing to the ER effect. The next question that arises is whether the Debye or the interfacial polarization would control the ER effect or if they would jointly control the ER effect. As is known, the Debye polarization is generated due to the dipole orientation in an electric field. For most solid materials, the dipole is almost unable to reorient because the solidification usually fixes the molecule with such rigidity in the solid lattice that there is little or no orientation of the dipoles even in an extremely strong electric field [34]. If the ER effect stems from the polarization of the solid particulate material, it seems unlikely that the Debye polarization would make any contribution, and the interfacial polarization will probably be responsible for the effect. Hao experimentally clarified this issue by measuring the ER response time [26] and temperature dependence of the dielectric property [35], as the ER response time and the dielectric dispersion peak would be determined by the relaxation time of the polarization that controls the ER effect.

As shown in Eq. (25) and Eq.(26) in Chapter 7, the relaxation times of the Debye and the interfacial polarizations are governed by different factors. The relaxation time of the Debye polarization, τ_d, is controlled by the potential barrier between two equilibrium positions where the dipole orientates from one to the other [34] and that of the interfacial polarization, τ_i , is controlled by the dielectric constants and conductivities of the two components [36,37]. Clearly, τ_d depends on the material state (liquid or solid), i.e., the molecule interaction and the environment, while τ_i depends on the dielectric properties of two components, especially on the particle conductivity. Hao [26] measured the response time of the aluminosilicate powder dispersing into the polydimethylsiloxane (PDMS) oil suspensions via the shear stress dependence on the applied electric field frequency. The results are shown in Figure 21 in Chapter 5. Two suspensions were made from the same powder but treated at different temperature 550°C and 750°C for 8 hours and named as AS1 and AS2, respectively. The response times of these two suspensions are very much different. The response time of the suspension with the particle conductivity 6.0×10^{-7} S/m (AS1) is 0.6 ms while that of the suspension with the particle conductivity 8.4×10^{-10} S/m (AS2) is 0.22 s. If the Debye polarization controls the ER effect, these two ER fluids should give, according to Eq.(25) in Chapter 7, almost the same response time, as the solid particulate material in these two suspensions are almost the same, and the energy barrier E_H should not largely depend on the water content on the particle surface. The quite different response time can only be reasonably understood through Eq.(26) in Chapter 7, indicating the interfacial polarization would probably govern the ER effect. Additional experimental evidence [38] showed that the measured response time is inversely proportional to the particle conductivity, agreeing well with Eq.(26) in Chapter 7. Wen [39] and Conrad [40] found that even the same particulate material will display different response time after adsorbing different amount of water; the larger amount of water does lead to the shorter response time. These are also consistent with Eq.(26) in Chapter 7; Since the oil is hydrophobic, the added water may adsorb on the particle surface and thus enhance the particle conductivity considerably. Whittle [41] also found the same inversely proportional relationship between the particle conductivity and the ER response time, further supporting that the ER effect may be controlled by the interfacial polarization.

Direct differentiation on the contribution of the Debye or the Wagner-Maxwell polarization to the ER effect was carried out by Hao [35]. The strategy employed in Hao's paper is to compare the temperature dependence difference of the dielectric loss tangent maximum values of commonly encountered three-type polarizations, the ionic polarization, the Debye polarization, and the interfacial polarization. As shown in Eq.(147) in Chapter 7, for the Maxwell-Wagner polarization (also called the interfacial polarization) the dielectric loss tangent can be expressed as follows if the particle conductivity σ_p is much larger than that of the medium

$$tg\delta = \frac{K\varpi t}{1+K+\varpi^2 t^2} \tag{1}$$

in this equation

$$K = \frac{9\Phi\varepsilon_{sm}}{2\varepsilon_{sm}+\varepsilon_{sp}} \tag{2}$$

$$t = \frac{\varepsilon_0(2\varepsilon_{sm}+\varepsilon_{sp})}{\sigma_p} \tag{3}$$

where ε_0, ε_{sm}, and ε_{sp} are the static dielectric constants of the vacuum, the medium and the particles; Φ is the particles volume fraction, ω is the frequency of alternating electric field. Differentiating Eq.(1) with respect to the frequency ω, one can easily obtain the maximum value of $tg\delta_{max}$

$$tg\delta_{max} = \frac{K}{2\sqrt{1+K}} = \frac{9\phi\varepsilon_{sm}}{2\sqrt{\left(9\phi\varepsilon_{sm}+2\varepsilon_{sm}+\varepsilon_{sp}\right)\left(2\varepsilon_{sm}+\varepsilon_{sp}\right)}} \tag{4}$$

Assuming both ε_{sm} and ε_{sp} are dependent on temperature T, differentiating Eq.(4) with respect to T leads to

$$\frac{dtg\delta_{\max}}{dT}=\frac{(2+K)\frac{dK}{dT}}{4(1+K)^{3/2}}$$

$$=\frac{(2+K)}{4(1+K)^{3/2}}\times\frac{9\phi\left(\varepsilon_{sp}\frac{d\varepsilon_{sm}}{dT}-\varepsilon_{sm}\frac{d\varepsilon_{sp}}{dT}\right)}{\left(2\varepsilon_{sm}+\varepsilon_{sp}\right)^{2}} \tag{5}$$

The sign of ($dtg\delta_{\max}/dT$) obviously depends on the sign of $\left(\varepsilon_{sp}\frac{d\varepsilon_{sm}}{dT}-\varepsilon_{sm}\frac{d\varepsilon_{sp}}{dT}\right)$, as the rest part always has a positive value. According to the Fröhlich's theory [34], for non-polar liquids, the static dielectric constant usually decreases as temperature increases. The most commonly used dispersing medium of ER fluids is the polydimethylsiloxane oil, an non-polar liquid, thus $\frac{d\varepsilon_{sm}}{dT}$ would be less than zero. If $\frac{d\varepsilon_{sp}}{dT}>0$, then in any cases, the term $\left(\varepsilon_{sp}\frac{d\varepsilon_{sm}}{dT}-\varepsilon_{sm}\frac{d\varepsilon_{sp}}{dT}\right)$ should be less than zero, thus $dtg\delta_{\max}/dT<0$; If $\frac{d\varepsilon_{sp}}{dT}<0$, the situation is somewhat complicated, and one needs to know all the parameters involved in the term $\left(\varepsilon_{sp}\frac{d\varepsilon_{sm}}{dT}-\varepsilon_{sm}\frac{d\varepsilon_{sp}}{dT}\right)$ for the purpose of determining the sign of ($dtg\delta_{\max}/dT$). These results feature the dielectric characteristics of an ER suspension in the Maxwell-Wagner polarization dominant case.

According to Eq. (58) and (59) in Chapter 7, the dielectric loss tangent of the Debye polarization can be expressed as:

$$tg\delta=\frac{\left(\varepsilon_{sp}-\varepsilon_{\infty p}\right)\omega\tau}{\varepsilon_{sp}+\varepsilon_{\infty p}\omega^{2}\tau^{2}} \tag{6}$$

The maximum value of the dielectric loss tangent, $tg\delta_{d\max}$, of the Debye polarization can be expressed:

$$tg\delta_{d\max} = \frac{\varepsilon_{sp} - \varepsilon_{\infty p}}{2\sqrt{\varepsilon_{sp}\varepsilon_{\infty p}}} \qquad (7)$$

where $\varepsilon_{\infty p}$ is the dielectric constant at very high frequency. Differentiating Eq.(7) with respect to T and assuming only ε_{sp} is temperature dependent, thus

$$\frac{dtg\delta_{d\max}}{dT} = \frac{\left(1 + \frac{\varepsilon_{\infty p}}{\varepsilon_{sp}}\right)\frac{d\varepsilon_{sp}}{dT}}{4\sqrt{\varepsilon_{\infty p}\varepsilon_{sp}}} \qquad (8)$$

Comparing Eq. (5) with Eq. (8), one will find that, if $\frac{d\varepsilon_{sp}}{dT}>0$, the observed ($dtg\delta_{\max}/dT$) determined by the Maxwell-Wagner polarization will give an opposite sign to that determined by the Debye polarization. In this case, it should be easier to distinguish the polarization type; If $\frac{d\varepsilon_{sp}}{dT}<0$, one has to know the exact value of $\frac{d\varepsilon_{sp}}{dT}$. For a homogeneous ER system such as liquid crystalline polymer (LCP) system, the dielectric constant of the liquid crystalline polymer at the isotropic state would usually decrease as temperature increase [42], similar to a polar liquid. In certain exceptional cases, for example, if the LCP material has a cyano group, then its dielectric constant at the isotropic state would increase as temperature increases [42]. In this case $dtg\delta_{\max}/dT$ will be larger than zero in the Debye polarization dominant case, while it will be less than zero in the Maxwell-Wagner polarization dominant case. These criteria are only applicable to the immiscible LCP ER systems that can only display detectable ER effect in the isotropic phase. For a perfect miscible isotropic LCP ER system, there should be no the interfacial polarization and ionic polarization, the sign of ($dtg\delta_{\max}/dT$) should be same as that of the ($d\varepsilon_{sp}/dT$), usually negative for the Debye polarization dominant case. The pure LCP system usually displays the ER effect in the nematic phase, and the sign of ($dtg\delta_{max}/dT$) should depend on the applied electric field either perpendicular or parallel to the director of nematic LCP, i.e., the sign of $d\varepsilon_{//}/dT$ and $d\varepsilon_{\perp}/dT$.. If the external electric field is parallel to the director, and ($d\varepsilon_{//}/dT$) is positive, then $dtg\delta_{max}/dT>0$ for the Debye polarization dominance case. Due to the complexity of the LCP system, one should at first determine the exact value of the parameters involved in Eq.(5) and then can determine the polarization type.

The ion displacement polarization usually takes place in an ER suspension having ionic solid particulate. If the conductivity of the solid material is not too high, then the maximum of the dielectric loss tangent $tg\delta_{i\max}$, due to the ion displacement polarization, can be expressed [28,43]

$$tg\delta_{i\max} = \frac{\varepsilon_{sp} - \varepsilon_{\infty p}}{2\sqrt{\varepsilon_{sp}\varepsilon_{\infty p}}} \qquad (9)$$

Note that Eq. (9) and (7) have a same form. Differentiating Eq.(9) with respect to T should lead to a same equation as shown in Eq. (8), if only ε_{sp} is also assumed to be dependent of temperature. This means that the Debye and ionic polarizations are undifferentiated using this method, and only Debye or ionic polarization with the Maxwell-Wagner polarization is differentiable. Clearly, for unambiguously distinguishing the polarization type in both heterogeneous and homogeneous ER systems, one may need to exactly obtain the parameter values involved in Eq. (5) and (8) through experimental measurements.

Two types of ER suspensions were employed by Hao for experimentally differentiating the polarization type in ER suspensions [35]. One is the heterogeneous suspension composed of molecular sieve 3A dispersed into polydimethylsiloxane (PDMS). Anther is the homogeneous suspension composed of the liquid crystalline polymer of the polysiloxane main chain (the number of silicon is 31) and 4-cyanophenyl-4-propyloxy benzoate side chain (the number is 14, this polymer is named LCPP) dissolved in phenyl-substituted polydimethylsiloxane (Ph-PDMS, 15% phenyl content; this sample is named LCPPh). The dielectric loss tangent of the molecular sieve 3A dispersed into the polydimethylsiloxane (PDMS) suspension vs. frequency at different temperatures ranging from 20 to 120 ^{o}C is shown in Figure 3. The peak value of the dielectric loss tangent clearly decreases as temperature increases.

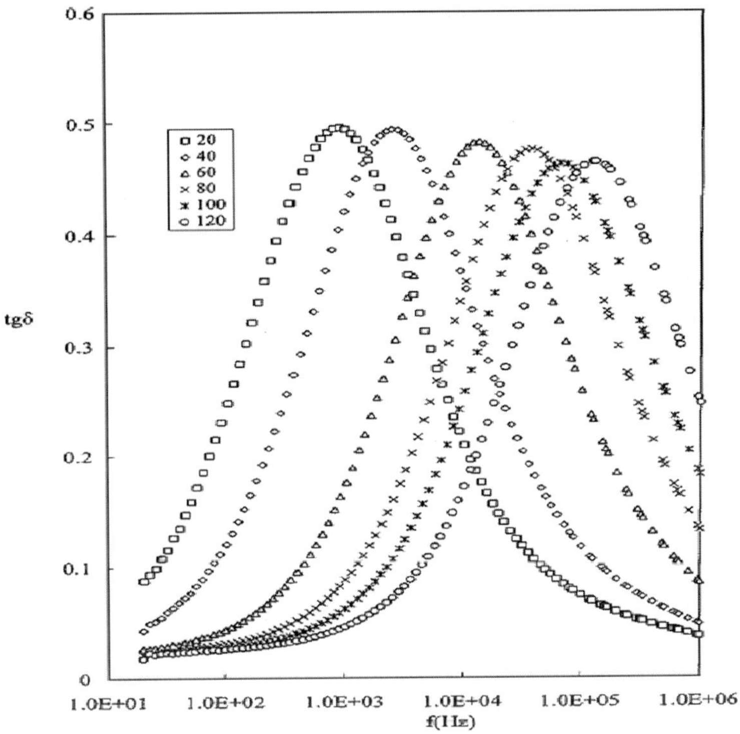

Figure 3 The dielectric loss tangent of molecular sieve 3A dispersed into the polydimethylsiloxane (PDMS) suspension vs. frequency at different temperature. Reproduced with permission from T. Hao, A. Kawai, and F. Ikazaki, J. Colloid Interface Sci., 239(2001)106

For the homogeneous solution made of a liquid crystalline polymer of the polysiloxane main chain (the number of silicon is 31) and 4-cyanophenyl-4-propyloxy benzoate side chain (the number is 14) dissolved in phenyl-substituted polydimethylsiloxane (Ph-PDMS, 15% phenyl content), the dielectric loss tangent peak value increases as temperature increases, which is shown in Figure 4. Clearly, for the liquid crystal polymer the temperature dependence of the maximum value of the dielectric loss tangent presents an opposite trend to that of the molecular sieve/PDMS suspension. The static dielectric constants of two dispersing media, PDMS and Ph-PDMS, are plotted against temperature in Figure 5 and Figure 6. Ph-PDMS has a relatively large static dielectric constant than that of PDMS, and both liquids show that the static dielectric constant decreases with temperature.

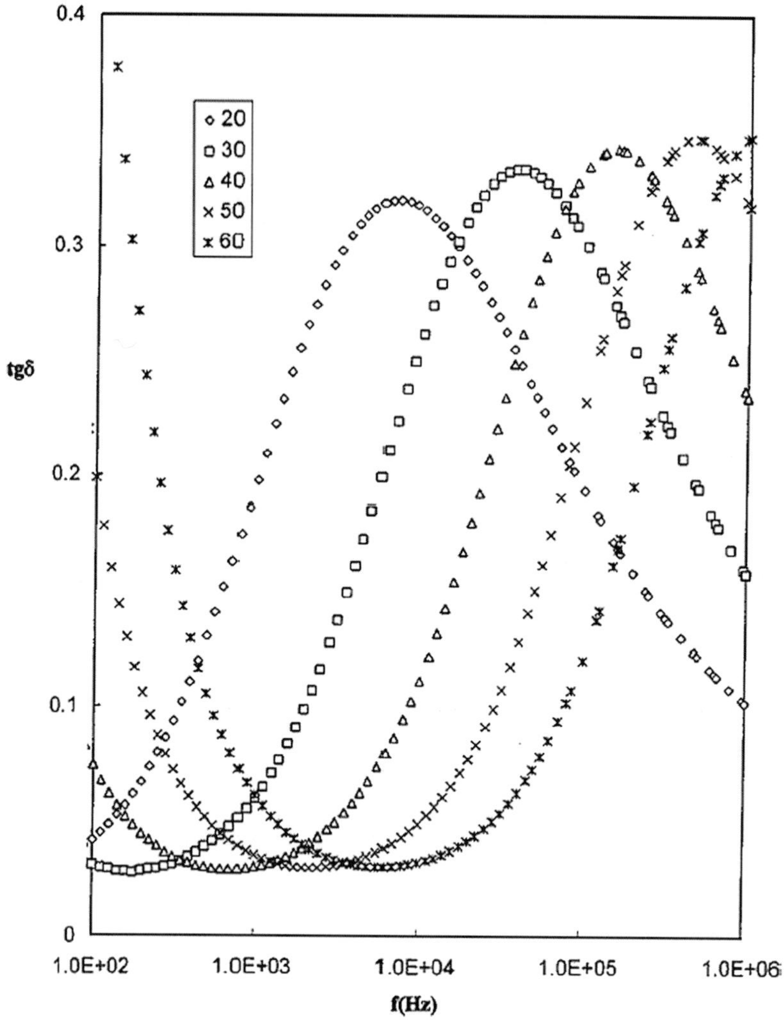

Figure 4 The dielectric loss tangent of the liquid crystalline polymer of the polysiloxane main chain (the number of silicon is 31) and 4-cyanophenyl-4-propyloxy benzoate side chain (the number is 14, this polymer is named LCPP) dissolved in phenyl-substituted polydimethylsiloxane (Ph-PDMS, 15% phenyl content) as a function of frequency obtained at 20–60 °C. Reproduced with permission from T. Hao, A. Kawai, and F.Ikazaki, J. Colloid and Interface Sci., 239(2001)106

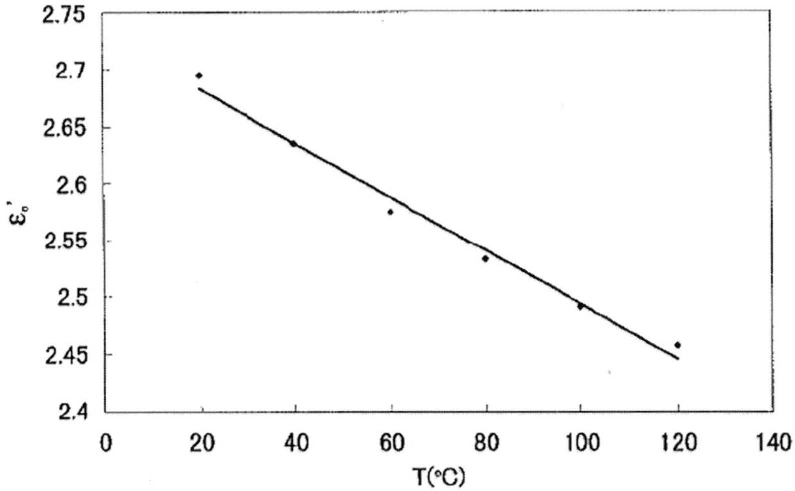

Figure 5 The static dielectric constant of polydimethylsiloxane (PDMS) vs. temperature. Reproduced with permission from T. Hao, A. Kawai, and F. Ikazaki, Langmuir 16(2000)3058

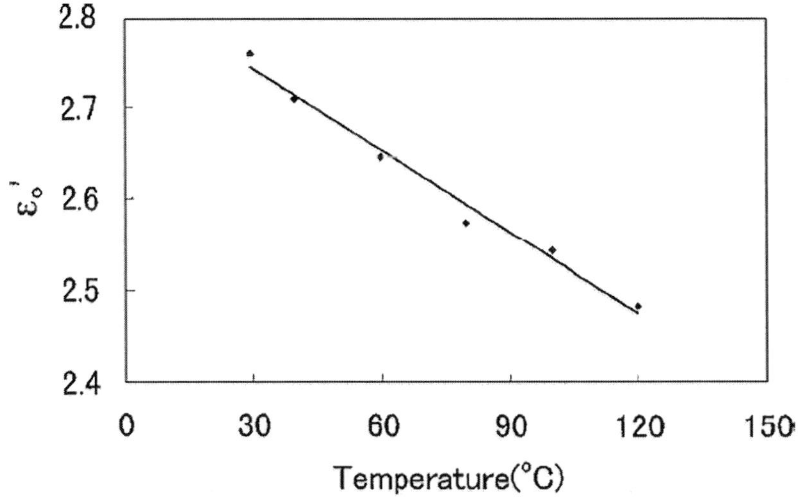

Figure 6 The static dielectric constant of phenyl-substituted polydimethylsiloxane (Ph-PDMS) oil vs. temperature. Reproduced with permission from T. Hao, A. Kawai, and F. Ikazaki, J. Colloid Interface Sci., 239(2001)106

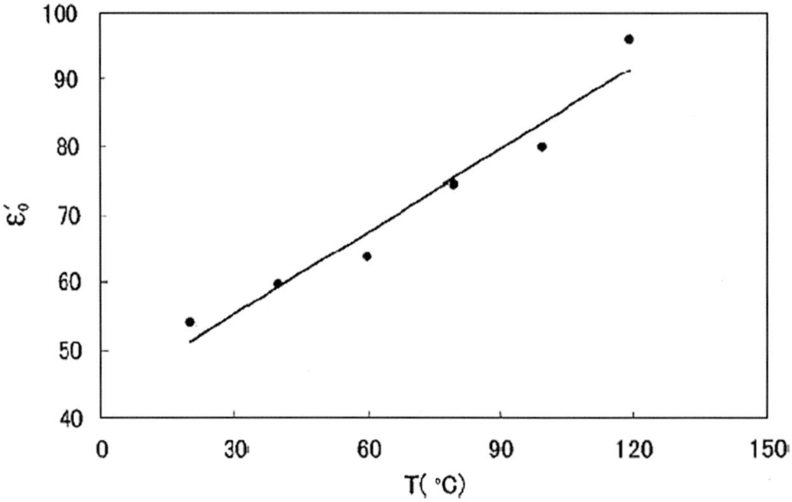

Figure 7 The static dielectric constant of pure MS3A vs. temperature. Reproduced with permission from T. Hao, A. Kawai, and F. Ikazaki, J. Colloid Interface Sci., 239(2001)106

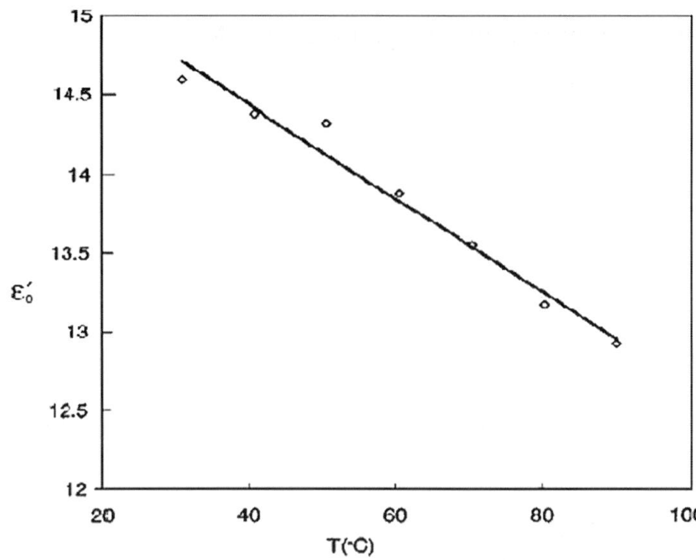

Figure 8 The static dielectric constant of the liquid crystalline polymer of the polysiloxane main chain (the number of silicon is 31) and 4-cyanophenyl-4-propyloxy benzoate side chain (the number is 14) vs. temperature. Reproduced with permission from T. Hao, A. Kawai, and F. Ikazaki, J. Colloid Interface Sci., 239(2001)106

For the dispersed phase the static dielectric constants of molecular sieve 3A and the liquid crystalline polymer of the polysiloxane main chain (the number of silicon is 31) and 4-cyanophenyl-4-propyloxy benzoate side chain (the number is 14) are plotted against temperature in Figures 7 and 8, respectively. Unlike the liquid crystalline polymer, the static dielectric constant of the molecular sieve increases with temperature. So ${d\varepsilon_s}/{dT}$ (ε_s is the static dielectric constant) is less than zero for PDMS, Ph-PDMS and the liquid crystalline polymer, while ${d\varepsilon_s}/{dT}$ for molecular sieve 3A is larger than zero. The value of ${d\varepsilon_s}/{dT}$ can be easily determined using the linear regression method based on the experimental data points, for example ${d\varepsilon_s}/{dT}$ is -2.4 x10^{-3}(1/ °C) for PDMS and 0.40 for the molecular sieve 3A. Thus the term $\left(\varepsilon_{sp}\frac{d\varepsilon_{sm}}{dT}-\varepsilon_{sm}\frac{d\varepsilon_{sp}}{dT}\right)$ in Eq.(5) thus can be calculated as -1.21, which means that ($dtg\delta_{\max}/dT$) is less than zero. According to the criteria given above, it easily comes to the conclusion that the Maxwell-Wagner polarization is dominant in the particle-type MS3A suspension, as the dielectric loss tangent peak value of the MS3A suspension obviously decreases as the temperature increases from 20 to 120°C. The same calculation procedure can be applied to the liquid crystalline polymer system, and the final conclusion is that the Maxwell-Wagner polarization is still dominant even in a homogeneous ER system [35]

In addition, the Wagner-Maxwell equation is able to adequately describe the dielectric property of ER fluids. Weiss [44] and Filisko [45] measured the dielectric constant and dielectric loss of ER fluids, and found that the Wagner-Maxwell equation is a suitable model to describe the dielectric property. The Wagner-Maxwell equation can also explain the frequency dependence of yield stress of ER fluids. As given in Figure 9 in Chapter 5, the frequency dependence of the yield stress and the dielectric constant follow a similar trend, further indicating that it is the Wagner-Maxwell polarization that controls the dielectric property of the ER suspension and then the ER effect.

As the experimental facts indicated above, the importance of the Wagner-Maxwell polarization for the ER effect is obvious. However, it is still not clear how the interfacial polarization controls the ER effect, and how the material dielectric constant and dielectric loss correlate with the ER effect. From the Wagner-Maxwell equations shown in Chapter 7, one can see that the interfacial polarization greatly depends on the conductivity of the solid particulate material. Figure 9 shows how the conductivity of dispersed particle shifts the dielectric loss tangent peak [25]. The dielectric loss tangent peaks at low frequencies if the conductivity of the dispersed particle is low. A material of the conductivity that can be easily adjusted and thus the ER performance of those materials dispersed in an insulating oil would be different between each other due to the different interfacial polarization contribution. This set of materials of different conductivity would be very useful for determining whether the dielectric constant or the dielectric loss would be dominant for the ER effect. Hao [27] selected the oxidized polyacrylonitrile (OP) and aluminosislicate (AS) materials as the candidates for this purpose, as the conductivity of the OP can be easily adjusted at different carbonating temperatures, and that of the AS can also be easily controlled using different heat treatment temperatures. The heating process is controlled to only change the material conductivity rather than the material itself. The shear stress against the particle dielectric constant and dielectric loss tangent of these two materials are shown in Figures 27 and 28 in Chapter 5 already. The shear stress does not monotonically change with the dielectric constant or dielectric loss tangent even for the same series samples. From these experimental results, Hao concluded that a material having a large dielectric constant and a large dielectric loss will display a strong ER effect; The ER effect will be weak if the material only has a large dielectric constant but a small dielectric loss. The dielectric loss tangent of the solid particulate material should be about 0.10 at 1000 Hz, which is required for a detectable ER effect. Furthermore, the larger the dielectric constant is, then the stronger ER effect will be. This means the dielectric loss is also an important parameter for the ER effect. The function of the dielectric loss in the ER effect was detailed in ref.[26], and a dielectric loss mechanism was thus proposed, which will be further discussed in a future chapter

3. THEORETICAL TREATMENT ON THE DIELECTRIC CRITERIA FOR HIGH PERFORMANCE ER SUSPENSIONS

The physical origins of the empirical criteria described above were explored theoretically by Hao [28]. As stated earlier, one typical feature of the ER fluid is that the ER particulates can form the fibrillated structure under the stimulation of an external electric field, while non-ER particulates do not have such a capability.

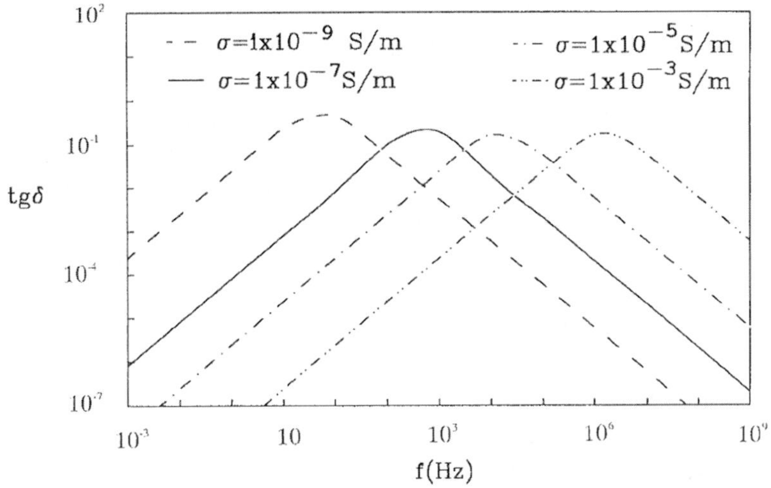

Figure 9. The dielectric loss tangent as a function of frequency predicted by the Wagner model under the assumptions of $\varepsilon_m = 2$, $\varepsilon_p = 10$, and $\Phi = 0.35$. Reproduced with permission from T. Hao, J. Colloid Interface Science, 206(1998)240.

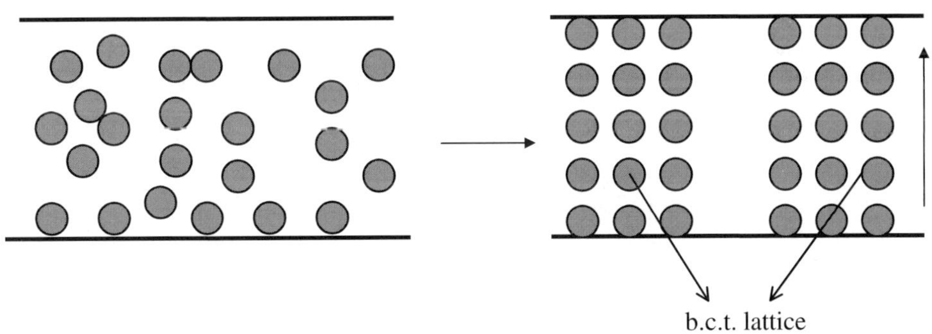

Figure 10 Schematic illustration of the structure change of an ER fluid before (a), and after (b) an external electric field is applied. The two parallel dark lines stand for two electrodes. Reproduced with permission from T. Hao, A. Kawai, and F. Ikazaki, Langmuir, 15(1999)918

The ER fluids can be solidified under an electric field, changing from a liquid state to a solid state. This process, schematically illustrated in Figure 10, has been comprehensively studied and is recognized as a second-order phase transition [46,47]. At the liquid state (before an electric field is applied), the ER particles are randomly distributed in the medium, and at the solid state (after an electric field is applied), the dispersed particles form a fibrillation structure (columns). Within a column the particles are arrayed as a crystallite of a body-centered tetragonal (bct) lattice, which was determined with a laser diffraction technique [48].

Since the order in ER systems increases dramatically after an electric field is applied, the entropy of these systems would markedly reduce; that is, the ΔS would be obviously less than zero. From this very basic fact, the dielectric criteria were theoretically obtained by Hao [28].

For a general ER system, assuming that the volume of the ER fluid is always kept constant, and that the temperature T is the only parameter besides the electric field E considered to be varied, after an electric field is applied, the increase dU of the internal energy U per unit volume of the ER suspension, can be expressed as follows, according to the first thermodynamic law and the electromagnetic theory [34]:

$$dU = dQ + \frac{E}{4\pi} dD \tag{10}$$

where dQ is the influx of heat per unit volume, D is the electric displacement, and $\frac{E}{4\pi}dD$ represents the influx of energy into the ER fluid. For the ER fluids, it is already found that, both theoretically and experimentally, there seems to be a critical electric field at which the liquid-solid (disorder-order) transition takes place [47,49,50]. The critical electric field is usually less than 200 V/mm, and the typical value is about 45 V/mm for many ER systems. In the case of the applied electric field slightly larger than the critical field, one may assume that the static dielectric constant of an ER suspension, ε_s, is independent of E, as the applied electric field is not so strong. Even if in a very strong field, the static dielectric constant of many materials would not be substantially changed [51]. Then $D=\varepsilon_s E$, thus

$$dD = d(\varepsilon_s E) = \varepsilon_s\, dE + E d\varepsilon_s = \varepsilon_s\, dE + E\frac{\partial \varepsilon_s}{\partial T} dT \tag{11}$$

as ε_s just depends on T. It will be useful to take T and E^2 as independent variables, thus Eq.(10) can be rewritten as:

$$dU = dQ + \frac{\varepsilon_s}{8\pi} d(E^2) + \frac{E^2}{4\pi}\frac{\partial \varepsilon_s}{\partial T} dT = \frac{\partial U}{\partial (E^2)} d(E^2) + \frac{\partial U}{\partial T} dT \qquad (12)$$

One also may assume the entropy S is a function of T and E^2, as $dS = \frac{dQ}{T}$, the second thermodynamic law for a reversible process. So that

$$dS = \frac{\partial S}{\partial T} dT + \frac{\partial S}{\partial (E^2)} d(E^2) \qquad (13)$$

Inserting dQ from Eq. (12) and (13)

$$dS = \frac{1}{T}\left(\frac{\partial U}{\partial T} - \frac{E^2}{4\pi}\frac{\partial \varepsilon_s}{\partial T}\right) dT + \frac{1}{T}\left(\frac{\partial U}{\partial (E^2)} - \frac{\varepsilon_s}{8\pi}\right) d(E^2) \qquad (14)$$

The condition that dS is a total differential requires that

$$\frac{\partial}{\partial (E^2)}\left\{\frac{1}{T}\left(\frac{\partial U}{\partial T} - \frac{E^2}{4\pi}\frac{\partial \varepsilon_s}{\partial T}\right)\right\} = \frac{\partial}{\partial T}\left\{\frac{1}{T}\left(\frac{\partial U}{\partial (E^2)} - \frac{\varepsilon_s}{8\pi}\right)\right\} \qquad (15)$$

Carrying out the differentiation one finds

$$\frac{\partial U}{\partial (E^2)} = \frac{1}{8\pi}\left(\varepsilon_s + T\frac{\partial \varepsilon_s}{\partial T}\right) \qquad (16)$$

Comparing Eq. (12) with (13) and inserting $\frac{\partial U}{\partial (E^2)}$ from Eq. (16)

$$\frac{\partial S}{\partial (E^2)} = \frac{1}{8\pi}\frac{\partial \varepsilon_s}{\partial T} \qquad (17)$$

Integrating

$$S = S_0(T) + \frac{\partial \varepsilon_s}{\partial T}\frac{E^2}{8\pi} \qquad (18)$$

where $S_0(T)$ is the entropy in the absence of a field. Thus

$$\Delta S = \frac{\partial \varepsilon_s}{\partial T}\frac{E^2}{8\pi} \tag{19}$$

For an ER fluid, $\Delta S < 0$, thus obviously requires $\frac{\partial \varepsilon_s}{\partial T} < 0$. The static dielectric constant ε_s of an ER fluid is the dielectric constant at $\omega=0$. Since the dielectric property of ER fluids can be described using the Wagner-Maxwell polarization equations. $\frac{\partial \varepsilon_s}{\partial T}$ thus can be derived from the Wagner-Maxwell equation. As stated earlier, If the particle conductivity σ_p is much larger than that of the medium, a simplified Maxwell-Wagner equation can be expressed by Eq. (146) in Chapter 7 as:

$$\varepsilon = \varepsilon_\infty\left[1+\frac{K}{(1+\varpi^2 t^2)}\right] \tag{20}$$

in this equation

$$\varepsilon_\infty = \varepsilon_{sm}\left[1+\frac{3\Phi(\varepsilon_{sp}-\varepsilon_{sm})}{2\varepsilon_{sm}+\varepsilon_{sp}}\right] \tag{21}$$

where ε_0, ε_{sm}, and ε_{sp} are the static dielectric constants of the vacuum, the medium and the particles; Φ is the particles volume fraction, ω is the frequency of alternating electric field. The parameters K and t are given in Eq. (2) and (3), respectively. Assuming $\omega=0$, Eq. (20) becomes

$$\varepsilon_s = \varepsilon_\infty(1+K) \tag{22}$$

Assuming that both ε_{sm}, and ε_{sp} are the function of T, thus

$$\frac{d\varepsilon_s}{dT} = \left[1+3\Phi+\frac{54\Phi^2\varepsilon_{sm}\left(\varepsilon_{sp}^2-1.5\varepsilon_{sm}\varepsilon_{sp}-\varepsilon_{sm}^2\right)}{\left(2\varepsilon_{sm}+\varepsilon_{sp}\right)^3}\right]\frac{d\varepsilon_{sm}}{dT} - \frac{27\varepsilon_{sm}^2\Phi^2\left(\varepsilon_{sp}-4\varepsilon_{sm}\right)}{(2\varepsilon_{sm}+\varepsilon_{sp})^3}\frac{d\varepsilon_{sp}}{dT} \tag{23}$$

For the non-polar liquid materials, $\frac{d\varepsilon_{sm}}{dT}$ is always less than zero [34]. So if $\varepsilon_{sp} > 4\varepsilon_{sm}$ and $\frac{d\varepsilon_{sp}}{dT}$>0, $\frac{d\varepsilon_s}{dT}$ would be less than zero in any conditions; if $\varepsilon_{sp} > 4\varepsilon_{sm}$, $\frac{d\varepsilon_{sp}}{dT}$<0, the situation becomes complicated, and the ER effect is still theoretically possible if

$$\left(\frac{d\varepsilon_{sp}}{dT}\right)/\left(\frac{d\varepsilon_{sm}}{dT}\right)<\frac{(1+3\Phi)\left(2\varepsilon_{sm}+\varepsilon_{sp}\right)^3+54\Phi^2\varepsilon_m\left(\varepsilon_{sp}^2-1.5\varepsilon_{sm}\varepsilon_{sp}-\varepsilon_{sm}^2\right)}{27\Phi^2\varepsilon_{sm}^2\left(\varepsilon_{sp}-4\varepsilon_{sm}\right)} \tag{24}$$

However, the possibility is not high, as the entropy decrease is not large (the particle contributes positively), and the ER solidification process will need a much larger entropy decrease for forming the b.c.t crystalline structure. So only the weak ER effect or even no ER effect at all would be generated. In another case, if

$$\left(\frac{d\varepsilon_{sp}}{dT}\right)/\left(\frac{d\varepsilon_{sm}}{dT}\right)>\frac{(1+3\Phi)\left(2\varepsilon_{sm}+\varepsilon_{sp}\right)^3+54\Phi^2\varepsilon_m\left(\varepsilon_{sp}^2-1.5\varepsilon_{sm}\varepsilon_{sp}-\varepsilon_{sm}^2\right)}{27\Phi^2\varepsilon_{sm}^2\left(\varepsilon_{sp}-4\varepsilon_{sm}\right)} \tag{25}$$

$\frac{d\varepsilon_s}{dT}$ would be larger than zero, a negative ER effect, the viscosity of fluid decreases as the applied electric field increases, may be generated; If $\varepsilon_{sp} < 4\varepsilon_{sm}$, physically $\frac{d\varepsilon_{sp}}{dT}$ would unlikely be less than zero, as the small dielectric constant is generally due to the polar group orientation, as well as the electronic and atomic polarization. For a polar material of comparative small dielectric constant, the term $\frac{d\varepsilon_{sp}}{dT}$, likely of the same order as $\frac{d\varepsilon_{sm}}{dT}$, would be larger than zero [34]. So the ER effect could also be possibly generated, but it should still be very weak. In this section, the concentration is focused on the positive ER situation, i.e., $\frac{d\varepsilon_{sp}}{dT}$>0.

According to the Clausius-Mossotti formula, the static dielectric constant of one pure material ε can be expressed as [43]

$$\frac{\varepsilon-1}{\varepsilon+2}=\frac{4}{3}\pi N(\alpha_e+\alpha_a+\alpha_i+\alpha_d) \tag{26}$$

where $\alpha_e, \alpha_a, \alpha_i$ and α_d are the polarizabilities of the electron, atom, the ion displacement, and the dipole rotation, N is the number of molecule per unit volume. For a solid material, the dipole polarization is negligible as the solidification usually fixes the molecular with such rigidity in the lattice that little or no orientation of the dipole in an electric field [34]. Thus for the ER particle

$$\frac{\varepsilon_{sp}-1}{\varepsilon_{sp}+2}=\frac{4}{3}\pi N(\alpha_e+\alpha_a+\alpha_i) \tag{27}$$

At a very high frequency region, Eq.(27) would lead to

$$\frac{\varepsilon_{\infty p}-1}{\varepsilon_{\infty p}+2}=\frac{4}{3}\pi N(\alpha_e+\alpha_a) \tag{28}$$

Differentiating Eq.(27) with respect to T

$$\frac{3}{(\varepsilon_{sp}+2)^2}\frac{d\varepsilon_{sp}}{dT}=\frac{4}{3}\pi(\alpha_e+\alpha_a+\alpha_i)\frac{dN}{dT}+\frac{4}{3}\pi N\frac{d\alpha_i}{dT} \tag{29}$$

Since $\frac{1}{N}\frac{dN}{dT}=-3\beta$, where β is the material linear coefficient of expansion, Eq.(29) can be rewritten as:

$$\frac{d\varepsilon_{sp}}{dT}=\left[-3\frac{\varepsilon_{sp}-1}{\varepsilon_{sp}+2}\beta+\frac{4\pi N}{3}\frac{d\alpha_i}{dT}\right]\frac{(\varepsilon_{sp}+2)^2}{3} \tag{30}$$

where $\alpha_i=\frac{2q^2}{k}$, and $k=\frac{2M(n-1)q^2r_0^{n-1}}{3r^{n+2}}$ [43], q is the ion charge, k is the elastic bonding coefficient between two ions of opposite sign of charge, M and n are constants, r_0 and r are the equilibrium distance, and distance between the two ions. Assuming the α_i and k are function of temperature due to the fact that r would change with temperature, thus

$$\frac{d\alpha_i}{dT} = -\frac{\alpha_i}{k}\frac{dk}{dT} \tag{31}$$

$$\frac{dk}{dT} = -(n+2)k\frac{1}{r}\frac{dr}{dT} \tag{32}$$

According to definition, $\frac{1}{r}\frac{dr}{dT} = \beta$, from Eq. (28-32)

$$\frac{d\alpha_i}{dT} = \alpha_i(n+2)\beta = \frac{3}{4\pi N}\left(\frac{\varepsilon_{sp}-1}{\varepsilon_{sp}+2} - \frac{\varepsilon_{\infty p}-1}{\varepsilon_{\infty p}+2}\right)(n+2)\beta \tag{33}$$

Inserting Eq. (33) into Eq. (30), and assuming $\frac{\varepsilon_{sp}-1}{\varepsilon_{sp}+2} = p_s$, and $\frac{\varepsilon_{\infty p}-1}{\varepsilon_{\infty p}+2} = p_\infty$

$$\frac{d\varepsilon_{sp}}{dT} = \left[(n-1)p_s - (n+2)p_\infty\right]\frac{(\varepsilon_{sp}+2)^2\beta}{3} \tag{34}$$

Only when $\frac{p_s}{p_\infty} > \frac{n+2}{n-1}$, $\frac{d\varepsilon_{sp}}{dT}$ would be larger than zero. This requirement would lead to

$$\frac{\varepsilon_{sp}-\varepsilon_{\infty p}}{\varepsilon_{sp}\varepsilon_{\infty p}} > \frac{1}{n} \approx 0.1 \tag{35}$$

due to for most materials, n≈10 [43]. Eq.(35) is the dielectric requirement for an ER particle material.

Since the conductivity of most ER particle materials is comparatively low, one may assume that the particle dielectric loss just results from the ion displacement polarization. The dielectric loss due to the ion displacement polarization can be calculated from the absorption current. Assuming that the initial conductivity of an ER particle material is σ_0, and the absorption current can be expressed as $\sigma_0 e^{-\frac{t}{\theta}}$, t is time, θ is a time constant, $\theta = \frac{\varepsilon_{sp}+2}{\varepsilon_{s\infty}+2}\tau$, τ is the relaxation time, thus under an oscillatory electric field $E=E_0 e^{i\varpi t}$, the induced current density d_i due to the ion displacement polarization can be expressed as [43]

$$d_i = \sigma_0 \int_{-\infty}^{t_1} \frac{dE}{dt} e^{-\frac{t_1-t}{\theta}} dt = i\varpi\sigma_0 E_0 \int_{-\infty}^{t_1} e^{i\varpi t-\frac{t_1-t}{\theta}} dt = \frac{i\varpi\theta\sigma_0}{1+i\varpi\theta} E_0 e^{-\frac{t_1}{\theta}} e^{\frac{1+i\varpi\theta}{\theta}}$$

$$= \frac{\varpi^2\theta^2\sigma_0}{1+\varpi^2\theta^2} E_0 e^{i\varpi t_1} + i\frac{\varpi\theta\sigma_0}{1+\varpi^2\theta^2} E_0 e^{i\varpi t_1} \quad (36)$$

where t_1 is a constant. The first term of the above-equation has a same phase with the applied electric field, can be called the ohmic component d_{io}, which would contribute to the dielectric loss; while the second term has a (π/2) phase difference with the electric field, can be called the capacitive component d_{ic}, which just contributes to the polarization. Because the atomic and electronic polarizations also will contribute to the capacitive current density and the general relation between the induced current density d and the dielectric constant ε is $d = \frac{\varpi\varepsilon E_0}{4\pi}$, the current density due to these two kinds of polarizations d_{ae} can be expressed as $d_{ae} = \frac{\varpi\varepsilon_{\infty p} E_0}{4\pi}$, then the total capacitive current density d_c can be written as

$$d_c = d_{ic} + d_{ae} = \varpi\left(\frac{\varepsilon_{\infty p}}{4\pi} + \frac{\theta\sigma_0}{1+\varpi^2\theta^2}\right)E_0 \quad (37)$$

According to the definition, the dielectric loss tangent tgδ

$$tg\delta = \frac{d_{io}}{d_c} = \frac{\varpi\theta^2\sigma_0}{\frac{\varepsilon_{\infty p}}{4\pi}\left(1+\varpi^2\theta^2\right)+\theta\sigma_0} \quad (38)$$

the total dielectric constant ε

$$\varepsilon = \frac{d_c}{\varpi E_0} 4\pi = \varepsilon_{\infty p} + \frac{4\pi\theta\sigma_0}{1+\varpi^2\theta^2} \quad (39)$$

At a dc field (ω=0),

$$\varepsilon_{sp} = \varepsilon_{\infty p} + 4\pi\theta\sigma_0 \tag{40}$$

Inserting Eq. (40) into Eq. (38)

$$tg\delta = \frac{\varpi\theta\left(\varepsilon_{sp} - \varepsilon_{\infty p}\right)}{\varepsilon_{sp} + \varepsilon_{\infty p}\varpi^2\theta^2} \tag{41}$$

Eq. (41) is a general expression for a solid material of a very low conductivity and marked ionic displacement polarization. Differentiating Eq. (41) with respect to ω, one would find the tgδ has a maximum value,

$$tg\delta_{\max} = \frac{\varepsilon_{sp} - \varepsilon_{\infty p}}{2\sqrt{\varepsilon_{sp}\varepsilon_{\infty p}}} \tag{42}$$

Comparing with Eq. (35), one would conclude

$$tg\delta_{\max} > \frac{\varepsilon_{sp} - \varepsilon_{\infty p}}{\varepsilon_{sp}\varepsilon_{\infty p}} > 0.1 \tag{43}$$

as for many ER particle materials, $\sqrt{\varepsilon_{sp}\varepsilon_{\infty p}}$ is always larger than 2. Eq. (43) indicates that the suspended particles can become order under an electric field only when the maximum value of the dielectric loss tangent is larger than 0.1, which is exactly the empirical criterion obtained experimentally.

In conclusion, from a basic fact that the entropy of ER fluids should greatly reduce after an electric field is applied, it is theoretically demonstrated that the maximum value of dielectric loss tangent of the dispersed particle should be larger than 0.1, which agrees well with the experimental results. the negative ER effect only become possible if $\frac{d\varepsilon_{sp}}{dT}$<0.

4. THE YIELD STRESS EQUATION

Why the large dielectric loss tangent is necessary for the ER effect? Hao proposed a qualitative model on the assumption that the particle turning process and particle polarization process are both important to the ER effect, and the interfacial polarization would be responsible for the ER effect [26], which is called the dielectric loss model and will be described in detail in next chapter. A large interfacial polarization (also called the Maxwell-Wagner polarization) would facilitate the particle to attain a large amount of charges on the surface, leading to the turning of particle along the direction of the applied electric field to form a fibrillation structure; the strength of the fibrillation chains is thus determined by the particle polarization ability, i.e., the particle dielectric constant. In order to generate sufficient large interfacial polarization, the particulate material of large dielectric loss should be necessary, and an empirical criterion that the dielectric loss tangent of dispersed particulate material must be larger than 0.1 at 1000 Hz is thus needed. The non-ER particles were assumed to unable to turn along the direction of the external electric field for forming the fibrillation structure, thanks to the very small dielectric loss. The currently observed ER phenomena could be reasonably explained with this framework. The yield stress equation derived on the basis of this successful model thus was made by Hao [29], who materialized this ER mechanism model into a theoretical tool for describing the ER rheological behaviors quantitatively. The internal energy and entropy change of an ER fluid under a static electric field was used to estimate the inter-particle force then the yield stress of whole system. It was found that the derived equation can reasonably describe the ER phenomena observed to date, and agree well with the experimental data

According to the ER mechanism mentioned above, the entropy change of the ER system obviously includes two parts, the one is the particle configuration entropy, which represents the entropy change from the randomly distributed particle state to the body-centered tetragonal (bct) crystalline state; the other is the entropy change from the very weak interparticle force state to the exceptionally strong interparticle force state. The former part would contribute to the particle arrangement, while the latter part would contribute to the ER effect, which is thought to be originally induced by the interfacial polarization. At the first part, the interfacial polarization is presumably thought not to take place, and the inter-particles force could be negligible since in this state it is extremely weak; at the second part, the interfacial polarization takes place, making the already well-arranged particles become strongly correlated. This assumed process is schematically illustrated in Figure 11. In order to determine the yield stress of the ER suspension, one has to know the internal energy and entropy change of the second step, ΔS_2, ΔU_2, respectively. Obviously

$$\Delta S_2 = \Delta S - \Delta S_1 \tag{44}$$

$$\Delta U_2 = \Delta U - \Delta U_1 \tag{45}$$

The above-equations mean that one can easily know the second step internal energy and entropy change if the total internal energy and entropy change, as well as the first step internal energy and entropy change, are determined. Integrating Eq. (16) leads to:

$$U = U_0(T) + \frac{E^2}{8\pi}(\varepsilon_s + T\frac{\partial \varepsilon_s}{\partial T}) \tag{46}$$

where $U_0(T)$ represents the internal energy of the ER suspension in the absence of an electric field. Thus the internal energy change ΔU due to the applied electric field is

$$\Delta U = \frac{E^2}{8\pi}(\varepsilon_s + T\frac{\partial \varepsilon_s}{\partial T}) \tag{47}$$

Eq. (19) shows the entropy change ΔS can be expressed as:

$$\Delta S = \frac{\partial \varepsilon_s}{\partial T}\frac{E^2}{8\pi} \tag{48}$$

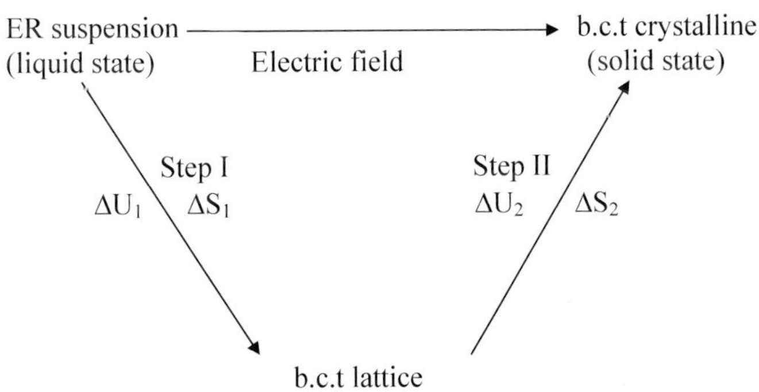

Figure 11 Schematic illustration of the assumed two-step process during the solidification transition of the ER suspension under an external electric field. Reproduced with permission from T. Hao, A. Kawai, and F.Ikazaki, Langmuir, 16(2000)3058

For the purpose of determining ΔU and ΔS, one may need to determine ε_s and $\frac{\partial \varepsilon_s}{\partial T}$ first. The Wagner-Maxwell equations are still used to derive the ε_s and $\frac{\partial \varepsilon_s}{\partial T}$, though the Wagner-Maxwell equations just hold for the diluted suspensions with the particle volume fraction less than 0.1. The Hanai equations is suitable for the concentrated suspensions [52]. However, note that there is no significant difference the Wagner-Maxwell and Hanai equations until the particle volume fraction is larger than 0.4. The Wanger-Maxwell equation is thus still used for computing ε_s and $\frac{\partial \varepsilon_s}{\partial T}$ in Hao's work [29]. According to Eq. (22), the static dielectric constant of the whole suspension ε_s can be expressed as:

$$\varepsilon_s = \varepsilon_{sm}\left(1 + 3\phi + \frac{27\phi^2 \varepsilon_{sm}\left(\varepsilon_{sp} - \varepsilon_{sm}\right)}{\left(2\varepsilon_{sm} + \varepsilon_{sp}\right)^2}\right) \quad (49)$$

Eq. (23) already gives the expression for $\frac{\partial \varepsilon_s}{\partial T}$. Substitute Eq.(23) and (49) into Eq.(46) leads to

$$U = U_0 + \frac{E^2}{8\pi}\left[\begin{array}{l}(1+3\Phi)\left(\varepsilon_{sm} + T\frac{d\varepsilon_{sm}}{dT}\right) + \frac{27\Phi^2\varepsilon_{sm}^2\left(\varepsilon_{sp} - \varepsilon_{sm}\right)}{\left(2\varepsilon_{sm} + \varepsilon_{sp}\right)^2} + \\ \frac{54\Phi^2\,\varepsilon_{sm}T\left(\varepsilon_{sp}^2 - 1.5\varepsilon_{sm}\varepsilon_{sp} - \varepsilon_{sm}^2\right)}{\left(2\varepsilon_{sm} + \varepsilon_{sp}\right)^3}\frac{d\varepsilon_{sm}}{dT} - \frac{27\Phi^2\varepsilon_{sm}^2 T\left(\varepsilon_{sp} - 4\varepsilon_{sm}\right)}{\left(2\varepsilon_{sm} + \varepsilon_{sp}\right)^3}\frac{d\varepsilon_{sp}}{dT}\end{array}\right] \quad (50)$$

For the step I, the interfacial polarization was assumed to be inactive, which is only physically likely under the condition of $\varepsilon_{sm} = \varepsilon_{sp}$. In such a condition, the static dielectric constant of this assumed system, according to Eq. (22), is $\varepsilon_{sm}(1+3\Phi)$. The influx electric energy into the assumed system for arraying the disordered particles to the very loose b.c.t. lattice structure can be expressed, analogue to Eq. (46)

$$\Delta U_1 = (1+3\Phi)\left(\varepsilon_{sm} + T\frac{d\varepsilon_{sm}}{dT}\right)\frac{E^2}{8\pi} \quad (51)$$

and the entropy change

$$\Delta S_1 = (1+3\Phi)\frac{E^2}{8\pi}\frac{\partial \varepsilon_{sm}}{\partial T} \tag{52}$$

Therefore according to Eq. (44) and (45)

$$\Delta U_2 = \frac{E^2}{8\pi}\left[\frac{27\Phi^2\varepsilon_{sm}^2(\varepsilon_{sp}-\varepsilon_{sm})}{(2\varepsilon_{sm}+\varepsilon_{sp})^2} + \frac{54\Phi^2\ \varepsilon_{sm}T(\varepsilon_{sp}^2-1.5\varepsilon_{sm}\varepsilon_{sp}-\varepsilon_{sm}^2)}{(2\varepsilon_{sm}+\varepsilon_{sp})^3}\frac{d\varepsilon_{sm}}{dT} - \frac{27\Phi^2\varepsilon_{sm}^2T(\varepsilon_{sp}-4\varepsilon_{sm})}{(2\varepsilon_{sm}+\varepsilon_{sp})^3}\frac{d\varepsilon_{sp}}{dT}\right] \tag{53}$$

$$\Delta S_2 = \frac{E^2}{8\pi}\left[\frac{54\Phi^2\ \varepsilon_{sm}T(\varepsilon_{sp}^2-1.5\varepsilon_{sm}\varepsilon_{sp}-\varepsilon_{sm}^2)}{(2\varepsilon_{sm}+\varepsilon_{sp})^3}\frac{d\varepsilon_{sm}}{dT} - \frac{27\Phi^2\varepsilon_{sm}^2T(\varepsilon_{sp}-4\varepsilon_{sm})}{(2\varepsilon_{sm}+\varepsilon_{sp})^3}\frac{d\varepsilon_{sp}}{dT}\right] \tag{54}$$

The ΔU_2 should be less than zero, as the inter-particle force in the ER crystalline lattice is attractive. Thus Eq. (53) should be expressed

$$\Delta U_2 = -\frac{E^2}{8\pi}\left[-\frac{27\Phi^2\varepsilon_{sm}^2(\varepsilon_{sp}-\varepsilon_{sm})}{(2\varepsilon_{sm}+\varepsilon_{sp})^2} - \frac{54\Phi^2\ \varepsilon_{sm}T(\varepsilon_{sp}^2-1.5\varepsilon_{sm}\varepsilon_{sp}-\varepsilon_{sm}^2)}{(2\varepsilon_{sm}+\varepsilon_{sp})^3}\frac{d\varepsilon_{sm}}{dT} + \frac{27\Phi^2\varepsilon_{sm}^2T(\varepsilon_{sp}-4\varepsilon_{sm})}{(2\varepsilon_{sm}+\varepsilon_{sp})^3}\frac{d\varepsilon_{sp}}{dT}\right] \tag{55}$$

ΔS_2 should also be less than zero, as in the step II, the entropy obviously decreases substantially, requiring $\varepsilon_{sp} > 4\varepsilon_{sm}$ at $\frac{d\varepsilon_{sp}}{dT}>0$ and $\varepsilon_{sp} < 4\varepsilon_{sm}$ at $\frac{d\varepsilon_{sp}}{dT}<0$. These are the preconditions of Eq.(55).

The Eq. (55) gives the internal energy change per unit volume of the whole suspension. For an ER suspension of volume V, particle volume fraction Φ and particle diameter r, the particles number N in this volume

$$N = \frac{V\Phi}{\frac{4}{3}\pi r^3} = \frac{3V\Phi}{4\pi r^3} \tag{56}$$

then the inter-particle energy U_{ip}

$$U_{ip} = \frac{\Delta U_2 V}{N/2} = \frac{8\pi r^3 \Delta U_2}{3\phi} \tag{57}$$

There are two ways to express the free volume. The first is to assume that the free volume is the volume unoccupied by the particles in a system, and then the free volume per particle V_{fp} under off-state electric field is

$$V_{fp} = \frac{V - V\Phi}{N} = \frac{4\pi r^3(1-\Phi)}{3\Phi} \tag{58}$$

The second is to assume that the free volume is the maximum particle packing volume minus the particle volume, and then the free volume per particle V_{fp} under off-state electric field is:

$$V_{fp} = \frac{V\Phi_m - V\Phi}{N} = \frac{4\pi r^3(\Phi_m - \Phi)}{3\Phi} \tag{59}$$

where Φ_m is the maximum packing fraction. It is known that under an electric field the ER particles will form the fibrillation chains of the b.c.t. lattice. The unit cell of the b.c.t. lattice is schematically illustrated in Figure 12. One center-particle would have 8 nearest neighbor particles of the inter-particle distance 2r, which belongs to the different chain class; besides, one center-particle also has another 6 nearest neighbor center-particles also of the inter-particle distance 2r, which belong to the same chain class. So the total energy for one particle in the unit cell U_{it}

$$U_{it} = (8+6)\times U_{ip} = \frac{112\pi r^3 \Delta U_2}{3\Phi} \tag{60}$$

Since the particle can not freely go out of the cell of the lattice, the free volume per particle under an electric field, V_{fE} should become

$$V_{fE} = V_{fp} - \left(\frac{2r\times\sqrt{6}r\times\sqrt{6}r}{2}\right) = \frac{4\pi r^3(1-\Phi) - 18\Phi r^3}{3\Phi} \tag{61}$$

or from Eq. (59)

$$V_{fE} = \frac{4\pi r^3(\Phi_m - \Phi) - 18\Phi r^3}{3\Phi} \tag{62}$$

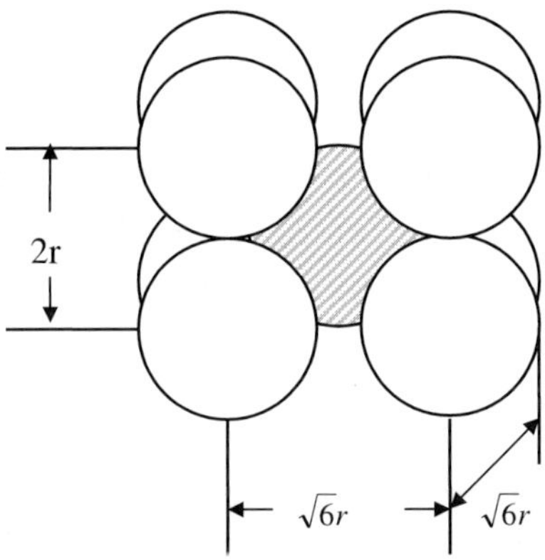

Figure 12 Schematic illustration of the unit cell of the b.c.t lattice formed by the ER particles under an electric field. The radius of the particle is r. Reproduced with permission from T. Hao, A. Kawai, and F. Ikazaki, Langmuir, 16(2000)3058

Note that in one unit cell there are two equivalent particles. Assuming that the yield stress τ_y is the force that can drive one particle to move within the free volume per particle, thus

$$\tau_y = \frac{|U_{it}|}{V_{fE}} = \frac{112\pi}{4\pi(1-\Phi)-18\Phi}|\Delta U_2| =$$

$$\frac{14E^2}{4\pi(1-\Phi)-18\Phi}\times \left[-\frac{27\Phi^2\varepsilon_{sm}^2(\varepsilon_{sp}-\varepsilon_{sm})}{(2\varepsilon_{sm}+\varepsilon_{sp})^2} - \frac{54\Phi^2\ \varepsilon_{sm}T(\varepsilon_{sp}^2-1.5\varepsilon_{sm}\varepsilon_{sp}-\varepsilon_{sm}^2)}{(2\varepsilon_{sm}+\varepsilon_{sp})^3}\frac{d\varepsilon_{sm}}{dT} + \frac{27\Phi^2\varepsilon_{sm}^2T(\varepsilon_{sp}-4\varepsilon_{sm})}{(2\varepsilon_{sm}+\varepsilon_{sp})^3}\frac{d\varepsilon_{sp}}{dT}\right] \quad (63)$$

or

$$\tau_y = \frac{14E^2}{4\pi(\Phi_m - \Phi) - 18\Phi} \times \left[-\frac{27\Phi^2\varepsilon_{sm}^2(\varepsilon_{sp} - \varepsilon_{sm})}{(2\varepsilon_{sm} + \varepsilon_{sp})^2} - \frac{54\Phi^2 \varepsilon_{sm}T(\varepsilon_{sp}^2 - 1.5\varepsilon_{sm}\varepsilon_{sp} - \varepsilon_{sm}^2)}{(2\varepsilon_{sm} + \varepsilon_{sp})^3}\frac{d\varepsilon_{sm}}{dT} + \frac{27\Phi^2\varepsilon_{sm}^2 T(\varepsilon_{sp} - 4\varepsilon_{sm})}{(2\varepsilon_{sm} + \varepsilon_{sp})^3}\frac{d\varepsilon_{sp}}{dT} \right] \tag{64}$$

The Eq. (63) and (64) present the correlation between the yield stress of the ER suspension and the physical parameters of the suspension materials. Principally, one can easily calculate the yield stress value from Eq. (63) or (64) if all the parameters are already known. Eq. (63) and (64) can be further developed to a more meaningful form if $d\varepsilon_{sm} / dT$ and $d\varepsilon_{sp} / dT$ are known. The Clausius-Mossotti equation provides a means of correlating the macroscopic dielectric property with the microscopic property—the number of molecules per unit volume and the molecule polarizability. The dielectric property varies with temperature is because the number of the molecule per unit volume changes due to the material thermal expansion. For non-polar liquid materials used in ER fluids, it is acceptable to assume that only the static dielectric constant ε_{sm} and the number of the molecules per unit volume N are dependent of temperature. Differentiating the Clausius-Mossotti equation shown in Eq. (26) with respect to temperature T for both sides, one may obtain

$$\frac{d}{dT}\left(\frac{\varepsilon_{sm} - 1}{\varepsilon_{sm} + 2}\right) = \frac{4}{3}\pi(\alpha_e + \alpha_a + \alpha_i + \alpha_d)\frac{dN}{dT} \tag{65}$$

or

$$\frac{3}{(\varepsilon_{sm} + 2)^2}\frac{d\varepsilon_{sm}}{dT} = \frac{4}{3}N\pi(\alpha_e + \alpha_a + \alpha_i + \alpha_d)\frac{1}{N}\frac{dN}{dT} \tag{66}$$

Since $\frac{1}{N}\frac{dN}{dT} = -\beta_m$, where β_m is the liquid linear expansion coefficient, Eq. (26), (65) and (66) lead to:

$$\frac{d\varepsilon_{sm}}{dT} = -\frac{(\varepsilon_{sm}-1)(\varepsilon_{sm}+2)}{3}\beta_m \tag{67}$$

For solid particulate materials, $d\varepsilon_{sp}/dT$ is already given in Eq. (34) and can be simplified as:

$$\frac{d\varepsilon_{sp}}{dT} = [(n-1)p_s-(n+2)p_\infty]\frac{(\varepsilon_{sp}+2)^2\beta_p}{3} = \frac{p(\varepsilon_{sp}+2)^2\beta_p}{3} \tag{68}$$

where β_p is the particle material linear coefficient of expansion, n is a constant of the value between 4 and 19.7, depending on the solid crystalline state [43]. $p=[(n-1)p_s-(n+2)p_\infty]$ with $p_s = \frac{\varepsilon_{sp}-1}{\varepsilon_{sp}+2}$, and $p_\infty = \frac{\varepsilon_{\infty p}-1}{\varepsilon_{\infty p}+2}$.

Assuming $\xi = \varepsilon_{sp}/\varepsilon_{sm}$, substituting Eq. (67) and (68) into Eq. (63) and (64):

$$\tau_y = \frac{126\Phi^2E^2}{4\pi(1-\Phi)-18\Phi}\left\{\begin{array}{l}-\frac{3\varepsilon_{sm}(\xi^2+\xi-2)}{(2+\xi)^3}+\frac{2\beta_mT(\xi^2-1.5\xi-1)(\varepsilon_{sm}^2+\varepsilon_{sm}-2)}{(2+\xi)^3}+\\ \frac{T\beta_p p(\xi-4)(\varepsilon_{sm}\xi+2)^2}{(2+\xi)^3}\end{array}\right\} \tag{69}$$

and

$$\tau_y = \frac{126\Phi^2E^2}{4\pi(\Phi_m-\Phi)-18\Phi}\left\{\begin{array}{l}-\frac{3\varepsilon_{sm}(\xi^2+\xi-2)}{(2+\xi)^3}+\frac{2\beta_mT(\xi^2-1.5\xi-1)(\varepsilon_{sm}^2+\varepsilon_{sm}-2)}{(2+\zeta)^3}+\\ \frac{T\beta_p p(\xi-4)(\varepsilon_{sm}\xi+2)^2}{(2+\xi)^3}\end{array}\right\} \tag{70}$$

Once the parameters are known, Eq. (69) and (70) will give the yield stress of the ER fluids. The yield stress of the ER fluids depends on many parameters such as the square of the applied electric field strength, the particle volume fraction, the dielectric constants of both the dispersing medium and dispersed particle, temperature, parameter p, and the thermal expansion coefficient of both the dispersing medium and dispersed particle.
If p is of a positive value, the yield stress thus will increase as ξ increases, as the numerator increases much faster than the denominator with ξ, which is in consistent with the prediction given by the polarization model [10,53].

However, if p is of a negative value, then the yield stress will decrease as ξ increases, which is unable to be explained by the polarization model but was experimentally observed. The parameter p only becomes positive when the dielectric loss tangent of the dispersed solid material is larger than 0.1, as we demonstrated before both experimentally and theoretically. Eq. (69) and (70) incorporates the dielectric criteria proposed earlier. Figure 13 shows the calculated yield stress using Eq.(69) as the function of the particle volume fraction at the particle-to-oil dielectric ratio ξ=10, $d\varepsilon_{sm}/dT = -2.4\times10^{-3}\,{}^{o}C^{-1}$ experimentally determined for the silicone oil, and $d\varepsilon_{sp}/dT = 0.4^{o}C^{-1}$ as $d\varepsilon_{sp}/dT$ should be larger than zero for the ER particulate material. When the particle volume fraction is small, the yield stress slightly increases as the particle volume fraction increases, which is in a good agreement with the experimental results [7, 54]. When the particle volume fraction is about 0.35, the yield stress dramatically increases, indicating that a critical volume fraction would exist in the ER suspension. Hao's previous work [46] shows that there is a critical volume fraction value in the ER suspension, and once the particle volume fraction is larger than this value, the rheological properties of the ER suspension would dramatically increase. This critical value was experimentally found to be about 0.37 on the basis of rheological investigation [46] and was calculated to be around 0.4 in aid of the Flory's gelation theory and the percolation concept [33]. The critical volume fraction value given by the present yield stress equation is quite close to the previous data, indicating that Eq. (69) gives a pretty good prediction.

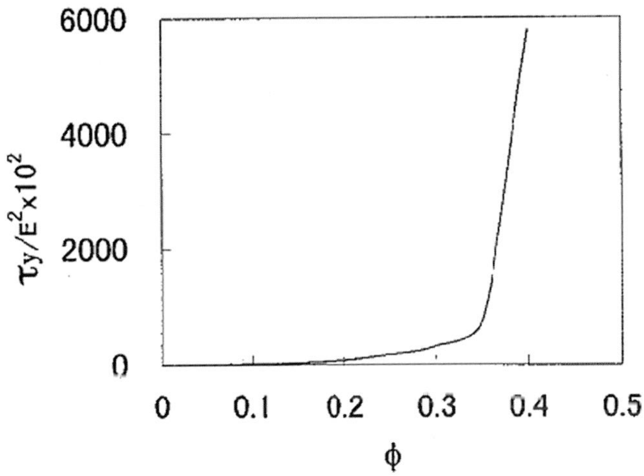

Figure 13 The normalized yield stress, $\tau_y / E^2 x10^2$, vs. particle volume fraction at the particle-to-oil dielectric constant ratio ξ=10, and $d\varepsilon_{sp} / dT = 0.4$. This is computed from Eq. (69). Reproduced with permission from T. Hao, A. Kawai, and F. Ikazaki, Langmuir, 16(2000)3058.

The effect of particle-to-oil dielectric constant ratio, ξ, on the yield stress of the ER fluids should be another interesting topic to evaluate. Figure 16 shows the yield stress of the ER fluid calculated from Eq. (69) as a function of the particle-to-oil dielectric constant ratio, ξ, under the assumption that $\frac{d\varepsilon_{sp}}{dT}$ is 0.40, and $\beta_p p$=3.82x10^{-4}, both of those are typical values for ER systems based on the experimental data. At low particle-to-oil dielectric constant ratio, the yield stress dramatically increases as ξ increases. However, the yield stress gradually levels off after ξ is more than 60. The sharp increase of the yield stress takes place in the range of ξ <50, indicating that the materials of the static dielectric constant around 150 would potentially display the best ER effect. The materials of the large static dielectric constant would not always display a strong ER effect if other properties are unfavorable to the ER effect.

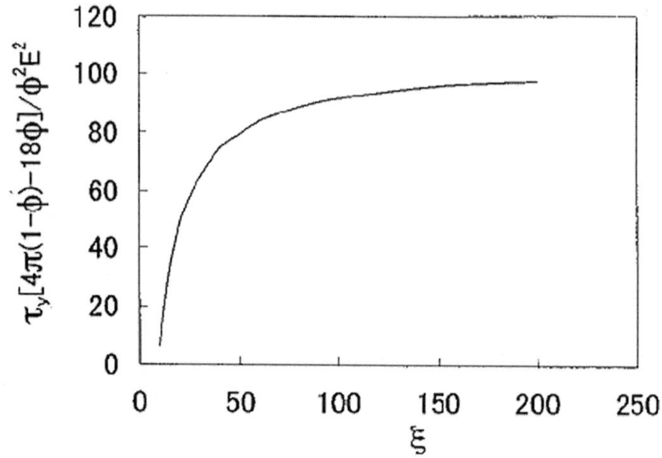

Figure 14 The numerical relationship between the normalized yield stress, $\tau_y\left[4\pi(1-\Phi)-18\Phi\right]/\Phi^2E^2$, and the particle-to-oil dielectric constant ratio, ξ, of the silicone oil ER system, based on the Eq.(69) at room temperature. Reproduced with permission from T. Hao, A. Kawai, and F. Ikazaki, Langmuir, 16(2000)3058

Since p is a very important parameter in the yield stress equation (69), it will be interesting to further explore which physical parameters would greatly affect p value and then the yield stress. According to the definition,

$$p=(n-1)\frac{\varepsilon_{sp}-1}{\varepsilon_{sp}+2}-(n+2)\frac{\varepsilon_{\infty p}-1}{\varepsilon_{\infty p}+2}$$

$$=(n-1)\left(\frac{\varepsilon_{sp}-1}{\varepsilon_{sp}+2}-\frac{\varepsilon_{\infty p}-1}{\varepsilon_{\infty p}+2}\right)-3\frac{\varepsilon_{\infty p}-1}{\varepsilon_{\infty p}+2} \tag{71}$$

and assuming $\varepsilon_{sp}-\varepsilon_{\infty p}=A$

$$
\begin{aligned}
p &= 3(n-1)\frac{A}{(\varepsilon_{\infty p}+2+A)(\varepsilon_{\infty p}+2)} - 3\frac{\varepsilon_{\infty p}-1}{\varepsilon_{\infty p}+2} \\
&= 3\left[\frac{(n-\varepsilon_{\infty p})A - \varepsilon_{\infty p}(\varepsilon_{\infty p}+1)+2}{(\varepsilon_{\infty p}+2)(\varepsilon_{\infty p}+2+A)}\right] = \frac{3(n-\varepsilon_{\infty p})A - 3\varepsilon_{\infty p}(\varepsilon_{\infty p}+1)+6}{(\varepsilon_{\infty p}+2)^2+(\varepsilon_{\infty p}+2)A}
\end{aligned}
\tag{72}
$$

If $\varepsilon_{\infty p}$ is not very large (less than 10 for most small dielectric constant materials), $3(n-\varepsilon_{\infty p})$ is always larger than $(\varepsilon_{\infty p}+2)$, thus p will increase as A increases, i.e., the ER effect will approximately increase with the difference between the dielectric constants below and above the relaxation frequency, which was already experimentally found by Ikazaki and Kawai [21,31]. For a solid material of a leak conductivity σ, an initial conductivity σ_0, and the relaxation time related constant θ, it is known that [43]

$$
\varepsilon_{sp} - \varepsilon_{\infty p} = 4\pi\theta\sigma_0 \tag{73}
$$

and

$$
tg\delta = \frac{\sigma(1+\varpi^2\theta^2)+\varpi^2\theta^2\sigma_0}{\frac{\varepsilon_{\infty p}\varpi}{4\pi}(1+\varpi^2\theta^2)+\varpi\theta\sigma_0} \tag{74}
$$

thus

$$
A = 4\pi\theta\sigma_0 = \frac{4\pi(1+\varpi^2\theta^2)\left(\frac{\varepsilon_{\infty p}\varpi tg\delta}{4\pi}-\sigma\right)}{\varpi^2\theta - \varpi tg\delta} \tag{75}
$$

where ϖ is the field frequency, and $tg\delta$ is the dielectric loss tangent of the solid material. Eq (75) indicates that the parameter A would obviously increase as $tg\delta$ increases, provided that $\varpi\theta > tg\delta$ (this always holds, as A should have a positive value). Differentiating Eq. (75) with respect to σ, one would find there is a maximum value for A at

$$\sigma = \frac{\left(\frac{\varepsilon_{\infty p}\varpi}{4\pi}\frac{dtg\delta}{d\sigma} - 1\right)\varpi^2\theta + \varpi tg\delta}{\varpi\frac{dtg\delta}{d\sigma}} \tag{76}$$

as the second derivative of A with respect to σ, $\frac{d^2A}{d\sigma^2}$, is negative. Considering that $\varpi\theta > tg\delta$, and $\frac{\varepsilon_{\infty p}\varpi}{4\pi}\frac{dtg\delta}{d\sigma} >> 1$, thus $\sigma \approx \frac{\varepsilon_{\infty p}\varpi^2\theta}{4\pi}$. For solid materials, one would assume the Debye polarization is unable to occur, and the dielectric adsorption phenomenon just stems from the ion polarization, and the ion polarization-induced $tg\delta$ would reach at a maximum value at $\varpi = \frac{1}{\theta}\sqrt{\frac{\varepsilon_{sp}}{\varepsilon_{\infty p}}}$ [43]. In such a case, the parameter A also reaches at a maximum value, thus $\sigma \approx \frac{\varepsilon_{sp}}{4\pi\theta}$, which indicates that for different dispersed solid material, the yield stress would peak at different particle conductivity. In the poly(acenequinones)/silicone oil system, the yield stress was found to peak at the particle conductivity around 10^{-5}S/m [7]. However, in the oxidized polyacrylonitrile/silicone oil system, the yield stress maximum value was found to occur nearby 10^{-7} S/m [27]. According to the dielectric data presented in each paper, the optimal conductivity ($\sigma \approx \frac{\varepsilon_{sp}}{4\pi\theta}$) is crudely estimated at 0.22 x 10^{-5} S/m for the former system, and 0.88x10^{-7} S/m for the latter system, we may say, agreeing well with the experimental results.

The temperature dependence of the yield stress can also be qualitatively analyzed using Eq. (69). Since σ and $tg\delta$ are much sensitive than the dielectric constant to temperature, one would still center on the temperature dependence of the parameter p. For most solid dielectric materials the conductivity will exponentially increase with temperature, thus the conductivity rather than other parameters will be a main variable and surely make a big contribution to the yield stress, as shown in Eq. (69). The yield stress would also go through a maximum value at the temperature where the conductivity reaches at the optimal value. The yield stress first increases and then decreases with temperature were found experimentally [55,56]. Accordingly, the yield stress would decrease with temperature if the conductivity of the solid particles is already larger than the optimal value, while the yield stress would increase with temperature if the conductivity is lower than the optimal value, which were also experimentally found previously [56].

The relaxation time constant θ would also influence the parameter A, finally p substantially. From either Eq.(73) and (75), one will find that p increases as θ increases, that is, the ER effect will be stronger if the dielectric relaxation is slower. However, too slow relaxation time (then the slow response time) would make ER fluids useless. Generally, the ER response time around 1 millisecond is favorable, thus requiring the relaxation time be of the same time scale, i.e., the dielectric relaxation frequency around 10^3Hz. Block presumably thought the polarization rate would be important in the ER response process, and too fast or too slow polarization is unfavorable to the ER effect [7]. Ikazaki and Kawai experimentally found that the ER fluids of the relaxation frequencies within the range 100-10^5 Hz would exhibit a large ER effect [21,31], supporting the derivation from Eq. (69).

Note that the conclusions above are derived under the assumption that the parameter p is of a positive value. The parameter *p* only becomes positive when the dielectric loss tangent of the dispersed solid material is larger than 0.1. If p is negative, a large particle-to-oil dielectric constant ratio would not generate a large yield stress. An excellent example is $BaTiO_3$ material, as $BaTiO_3$ usually has a very large dielectric constant (around 2000, depending on its crystalline state), and should have had a strong ER effect. However, the ER effect of the barium titanate/insulating oil suspensions was found to be inactive under a dc field [57]. and active after adsorbing a small amount of water[58] or being stimulated by an ac field [9]. The static dielectric constant of pure $BaTiO_3$ were measured by Hao at different temperatures (20—120°C) [29] and is plotted against temperature in Figure 15. As $BaTiO_3$ is a ferroelectric material and the dielectric property greatly depends on the crystalline state and manufacture method, the static dielectric constant of the $BaTiO_3$ displays an unusual temperature dependence: Within temperature range 20-40°C, it sharply decreases, while it slowly decreases between 60-120°C. At low temperature range, $\frac{d\varepsilon_{sp}}{dT}$ is -47.5, and at high temperature range, it values as -1.08. This clearly indicates that p is negative for $BaTiO_3$, see Eq. (68). Based on the static dielectric constant data obtained at different temperatures, one may calculate the yield stress of $BaTiO_3$/silicone oil suspension according to Eq. (69). The computed yield stresses at all temperatures (20, 40, 60, 80, 100, 120 °C) are negative, indicating that the $BaTiO_3$/silicone oil suspension would not display the ER effect, though $BaTiO_3$ has a very large dielectric constant. The predicted negative yield stress means that the inter-particles force is not attractive, instead repulsive. However, whether the $BaTiO_3$/silicone oil suspension displays a negative ER effect would largely depend on which interaction, the field-induced repulsive force or the off-field original inter-particle force, is a main contribution to the apparent mechanical properties of whole suspension. Note that Eq.(69) is inappropriate to predict the negative ER effect, as it involves the assumption that the particle will form the b.c.t. lattice under an electric field (p>0). For accurate reason, Hao conducted the experimental measurement of the yield stress of a $BaTiO_3$/silicone oil suspension with the particle volume fraction 0.20. Figure 16 shows the yield stress of $BaTiO_3$/silicone oil suspension as a function of temperature, experimentally determined under zero and 2kV/mm electric field.. The $BaTiO_3$/silicone oil suspension indeed does not show any positive ER effect, in contrast, it gives a slight negative ER effect. As a result, the positive p would be very important for an ER material.

A quantitative comparison between the prediction derived from the yield stress equation and the experimental results was made by Hao for the zeolite/silicone oil system [29]. The static dielectric constant of the pure zeolite material and the yield stress of zeolite/silicone oil suspension of the particle volume fraction 0.23 were experimentally measured at different temperatures. The calculated yield stress values from Eq. (69) vs. temperature is shown in Figure 17 as a solid line. For comparison, the experimentally measured data are also shown in Figure 17 as black points. As we can see, the predicted values agree very well with the experimental ones, indicating that Eq. (69) is able to predict the yield stress of ER fluids, indeed.

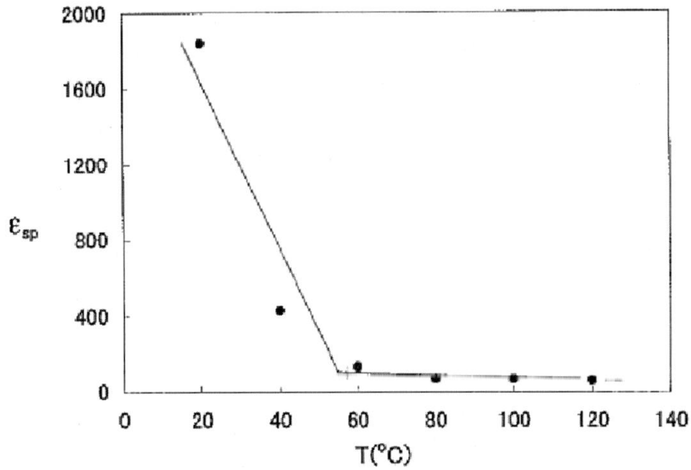

Figure 15 Static dielectric constant of $BaTiO_3$ as a function of temperature. Reproduced with permission from T. Hao, A. Kawai, and F. Ikazaki, Langmuir, 16(2000)3058.

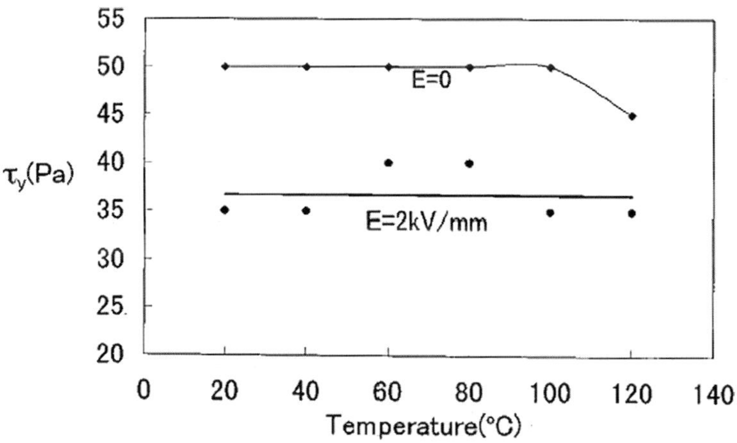

Figure 16 Yield stress of $BaTiO_3$/silicone oil suspension as a function of temperature. The particle volume fraction is 0.20. Reproduced with permission from T. Hao, A. Kawai, and F. Ikazaki, Langmuir, 16(2000)3058.

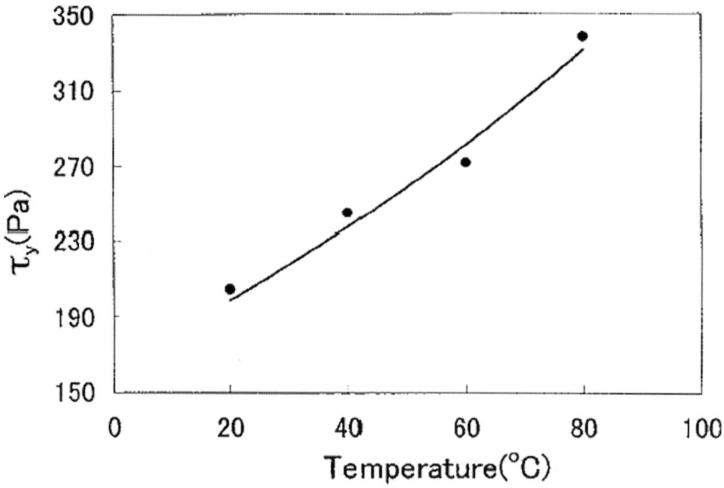

Figure 17 Yield stress of the zeolite/silicone oil suspension vs. temperature at the electric field E =2 kV/mm: The solid line is calculated with Eq. (69); The black points are experimental data. The particle volume fraction is 0.23. Reproduced with permission from T. Hao, A. Kawai, and F. Ikazaki, Langmuir, 16(2000)3058.

5 PARTICLE SHAPE EFFECT ON THE DIELECTRIC PROPERTIES OF ER SUSPENSIONS AND THE ER EFFECT

The discussions above are for the spherical or quasi-spherical particles dispersed in an insulating medium. As stated in Chapter 7, the particle shape play an important role in the dielectric property of whole suspension. In this section, the influence of particle shape on the dielectric properties and then the ER effect is discussed. Kanu [59] investigated the cylindrical poly(p-phenylene-2,6-benzobisthiazole) particle of the repeat unit shown in Figure 18 dispersed in the mineral oil. The fibers of aspect ratios (L/a) of 2, 4, 10 were used for this study. The average diameter of the fiber is 16 μm, and the particle volume fraction is 15 vol%. The dielectric constants of those suspensions are plotted against frequency at 30 °C in Figure 19. It shows that there is a small difference in dielectric constants of the suspensions with randomly dispersed particles, though those particles have different aspect ratio. However, when the suspended particles are aligned, the dielectric constants of the suspensions increase as particle geometric aspect ratio increases. The dynamic modulus of those suspensions obtained experimentally at strain amplitude 10^{-2} vs. the aspect ratio of the particle is

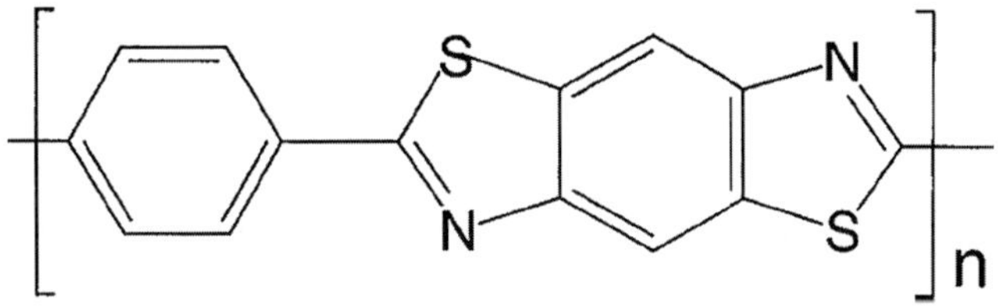

Figure 18 Molecular structure of poly(p-phenylene-2,6-benzobisthiazole). Reproduced with permission from R.C Kanu, and M.T. Shaw, J. Rheol., 42(1998)657.

shown in Figure 16, in Chapter 5. The dynamic modulus increases as the particle aspect ratio increases, roughly following the same trend that the dielectric constant increases with the aspect ratio, as seen in Figure 19. Those findings suggest that the particle with the largest geometric aspect ratio will be polarized to the greatest extent when the major axis of the particle is aligned to the direction of the external electric field, and therefore produce the strongest particle interaction or the ER effect. However, an opposite trend was observed in glass fiber particle/silicone oil system [60]. Figure 20 shows the dielectric constant and shear stress of glass microspheres and fibers dispersed in silicone oil vs. the particle aspect ratio. The dielectric constant obtained at 10 kHz decreases as the aspect ratio increases, as does the shear stress obtained at 2.2 kV /mm and shear rate 5 s^{-1}. Those results are surprising and contradict the dielectric theory described in Chapter 7. According to the calculation shown in Figure 22 in Chapter 7 [61], those results only become possible when the b-axis of the particle is aligned with the electric field. In addition, those ER suspensions are water-activated, and it should be hard to keep water content the same in all suspensions of different particle aspect ratios.

Monte Carlo calculations [62] show that both prolate and oblate ER particles align with their longest axis along the applied field allowing for largely induced dipole moments than those obtained for spherical ER particles of equal volume. Even slightly nonspherical ER particles are highly ordered by an electric field that is weaker than that required to induce chain formation for spherical particles. A very weak field is sufficient to generate a strong orientational order in prolate ER particle with the moderate aspect ratio. Further, field-ordered oblate ER particles tend to align their symmetry (short) axes to form a biaxial phase at high densities.

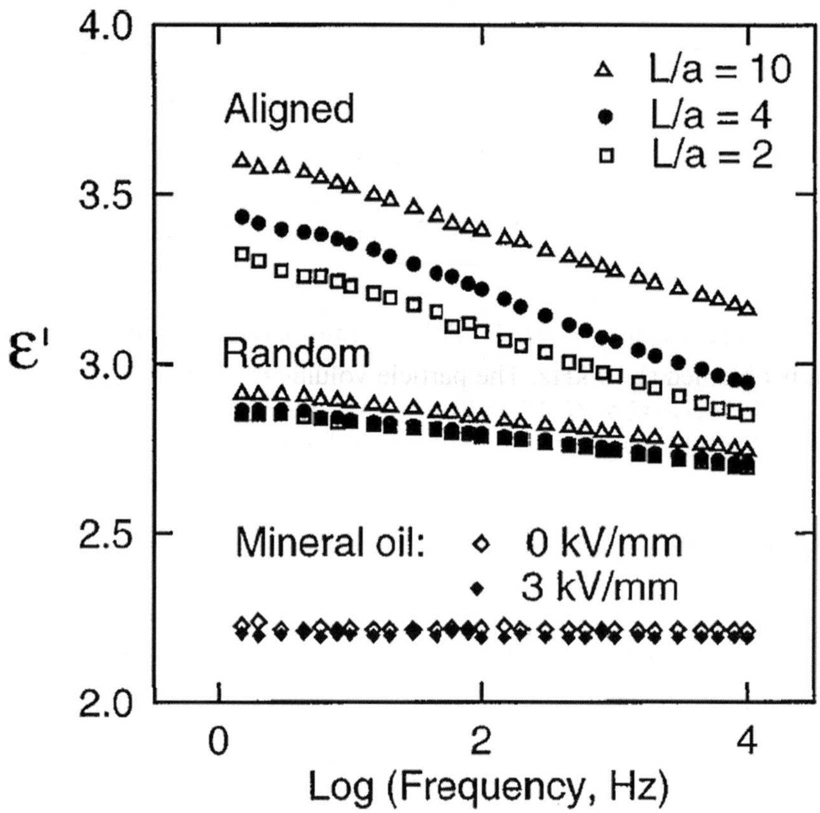

Figure 19 The static dielectric constant of aligned and randomly dispersed cylindrical poly(p-phenylene-2,6-benzobisthiazole) particle of different aspect ratio in mineral oil. The average diameter of the fiber is 16 μm, and the particle volume fraction is 15 vol%. Reproduced with permission from R.C. Kanu, and M.T. Shaw, J. Rheol., 42(1998)657.

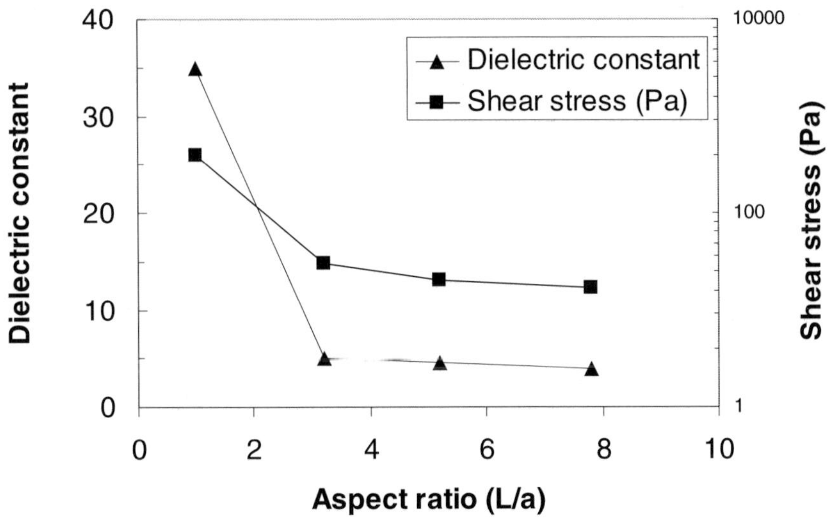

Figure 20 The dielectric constant and shear stress of glass microspheres and fibers dispersed in silicone oil vs. the particle aspect ratio. The dielectric constant is obtained at 10 kHz. The particle volume fraction is 20 vol%. The shear stress is obtained at 2.2 kV /mm and shear rate 5 s^{-1}. Redrawn from Y. Qi, and W. Wen, J. Phys. D: Appl. Phys. 35(2002)2231

6 THE RESPONSE TIME OF ER SUSPENSIONS

Since the ER effect is determined by the Wagner-Maxwell polarization, the response time of the ER suspensions should be identical to the relaxation time of the Wagner-Maxwell polarization. That is, for a system where the conductivity of the dispersed particle is much larger than that of the dispersed medium and the conductivity contribution from the dispersed medium is negligible, the relaxation time can be described by a simplified form as shown in Eq. (144) in Chapter 7.

$$\tau_i = \frac{\varepsilon_0 \left(2\varepsilon_m + \varepsilon_p\right)}{\sigma_p} \tag{77}$$

However when the conductivity of the dispersed medium is large enough in comparison with that of the dispersed particle, the relaxation time have a more complicated form as shown in Eq. (142) in Chapter 7

$$\tau_i = \frac{2\varepsilon_m + \varepsilon_p + \phi(\varepsilon_m - \varepsilon_p)}{2\sigma_m + \sigma_p + \phi(\sigma_m - \sigma_p)}\varepsilon_0 \quad (78)$$

The experimental evidences for supporting Eq. (76) or (77) are given previously. Eq. (77) and (78) provide a means for controlling the response time of ER fluids.

7 DIELECTRIC PROPERTY OF ER FLUIDS UNDER A HIGH ELECTRIC FIELD

The dielectric constant and loss typically are independent of the applied electric field strength [51]. However, many materials, for example, ferroelectric ceramics [63], do show the dielectric property dependence on the applied field strength. Since the rheological properties continuously increase with the electric field, the degree of particle re-orientation order are supposed to increase with the electric field, too, as do the dielectric properties. Klass [64] found that the dielectric constant of silica/silicone oil system increases with the applied field strength only when the particle volume fraction is less than 10 vol%, as shown in Figure 24, in Chapter 5. Deinega [65] found that the dielectric constant initially increases with the applied electric field (0.4-4 kV/mm), and then levels off at high electric fields. The dielectric property of polyurethane particle/silicone oil ER suspension with particle volume fraction 60 vol% was investigated under electric field from 0 to 400 V/mm [66], and the relaxation strength, defined as the dielectric constant difference before and after the relaxation, was found to increase with the applied electric field, as shown in Figure 21. The dielectric relaxation strength saturates at E=250 V/mm, and the decrease at E>250 V/mm was attributed to the diminishing of the Maxwell-Wagner polarization at the particle-liquid interface.

Besides dielectric constant, the dielectric loss also increases with the applied electric field strength substantially for polyurethane particle/silicone oil ER suspension [67].

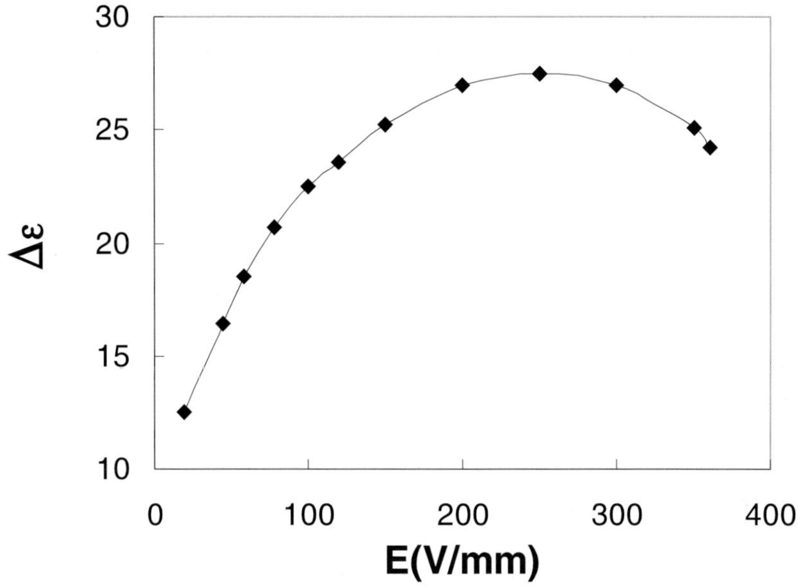

Figure 21 The dielectric relaxation strength of polyurethane particle/silicone oil ER suspension with particle volume fraction 60 vol% vs. the electric field from 0 to 400 V/mm at T=60 °C. Redrawn from P. Placke, R. Richert, E.W. Fischer, Colloid Polym. Sci., 273(1995)1156.

8 SUMMARY

Among the polarizations taking place in ER fluids under an electric field, including the electronic, the atomic, the Debye the electrode, and the interfacial polarizations, the interfacial polarization is experimentally found to contribute to the ER effect, and the Wanger-Maxwell Equation, which deals with the interfacial polarization in the heterogeneous system, is found to be capable of describing the dielectric phenomena observed in ER fluids. The dielectric loss of the particulate material is found to play a important role in the ER effect, and the dielectric constant becomes dominant only when the dielectric loss tangent is larger than 0.10 at 1000Hz. Those empirical criteria for screening high performance ER solid materials have a solid physical basis—They are the preconditions for the entropy decrease commonly observed in ER systems. Theoretically, one also can reach the same conclusion if the entropy of ER systems is assumed to greatly decrease, as the ER fluids change from the randomly distributed colloidal suspension state to a body-center-tetragonal crystal state after the application of an external electric field. The mechanism of the ER effect is suggested to have two steps: The first step is the particle turning process, which is controlled by the particle dielectric loss; the second step is the particle strongly correlating process, which is controlled by the dielectric constant. The entropy change in the ER system obviously includes two parts, the one is the particle configuration entropy, which represents the entropy change from the randomly distributed particle state to the b.c.t. lattice state; the other is the entropy change from the very weak inter-particle force state to the exceptionally strong inter-particle force state. The former part would contribute to the particle re-arrangement, while the latter part would contribute to the ER effect, which is thought to be originally induced by the interfacial polarization. On the basis of this ER mechanism, a general form yield stress equation is derived from the internal energy and entropy changes of an ER fluid under an external static electric field. The yield stress equation can give very good predictions in accordance with the experimental results obtained to date.

The yield stress equation involves a very important parameter, p, which gives an appropriate expression for the dielectric loss tangent criterion. When p is positive, the yield stress would increase with the particle-to-medium dielectric ratio ξ, as shown by the polarization model and other models. However, the yield stress gradually levels off after $\xi>60$. The sharp increase of the yield stress takes place in the range of $\xi<50$, indicating that the materials of the static dielectric constant about 150 would potentially display the best ER effect (in the PDMS oil system), and the materials of a large static dielectric constant would not always display a strong ER effect if other properties are unfavorable to the ER effect; When p is negative, the suspension would give a weak or even no ER effect at all. For the purpose of making the value of p as large as possible, the very slow relaxation time and suitable particle conductivity are essential. For making a good ER fluid, from Eq. (69), one should use the liquid medium and solid material that have high dielectric constants, i.e., a large dielectric constant ratio of the solid material to the liquid.. In addition, the solid material must have a large dielectric loss tangent, at least the maximum value larger than 0.1 for keeping the parameter p positive.

REFERENCES

[1] W.M. Winslow, U.S. Patent 2417850, 1947
[2] D.L Klass, and T.W Martinek, J. Appl. Phys. 38(1967)67
[3] H. Uejima, Jpn. J. Appl. Phys., 11(1972)319
[4] J. E Stangroom,. Phys. Technol., 14(1983)290
[5] H. Block, and J. P Kelly,. Pro. IEE Colloq., 14(1985)1
[6] F.E. Filisko, L. H. Radzilowski, J. Rheol, 34(1990)539
[7] H. Block, J.P Kelly, A. Qin and T. Watson, Langmuir, 6(1990)6
[8] R.A. Anderson, Langmuir, 10(1994)2917
[9] T. Garino, D. Adolf, and B. Hance, Proc. Int. Conf. on ER fluids, R. Tao, Ed.; World Scientific, Singapore, P.167, 1992
[10] L.C. Davis, J. Appl. Phys. 72(1992)1334
[11] L.C. Davis, J. Appl. Phys., 73(1993)680
[12] L.C. Davis, Appl. Phys.Lett, 60(1992)319
[13] Y Otsubo, and K. Watanabe, J. Soc. Rheol. Japan, 18(1990)111
[14] C. F Zukoski, Ann. Rev. Mater. Sci. 23(1993)45
[15] B. Khusid, and A. Acrivos, Phys. Rev. E 52(1995)1669
[16] F.E. Filisko, L.H. Radzilowski, J. Rheol, 34(1990)539
[17] U. Treasurer, L.H. Radzilowski, and F.E. Filisko, J. Rheol. 35(1991)1051;
[18] F. E. Filisko, Prog. in Electrorheology, K. O.Havelka, and F.E. Filisko, Eds. Plenum Press: New York, P.3, 1994
[19] A. W. Schubring, and F.E. Filisko, in Prog. in Electrorheology, K.O Havelka, and F.E. Filisko, Eds.; Plenum Press: New York, P.215, 1994
[20] A. Kawai, K. Uchida, K. Kamiya, A. Gotoh, S. Yoda, K. Urabe, and F. Ikazaki, Intern.J.Modern Phys. B 10(1996)2849
[21] F. Ikazaki, A. Kawai, T. Kawakami, K. Edamura, K. Sakurai, H. Anzai, and Y. Asako, J. Appl. Phys. *D,* 31(1998)336
[22] F. Ikazaki, A. Kawai, T. Kawakami, M. Konishi, and Y. Asako, Proc. Intern. Conf. on Electrorheological and Magnetorheological Fluids, K. Koyama and M. Nakano, Eds; World Scientific: Singapore, p205, 1998
[23] Z.Y Qiu, H. Zhang, Y.Tang, L.W. Zhou, C. Wei, S.H. Zhang, E.V. Korobko, Proc. Intern. Conf. on Electrorheological and Magnetorheological Fluids,. K. Koyama and M. Nakano, Eds; World Scientific: Singapore, p197, 1998
[24] T. Hao, Appl. Phys. Lett. **70** (1997)1956
[25] T.Hao, J. Coll. Interface. Sci., 206(1998)240;
[26] T. Hao. A.Kawai, and F. Ikazaki, Langmuir 14(1998)1256
[27] T. Hao, Z. Xu, and Y.Xu, J.Colloid Interface Sci., 190(1997)334
[28] T. Hao. A. Kawai, and F. Ikazaki, Langmuir 15(1999) 918
[29] T. Hao. A. Kawai, and F. Ikazaki, Langmuir 16(2000)3058
[30] T. Hao. A. Kawai, and F. Ikazaki, J. Colloid Interface Sci., 239(2001)106
[31] A. Kawai, K. Uchida, K. Kamiya, A. Gotoh, S. Yoda, K. Urabe, and F. Ikazaki, Int. J. Modern Phys. B 10(1996)2849
[32] F.E. Filisko, Proc. Int. Conf. on ER fluids, R. Tao, Ed.; World Scientific: Singapore, P.116. 1992
[33] T. Hao, and Y. Xu, J. Colloid Interface Sci., 181(1996)581

[34] H. Fröhlich, Theory of Dielectrics, Clarendon Press, Oxford, 1958
[35] T. Hao, A. Kawai, and F. Ikazaki, J. Colloid Interface Sci., 239(2001)106
[36] S.O. Morgan, Trans. Am. Electrochem. Soc. 65(1934)109
[37] R.W. Sillars, J.I.E.E., 80(1937) 378
[38] K.D. Weiss, D.A. Nixon, J.D. Carlson, and A.J. Margida, in Prog. in Electrorheology, K.O. Havelka, and F.E. Filisko, Eds.; Plenum Press: New York, P.207, 1994
[39] W. Wen, H. Ma, W.Y. Tam, and P. Sheng, Phys. Rev.E. 55(1997)R1294
[40] H. Conrad, and Y. Chen, in Prog. in Electrorheology, K.O. Havelka, and F.E. Filisko, Eds.; Plenum Press: New York, P.55, 1994
[41] M. Whittle, W.A. Bullough, D.J. Peel, and R. Firoozian, Phys. Rev. E 49(1994)5249
[42] Jeu, W. H. de, "Physical Properties of Liquid Crystalline Materials," Gordon and Breach, London/New York/Paris, 1980
[43] G. I. Skanavi, "Dielectric Physics," translated by Y. H. Chen. The High Education Press, Beijing, 1958.
[44] K.D. Weiss, and J. D. Carlson, Proc.3rd Int. Conf. Electrorheological Fluids, R.Tao, Eds.; World Scientific, Singapore, p.264, 1992
[45] F.E. Filisko, Proc. Intern. Conf. on ER fluids, R. Tao, Ed.; World Scientific: Singapore,. 1992, P.116
[46] T. Hao, Y.Chen, Z. Xu, Y. Xu, Y. Huang, Chin. J. Polym. Sci. 12(1994)97
[47] R.Tao, J. T. Woestman, N. K. Jaggi, Appl. Phys. Lett. *55(*1989)1844
[48] T. J. Chen, R. N. Zitter, R. Tao, Phys. Rev. Lett. 68(1992)2555
[49] B. Khusid, and A. Acrivos, Phys. Rev. E 52(1995) 1669
[50] W. Wen, S.J. Men, K.Lu, Phys. Rev. E 55(1997)3015
[51] C.P. Smyth, Dielectric Behavior and Structure, McGRAW-Hill Book Company, Inc.: New York, Toronto, London, 1955
[52] T. Hanai, in Emulsion Science; P. Sherman, eds., Academic Press, Chapter 5, 1968
[53] D.J. Klingenberg, and C.F. Zukoski, Langmuir, 6(1990)15
[54] L. Marshall, C.F. Zukoski, and J. Goodwin, J. Chem. Soc. Faraday Trans. 85(1989)2785;
[55] H. Conrad, Y. Li, and Y. Chen, J.Rheol. 39(1995)1041
[56] T.Hao, H. Yu, and Y. Xu, J. Colloid Interf. Sci. 184(1996)542
[57] Y. Otsubo, K. Watanabe, J. Soc. Rheol. Jpn. 18(1990)111
[58] C.F. Zukoski. Annu. Rev. Mater. Sci. 23(1993)45
[59] R.C Kanu, and M.T. Shaw, J. Rheol., 42(1998)657
[60] Y. Qi, and W. Wen, J. Phys. D: Appl.Phys. 35(2002)2231
[61] S.B. Jones, and S.P.Friedman, Water Resour. Res. , 36(2000)2821
[62] M. J. Blair and G. N. Patey, J. Chem..Phys., 111(1999)3278
[63] A. von Hippel, Rev. Mod. Phys., 22(1950)221
[64] D.L. Klass, and T.W. Martinek, J. App. Phys. 38(1967)75
[65] Yu.F. Deinega, K.K. Popko, and N.Ya. Kovganich, Heat-Tranfer-Sov. Res., 10(1978)50
[66] P. Placke, R. Richert, E.W.Fischer, Colloid Polym. Sci., 273(1995)1156
[67] P. Placke, R. Richert, and E.W.Fischer, Colloid Polym. Sci., 273(1995)848

Chapter 9

Mechanisms of the electrorheological effect

The mechanism of the ER effect has been targeted since the ER effect was discovered in the 1930s. There are various models or mechanisms previously proposed for explaining observed ER phenomena. In this chapter, the physical mechanisms that relate the microscopic material-related properties to the macroscopic properties are addressed. The phenomenological models, such as rheological models that characterize the ER rheological performance, will not be focused on. The mechanisms will be chronologically presented to show how the ER mechanisms have evolved from early phenomenological description to more mature physical models.

1 FIBRILLATION MODEL

The fibrillation model was proposed by Winslow [1] based on his observation that the fibrillated chains were formed in ER suspensions The particles can be polarized and aligned as dipoles along the direction of the electric field. The interaction between the polarized particles will be dramatically increased, resulting in the remarked ER effect. This process has been discussed previously many times. The particle could bear some net charge, arising from the non-uniform polarization, thus dielectrophoresis may contribute to the particle motion for particle re-arrangement. The electrophoresis refers to the particle motion arising from the force generated by an uniform electric field on a charged body. Since most ER fluids don't contain any charging agent, there is no net charge on the particle surface. In an uniform electric field, a neutral particle will be polarized. However, as pointed out by Jordan [2], the induced opposite charges are of equal amount, and the net force exerted on the particle is zero. The particle may remain stationary in an electric field and is unable to form the fibrillated structure. In a non-uniform electric field, the polarized neutral particle may experience a net force on it due to the unequal field strength on the particle surface. So the dielectrophoresis phenomenon may contribute to the fibrillation process. Detailed discussion on the dielectrophoresis can be found in Pohl's book [3]. In addition, the surfactant or polar additives in ER suspensions may adsorb on the particle surface and make the particle charged. The electrophoresis may thus contribute to the particle fibrillation process, or compete with the particle fibrillation process, as the migration of particles to a single electrode may produce phase separation in the system, which in turn weaken the mechanical strength of the whole system. This is the reason that ac electric field sometimes is used for stimulating the ER effect [4].

Direct observation of the fibrillated chain formation process in ER fluids indicate that the chain length increases with the electric field strength and the particle volume fraction, and decreases with field frequency. These trends may be explained by considering the polarization forces between chained particles within a single chain and between fibrillated chains. The magnitude of the polarization force thus determines the mechanical strength of the fibrillated structure.

It is worth mentioning that when the particle volume fraction is high enough, the three-dimensional network structure instead of fibrillated chain structure may form in ER fluids [5]. Sometimes the three-dimensional dendritic structure may form [6-8]. Figure 1 shows the microstructure of a metal/oil, glass sphere/oil, and nickel coated PMMA/oil under an electric field. Both the metal and metal coated PMMA particles form the dendritic structure, and only the glass spheres form the fibrillated chain structure. The appearance of fractal structure is believed to arise from the huge conductivity difference between the dispersed particle and the dispersed medium.

In a word, the fibrillation model is based on the observation of fibrillated chain structure in ER fluids. This model was further expanded for quantitatively calculating the mechanical strength of ER suspensions, which is called the polarization model and will be addressed in detail in a future section. Sometimes the fibrillated chain model may be called a primary polarization model, because the particle polarization is emphasized.

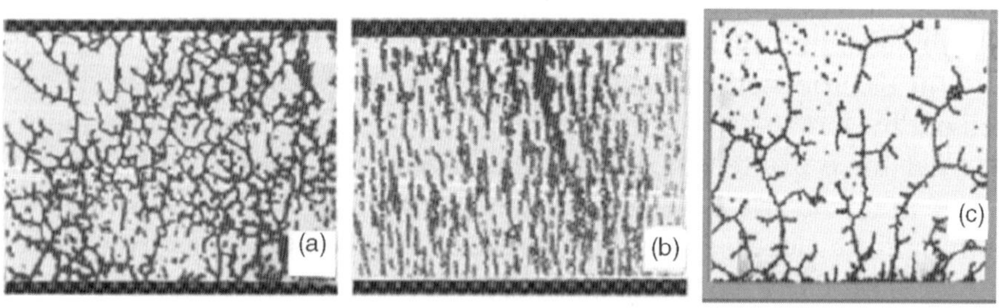

Figure 1 The microstructure of particle/oil suspensions. (a) a metal/oil; (b) glass sphere/oil; (c) nickel coated PMMA/oil. PMMA=polymethyl methacrylate. Reproduced with permission from W. Wen, and K. Lu, Phys. Fluids, 8(1996)2789; W. Wen, and K. Lu, Phys. Fluids, 9(1997)1828.

2 ELECTRIC DOUBLE LAYER (EDL) MODEL

The EDL model was primarily proposed by Klass [9,10] to explain why water plays a key role in the ER response and why the ER effect can take place on a millisecond time scale. Klass believed that the fibrillation process will be rather slow compared with the ER response time, and thus the fibrillation model is inadequate to describe ER phenomena. If water is in an ER suspension, each particle will be surrounded by the EDL. The EDL can be polarized and distorted. The neighboring distorted EDLs may overlap with each other, generating strong electrostatic forces and thus resulting in the ER effect. This process is schematically illustrated in Figure 2. A more detailed EDL model was developed by Uejima [11] and by Deinega [6] to explain the dependences of ER effect on water, field frequency and temperature. According to those explanations, the adsorbed water directly affects the surface charge density of the EDL. The adsorbed water is considered to have two types. One is to adsorb onto the surface of particle for forming the EDL; the other is to absorb in the inner pores of the particles. When the amount of water is in a small scale, all adsorbed water may be in the inner pores and there is no EDL around the particles. Thus there is no observable ER effect. As the water amount increases, the EDL may gradually form for all particles, and thus the ER effect increases. However, an excess amount of water may weaken the ER effect, as the EDL thickness may be reduced as the water amount increases, and the overlap interaction between the EDLs reduce too. The ER effect could be enhanced if the charge carriers only move within the diffuse region, while it could be weakened once the charge carriers migrate between particles. This is the reason that the ER effect undergoes a peak as the water amount increases. Nevertheless, the EDL model is unable to explain why the EDL overlap can result in an increase of rheological properties of several orders of magnitude. Also, for the water-free ER suspensions there is obviously no EDL surrounding each particle, however, a very strong ER effect is still observed.

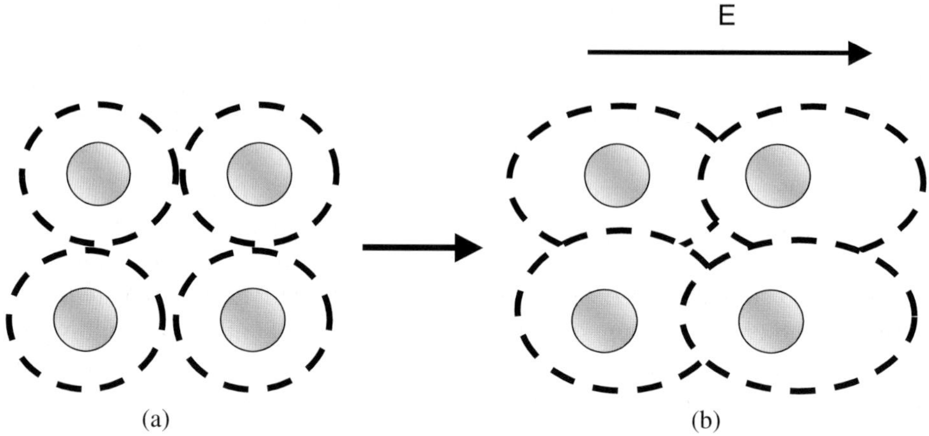

Figure 2 Schematic illustration of the electrical double layer polarization (a) before (b) after applying an electric field.

3. WATER/SURFACTANT BRIDGE MODEL

The water bridge mechanism was proposed by Stangroom [12] to explain the ER effect observed in hydrous ER fluids. He thought that a good ER fluid should meet the following requirements: a) the liquid medium must be hydrophilic; b) the solid particles must be water-like and porous materials with the capability of adsorbing and keeping some amount of water; c) the water amount on the particle surface will determine the ER effect. Under an electric field, ions from water may move out of pores and migrate from one particle to another. Thus an adhesive water bridge could be formed between particles, as shown in Figure 3. This water bridge is very strong due to the high surface tension of water. A similar surfactant bridge was observed by Kim [13], as shown in Figure 8, Chapter 4. See [14] slightly modified the water bridge model by considering the electrostatic energy distribution in the area surrounding particles. The force due to water or the surfactant bridge has already been addressed in section 1.6, Chapter 6. The weak point of the water or surfactant bridge model is obvious: If the ER effect is mainly determined by the surface tension of water or surfactant, then the ER effect of all ER fluids should be similar, however, this is not the truth. The particle polarization (or dielectric properties) should be more important, indicating that water or the surfactant bridge model is too simple for the ER effect.

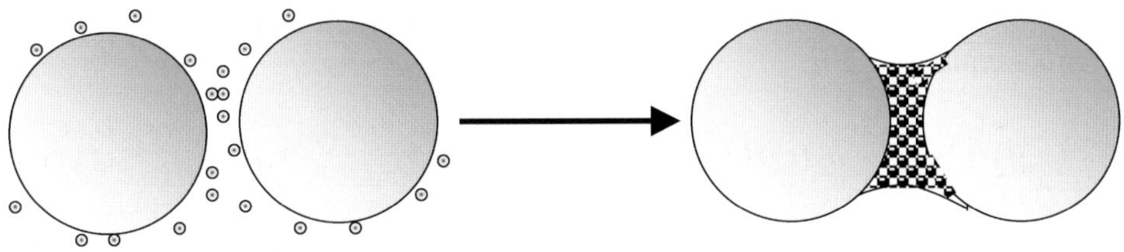

Figure 3 Schematic illustration of water (or surfactant) bridge formed between particles.

4 POLARIZATION MODEL

Both the EDL and water bridge mechanisms lost the physical ground when anhydrous ER fluids were invented in 1985. The fibrillation model received much attention again and many attempts were made to quantitatively calculate the electrostatic polarization force between particles within a chain structure. Several review articles have addressed the various slightly different polarization models and tried to compare the calculated results with the experimental data [2, 15-18]. Since the polarization model has strong limitations for explaining the ER phenomena, only a brief description will be presented in this section.

The objective of the polarization model is to relate the material parameters, such as the dielectric properties of both the liquid and solid particles, the particle volume fraction, the electric field strength, etc., to the rheological properties of the whole suspension, in combination with other microstructure features such as fibrillated chains. A idealized physical model ER system—an uniform, hard dielectric sphere dispersed in a Newtonian continuous medium, is usually assumed for simplification reason, and this model is thus also called the idealized electrostatic polarization model. The hard sphere means that the particle is uncharged and there are no electrostatic and dispersion interactions between the particles and the dispersing medium before the application of an external electric field. For the idealized electrostatic polarization model, there are roughly two ways to deal with the suspensions: One is to consider the Brownian motion of particle, and another is to ignore the Brownian motion and particle inertia. For both cases the anisotropic structure of such a hard sphere suspension is assumed to be represented by the pair correlation function $g(r,\theta)$, derived by

Hayter [19] from the mean spherical approximation closure to the Ornstein-Zernike equation [20]:

$$g^{eq}(r,\theta) = g_1(r) + h_2(r)P_2(\cos\theta) \tag{1}$$

where r is the center-to-center (or point-to-point) distance between dipoles, and θ is the angle between the electric field vector and the center-to-center dipole vector, see Figure 4. $g_1(r)$ is the isotropic part, and $P_2(\cos\theta)$ is the second Legendre polynomial,

$$P_2(\cos\theta) = \frac{1}{2}\left(3\cos^2\theta - 1\right) \tag{2}$$

The function $h_2(r)$ depends on the mean dipole strength μ that may be expressed as:

$$\mu = \mu_0\left(1 - \beta\phi - \frac{6}{5}\beta\phi\int_{2r}^{\infty}\frac{1}{\delta}h_2(r)dr\right)^{-1} \tag{3}$$

Where $\beta = (\xi - 1)/(\xi + 2)$ with $\xi = \varepsilon_p / \varepsilon_m$, called the polarizability, ϕ is the particle volume fraction, a is the radius of the particle, $\mu_0 = (\pi/2)\varepsilon_0\varepsilon_m(2a)^3\beta E$, is the dipole strength for an isolated sphere. Assuming that particles interact through both hard-sphere repulsion and point-dipole electrostatics corrected for multibody interactions in a mean-field manner, Adriani [21] calculated the shear modulus and the dynamic

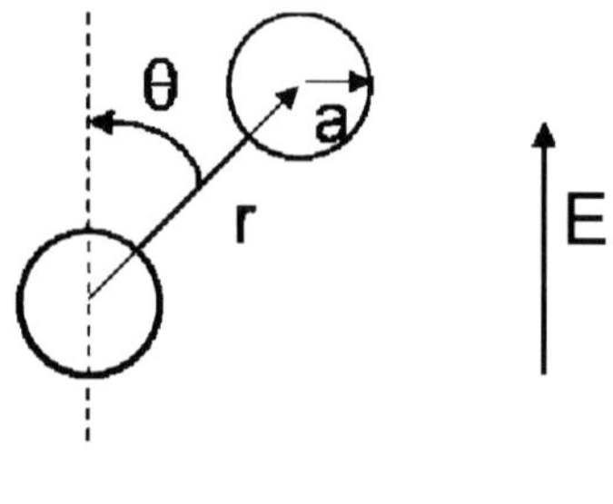

Figure 4 Coordinate system of the monodispersed sphere system with the radius a in an electric field.

viscosity from a small perturbation to the particle distribution described by the pair correlation function, in which the Brownian motion, the electrostatic force, and other interaction forces are considered. The predicted high frequency shear modulus is plotted vs. the relative magnitude of the electrostatic to thermal contributions in Figure 5. The shear modulus has a strong dependence on the particle volume fraction, the dipole strength, and the square of the electric field strength. However, the experimental results show that the real modulus of polystyrene bead fluid is drastically underestimated with the Adriani's model [22]. This model was late extended to examine the birefringence and dichroism of the ER suspensions under combined electric and shear fields [23]. Again, the experimental data on the dichroism of silica suspensions show that the calculated results are about one order of magnitude lower than the experimental data [24]. The discrepancy may be arisen from the facts that the point dipole assumption probably underestimates the interparticle forces and the theory is only valid for the situation that the relative magnitude of the electrostatic to the thermal forces is far less than 1 [15, 25]. No matter whether the Brownian force is considered, the dynamics simulation employing the idealized electrostatic polarization model will produce the fibrillated structure, as shown in Figure 6. If the shorter-ranged repulsive force is considered, much thicker chains are formed in Figure 6b in comparison with Figure 6a, where the shorter-ranged repulsive force is not considered.

A more direct way to calculate the rheological properties of ER suspensions is to directly estimate the interparticle force induced by the an electric field using the point-dipole approximation. As shown earlier, for uncharged particles in an electric field, Laplace's equation holds:

$$\nabla^2 \psi = 0 \tag{4}$$

where ψ is the electric potential on the particle surface. If an uniform electric field E is applied in the z direction shown in Figure 4, the solution to Eq. (4) is:

$$\psi = -\vec{E} \cdot \vec{r} \cos\theta \left(1 - \frac{\beta a^3}{r^3} \right) \tag{5}$$

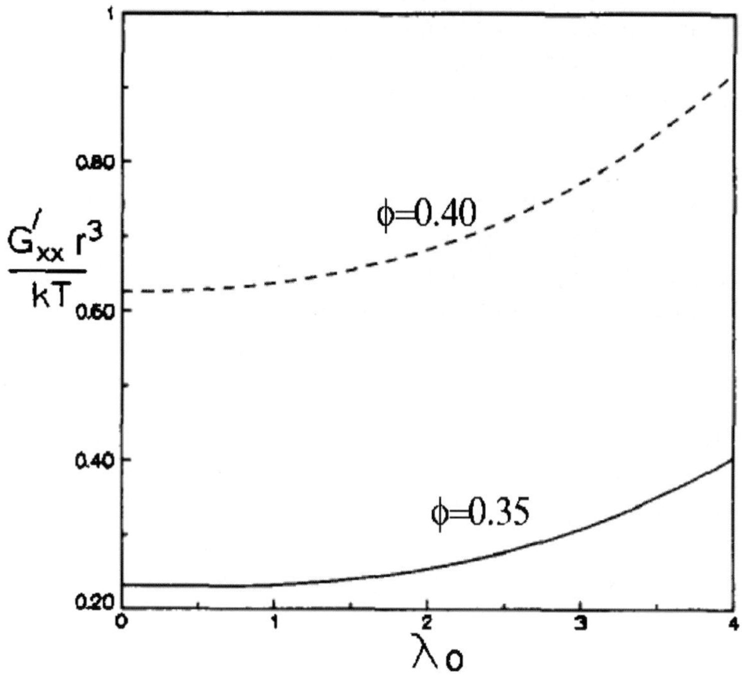

Figure 5 Prediction of high frequency modulus vs. the relative magnitude of the electrostatic to thermal contributions λ_0, $\lambda_0 = (\pi/2)\varepsilon_0\varepsilon_m(2a)^3\beta^2E^2/k_BT$ at the particle volume fraction 0.35 and 0.40. β=0.75. Reproduced with permission from P.M. Adriani, and A.P. Gast, Phys. Fluids, 31(1988)2757.

The dimensionless interaction energy between two point-dipoles can be expressed as [23, 26]

$$\frac{U(r,\theta)}{k_BT} = -\frac{\lambda}{2}\left[\frac{3\cos^2\theta - 1}{(r/2a)^3}\right] = -\lambda\left(\frac{2a}{r}\right)^3 P_2(\cos\theta) \qquad r/2a \geq 1 \qquad (6)$$

where $\lambda = \lambda_0/4 = \pi\varepsilon_0\varepsilon_m a^3\beta^2E^2/k_BT$, characterizing the relative importance of the electrostatic polarization energy to the thermal energy. This interaction energy is anisotropic and can be decomposed into two parts: An attraction force along with the direction of the electric field and a

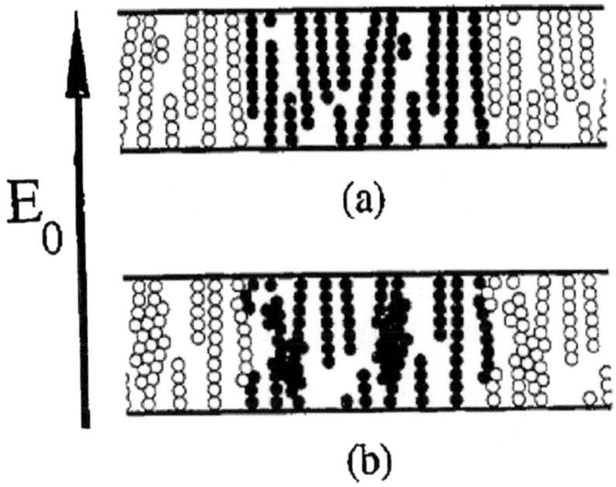

Figure 6 Fibrillated structure produced with the simulated method employing the idealized electrostatic polarization model without considering the Brownian motion. Thicker columns are formed under the shorter-ranged repulsive force (b). Reproduced with permission from D.J. Klingenberg, F. van Swol, and C.F. Zukoski, J. Chem.Phys., 91(1989)7888.

repulsive force perpendicular to the electric field. When two pair particles are far away from each other, differentiation of Eq. (6) leads to the interaction force of two point-dipoles [25, 26]:

$$\vec{F}_{pair} = 12\pi\varepsilon_0\varepsilon_m a^2\beta^2E^2\left(\frac{a}{r}\right)^4\left[\left(2\cos^2\theta - \sin^2\theta\right)\vec{e}_r + \sin 2\theta\vec{e}_\theta\right] \tag{7}$$

However, both the experimental and theoretical evidences show that the force on a sphere can significantly exceed that given by Eq. (7) [27-30]. The underestimation is attributed to the ignorance of multipole and multibody and the local electric field effects arisen from other particles and the particle chains nearby. Eq. (7) is only considered to be accurate if the conductivity of both the particle and the liquid medium is zero ($\beta \to 0$), or $(a/r) \to 0$. Many attempts have thus been made to include the multipole and multibody effects in the calculation [30-32]. For incorporating the multipole effect, the exact solution to Laplace's equation is required and the following equation is obtained using the multipole expansion method [30,32,33]:

$$\vec{F} = 12\pi\varepsilon_0\varepsilon_m a^2\beta^2 E^2 \vec{f} \tag{8}$$

with

$$\vec{f} = \left(\frac{a}{r}\right)^4 \left[\left(2f_{\parallel}\cos^2\theta - f_{\perp}\sin^2\theta\right)\vec{e}_r + f_{\Gamma}\sin 2\theta\vec{e}_{\theta}\right] \tag{9}$$

where $f_{\parallel}$, $f_{\perp}$, f_{Γ} are the parallel, perpendicular, and torsional force components, and $\vec{e}_r$, $\vec{e}_{\theta}$ are unit vectors in r and θ. In the point-dipole approximation, $f_{\parallel}$, $f_{\perp}$, f_{Γ} are all unity and Eq. (8) reduces to Eq.(7). In the exact calculation they are the function of the dielectric constant of both particle and liquid medium, the radius of particle, and the distance between the two dipoles. Figure 7 shows $f_{\parallel}$ defined in Eq.(9) against the interparticle center-to-center separation distance r/2a. A huge force between two spheres is generated when two particles becomes closer and the dielectric constant ratio of particle-to-liquid is larger, implying that the point-dipole approximation fails at such conditions. Note that the interparticle force becomes unity at large separations, where the point-dipole approximation holds. Bonnecaze [31] used the electrostatic energy method for calculating the interparticle force under assumption that the force on a given sphere is the derivative of the electrostatic energy with respect to the particle position. The energy method results are claimed to be in excellent agreement with the data shown in Figure 7 from ref.[15]. The calculated effective viscosity using the electrostatic energy method with or without near-field(NF) interaction vs. the Mason number, Ma, the ratio of the viscous to the electrostatic forces, is shown in Figure 8. A good agreement is obtained through the whole Ma number range, indicating that the electrostatic energy method is accurate and captures both the qualitative and quantitative features of ER suspensions. The influence of the third particle on the interaction force of a pair- dipole (many-body effect) is shown in Figure 9 for almost touching conducting spheres. It clearly shows that the rate of divergence of the force between two particles in the presence of third particle is almost two times than that of two particles alone. However, those calculation methods obviously overlook the local electric field correction, which should be important when the particles can be highly polarized. Chen [28] introduced the local electric field correction via mapping out the electric distribution on the particle surface, and then calculated the interaction force on the central particle using the Coulomb's Law Committee method [34].

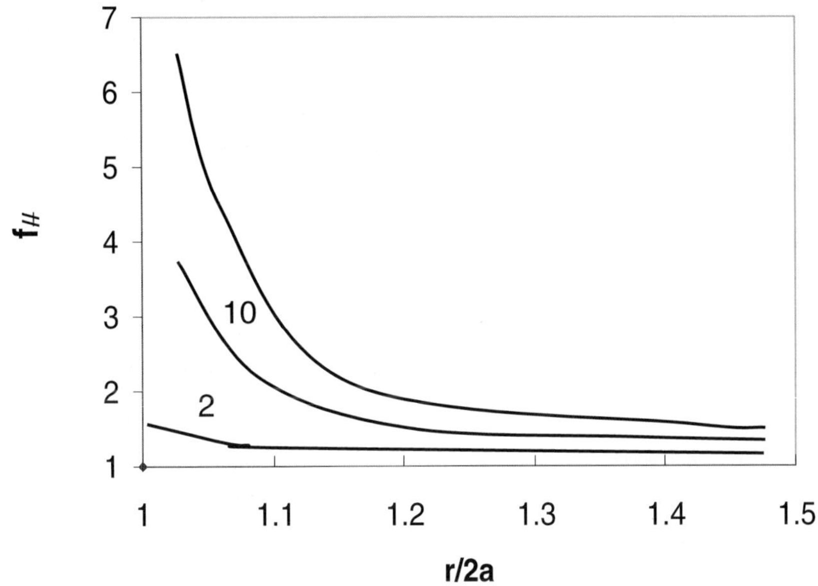

Figure 7 The force component $f_{\parallel}$ defined in Eq.(9) against the interparticle center-to-center separation r/2a for three values of $\varepsilon_p / \varepsilon_m = 2,10,\infty$. Redrawn from A.P. Gast, and C.F. Zukoski, Adv. Colloid Interface Sci., 30(1989)13 with permission. For point-dipole approximation, $f_{\parallel}$ =1 for all r/2a.

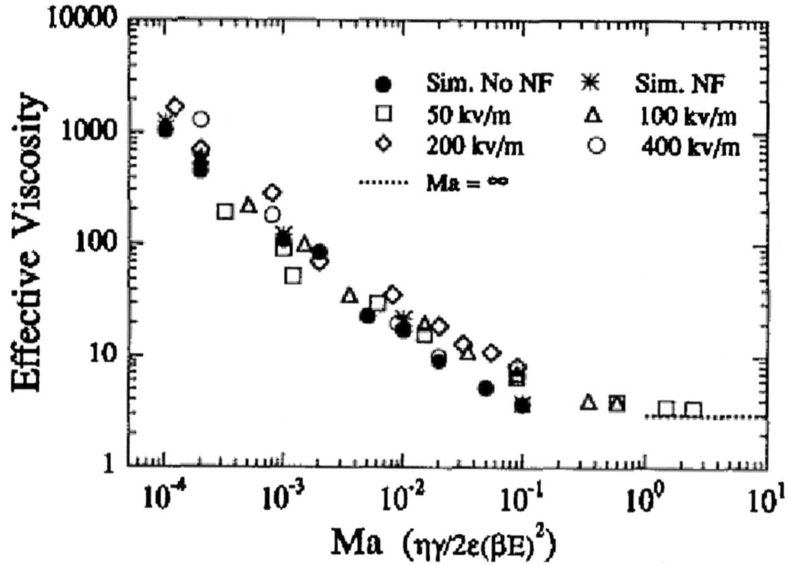

Figure 8. Comparison of simulated effective viscosity from the electrostatic energy method to the experimental results of Marshall et al. (L. Marshall, J. W. Goodwin, and C. F. Zukoski, J. Chem. Soc. Faraday I, 85(1989)2785) illustrated with the open symbols and the applied field used. Reproduced with permission from R.T. Bonnecaze, and J.F. Brady, J. Chem. Phys., 96(1992)2183.

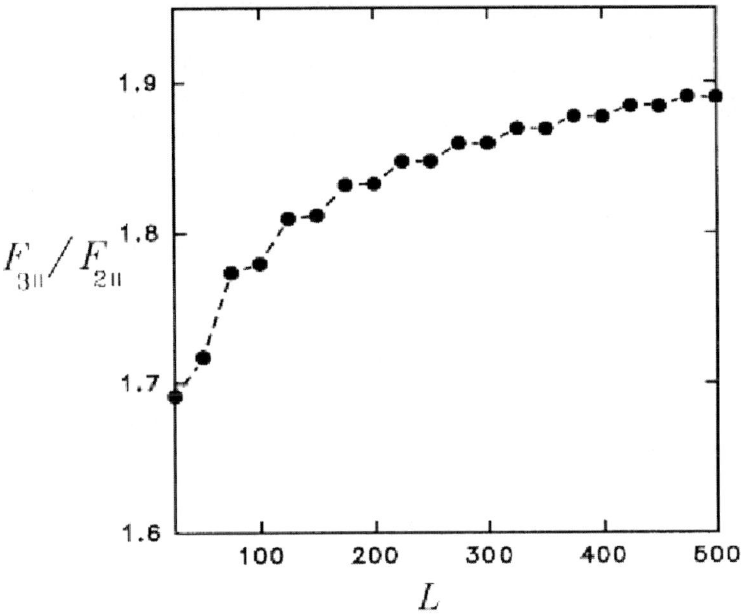

Figure 9 The normalized force function $F_{3||}/F_{2||}$ for touching conducting spheres as a function of the number of multipole moments L. Reproduced with permission from H.J.H. Clercx, and G.Bossis, Phys. Rev. E, 48(1993)2721

The component of the electric field can be obtained at any point by taking the gradient of the electric potential:

$$\vec{E}_r = -\frac{\partial \vec{E}}{\partial r} \tag{10}$$

and the interaction force is expressed as:

$$F_Z = \frac{\varepsilon_m}{\varepsilon_o} \int_{\Sigma} \sigma_t E_z(out) ds \tag{11}$$

where the integration is taken over the central particle surface and σ_t is the total charge density induced by the polarization and distributed on the particle surface. E_z is the Z component of the electric field due to the adjacent particles. Figure 10 shows normalized particle-particle interaction force vs. the interparticle distance R/a calculated with the Chen model given in Eq. (11) and the point-dipole approximation given in Eq. (7). The effects of nearby particles become significant when the interparticle distance is about one diameter. The inter-particle force is almost one order of magnitude greater than predicted with the point-dipole approximation when the two particles move much closer. Davis [35,36] introduced the finite-element analysis (FEA) method for calculating the inter-particle polarization forces for the reason that this method automatically includes two important effects: the local electric-field corrections and all multipole terms.

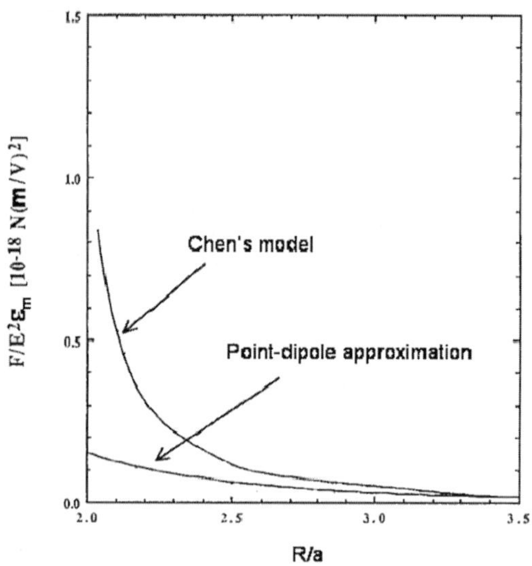

Figure 10 Normalized particle-particle interaction force vs. the separation distance R/a obtained using Chen's model and point-dipole approximation for $\varepsilon_p / \varepsilon_m = 10$. Reproduced with permission from Y. Chen, A. F. Sprecher, and H. Conrad, J. Appl. Phys. 70 (1991)6796.

The local electric field, E_{loc}, is the sum of the applied electric field, E_0, and the electric field E_p from all the other dipoles surrounded on the sphere under consideration:

$$\vec{E}_{loc} = \vec{E}_0 + \vec{E}_p \tag{12}$$

$\vec{E}_p$ may have two components parallel ($E_{p,\parallel}$) and perpendicular($E_{p,\perp}$) to the chain, and can be expressed as [36]:

$$E_{p,\parallel} = \frac{4.808 p_{\parallel}}{4\pi\varepsilon_0\varepsilon_m r^3} \tag{13}$$

and

$$E_{p,\perp} = \frac{-2.404 p_{\perp}}{4\pi\varepsilon_0\varepsilon_m r^3} \tag{14}$$

where r is the interparticle distance and p is the dipole moment. The effective force between particles can be estimated through the stretching portion of the shear modulus that can be given as [36]:

$$\begin{aligned} G_s &= 0.5\varepsilon_0\left(-r\frac{\partial\varepsilon_{eff}}{\partial r}\right)E_0^2 \\ &= 1.202 N_p F_d \end{aligned} \tag{15}$$

where ε_{eff} is the effective dielectric constant of the ER suspension and is the function of the local electric field, N_p is the number of chains per unit area, and F_d is the force between two dipoles. The ratio of the effective force calculated with Eq.(15) to the point-dipole force given in Eq.(7) with the simplification that the particles are aligned along the vector separating them, is plotted in Figure 11 against the particle separation distance, r/2a. When two particles almost touch each other, the interparticle force predicted with FEA method is nearly two times higher than that predicted with Eq.(7), which only considers the multipole effect. Figure 11 clearly indicates that not only the multiple-dipole effect but also the multiple-body effect are significant in ER systems.

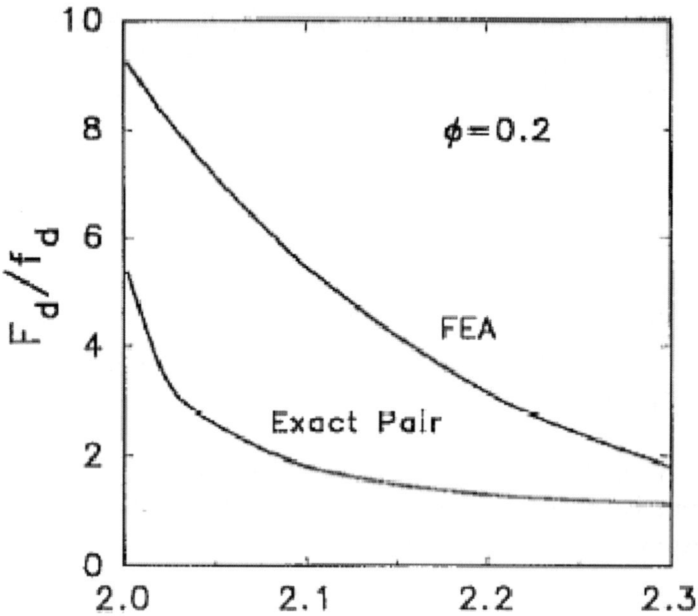

Figure 11 The ratio of the effective force calculated with Eq. (15) to the point-dipole force given in Eq. (7) for the case where θ=0 as a function of r/a. The "exact pair" is the force calculated with Eq. (8). Reproduced with permission from ref. L.C. Davis, J. Appl.Phys., 72(1992)1334.

There are many other attempts for incorporating the local electric field correction. Tao [37] used the finite-element analysis method for calculating the electrostatic interaction in a chain and found that the interaction force between a pair-particle can be expressed as:

$$f_{pair} \approx -\frac{1}{18}\varepsilon_m E_0^2 a^2 \left(\varepsilon_p / \varepsilon_m - 1\right)^2 \tag{16}$$

Eq. (16) is similar to the equation derived by Anderson as shown below [29]:

$$f_{pair} = -0.04138\varepsilon_m E_0^2 a^2 \left(\varepsilon_p / \varepsilon_m\right)^2 \tag{17}$$

Eq. (16) gives a slightly stronger force than that given by Eq. (17), and predicts 20% higher interparticle force than that given by the point-dipole approximation method with the local electric field correction [37]. Generally, a more universal form could be used to express the derived electrostatic force [38]:

$$F = k\varepsilon_m a^2 \beta^2 E^2 S \tag{18}$$

Where F is the electrostatic force, k a constant, $\beta = (\xi - 1)/(\xi + 2)$ with $\xi = \varepsilon_p / \varepsilon_m$, S is a structure factor related to the particle microstructure. Once the interparticle force is determined, the rheological parameters such as the shear modulus and the yield stress can be related to the physical properties of the materials through Eq. (18). For example, Klingenberg [39] equate the time average of the shear stress with the yield stress and obtain the yield stress τ_y as:

$$\tau_y = 18\phi\varepsilon_0\varepsilon_m\beta^2 E^2 f_m \left[1 - \frac{(\pi/6)^{1/2}}{(d/a)\tan\theta_m \phi^{1/2}}\right] \tag{19}$$

where f_m and θ_m are the maximum in the dimensionless restoring force and the angle at the maximum, respectively. d is the gap distance between two electrodes. Assuming the shear stress is equal to the change of the free energy under an electric field [40] in the absence of nonelectrostatic forces, Davis [35,36] obtained the yield stress as:

$$\tau_y \approx \frac{1}{4}\varepsilon_0 \left(-r\frac{\partial \varepsilon_{eff}}{\partial r}\right) E^2 \tag{20}$$

A basic derivation from Eq. (20) is that yield stress will increase linearly with the dielectric constant ratio $\varepsilon_p/\varepsilon_m$, indicating that a high particle dielectric constant will give a strong ER effect. The material of an extremely high dielectric constant was thus used experimentally as the solid particulate phase of the ER fluid. A barium titanate suspension ($BaTiO_3$, of the dielectric constant around 2000, depending on its crystallization state), however, presents a surprising result: inactive under a dc field [41], and active after adsorbing a small amount of water [42]or being stimulated by an ac field [43,44]. Those results suggest that the polarization model still has much room to improve. Obviously, the polarization model fails to describe other important ER experimental observations, such as the rheological property dependence on the electric field frequency and the particle conductivity. Klingenberg [25,45] proposed that in the interparticle force equations the polarizabilty β should be replaced by the "effective relative polarizabilty", β_{eff}, which can be obtained from the Maxwell-Wagner polarization model [46]:

$$\beta_{eff}^2 = \beta_d^2 \frac{\left[(\omega\tau_{MW})^2 + \beta_c/\beta_d\right]^2 + (\omega\tau_{MW})^2(1-\beta_c/\beta_d)^2}{\left[1+(\omega\tau_{MW})^2\right]^2} \tag{21}$$

where

$$\beta_d = \frac{\xi - 1}{\xi + 2} \qquad (\xi = \varepsilon_p/\varepsilon_m) \tag{22}$$

$$\beta_c = \frac{\sigma_p - \sigma_m}{\sigma_p + 2\sigma_m} \tag{23}$$

$$\tau_{MW} = \varepsilon_0 \frac{\varepsilon_p + 2\varepsilon_m}{\sigma_p + 2\sigma_m} \tag{24}$$

where ω is the angular frequency, σ_p and σ_m are the conductivity of the particle and medium, respectively. The value of β_{eff} depends on the frequency relative to the polarization relaxation time τ_{MW}. When the frequency is large enough that $\omega\tau_{MW} \to \infty$, then Eq. (21) becomes:

$$\beta_{eff}^2 = \beta_d^2 \tag{25}$$

The dielectric constants of the particle and the dispersing medium dominate the polarizability. While in the dc electric field, Eq. (21) becomes:

$$\beta_{eff}^2 = \beta_c^2 \tag{26}$$

thus the conductivities of both the particle and the dispersing medium dominates the polarizability. In other words, the conductivity mismatch between particle and liquid medium, rather than the dielectric constant mismatch, was thought to be a dominant factor for dc and low frequency ac excitation [29,35,36]. This modification on the polarization model lays the foundation for the generalized polarization model. Note that the generalized polarization model can only qualitatively describe some experimental facts, for example, why the high dielectric constant material cannot always give a good ER response. It is still very poor in many aspects; for example, it could not explain why the negative ER effect could occur in some suspensions. Quantitatively, there is still a huge discrepancy between the theoretical prediction and the experimental results.

4 CONDUCTION MODEL

Atten [46] and Foulc [47] proposed a conduction model in which the ER effect was thought to be determined by the particle-to-liquid conductivity ratio σ_p/σ_m if $\sigma_p>\sigma_m$; if $\sigma_p<\sigma_m$, a negative ER effect would be expected. This model was extensively extended by Tang [48] and Wu [49]. Atten [47] directly measured the attraction force between a macroscopic polyamid sphere of 2 cm in diameter and a flat electrode immersed in mineral oil. AOT (sodium [*bis*-2-ethylhexyl]-sulphosuccinate) was added to liquid medium for adjusting the conductivity of liquid phase. The measured attraction force is plotted against the applied electric field at different particle-to-medium conductivity ratio in Figure 12. When the particle-to-medium conductivity ratio, $\Gamma=\sigma_p/\sigma_m$, is small enough, the attraction force F almost linearly increases with E^2. Once $\Gamma>10$, the attraction force F only linearly increases with E^2 at weak electric field and with E at strong electric field. Those experimental observations can be well described with the following equations[46,47] :

$$F = 4\pi R^2\left[\pi \ln(R/R_b)\right]^{-2}\varepsilon_m E^2\Gamma^2 \tag{27}$$

for low electric fields, and

$$F = 2\pi R^2\varepsilon_m E E_c\left\{\ln\left[(10\Gamma/\pi)(2E/E_c)^{1/2}\right]\right\}^2 \tag{28}$$

for high electric fields. Where R_b is the radius of the area in which the current passes through from the sphere to the electrode, the contact zone, and E_c is a constant. Eq. (28) gives an approximately linear relationship between the force and the applied electric field. Eq. (27) can lead to an equation correlating the force with the current passing through the particle under the assumption that the current I_p is related to the radius of the contact zone [46]

$$F \cong 4\pi R^2 \varepsilon_m \sigma_p V^4 / I_p^2 \qquad (29)$$

where V is the potential difference between two adjacent spheres. Eq. (29) predicts a linear relationship between the forces F and (V^2/I_p), which was experimentally confirmed in the system of two polyamide spheres immersed in AOT/mineral oil. Again, AOT was used to adjust the conductivity of the liquid medium.

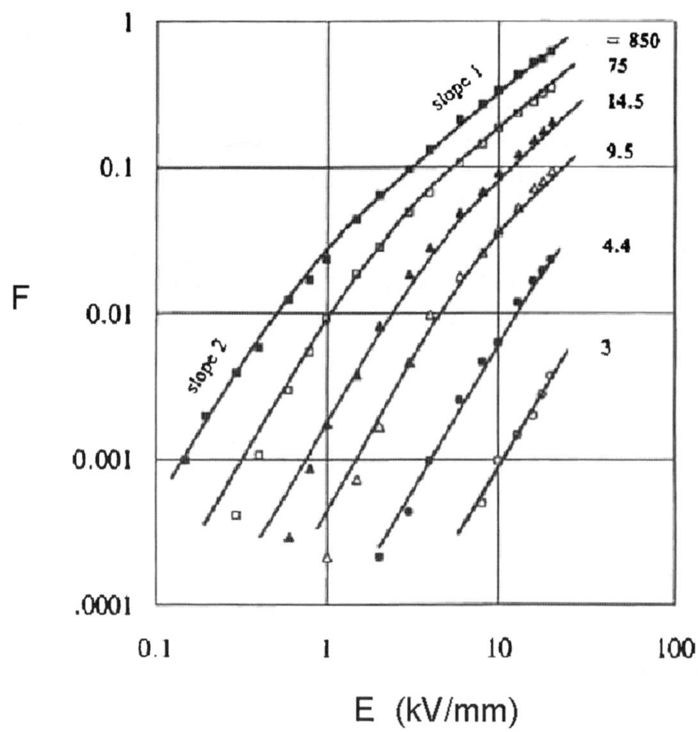

Figure 12 The attraction force between a polymeric sphere and a flat electrode vs. the applied electric field at various values of the particle-to-liquid conductivity ratio, $\Gamma = \sigma_p / \sigma_m$. Reproduced with permission from J-N. Foulc, P. Atten, N. Felici, J. Electrostatic, 33(1994)103.

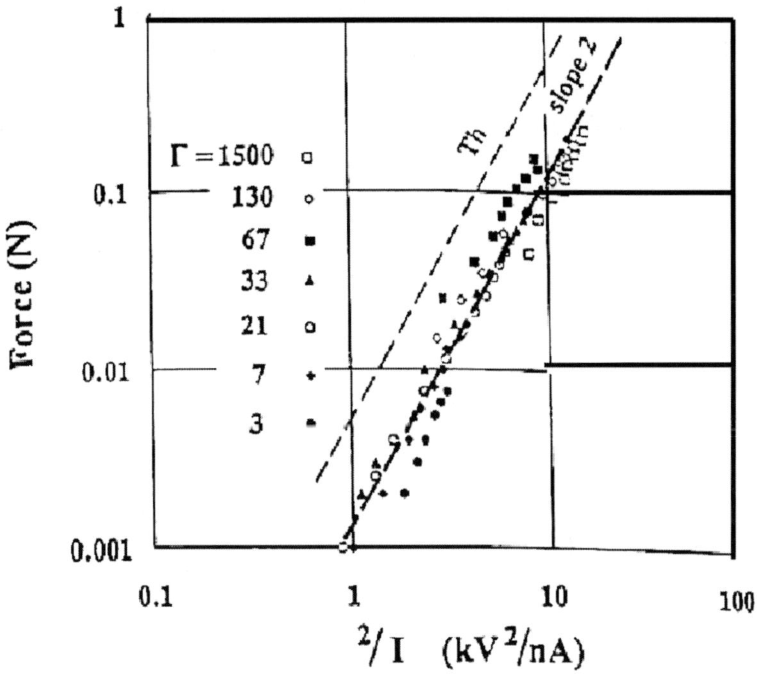

Figure 13 Experimental (dots) and theoretical prediction (Th) from Eq. (29) forces vs. (V^2/I_p) between two half polyamide spheres immersed in AOT/ mineral oil mixture at different particle-to-liquid conductivity ratio. Reproduced with permission from P. Atten, J-N. Foulc, and N. Felici, Int. J. Mod. Phys. B, 8(1994)2731.

Figure 13 shows the calculated force from Eq. (29) and experimentally measured force vs. (V^2/I_p) at different particle-to-liquid conductivity ratio for two polyamide spheres. Qualitatively there is a good agreement between theoretical predictions from Eq. (29) and the experimental results, though quantitatively Eq. (29) gives 5 times higher values than the experimental data. The discrepancies are ascribed to the overestimation of the conductance of the two spheres. An important implication of Eq. (29) is that a huge leakage current of an ER fluid and, therefore, the heating phenomenon that results, appear to be unavoidable if a strong ER effect is going to be expected.

In the conduction model, the surface conductivity of the particle is thought to function in a similar way as the bulk conductivity does. Atten [46] obtained the following equations when the surface conductivity of the sphere is considered:

$$F = 4\pi R^2 \varepsilon_m E^2 \exp\left[(2\Gamma_s / A)^{1/2}\right] \tag{30}$$

for low electric fields with

$$\Gamma_s = \sigma_{sp} / (R\sigma_m) \tag{31}$$

and

$$A = \frac{\ln(2R/R_b)}{\ln(R/R_b)} \tag{32}$$

where σ_{sp} is the surface conductivity of particle, R_b is the radius of conducting region. For high electric fields,

$$F \cong 2\pi R^2 \varepsilon_m E E_c \left\{\ln\left[(A_1\Gamma_s)(2E/E_c)^{1/2}\right]\right\}^2 \tag{33}$$

where

$$A_1 = \frac{5}{A[\ln(R/R_b)/R/R_b]} \tag{34}$$

Eq. (30) and (33) are quite similar to Eq. (27) and (28) in regard to the electric field dependence. Using a different approach, Tang [48] derives the force and current as a function of the separation of the particles, rather than dividing the contact zone between particles into two distinct regions as used by Atten. Since the surface conductivity of particle is as important as the bulk conductivity, Wu [50] addressed how a surface film on the particle influences the ER response and obtained following equation on the basis of the conduction model:

$$f = \pi R^2 \varepsilon_m E^2 F(d) \tag{34}$$

where F(d) is the normalized force, a function of the thickness and conductivity of the surface film, the separation distance of the two spheres (d), the applied electric field (E), and the radius of the particle. It can be expressed as:

$$F(d) = \int \left\{ \left[1 + H(x)\right] \left[\frac{\left\{1 - B + B\exp[\sqrt{E_m^{(k)} / E_c}]\right\} \times}{\left[\frac{1}{\Gamma_p} + \frac{1}{\sigma_r \left[1 - (x/R)^2\right]} \right] + H(x)} \right]^{-1} \right\} \frac{x}{R^2} dx \tag{35}$$

with

$$H(x) = \frac{1 + d/2R}{\sqrt{1 - (x/R)^2} - 1} \tag{36}$$

where $E_m^{(k)}$ is the local electric field at the liquid medium, k denotes the kth iteration. $\Gamma_p = \sigma_p / \sigma_m(0)$, $\sigma_r = (R/t_f)[\sigma_f / \sigma_m(0)]$, t_f and σ_f are the thickness and the conductivity of the surface film, $\sigma_m(0)$ is the conductivity of liquid medium at low electric field, and B and E_c are constant inherited from the modified Onsager's theory [47,51] for the non-ohmic conductivity of a nonpolar liquid:

$$\sigma_m = \sigma_m(0)\left[1 - B + B\exp\left(\sqrt{E_m / E_c}\right)\right] \tag{37}$$

Figure 14 shows the normalized attractive force F(d) between two particles vs. the normalized separation distance of particle d/2R for E =3 kV/mm, and the particle-to-liquid medium conductivity ratio, $\Gamma_p = \sigma_p / \sigma_m(0) = 10^{12}$ at different σ_r values. When the separation distance is large (d/2R>0.1), the attractive force between particles rapidly decreases, becoming almost independent of the surface film thickness and conductivity; When d/2R< 0.01, the attractive force between particles becomes independent of the separation distance, indicating that there is a field saturation phenomenon occurred in the liquid medium; however, in this region the attraction force does increase substantially with the increase of the surface film conductivity and the decrease of the surface film thickness. A reasonable conductivity and intermediate thickness of the surface film were thus suggested in the consideration of the unacceptably high current density of the suspension and the low electric breakdown strength of the surface film[50]. For increasing the electric breakdown strength of the surface film, a surface film material of a high dielectric constant is desired [52]. Note that Eq. (35) is only valid for dc field. A similar equation was derived for calculating the interaction force between particles under ac field [52]. A good agreement was found between the predictions and the experimental results for the surface oxidized aluminum particle/silicone oil system under both dc [50] and ac fields [52], in term of the yield stress dependence on the applied electric field and the thickness of the surface film. The shear stress of the ER suspension was easily derived on the basis of the interaction force between particles in a chain shown in Eq. (34). Suppose that there are N particles in a chain, and then the chain length is 2RN, as illustrated in Figure 15. Under a shear field with shear strain, γ, the chain length increases to $2RN + (N-1)d \approx (2R+d)N$, so

$$\sin\theta = \frac{\sqrt{(d/2R)^2 + 2(d/2R)}}{1+(d/2R)} \tag{38}$$

If the number of chain per area is N_{chain}, then the shear stress τ

$$\tau = N_{chain} f \sin\theta \tag{39}$$

In an unit volume and the particle volume fraction is ϕ,

$$\frac{1}{2R} \times \frac{4}{3}\pi R^3 \times N_{chain} = \phi \tag{40}$$

thus

$$N_{chain} = \frac{3\phi}{2\pi R^2} \tag{41}$$

Substituting Eq. (34), Eq. (38), and Eq. (41) into Eq. (39) leads to

$$\tau = \frac{3}{2}\phi\varepsilon_m E^2 F(d) \frac{\sqrt{(d/2R)^2 + 2(d/2R)}}{1+(d/2R)} \tag{42}$$

The yield stress τ_y is the maximum value of τ given in Eq. (42). Figure 16 shows the predicted yield stress with Eq. (42) and experimentally measured values from the reference [53] vs. temperature for zeolite particle/silicone oil suspension. A good agreement between the theoretical calculations and the experimental results is obtained for the entire temperature range from 25 to 160 °C. The yield stress passed through a maximum around 100 °C. A direct measurement of the interparticle force between two big polyamide spheres 7 mm in radius immersed in silicone oil against temperature shows that the interparticle force peaked at temperature around 40 °C [54]. The result is shown in Figure 17 as the percentage variation of the interparticle force compared to the room temperature value vs. temperature. The temperature dependence difference between those two systems may result from the different temperature dependence of the conductivity of the dispersed particle material. From Eq. (42) it is clear that the thickness of the coated film can have a great impact on the yield stress of the suspension. The yield stress of the oxidized silicon particle/silicone oil system vs. the thickness of the surface oxidized film is shown in Figure 18. As expected, the yield stress decreases as the thickness of the surface film increases. Again, a good agreement between the theoretical calculation and the experimental results is obtained for this ER system.

In addition to the thickness of the coated film, the conductivity of the coated film is also important. Figure 19 shows the directly measured interparticle force between two semispheres (polystyrene) immersed in the transformer oil vs. the electric field. The two semispheres were coated with a thin doped polyaniline layer with the thickness around 0.1 μm with different conductivities controlled via the doped level. The relation between the force and the applied voltage can be fitted with a quadratic law for the entire electric field range. For the spheres of the conducting film, the quadratic relation is only valid at low electric fields, and a linear relation is observed in high electric fields. For the conducting film the data could be well described by using the conduction model as shown in the following equation [55]:

$$F = \pi\varepsilon_m \frac{V_0^2}{\sqrt{2R\sigma_m/(t_f\sigma_f)}\left(\ln\frac{4}{\sqrt{2R\sigma_m/(t_f\sigma_f)}}\right)^2} \tag{43}$$

where V_0 is the applied voltage. For the insulating film the model is slightly underestimate the force, but still they are in the same order of magnitude. Eq. (43) indicates that for the given thickness of the surface film the force varies with the square root of the conductivity ratio (σ_f/σ_m). It is very much different from the particle without a coating film, in which the force varies with the square of the conductivity ratio at low fields. For a given film conductivity the force varies with the square root of film thickness, which is also different from that shown in Figure 18.

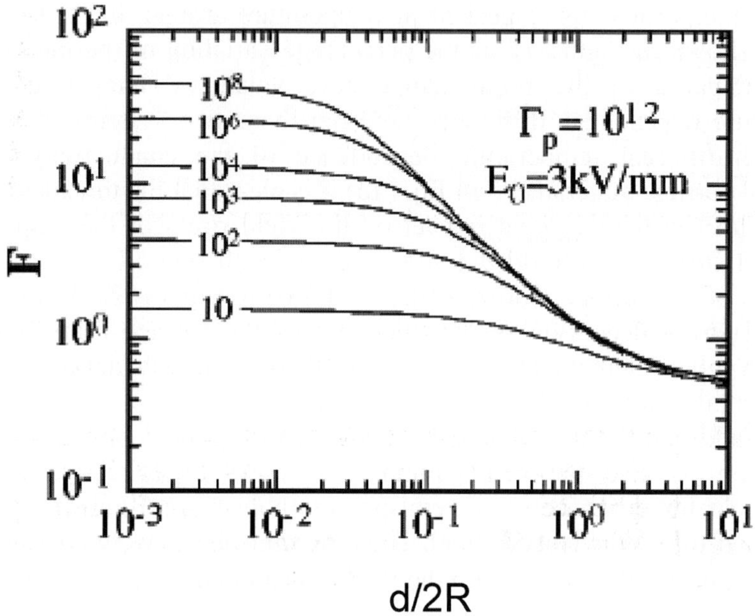

Figure 14 Normalized attractive force F(d) between two particles vs. the normalized separation distance of particle d/2R for E =3 kV/mm, and the particle-to-liquid medium conductivity ratio, $\Gamma_p = \sigma_p / \sigma_m(0) = 10^{12}$, $\sigma_m(0)$ is the conductivity of liquid medium at low electric field. $\sigma_r = (R/t_f)[\sigma_f / \sigma_m(0)]$, with t_f and σ_f are the thickness and the conductivity of the surface film, respectively. Reproduced with permission from C. Wu, and H. Conrad, J. Appl. Phys., 81(1997)383.

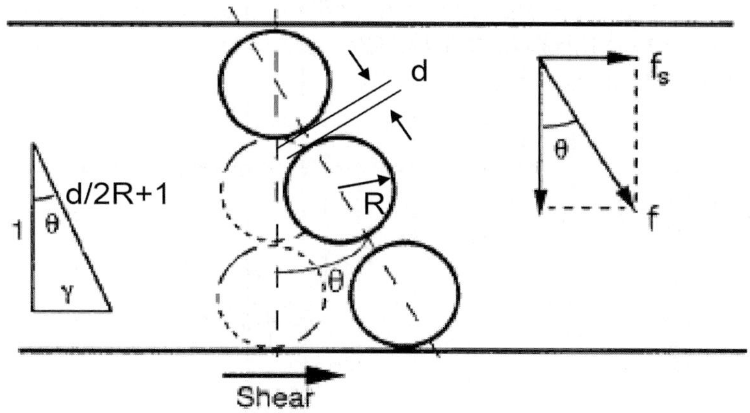

Figure 15 Schematic illustration of the fibrillated chain structure of an ER fluid under shear.

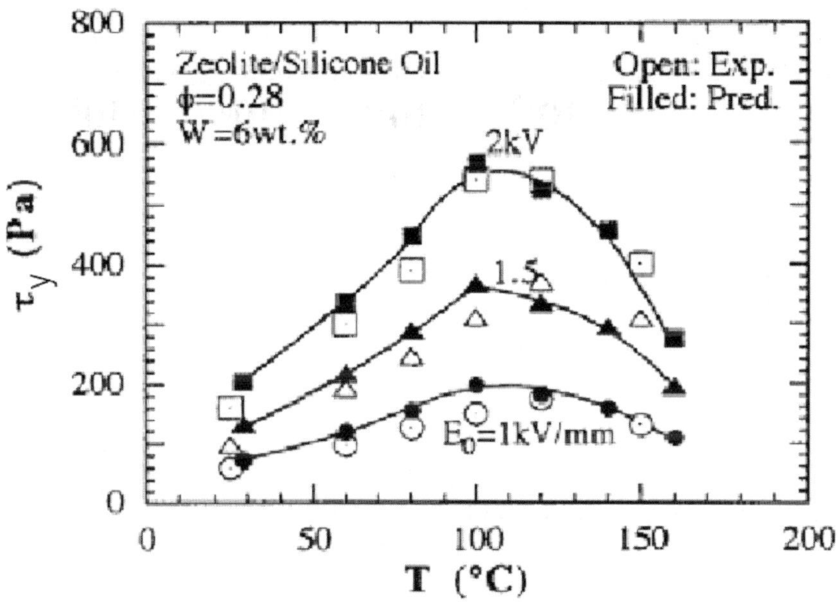

Figure 16 Comparison of the predicted yield stresses with that measured data from ref. [H. Conrad, and Y. Chen, Progress in Electrorheology, K O Havelka and F E Filisko, ed., New York: Plenum, pp 55–86, 1995] for a suspension of zeolite particles in silicone oil at different temperatures and applied fields. Reproduced with permission from C. Wu, and H. Conrad, J. Phys. D: Appl. Phys. 29 (1996) 3147.

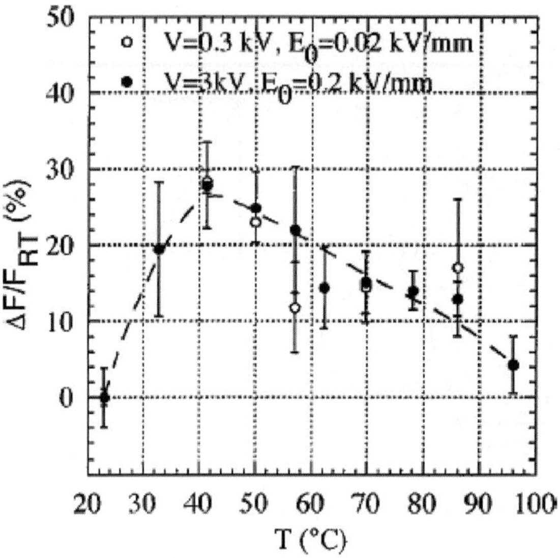

Figure 17 Percentage variation of the force ΔF/F(compared with the room temperature) vs. temperature. Measurements were performed at two electric field, E=0.02 kV/mm, and E=0.2 kV/mm. Error bars are representatives of the maximum and minimum force values obtained for four successive measurements in 5 min.. Reproduced with permission from P. Gonon, and J.-N Foulc, J. Appl.Phys. 87(2000)3563.

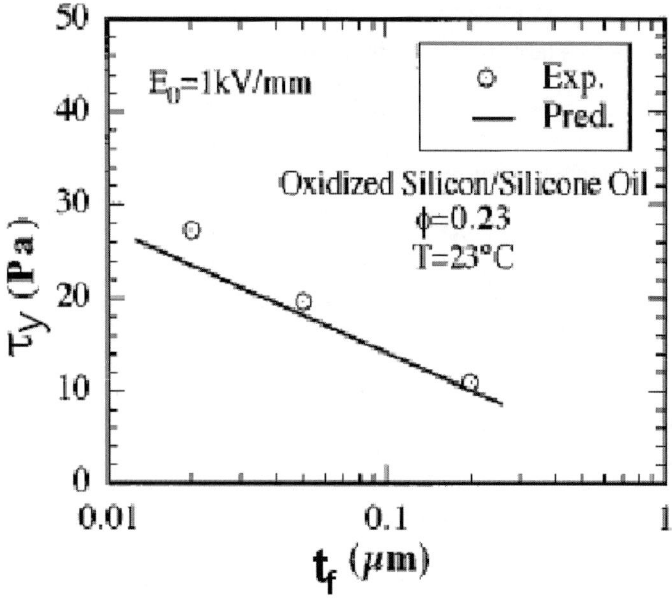

Figure 18 The yield stress of oxidized silicon particle/silicone oil vs. the thickness of the surface oxidized film. The particle volume fraction is 0.23, and the applied electric field is 1 kV/mm. Reproduced with permission from. C. Wu, and H. Conrad, J. Appl.Phys., 81(1997)383.

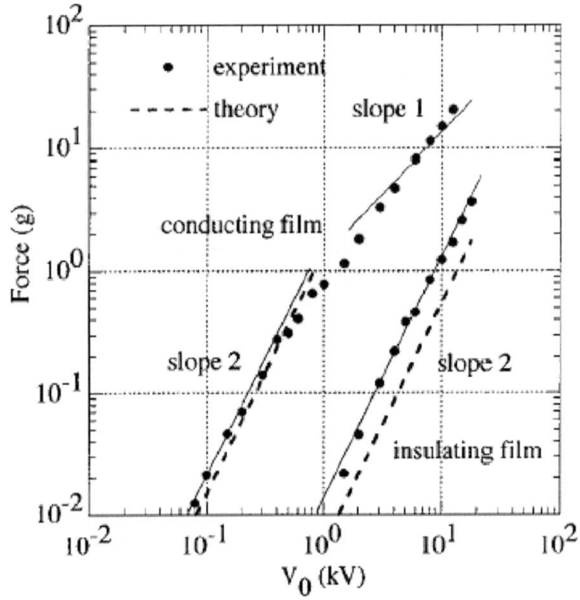

Figure 19 The interaction force of two coated polystyrene semispheres immersed in the transformer oil vs. the applied electric voltage V_0 for the spheres coated with the conducting (~10^{-6} S/m) and the insulating (~10^{-12} S/m) film. The dots are experimental points, and the dashed and solid lines are the theoretical calculation from Eq. (43) and the square or linear fits, respectively. Reproduced with permission from P. Gonon, J.-N Foulc, P. Atten, and C. Boissy, J. Appl. Phys., 86(1999)7160.

The conduction model is thought to be only valid for ER suspensions in reaction with dc or low frequency ac fields. For high frequency ac fields, the polarization model is dominant [55,56]. As shown in Eq. (25) and (26), once the Wagner-Maxwell polarization is taken into account, the parameter β is determined by the conductivity mismatch in dc or low frequency ac fields, and by the dielectric mismatch in high frequency fields (the low or high frequency is relative to the relaxation time of the Wagner-Maxwell polarization). The parameter β in the conduction model is:

$$\beta = \frac{\sigma_p - \sigma_m}{\sigma_p + 2\sigma_m} \tag{44}$$

So that when $\sigma_p < \sigma_m$, $\beta < 0$. In this case the dipole moment is opposite to the applied electric field, and the particles cannot form chains between two electrodes. A negative ER effect is expected in this case[57,58], and both the Teflon and PMMA particle/oil systems support this prediction.

The conduction model can successfully explain ER phenomena that are unexplainable by the polarization model. It could predict the current density, the yield stress and the temperature dependence of ER suspensions. However, as indicated in ref.[59], the conduction model can only be used for the situation where the suspension microstructure has been fully formed. The conduction model only considers the particle interaction, regardless of the microstructure change after an electric field is applied. It therefore could not give an explanation of the dynamic phenomena, such as the response time of ER fluid. More important, some experimental results provide evidence against this mechanism. For example, a magnesium hydroxide/poly(methylphenylsiloxane) suspension should have exhibited a positive ER effect according to the conduction model, however, it displays an obvious negative ER effect [60]. Figure 20 shows the viscosity of magnesium hydroxide (conductivity 5.8×10^{-7} S/m)/silicone oil (conductivity 1.0×10^{-12} S/m) vs. the electric field. A negative ER effect was clearly observed, which contradicts with the prediction from the conduction model.

The fatal shortcoming of the polarization and conduction models is that both of them are static, and do not take dynamic processes occurring in ER fluids into account. Khusid [59] considered dynamic events in ER fluids and examined the effects of the conductivity on both the field-induced particle aggregation process and the interfacial polarization process. An excellent qualitative theory was derived, and it is much more powerful than the polarization and conduction models, though some discrepancies with the experimental results still exist. This is because Khusid's two presumptions are not always valid in ER fluids. Both dispersed particles and the liquid medium were assumed to have no intrinsic dielectric dispersion, and the variation of the applied electric field was assumed to be very slow compared with the polarization rate, as stated earlier.

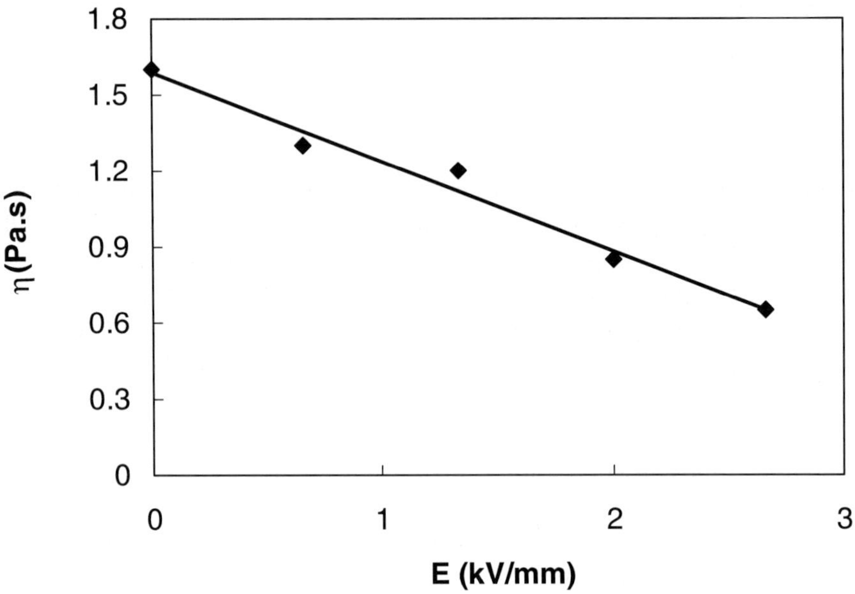

Figure 20 The viscosity of magnesium hydroxide/silicone oil vs. the electric field. The particle weight fraction is 30 wt%. Redrawn from ref. J. Trlica, O. Quadrat, P. Bradna, V. Pavlinek, and P. Saha, J. Rheol., 40(1996)943

There are other models available [61,62], however, none of them, including the polarization and conduction model, could explain all the current ER phenomena. They all suffer from a severe limitation: They cannot predict the yield stress based on the physical properties of ER suspension components and on the operating conditions (field strength, temperature, frequency, etc.). They could not provide a clear clue or implication on how to formulate a good ER suspension.

6 DIELECTRIC LOSS MODEL

Hao [63-65] proposed a dielectric loss model to explain the ER phenomena on the basis of experimental findings. Two dynamic processes were emphasized in this model. The first is the particle polarization process, in which the particle dielectric constant is dominant. The second is particle turning, i.e., the polarized particle should have the capability to align along the direction of the electric field, see Figure 21. This step was determined by the particle dielectric loss. The second step is the most important one, which distinguishes the ER particle from the non-ER particle. In other words, both the ER particle and non-ER particle can be polarized under an electric field, however, only the ER particle can re-orientate along the electric field direction, building up the fibrillated bridges between two electrodes. The non-ER particle does not have such ability. The possible reason is that the ER particle has a comparatively high dielectric loss tangent, around 0.1 at 1000 Hz, which may generate a large amount of bounded surface charge. Both the large particle dielectric loss and large dielectric constant are found, experimentally and theoretically, to be very important for the ER effect. The interfacial polarization is also found to be crucial. What is the relationship between the interfacial polarization and the particle dielectric constant and loss? Why can ER particulates form a fibrillated structure? Before those questions are answered, the issue where the dielectric constant and loss come from should be addressed.

As shown in Figure in Chapter 7, the total polarization of a heterogeneous system, P, can be expressed as:

$$P = P_{EL} + P_I + P_D + P_A + P_E \tag{45}$$

Correspondingly, the total dielectric constant,ε, can be expressed as,

$$\begin{aligned}\varepsilon &= \varepsilon_s + \varepsilon_{EL} + \varepsilon_I + \varepsilon_D + \varepsilon_A + \varepsilon_E \\ &= \varepsilon_s + \varepsilon_{EL} + \varepsilon_I + \varepsilon_D + \varepsilon_\infty\end{aligned} \tag{46}$$

where ε_{EL}, ε_I, ε_D, ε_A and ε_E are artificially regarded as the dielectric constant induced by the electrode, the interfacial, the Debye, the atomic and the electronic polarizations, respectively. ε_s is the static dielectric constant, $\varepsilon_\infty = \varepsilon_A + \varepsilon_E$, high-frequency dielectric constant. As stated before, for solid particulate materials, the dipole orientation contribution could be negligible, thus Eq. (46) can be written as:

$$\varepsilon = \varepsilon_s + \varepsilon_{EL} + \varepsilon_I + \varepsilon_\infty \tag{47}$$

Considering that most ER suspensions are non-aqueous systems that do not contain any charging agent, the contribution from the electrode polarization could be negligible, too. Eq. (47) can be further simplified as:

$$\varepsilon = \varepsilon_s + \varepsilon_I + \varepsilon_\infty \tag{48}$$

As we know, the dielectric constant and the dielectric loss are not independent, and the dielectric loss originally results from the slow polarization, i.e., the interfacial polarization in this case. For an ER suspension of a large dielectric loss tangent it probably means that the proportion of ε_I to ε is large. Obviously, the dispersed particle of a large dielectric loss may definitely result in a large dielectric loss of the whole suspension, then a large interfacial polarization. So the interfacial polarization physically stems from the dielectric loss of dispersed particles. There should be no obvious interfacial polarization if the particulate material does not have an appreciable dielectric loss.

One typical feature of the ER fluid is that the ER particle can fibrillate between two electrodes, however, the non-ER particle is unable to do so. At primary stage, the ER particle and the non-ER particle have an almost same microstructure—the particles randomly distribute in the medium, or stochastically form some clusters. After exposure under an external electric field, why do they behave quite differently? The major difference between the ER particle and the non-ER one is that the dielectric loss tangent of many ER particle materials is comparatively high, about 0.10 at 1000 Hz., i.e., the ER suspension usually will have a large interfacial polarization. Why does the large interfacial polarization produce the difference between the ER particle and the non-ER particle? Since the interfacial polarization is originally associated with the bounded surface charges, one has to presume that the large amount of surface charges can make the ER particles turn along the direction of an external electric field. As for the non-ER particles, however, they can not turn due to the shortage of surface charges. Although they still can be polarized, the total interparticle force should be canceled out owing to the diversity of particle dipole vectors, as shown in Figure 21.

As stated earlier, the surface charge is associated with the material dielectric loss, and can be produced through the following mechanisms [66, 67]: (i) the Debye polarization, associated with the dipole orientation; (ii) the interfacial polarization (Maxwell-Wagner polarization), associated with the displacement of the charge carriers over a microscopic distance; (iii) leakage conduction, associated with the formation of space charges. The two former effects do not involve any transfer of charge carriers between the dielectric and the electrodes, and would contribute to the dielectric absorption, and the last one would scale the space charge movement between the dielectric and the electrodes. So the surface net charge of a dielectric can be determined by the two former effects. Because the Debye rotation can be negligible in ER systems, the surface net charge can only be produced by the interfacial polarization and can be calculated through dc absorption current measurement, as shown earlier.

The next question is whether the strong interfacial polarization can induce particle tuning under an electric field. Hao [63] experimentally proved that even under a very weak electric field, $E_{cr.}$=12V/mm, the particle can turn to align along the direction of the applied electric field if a large interfacial polarization is generated. The ER particle turning under an electric field also was directly detected by using the X-ray diffraction method [68,69]. In Wen's work, the ER sample consists of silicon oil containing $BaTiO_3$ single-crystal spheres with a radius of 20 μm. These ferroelectric microspheres were carefully fabricated for the experiment so that their permanent dipoles are all polarized in the same direction. The relative difference of the X-ray diffraction amplitude without (I_n at E= 0 kV/mm) and with an electric field (E=2.5 kV/mm, I_j,) of the B_aTiO_3 crystal

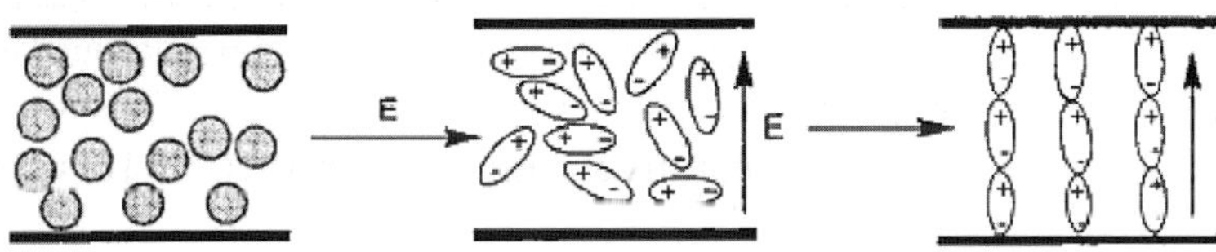

(a)

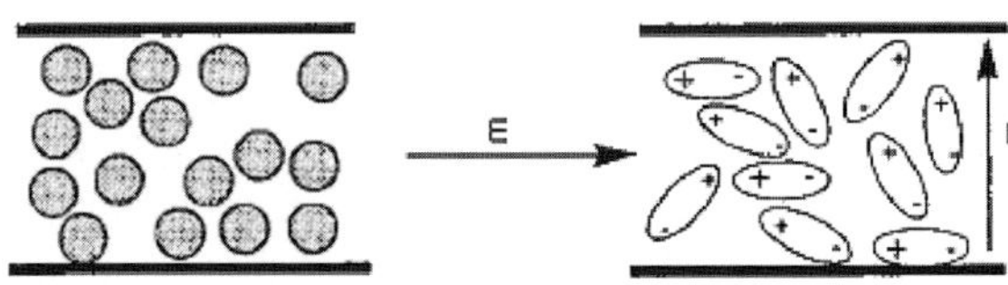

(b)

Figure 21 Schematic illustration of the behaviors of ER particles and the non-ER particles behaviors before and after an external electric field is applied. (a) ER particle; (b) non-ER particles. Reproduced with permission from T. Hao, A. Kawai, F. Ikazaki, Langmuir, 14(1998)1256 microspheres with a volume fraction of 31.5% are shown in Figure 22. Clearly, the diffraction from the (102) and (104,110) planes increases dramatically, while diffraction in the high-angles range remains unchanged, implying that the particles are re-orientated in the direction of the applied electric field. A further careful study on dielectric particle turning under an electric field was carried out by Lan [69]. In Lan's study, single crystal TGS (triglycine sulfate $(NH_3CH_2COOH)_3{\cdot}H_2SO_4)$ particles of 30 micron in diameter were mixed with wax with the particle volume fraction 10%. The reason that TGS particles were chosen is that TGS has a spontaneous polarization moment in the (010) direction at temperatures below 50 °C and becomes paraelectric without spontaneous polarization at temperatures above 50 °C [70]. Two kinds of waxes of different melting points, 46 °C for wax 1 and 58 °C for wax 2, were thus used for mixing with the TGS particle. There were three samples prepared: Two TGS/wax 1 mixtures and one TGS/wax 2 mixture, at the condition above the melting temperature for both waxes. One (sample *A*) of the TGS/wax 1 samples was cooled down to room temperature in the absence of an electric field, preserving the random distribution of TGS particle structure. Another TGS/wax 1 (sample *B*) and TGS/wax 2 (sample *C*) were cooled down under an ac field of 1500 V/mm, preserving the induced chains or columns structure in sample *B* and sample *C*.

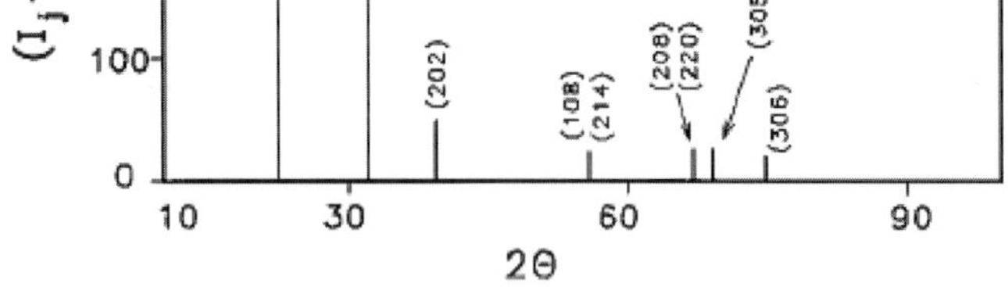

Figure 22 The relative difference of the x-ray diffraction amplitude without (I_n at E= 0 kV/mm) and with an electric field (E=2.5 kV/mm, I_j,) of the B_aTiO_3 crystal microspheres with a volume fraction of 31.5%. Reproduced with permission from W. Wen and K. Lu, Appl. Phys. Lett. 68(1996)1046

Those three solid samples were cut into pieces for x-ray-diffraction measurements at room temperature. The x-ray-diffraction spectra of samples *A, B*, and *C* are shown in Figure 23, in which the diffraction peaks of the solid wax were subtracted. For sample A prepared without an applied electric field, the dipoles can point in any direction, and the particles are randomly distributed without any preferred orientations. Thus the diffraction pattern of the sample A can be regarded as a normal powder diffraction pattern. For the sample *C*, TGS has no permanent dipole moment at temperature above 50 °C. The particles form the chain structure solely due to the induced dipole moment at high temperatures. Once the temperature goes below the Curie temperature, 50 °C, the permanent dipole moment comes back again, and the interaction between the permanent dipole moment cannot change the chain structure due to the solidification of the whole system. Thus the (010) plane of particles, the direction of the permanent dipole, is distributed randomly, and there is no preferred orientation in the diffraction pattern of the sample C, which is almost the same as that of sample *A*. The sample *B* was made at an electric field at 48 °C below the Curie temperature. At this condition the TGS particles of permanent dipole moment must interact with the external field, and the permanent dipoles should align along with the external electric field. Therefore, the diffraction pattern shows that the intensities of the (020) and (040) diffraction peaks increase dramatically, while that of (200) decreases in comparison with that of sample *A* and *C*. The relative intensity ratio of the diffraction peaks between sample *B* and sample *A* is plotted in Figure 24. I_A *and* I_B represent the x-ray-diffraction intensities of the sample A (E=0 kV/mm) and *B* (E=1.5 kV/mm). It is clear that the amplitudes of (020) and (040) increase dramatically under an external electric field, i.e., the preferred orientation of particles is in the direction of (010) along the electric field. For sample C, the particles can be polarized but cannot turn due to the lack of spontaneous polarization. The x-ray diffraction patterns of the microspheres of both $BaTiO_3$ and TGS single crystal are clear evidence that the particle does turn under an electric field for forming the orientated chain structure.

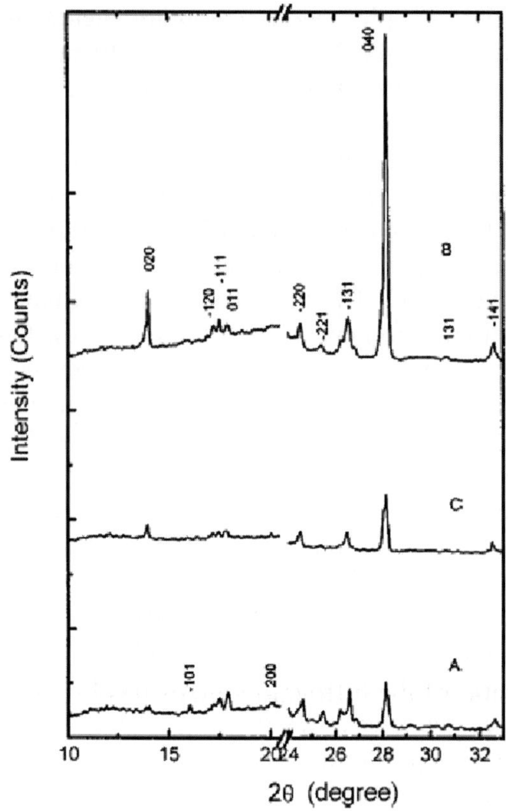

Figure 23. The x-ray-diffraction spectra of TGS (triglycine sulfate $(NH_3CH_2COOH)_3{\cdot}H_2SO_4)$/wax samples A (E=0 kV/mm), B(E=1.5kV/mm), and C(E=1.5 kV/mm). Reproduced with permission from Y. Lan, X. Xu, S. Men, and K. Lu, Phys. Rev. E., 60 (1999) 4336

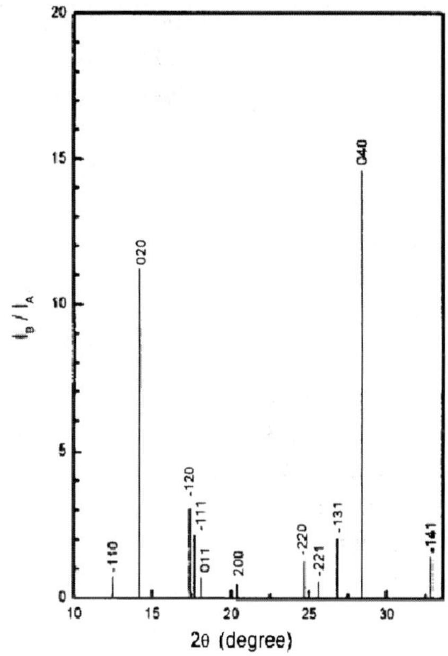

Figure 24 The ratio of the diffraction intensities between single crystal TGS (triglycine sulfate $(NH_3CH_2COOH)_3\ H_2SO_4$)/wax sample *B* and sample *A*. Reproduced with permission from Y. Lan, X. Xu, S. Men, and K. Lu, Phys. Rev. E., 60 (1999) 4336.

Obviously, once the ER particles turn along the direction of an electric field, the interparticle force will be mainly determined by how large the particles are polarized, i.e., the dielectric constant of particle. The dielectric loss mechanism therefore assumes that the ER effect should contain two steps: The first step is the particle turning along the direction of the applied electric field; the second is the particle binding together due to the polarization. The first step should be controlled by the dielectric loss of the dispersed particles, and the second step would be controlled by the dielectric constant. Both the dielectric loss and the dielectric constant are therefore important for the ER effect. Filisko [71,72] doubted the function of the interfacial polarization in the ER response based on the fact that the ER effect disappears when the water is removed even though the interfacial polarization still remains. Since the interfacial polarization and the large dielectric loss of dispersed particle are physically equal, as analyzed above, this fact can be attributed to the great decrease of the particle dielectric loss in the dry state and then the vanishing of the interfacial polarization.

For understanding why a large dielectric loss is experimentally found necessary for the ER response, a theoretical approach was developed on the assumption of the interfacial polarization responsible for the ER effect [64]. Since the ER fluid changes from a liquid state (the particle randomly distributed in the liquid medium) to a solid state (fibrillated chain of bct lattice), the entropy of the ER system may dramatically decrease. Based on this fact, Hao [64] theoretically came to the conclusion that the particle dielectric loss tangent maximum value should be larger than 0.10, which agrees well with the empirical criteria put forward earlier [73] . The criteria for the positive ER effect and negative ER effect were derived on the basis of this fact also. Those criteria are presented in the preceding chapter, and for emphasis they are shown here again under the dielectric loss mechanism umbrella. According to Hao [64] If the static dielectric constants of the liquid medium and the particle are ε_{sm}, and ε_{sp}, respectively, a strong positive ER effect will occur if $\varepsilon_{sp} > 4\varepsilon_{sm}$ and $\frac{d\varepsilon_{sp}}{dT}>0$. A weak or no ER effect is expected if $\varepsilon_{sp} > 4\varepsilon_{sm}$, $\frac{d\varepsilon_{sp}}{dT}<0$, and Eq.(49) is satisfied; A negative ER effect is anticipated if $\varepsilon_{sp} > 4\varepsilon_{sm}$, $\frac{d\varepsilon_{sp}}{dT}<0$, and Eq.(50) is satisfied.

$$\frac{d\varepsilon_{sp}/dT}{d\varepsilon_{sm}/dT} < \frac{(1+3\Phi)(2\varepsilon_{sm}+\varepsilon_{sp})^3 + 54\Phi^2\varepsilon_m(\varepsilon_{sp}^2 - 1.5\varepsilon_{sm}\varepsilon_{sp} - \varepsilon_{sm}^2)}{27\Phi^2\varepsilon_{sm}^2(\varepsilon_{sp} - 4\varepsilon_{sm})} \tag{49}$$

$$\frac{d\varepsilon_{sp}/dT}{d\varepsilon_{sm}/dT} > \frac{(1+3\Phi)(2\varepsilon_{sm}+\varepsilon_{sp})^3 + 54\Phi^2\varepsilon_m(\varepsilon_{sp}^2 - 1.5\varepsilon_{sm}\varepsilon_{sp} - \varepsilon_{sm}^2)}{27\Phi^2\varepsilon_{sm}^2(\varepsilon_{sp} - 4\varepsilon_{sm})} \tag{50}$$

If $\varepsilon_{sp} < 4\varepsilon_{sm}$, physically $\frac{d\varepsilon_{sp}}{dT}$ would unlikely be less than zero, thus a weak or negative ER effect would become possible at this condition. An interesting experimental result is related to the poly(methyl methacrylate) (PMMA)dispersions stabilized by polystyrene-*block*-poly- (ethylene-*co*-propylene) in decane [74]. This suspension could switch from positive ER effect to the negative ER effect via merely varying the particle volume fraction. Eq. (49) and (50) do indicate that the particle volume fraction could have such a profound impact on the ER effect.

A yield stress equation was also derived on the basis of the dielectric loss mechanism, as described in the preceding chapter. Under the assumption that only interfacial polarization would contribute to the ER effect and the ER particle would form the bct structure under an electric field, a yield stress equation could be expressed in Eq. (69) or Eq. (70) in Chapter 8, which obviously indicates that the yield stress of an ER fluid would increase with the square of the applied electric field, the particle volume fraction and the dielectric constant of the liquid medium. Those predictions agree very well with previous experimental results [75-77].

Note that the yield stress equation contains an important parameter p. If p is positive, the yield stress will increase with ξ, which is consistent with the prediction given by the polarization model [35,39]. However, if p is negative, then the yield stress will decrease with ξ, which can not be explained by the polarization model. The parameter p only becomes positive when the dielectric loss tangent of the dispersed solid material is larger than 0.1 [64,73,78].

REFERENCES

[1] W.M.Winslow, J. Appl.Phys. 20 (1949)1137
[2] T. C. Jordan and M.T.Shaw, IEEE Trans. Electric Insul., 24(1989)849
[3] H.A. Pohl, Dielectrophoresis: The Behavior of Neutral Matter in Non-uniform Electric Fields, Cambridge Univ. Press, Cambridge, 1978
[4] N.I.Gamayunov and V.A.Murtsovkin, J.Eng. Phys., 43(1982)963
[5] T. Hao, Y. Chen, Z. Xu, Y. Xu, Y. Huang, Chin. J. Polym.Sci., 12(1994)97
[6] Yu.F.Deinega and G.V. Vinogradov, Rheo.Acta, 23(1984)636
[7] W.Wen and K. Lu, Phys. Fluids, 8(1996)2789
[8] W.Wen and K. Lu, Phys. Fluids, 9(1997)1828
[9] D.L.Klass, T.W. Martinek, J. Appl.Phys. 38(1967)67
[10] D.L.Klass, T.W. Martinek, J. Appl.Phys. 38(1967)75
[11] H. Uejima, Jpn. J. Apply. Phys. 11(1972)319
[12] J.E. Stangroom, Phys. Technol., 14(1983)290
[13] Y.D. Kim, D.J.Klingenberg, J. Colloid Interface Sci., 183(1996)568
[14] H. See, H.Tamura, M.Doi, J.Phys. D: Appl.Phys. 26(1993)746
[15] A.P. Alice and C.F. Zukoski, Adv. Colloid Interf. Sci., 30(1989)153
[16] C.F. Zukoski, Annu. Rev. Mater.Sci., 23(1993)45
[17] M. Parthasarathy and D. J. Klingenberg, Mat. Sci. Eng., R17(1996)7
[18] D. J. Klingenberg, MRS Bulletin, 23(1998)30
[19] J.B. Hayter and R. Pynn, Phys. Rev. Lett., 49(1982)1103
[20] L.S. Ornstein and F. Zernike, Proc. Akad. Sci. (Amsterdam), 17(1914)793
[21] P.M. Adriani and A.P.Gast, Phys. Fluids, 31(1988)2757

[22] D. Brooks, J.Goodwin, C.Hjelm, L.Marshall, and C.Zukoski, Colloids and Surfaces, 18(1986)293
[23] P.M. Adriani and A.P.Gast, J. Chem. Phys. 91(1989)6282
[24] K.L.Smith,and G.G. Fuller, J. Colloid Interf. Sci., 155(1993)183
[25] M. Parthasarathy, D. J. Klingenberg, Mat. Sci. Eng., R17(1996)7
[26] C.A. Coulson, Electricity, Wiley-interscience, New York, 1961
[27] A.F.Sprecher, J.D. Carlson, and H. Conrad, Mat. Sci. Eng., 95(1987)187
[28] Y. Chen, A.F. Sprecher, and H. Conrad, J. Appl.Phys.70(1991)6796
[29] R.A. Anderson, Langmuir, 10(1994)2917
[30] H.J.H. Clercx, and G. Bossis, Phys. Rev.E, 48(1993)2721
[31] R.T. Bonnecaze, and J.F. Brady, J. Chem. Phys., 96(1992)2183
[32] D. J. Klingenberg, F. van Swol, and C. F. Zukoski, J. Chem. Phys., 94(1991)6170
[33] D.K. Ross, Aust. J. Phys., 21(1968)817
[34] Coulomb's Law Committee, Am. J. Phys. 18 (1950)1
[35] L.C. Davis, Appl.Phys. Lett., 60(1992)319
[36] L.C. Davis, J. Appl.Phys., 72(1992)1334
[37] R.Tao, Q.Jiang, and H.K.Sim, Phys. Rev. E, 52(1995)2727
[38] T.Hao, Adv. Colloid Interface Sci., 97(2002)1
[39] D.J.Klingenberg and C.F. Zukoski IV, Lanmguir, 6(1990)15
[40] R.Tao, J.T.Woestman, and N.K.Jaggi, Appl.Phys. Lett., 55(1989)1844
[41] O.Otsuba and K.Watanabe, J.Soc., Rheol. Jpn., 18(1990)111
[42] C.F.Zukoski, Annu. Rev. Mater. Sci., 23(1993)45
[43] T.Garino, A.Adolf, B.Hance, Proc. Int. Conf.on ER Fluids, R.Tao, Ed.,World Scientific, 1992, p.167
[44] P. J. Rankin, D. J. Klingernberg, J.Rheol. 42(1998)639
[45] D.J.Klingenberg, MRS Bulletin, 23(1998)30
[46] P.Atten, J-N.Foulc, and N. Felici, Int. J. Mod. Phys. B, 8(1994)2731
[47] J-N. Foulc, P. Atten, N. Felici, J. Electrostatic, 33(1994)103
[48] X. Tang, C. Wu, H. Conrad, J.Rheol. 39(1995)1059
[49] C. Wu, H. Conrad, , J. Phys. D: Appl. Phys. 29(1996)3147
[50] C.Wu and H. Conrad, J. Appl. Phys., 81(1997)383
[51] L.Onsager, J.Chem.Phys. 2(1934)599
[52] C.Wu, and H. Conrad, J. Appl. Phys., 81(1997)8057
[53] H. Conrad, and Y. Chen, Progress in Electrorheology, K O.Havelka and F E Filisko, ed., New York: Plenum, pp 55–86, 1995
[54] P. Gonon, and J.-N Foulc, J. Appl.Phys. 87(2000)3563
[55] P. Gonon, J.-N Foulc, P.Atten, and C. Boissy, J. Appl. Phys., 86(1999)7160
[56] C. Wu, and H. Conrad, Phys. Rev. E, 56(1997)5789
[57] C. Wu, and H. Conrad, J. Rheol., 41(1997)267
[58] C. Boissy, P.Atten, and J.-N. Foulc, J. Electrostatics, 35(1995)13
[59] B. Khusid and A. Acrivos, Phys. Rev. E, 52(1995)1669
[60] J. Trlica, O.Quadrat, P.Bradna, V.Pavlinek, and P.Saha, J. Rheol., 40(1996)943
[61] H.See, T.Saito, Rheol.Acta 35(1996)233
[62] H. Ma, W.Wen, W.Y.Tam, P.Sheng, Phys. Rev. Lett, 77(1996)2499
[63] T. Hao, A. Kawai, F. Ikazaki, Langmuir, 14(1998)1256
[64] T. Hao, A. Kawai, F. Ikazaki, , Langmuir, 15(1999)918

[65] T. Hao, A. Kawai, F. Ikazaki, , Langmuir, 16(2000)3058
[66] B.Gross, J. Chem. Phys. 17(1949)866
[67] C.P. Smyth,. Dielectric Behavior and Structure, McGraw-Hill, New York, Toronto, London, 1955; p 201
[68] W.Wen, and K.Lu, Appl. Phys. Lett. 68(1996)1046
[69] Y. Lan, X. Xu, S. Men, and K. Lu, Phys. Rew. E., 60 (1999) 4336
[70] T. Mitsui, I. Tatsuzaki, and E. Nakamura, An Introduction to the Physics of Ferroelectrics, Gordon and Breach, Tokyo, 1974
[71] F.E. Filisko, In Progress in Electrorheology; K.O. Havelka, F.E.Filisko, Eds., Plenum Press: New York, 1994; p 3.
[72] A.W. Schubring, and F.E. Filisko, In Progress in Electrorheology; K.O. Havelka, F.E.Filisko, Eds., Plenum Press: New York, 1994, p215
[73] T.Hao, Z.Xu, and Y.Xu, J. Colloid Interface Sci., 190(1997)334
[74] V. Pavlinek, P. Saha, O. Quadrat, J. Stejskal, Langmuir, 16(2000)1447
[75] H. Block, J.P. Kelly,A. Qin, T. Watson, Langmuir, 6(1990)6;
[76] T. Hao, Y. Chen, Z. Xu, Y. Xu, and Y. Huang, Chin. J. Polym. Sci., 12(1994)97
[77] L. Marshall, C. F. Zukoski, J.W. Goodwin, J. Chem. Soc. Faraday Trans. 85(1989)2785
[78] T.Hao, Appl. Phys. Lett., 70(1997)1956

Chapter 10

Applications of electrorheological fluids

An unique feature of ER suspensions is that mechanical strength can be continuously and reversibly adjusted from the liquid to solid state by simply adjusting the applied electric field strength. The ER suspension is therefore a perfect electro-mechanical interface, stimulating a great deal of interest in a wide range of industrial areas. Since the positive ER effect was reported almost 60 years ago, ER devices have been designed mainly on the basis of the positive ER effect. The applications of the positive ER effect are thus focused upon in this chapter.

The requirements for ER fluids applied to automotive devices including engine dampers, vehicle shock absorbers, alternator variable speed drives, and engine accessory variable speed drives, are discussed in ref. [1]. Basically, ER fluids should be nontoxic to humans and the environment, chemically durable and physically stable without particle settling issues, low power consumption and manufacturing cost, wide working temperature range, preferably between -40 to 200 ^{o}C, compatible with sealing materials, nonabrasive and noncorrosive to the device, and rapid on/off characteristics. The key physical parameters are the maximum viscosity at off-electric field state, the minimum yield stress at the maximumlly applicable field, and the maximum current density, which determines the power consumption.

The application areas can be categorized into three main fields. a) ER fluid is used as a smart medium of adjustable mechanical strength for controlling other devices such as car clutches, fluid brakes, shock adsorbers, damping devices, grippling devices, etc.; b) Instead of functioning as a mechanical strength transferring medium, the ER fluids are used as a sensor for monitoring the mechanical vibration signal; c) the ER effect rather than the unique mechanical strength of ER fluids is used for fabricating specially patterned or anisotropical materials such as phontonic crystals, polymer membrance for fuel cell batteries, etc.

1. MECHANICAL FORCE TRANSFERRING AND CONTROLLING DEVICES

The ER clutches, brakes, damping devices, hydraulic valves, shock absorbers, and robotic controlling systems [2-7] gripping devices [8], seismic controlling frame structures [9], human muscle stimulator [10,11], and spacecraft deployment damper [12] have been proposed. The basic working mechanisms of some ER devices are described below.

1.1 ER valves

An ER valve is a device in which an electric field is applied to control the flow rate. In regular hydraulic devices, the pressure is the driving force for fluid to flow through the tubes. ER valve can be fabricated to be a much smaller and much faster device that may modulate the flow continuously from fully open to fully shut. The working mechanism of an ER valve is illustrated in Figure 1. An ER fluid flows through a tube from the left side to the right, and the electrodes are applied on either side of the tube. The fibrillated chains under the electric field create a resistance force to the flow, resulting in the pressure increase in the tube. According to Yoshida [13], the differential pressure ΔP caused by the viscosity increase under an electric field E can be expressed as:

$$\Delta P = \frac{64 L_{electr}}{W_{electr} d_{ER}^{3} k} \eta(E) Q = R(E) Q$$

where L_{electr}, W_{electr}, and d_{ER} are the length, width, and gap size of electrodes, respectively, Q is the flow rate, k is a parameter dependent on channel section shape. η(E) and R (E) is the viscosity and channel resistance strongly dependent on the applied electric field. A two-port ER valve of only one flow in and flow out opening was fabricated by Yoshida [13], and the relationship among the differential pressure, the flow rate, and the applied electric field is shown in Figure 2. As the flow rate and the applied electric field increase, the differential pressure also increases. Importantly, the differential pressure strongly depends on the applied electric field, reaching the purpose of controlling the flow through simply changing the electric field.

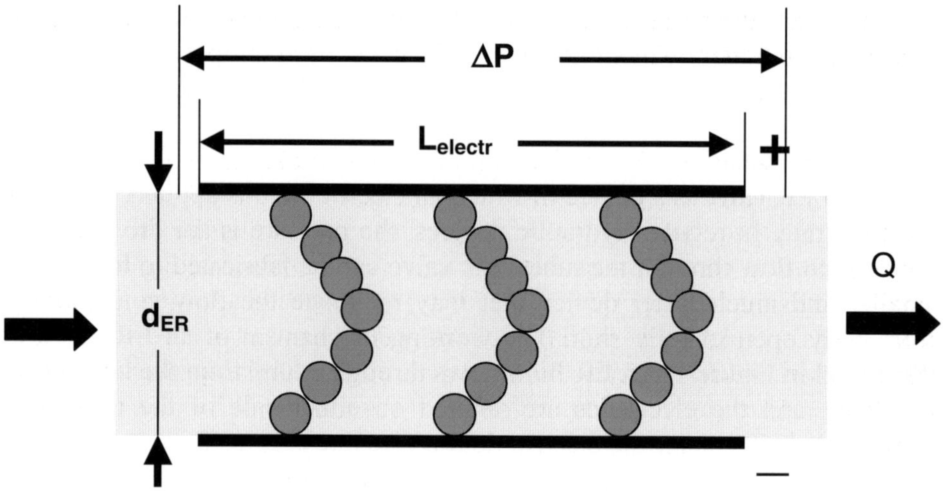

Figure 1 An illustration of an ER hydraulic valve. The electrodes are applied on the top and bottom sides of a tube through which an ER fluid is flowing and the pressure drop across the tube can be regulated via. the applied electric field strength.

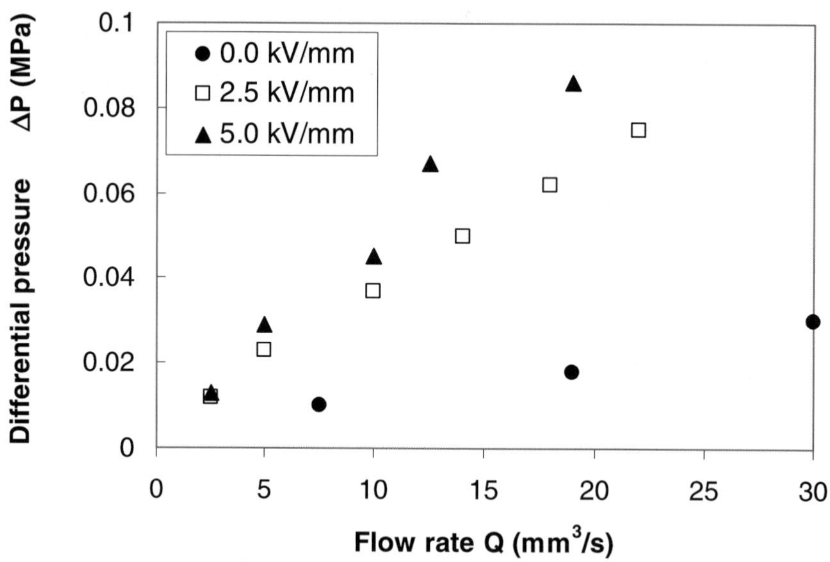

Figure 2 The differential pressure of a two-port (only one flow in and flow out opening) ER valve vs. the flow rate of the ER fluid (nematic liquid crystal) under different electric field strength. The electric gap is 0.15 mm.

The data are from K. Yoshida, M. Kikuchi, J.H. Park, and S. Yokota, Sensors Actuators, 95(2002)227.

Using such an ER valve, an anti-lock brake system (ABS) for passenger vehicles was designed by Choi [14], as shown in Figure 3. There are two ER valves, one for the braking mode and another for the releasing mode, replacing conventional solenoid valves. The ABS braking force for a small sized vehicle is approximately 10 MPa, and the size of the ER valves can be easily calculated on the basis of yield stress of the ER fluid. On various road condition tests this new type of ER brake system performs excellently, opening a possibility of new generation ABS for passenger vehicles.

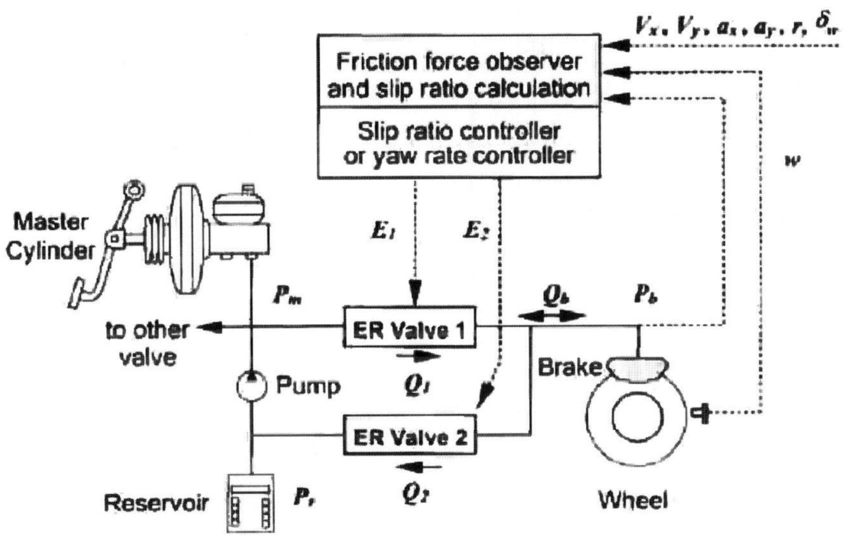

Figure 3 Anti-lock brake systems (ABS) of passenger vehicle using ER valve. Reproduced with permission from S.B. Choi, J.H. Bang, M.S.Cho, and Y.S.Lee, Proc. Inst. Mech. Engrs. Part D, 216(2002)897.

1.2 Damping devices and clutches

According to Filisko [15], the damping device employing the ER fluids can be roughly categorized as three modes: The flow mode, shear mode, and the mixed mode that combines both the flow and shear modes. These three modes are illustrated in Figure 4. In the flow mode, the piston moves up, driving the ER fluid to flow out of the chamber to the ER control gap, where an electric field is applied. The highly chained ER fluid under the electric field prevents the ER fluid flowing from the top to the bottom, thus damping the piston to move up. In the shear mode, the piston functions as a positive electrode and creates a shear field when it moves. In the meanwhile, the ER fluid will form the fibrillated chain structure between the piston and the wall of the chamber, thus directly preventing the piston from moving up. In the mixed mode, the ER fluid can flow through the channels within the piston, and the piston functions as a positive electrode, hence still creates a shear field when it moves.

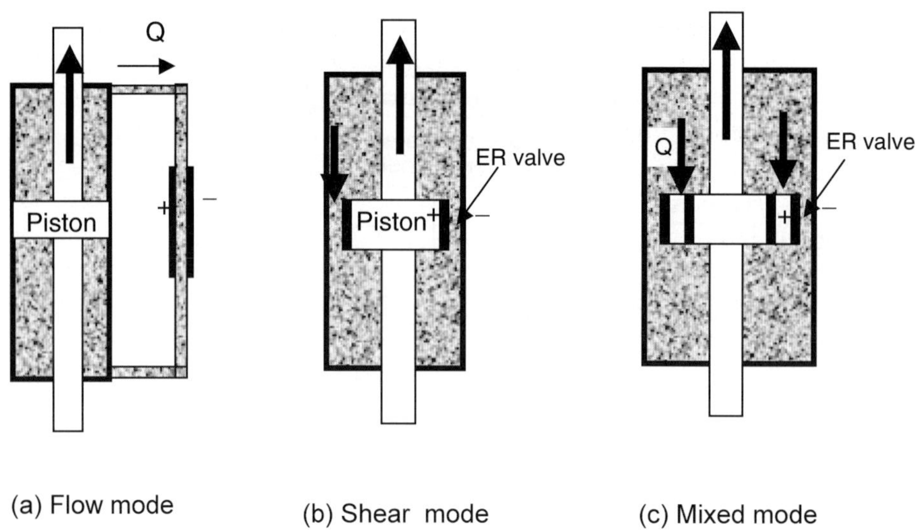

Figure 4 Illustration of three modes of ER dampers. a) flow mode; b) shear mode; and c) mixed mode.

Actual ER damping devices have been designed for meeting practical needs such as shock absorbers [16], full-car suspension system [17,18], automotive engine mounts [1], bearing damping with an ER fluid [19], etc. A concentric cylinder shock absorber using the Bayer Rheobay 3565 ER fluid was constructed and tested [16], and the experimental data are depicted in Figure 5. With the increase of the applied electric field strength the hysteresis force area increases with the displacement cycle, indicating an increase of the damping capability. In the plot of the force vs. the velocity, the pre-yield and post-yield behaviors of the ER fluid are very clear. In very small velocities the damping force almost linearly increases with the velocity, and reaches a constant value at a certain velocity that increases with the applied electric field strength. At the pre-yield region this shock absorber behaves as a viscous damper and at the post-yield region it behaves as a linear elastic damper.

There is a lot of literature addressing how to design a high-force ER damper [20-23]. The finite element analysis method is also used for addressing theoretical and practical issues related to ER dampers [24].

As for the ER clutches, the concentric cylinders and parallel plates clutches are proposed, as shown in Figure 6. In the concentric cylinder clutch the ER fluid is placed in the gap between two cylinders, and an electric field is applied across the gap. In the parallel-plate clutch the ER fluid is placed in between two plates that also serve as two electrodes. When one cylinder or plate rotates the created force will be transferred through the ER fluid to another cylinder or plate. The force transferred is determined by how strong the ER fluid is, and it thus can be adjusted by varying the applied electric field. To be used in the ER clutch the ER fluid must have a very low viscosity under zero electric field for minimizing the drag force, and a very high yield stress under an electric field for maximizing torque transfer. Several ER fluids of giant ER effect may be suitable for being used in the ER clutch [25-27]. There are plenty of ER clutches being patented [28-30]

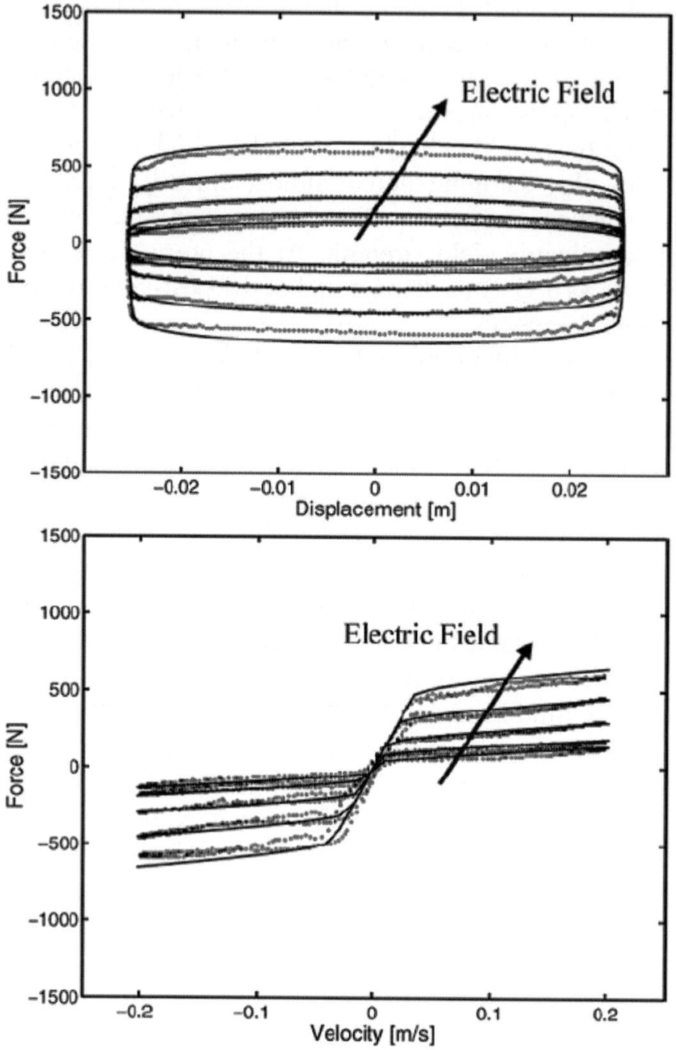

Figure 5. The damping force of a monotube ER shock absorber vs. the displacement or velocity with an electrode gap of d = 0.9 mm under an electric field varying from 0 to 3.5 kV/mm. Reproduced with permission from N.M. Wereley, J. Lindler, N. Rosenfeld, and Y. Choi, Smart. Mater. Struct. 13 (2004)743

(a)

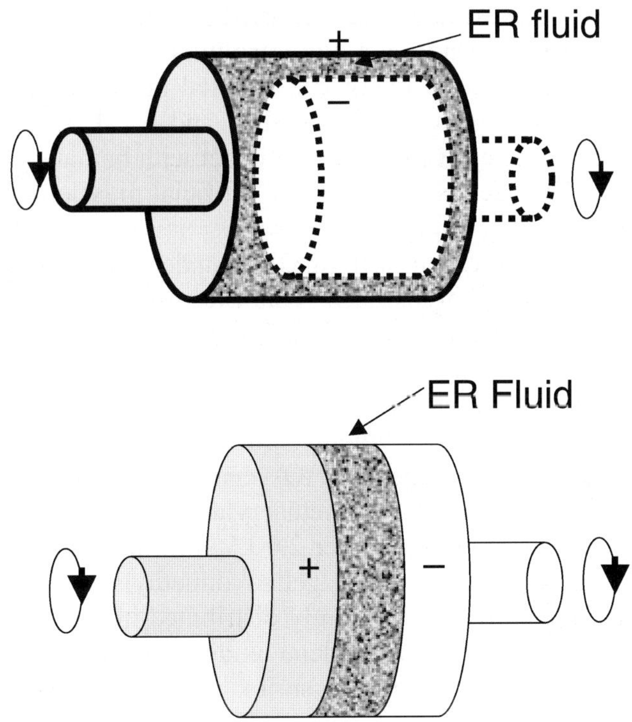

(b)

Figure 6 Illustration of the concentric ER clutch (a), and the parallel plates clutch (b).

1.3 Actuator and artificial muscle

An actuator is a device that incorporates an energy source to actuate and move certain components of the machines for realizing the automation of motion. It is the core component of the robotic and automation industry. Conventional actuators include the hydraulic, pneumatic, and electric ones, and they constitute the majority of robotic systems. The hydraulic and pneumatic actuators use the fluid to create the motion, while the electric actuator converts the electric energies to the mechanical motion. The advantages and disadvantages of those actuators can be found in ref.[31]. The ER actuator obviously can combine fluidic and electric energy together for automation purposes, and it thus has a lot of advantages, such as relatively large force, small size, and high bandwidth. A stepper motor controlled by an ER actuator was reported by Chu [32]. This stepper motor can travel in a large area with a high resolution of less than 0.36 μm on the base plate. Such a precise motor can be used in the fields such as MEMs, optical manipulator, manipulator in SEM and STM, laser adjustor, micro machining, etc. Since ER actuators can react with human and environmental stimuli, it can be potentially used as an artificial muscle or a haptic system, too.

The ER particle can still fibrillate under an electric field even the dispersing medium is polymerized. This ER system is commonly called the ER gel, which underpins the foundation for fabricating the ER artificial muscle. This type of ER gels doesn't have particle sedimentation problem that is commonly observed in regular ER systems. There are several examples reported on polydimethylsiloxane (PDMS) ER gels. Figure 7 shows a microscopic image of TiO_2 coated copolymer sphere of butyl acrylate and 1,3-butanediol dimethacrylate dispersed in hydrosilylated dimethylsilicone oil [33,34]. Under 2.0 kV/mm, this particle still forms the fibrillated chain structure in the gelled dimethylsilicone oil. Under zero electric field, the particle is randomly distributed in the gel. Another PDMS ER gel is the crosslinked polyethylene oxide dispersed in cured PDMS [35]. This kind of PDMS ER gel can shrink under an electric field, and the displacement vs. the applied electric field is shown in Figure 8. The compressional displacement increases with the applied electric field. The shrinkage takes place within 100 milliseconds, however, once the electric field is turned off, it takes from seconds up to minutes for such a gel to relax back to the original state.

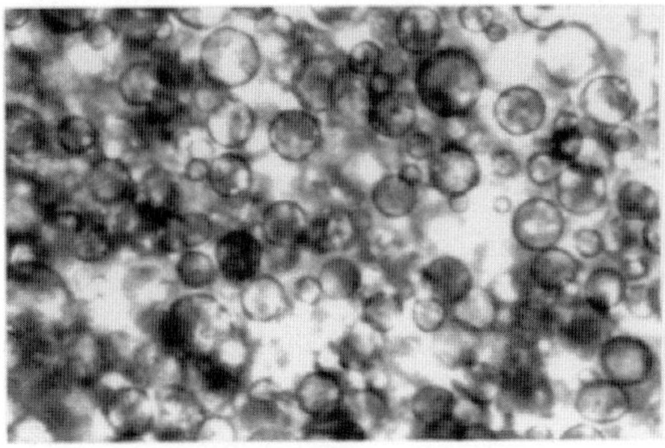

(a) $E = 0$

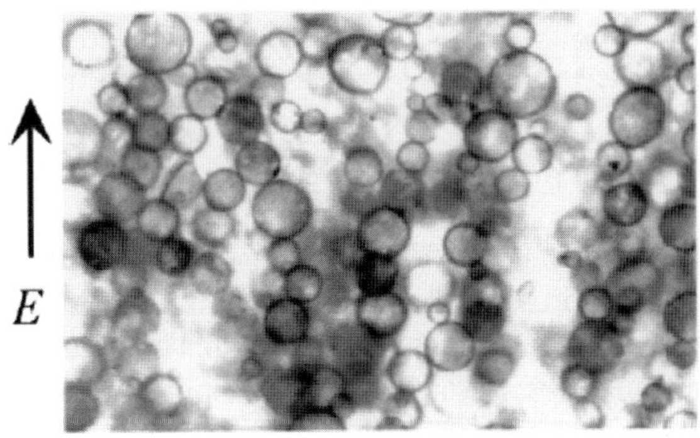

(b) $E = 2.0$kV/mm

Figure 7 The microscopic image of TiO2 coated copolymer sphere of butyl acrylate and 1,3-butanediol dimethacrylate dispersed in hydrosilylated dimethylsilicone oil. The particle diameter is about 10 μm. Reproduced with permission from R. Hanaoka, S. Tanaka, Y. Nakazawa, T. Fukami, and K. Sakurai, Electr. Eng. Jpn, 142(2003)1.

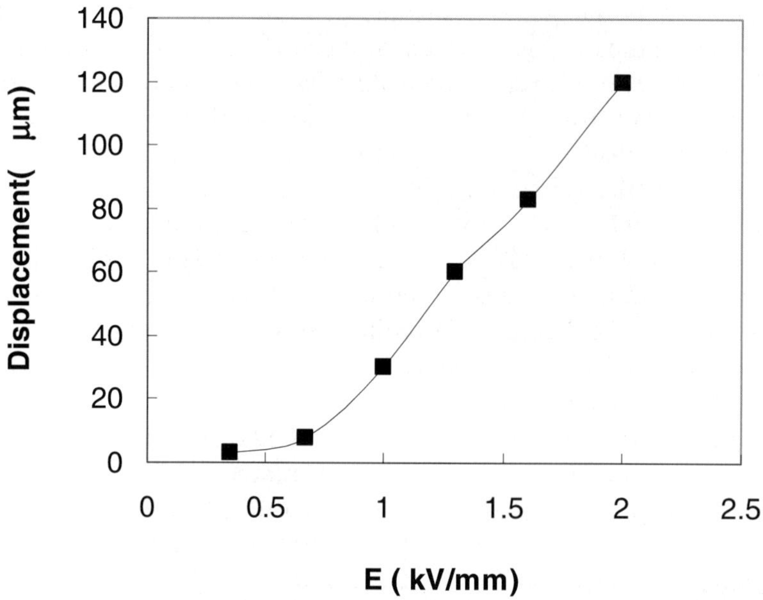

Figure 8 the compressional displacement of polyethylene oxide particle/PDMS gel vs. the applied electric field. The data are from K. Bohon, and S. Krause, J. Polym. Sci.: Part B: Polym. Phys., 36(1998)1091.

2. ER COMPOSITE MATERIALS

Through utilizing the principle of the ER effect or directly using the ER fluids, many new type materials can be manipulated with outperformed characters or specifications [35-38]. An experiment shown in Figure 9 indicates the vibration amplitude of a PMMA plate embodying a molecular sieve/silicone oil ER fluid decreases under an electric field, 0.5kV/mm. The intrinsic vibration frequency of this ER composite material was found to also shift from about 15 Hz (without an electric field) to 21 Hz at 0.5 kV/mm [35]. The ER composites thus have a potential to be used as damping materials, for example, in airplanes. A similar damping performance of a laminated composite material containing electrodes, flexible laminae that may serve as electrodes, and an ER fluid located between the laminae, was disclosed in a patent [36]. Theoretical treatment on a sandwich plate with a constraining layer and an electrorheological (ER) fluid core indicates that the increased electric field will increase the natural frequency of the sandwich plate and change the stiffness of the sandwich plate [39]. Additionally, the increase in thickness of the ER layer will decrease the resonance frequencies of the sandwich plate. Park [37] reported that a series of particulates ranging from sub-micrometers to tens of micrometers can be aligned under an electric field in photopolymerizable liquids. The particulates include silica-zirconia, glass sphere, glass fiber, aluminosilicate, barium titanate, and mica. The polymerizable liquid is urethane dimethacrylate mixed with 1,6-hexanediol dimethacrylate hardened with camphorquinone as the photosensitive initiator and *N,N*-dimethylaminoethyl methacrylate as the accelerator. Figure 10 shows scanning electron micrographs of aligned glass spheres (20 vol%) in urethane dimethacrylate mixed with 1,6-hexanediol dimethacrylate polymerized after 5 min under 0.6 kV/mm, 60 Hz. The average particle size of the glass sphere is 18 μm, and the spheres are well aligned in the polymer matrix. Interestingly, the big particles tend to aggregate together, while the small particles tend to form more close clusters. Such anisotropically structured materials should present anisotropically mechanical and dielectric properties.

Iron particle was aligned in a polydimethyl siloxane prepolymer under a magnetic field and then cured for making a soft gel. The ER property of such a soft gel was examined by An [37,40]. Figure 11 shows the measured shear stress of a iron/silicone soft gel with chains tilted at different angles to the normal of the sample surface. It was found that the shear stress of this iron/silicone soft gel increases with the applied electric field, and maximizes at the tilt angle, about 30 degree to the normal of the sample surface. The authors claimed that the point-dipole approximation can well explain the tilt angle dependence of the ER property. The real modulus and the normal stress of the whole composite do vary with the applied electric field, too, implying that this soft gel can function as a smart composite material.

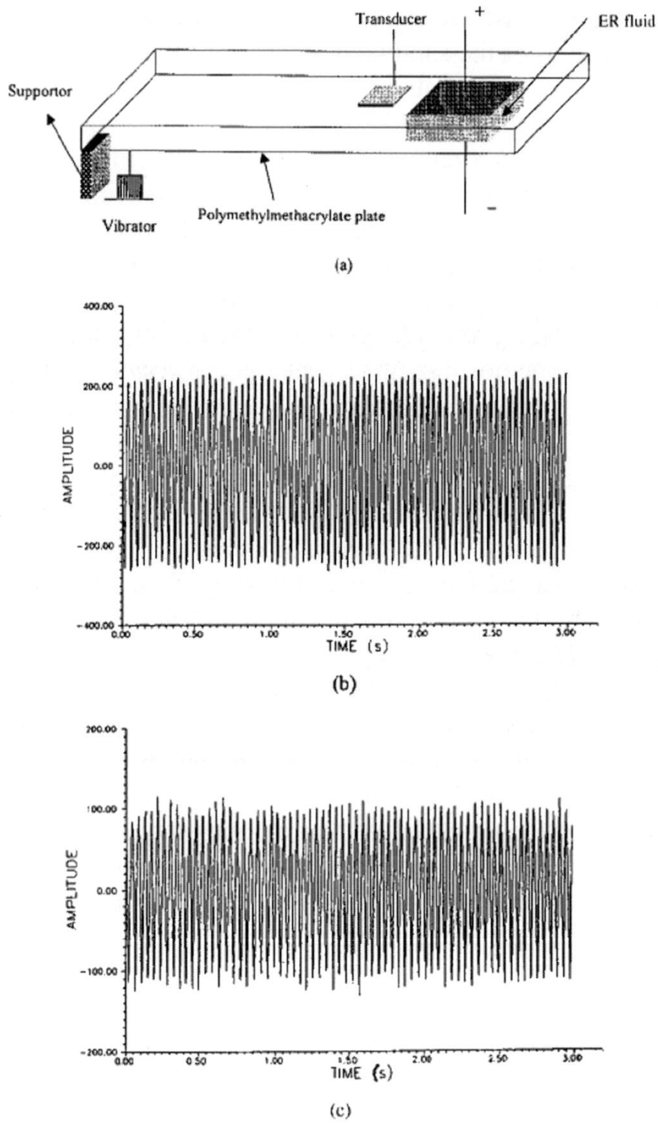

Figure 9 . PMMA and molecular sieve/silicone oil ER composite material. a) The experimental setup. One end of the PMMA plate is fixed, and at the other end there is a very thin small slot, containing the ER fluid composed of molecular sieve particles dispersed in silicone oil. The two electrodes are horizontal. b) The vibration amplitude of the plate measured with the sensor against time when the plate is stimulated with a vibrator and no electric field is applied to the ER fluid. c) The vibration amplitude against time when an electric field of 0.5 kV/mm is applied to the ER fluid. The amplitude is half that obtained with no applied electric field. Reproduced with permission from T. Hao, Adv. Mat., 13(2001)1847.

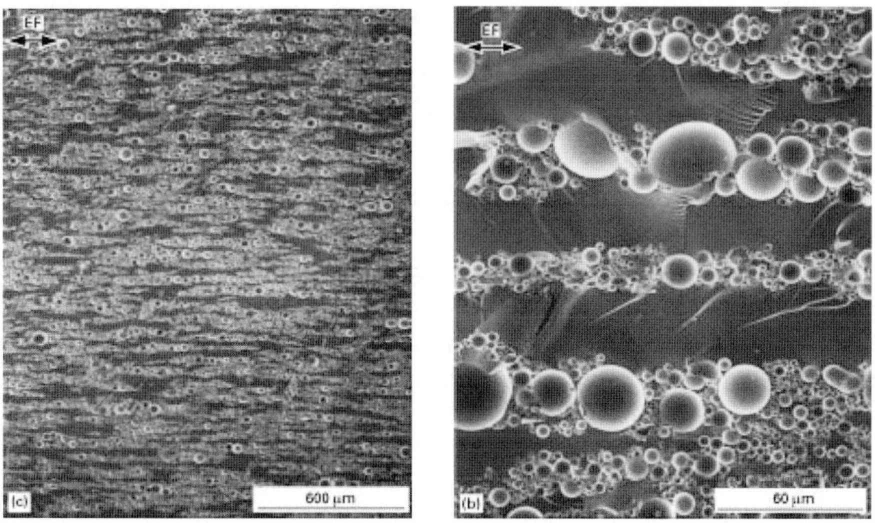

Figure 10 Scanning electron micrographs of aligned glass spheres (20 vol%) in urethane dimethacrylate mixed with 1,6-hexanediol dimethacrylate polymerized after 5 min under 0.6 kVmm, 60 Hz. Reproduced with permission from C. Park, and R.E. Robertson, J. Mat. Sci., 33(1998)3541

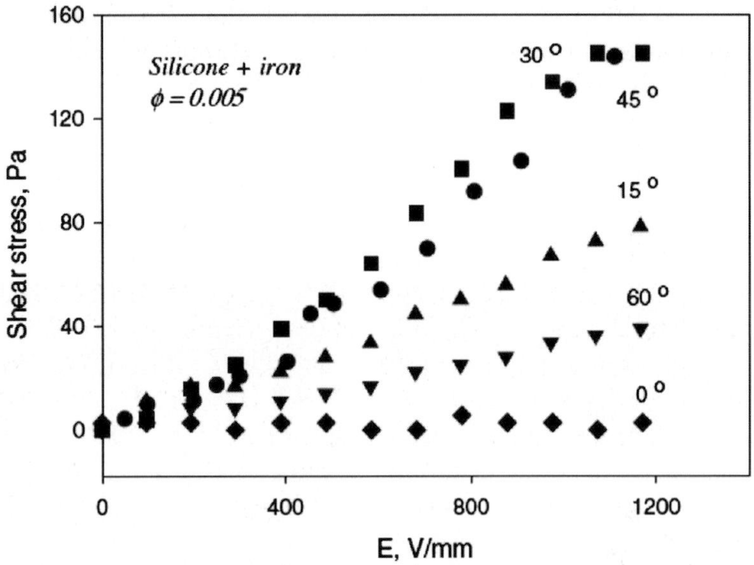

Figure 11 Shear stress produced by a iron/silicone soft gel with chains tilted at different angles to the normal of the sample surface. The particle volume fraction is 0.5 vol%. Reproduced with permission from Y. An, and M. T. Shaw, Smart Mater. Struct. 12 (2003)157.

3. ER INKS AND PIGMENTS

ER fluids can be used as smart ink materials whose droplets may be exactly controlled via an electric field, thus eliminating image artifacts, speeding up the printing process and improving the printing resolution. Apparatus based on controlling the droplet ejection [41], controlling ejected droplet amount [42], and accurately placing the droplet on the receiver medium [43], were reported for fabricating digital ink-jet printers. Approaches of manipulating ER pigment powder [44], liquid ER colorant toner [45], and ER dye materials [46] are already developed, creating a high possibility for ER materials becoming one kind of excellent electronic ink materials that underpin the paperlike display technology, too [47]. Figure 12 shows enlarged images of the intersection of two perpendicular lines produced by conventional (A) and ER (B) toners. The ER toner is the composite particle consisted of color pigments (phthalocyanine blue and brilliant carmine 6B), titanium dioxide, binder resin (copolymer of ethylene and vinyl acetate), and dispersed in low molecular weight hydrocarbon liquid (the isopar L from the Exxon Chemicals). Clearly, the ER toner presents a sharp, homogeneous, and well-defined cross shape image, while the conventional toner delivers a fuzzy, heterogeneous, and unsmooth image with a lot of areas without inks. The ER toner can produce a high quality image. Figure 13 shows another example that the ER toner (the cornstarch/silicone oil with particle concentration 50 wt %) is used and an ER valve is designed for controlling the ink droplet size and ejection velocity. Under 100 V/mm, the ink droplet is still ejected from the orifice with a spherical shape. When the electric field increases to 250 V/mm, the droplet cannot be ejected away from the orifice and a big egg-shaped droplet forms at the exit of the orifice. With the electric field further increased to 300 V/mm, there is no droplet at the opening of the orifice, and the electric field fully blocks the ejection of the ink droplet. The effect of the applied electric field to the ER valve on the pressures of the fluid chambers and the flying velocity of the ejected droplet in that experiment is shown in Figure 14. With the increase of the applied electric field, the ER valve gradually shut down, resulting in the pressure of the ink input chamber, P_1, to increase with the electric field. In the meanwhile, the pressure of the ink output chamber, P_2, decreases with the electric field. The flying velocity of ink droplet should depend on the pressure of P_2, thus decreasing with the electric field, too. The flying velocity becomes zero when the electric field is increased to 166 V/mm, at which the ink ejection is fully stopped. Advantages of ER inks over conventional inks are obvious.

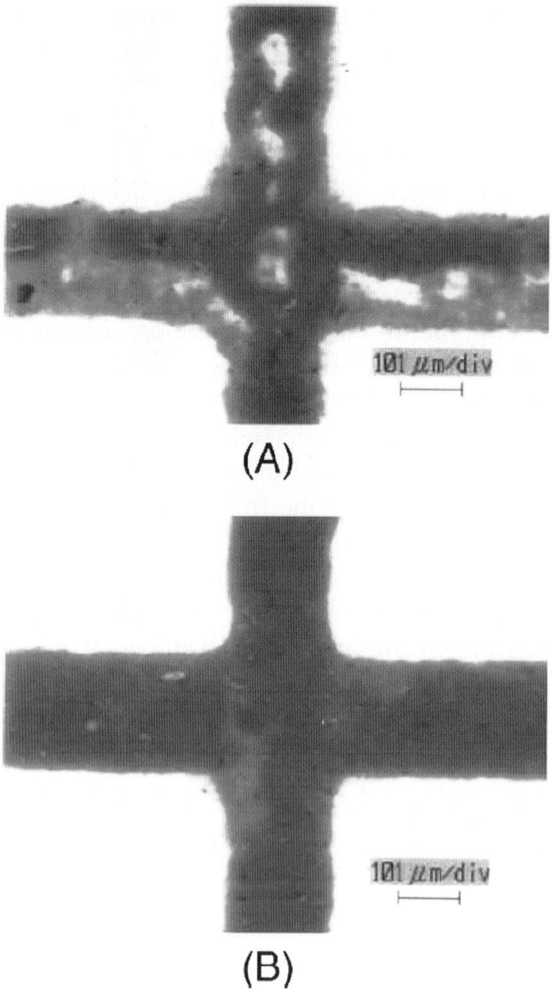

Figure 12 Enlarged images of the intersection of two perpendicular lines produced by conventional (A) and ER (B) toners. The ER toner is the composite particles (toner particles) consisting of color pigments (phthalocyanine blue and brilliant carmine 6B), titanium dioxide, and binder resin (copolymer of ethylene and vinyl acetate), and dispersed in isopar L (from Exxon Chemicals). Reproduced with permission from Y. Otsubo, and Y. Suda, J. Colloid Interf. Sci., 253 (2002)224

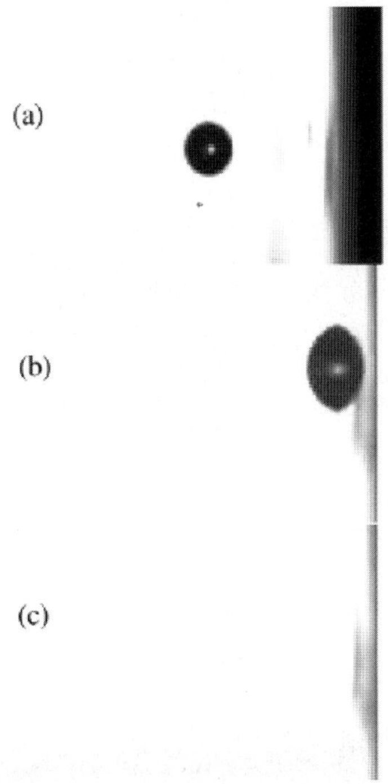

Figure 13 The images of the ink droplets in front of 0.15 mm diameter orifice when the ER toner is used and the applied electric field electrodes of the ER valve is : (a) 100 V/mm; (b) 250 V/mm; and (c) 300V/mm.the ejection of the ink droplet is fully stopped at the electric field 300 V/mm. The ER toner mainly is the cornstarch/silicone oil with particle concentration 50 wt %. Reproduced with permission from C. Lee, and C. Tseng, Materials and Design 23 (2002)727

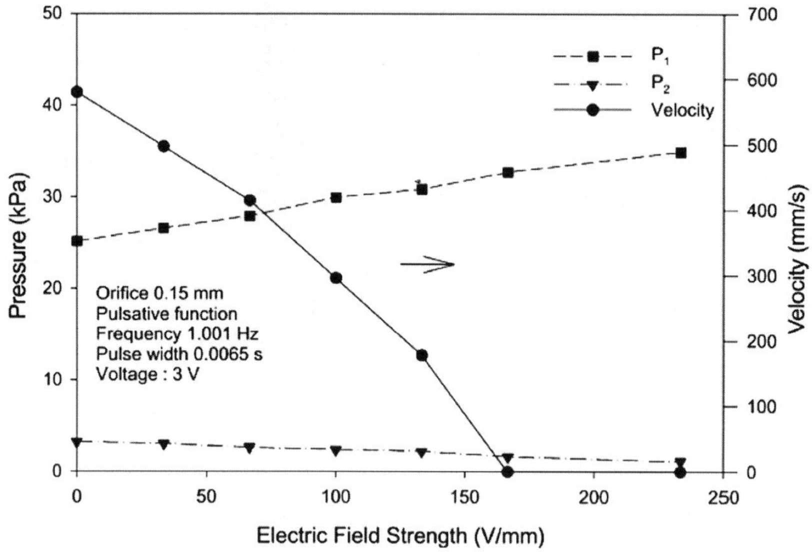

Figure 14 The effect of the applied electric field to the ER valve on the pressures of the fluid chambers and the flying velocity of the ejected droplet. P_1 and P_2 are the pressures before and after the ER valve that controls the toner flowing to the orifice. The ER toner mainly is the cornstarch/silicone oil with particle concentration 50 wt %. Reproduced with permission from C. Lee, and C. Tseng, Materials and Design 23 (2002)727

4. PHOTONIC CRYSTALS

The photonic crystals (also known as photonic band-gap materials) are a new type of materials in which the periodic dielectric structures with a band gap forbid propagation of a certain frequency range of light. This property enables one to control light with amazing facility and produce effects that are impossible with conventional optics. ER technology may be used to fabricate such photonic crystals. One example was reported of making 3-D photonic crystal using polyester resin [48]. The polyester resin was placed between two electrodes, and the air bubbles were generated within the materials as a result of chemical reaction. The air bubbles were aligned along the direction of the applied electric field due to the dielectric constant/conductivity mismatch between the resin and bubbles, finally forming a photonic crystal material. A reversible colloidal crystal was found to form in polystyrene colloidal suspension, as shown in Figure 15 [49]. Once an electric field is applied to the polystyrene colloidal suspension, the colloidal crystal is formed immediately of excellent optical properties, diffracting visible light with the differing crystal orientations visible. Upon removal of the electric field, the ordered structure melts into an isotropic colloidal suspension. It reforms once the electric field is re-applied. By manipulating the interaction force between the particles, the colloidal particle can be crystallized in various systems [50-52]

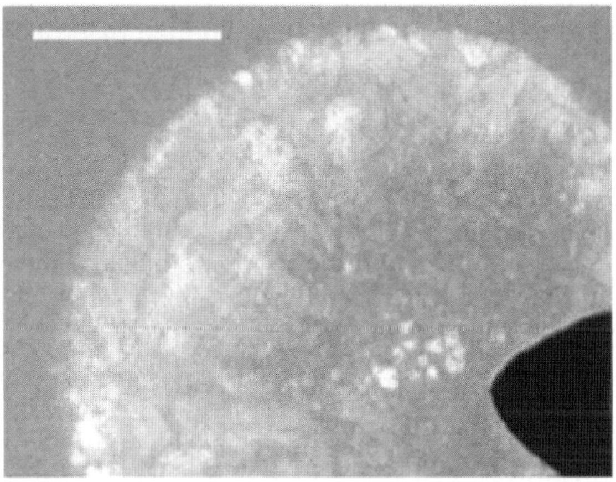

Figure 15 The 450 nm polystyrene colloidal particles were ordered to form the colloidal crystal under 12 V electric field across the electrode gap 7 μm. The scale bar is 100 μm. the dark area is the air bubble. Reproduced with permission from T. Gong, D. Wu, and D. M. Marr, Langmuir 19(2003)5967.

5. MECHANICAL POLISHING

An ER fluid composed of diamond particles was used to polish the ceramic disk under the application of low frequency ac field for inducing a reciprocating motion of the diamond particles [53]. This primary experiment is inspiring, which implies that the ER fluids can be used as very powerful slurries in the polishing industry. The optical micro-aspherical lens was also precisely polished with an ER fluid [54]. Kim [55] investigated the basic principle of the electromechanical polishing with the ER fluid. Figure 16 shows the atomic force microscope images of the borosilicate glass surface before and after polished with the 30 wt% 10μm-in-diam starch particle mixed with 5 wt% 0.3 μm-in-diam Al_2O_3 particle dispersed in the silicone oil under 1.0 kV/mm for 10 min. The starch particle was fund to have a relatively strong ER effect, 35 wt% starch/silicone oil suspension having the yield stress 1.5 kPa at 3.0 kV/mm. The starch particle functions as a soft particle for avoiding surface scratch and Al_2O_3 particle functions as an abrasive particle. After 10 min. polishing under 1.0 kV/mm, the average and maximum roughness of the borosilicate glass surface changes from 28.0 nm and 151 nm to 2.8 nm and 17.5 nm, respectively. The depth of surface removed was found to be a function of the applied electric field. Figure 17 shows the depth of borosilicate glass surface removed as a function of the applied electric field using the 30 wt% 10μm-in-diam starch particle mixed with 4 wt% 2 μm-in-diam diamond particle dispersed in the silicone oil, and the diamond/silicone oil suspension, respectively. The depth of surface removed is strongly dependent on the electric field strength when the 30 wt% 10 μm-in-diam starch particle mixed with 4 wt% 2 μm-in-diam diamond particle dispersed in the silicone oil is used for polishing. At 1.5 kV/mm, almost 1.5 μm material is removed, while only 0.25 μm material is removed when the diamond/silicone oil suspensions is used as a polishing slurry. The ER fluid is an excellent polishing slurry for material removal.

The ER polishing slurries are obviously superior to the conventional colloidal slurries, as the mechanical properties of ER fluids can be easily adjusted via the applied electric field. The semiconductor wafers, even more complex semiconductor devices, can be efficiently and effectively polished using the cheap and smart ER slurries. We believe the ER slurries may revolutionize the chemical-mechanical-polishing (CMP) technology, which now plays a key role in the polishing industry. Of course the magnetorheological fluid can also function as a smart fluid for polishing purposes [56,57]

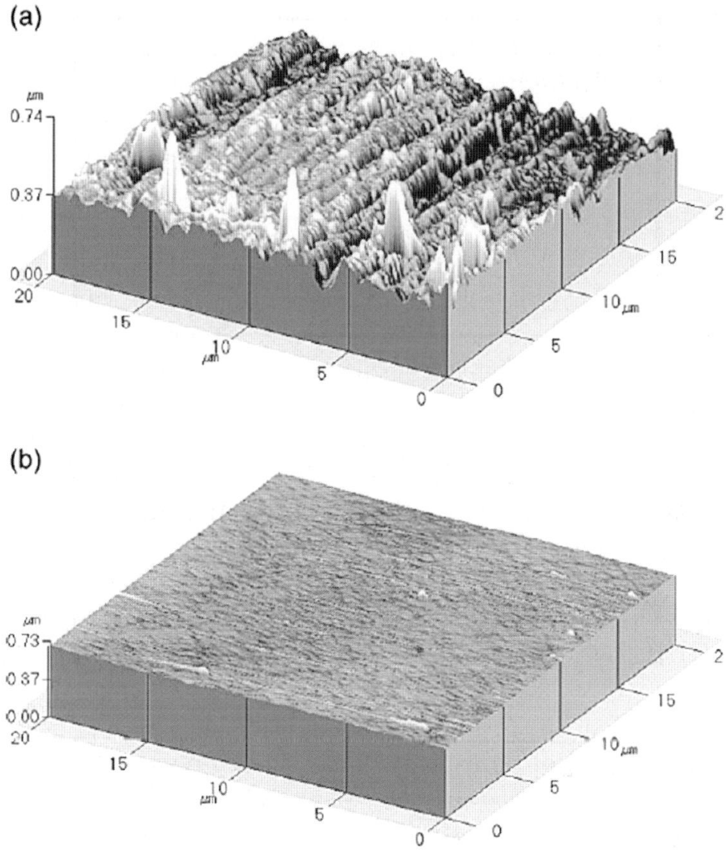

Figure 16 Atomic force microscope images of the glass surface before and after polished with the 30 wt% 10μm-in-diam starch particle mixed with 5 wt% 0.3 μm-in-diam Al_2O_3 dispersed in the silicone oil under 1.0 kV/mm for 10 min.. (a) Initial surface with the average surface roughness 28.0 nm and maximum surface roughness 0.151μm; (b) Polished surface with the average surface roughness 2.8 nm, and maximum surface roughness 17.5 nm. Reproduced with permission from W.B. Kim, S.J. Lee, Y.J. Kim, E.S. Lee, Int. J. Machine Tools Manufacture, 43 (2003) 81.

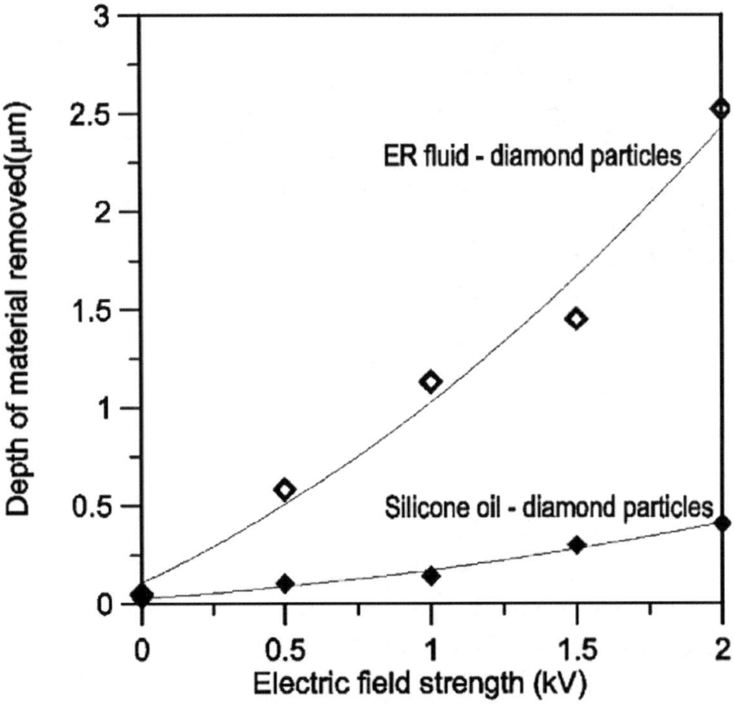

Figure 17 The depth of borosilicate glass surface removed as a function of the applied electric field using the 30 wt% 10μm-in-diam starch particle mixed with 4 wt% 2 μm-in-diam diamond particle dispersed in the silicone oil, and the diamond/silicone oil suspension, respectively The diamond/silicone oil suspension shows a very weak ER effect. Reproduced with permission from W.B. Kim, S.J. Lee, Y.J. Kim, E.S. Lee, Int. J. Machine Tools Manufacture, 43 (2003) 81.

6. ER TACTILE AND OPTICAL DISPLAYS

A tactile display is a device detectable by human touch as a finger is pressed on the surface, creating the illusion of viscoelasticity or as an aid to the visually handicapped [58]. Since the viscosity of an ER fluid greatly depends on the applied electric field (in other words, depends on the distance between two electrodes), the ER fluid can thus be used in the tactile display once the gap of two electrodes that hold the ER fluid can be easily changed via human touch. In comparison with other technologies used in the tactile display, the ER-based tactile display has many advantages such as low power consumption (low current), much safe for user, high force profiles, etc. A finger-touch tactile display was designed by Taylor [59], and the cross section of such a tactile array is schematically shown in Figure 18. Two different ER fluids, one a 24 vol% zeolite/silicone oil and the other a nematic liquid crystal mixed with 33 vol% lithium polymethacrylate, were tested in this tactile display, and both ER fluids show acceptable sensitivity and resolution. A single tactile cell is analogous to a piston freely moving through an ER fluid between two electrodes. The response of such a piston to an external mechanical force and the applied electric field strength were numerically simulated and experimentally verified in ref. [60]. Figure 19 shows the viscous force of a single tactile piston vs. the piston moving velocity at different electric fields from 0 to 2.0 kV/mm. The piston moving velocity can be assumed as the finger pressing speed. The viscous force generated during piston movement is strongly dependent on the applied electric field, and slightly increases with the velocity at high electric field strengths. This is exactly desired for the tactile display.

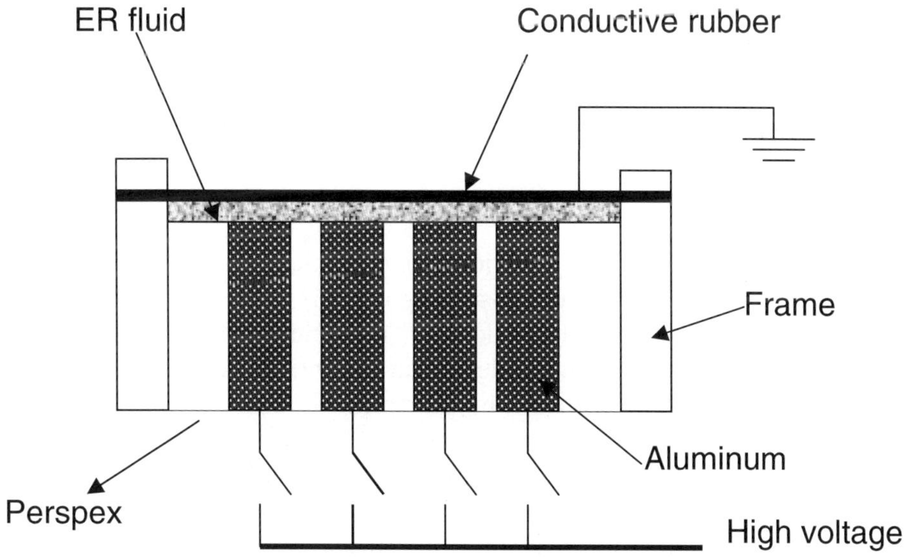

Figure 18 Cross section of tactile array system. Redrawn from P. Taylor, D. Poller, A. Sianaki, and C. Varley, Displays, 18(1998)135

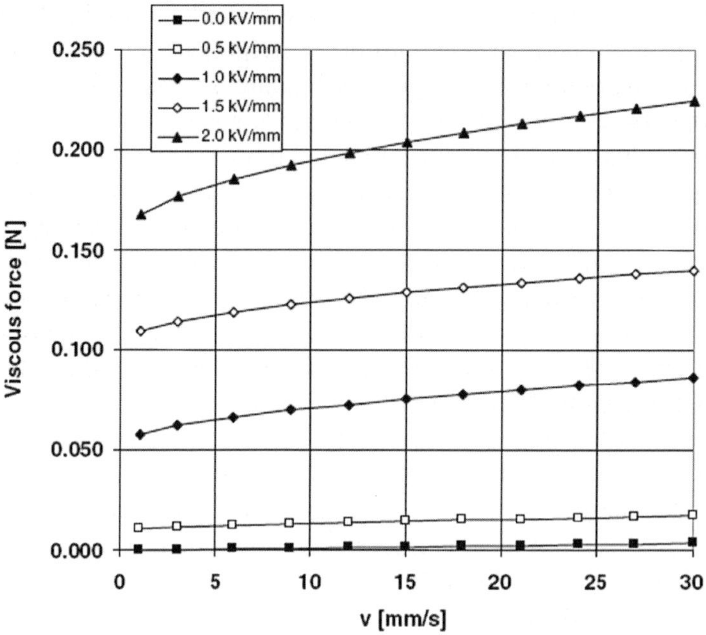

Figure 19 The viscous force of a single tactile piston vs. the piston moving velocity at different electric fields from 0 to 2.0 kV/mm. The ER fluids is Rheobay TP AI 3565 ERF from Bayer AG. Reproduced with permission from G. J. Monkman, S. Egersdorfer, H. Bose, and M. Baumann, J. Phys. D: Appl. Phys. 37 (2004)794

Another kind of display device using the chain-like structures of the ER fluids as optical shutters was also proposed [61]. The ER fluid is placed in between two transparent electrodes. When no electric field is applied, the ER particles are randomly suspended in an insulating medium. In this case the incident light will be scattered by the randomly distributed ER particles. After an electric field is imposed upon the ER fluid, the particles form a fibrillated chain structure, and thus the ER fluid becomes optically transparent. The increment of the light intensity of such an electro-optical display using the ER fluid against the particle weight fraction and the electric field strength is shown in Figure 20. In this figure the increase in light intensity 3.2 dBm corresponds to a 2.09 times increase in optical intensity, and the increase of 4.2 dBm corresponds to a 2.63 times increase in optical intensity. Figure 20 shows that the increase in light intensity passes through a peak against the particle concentration and monotonically increases with the increase of the applied electric field. The maximum intensity appears at particle concentration 5 wt%. When the particle concentration is higher than 5 wt%, there may be more spots occupied by fibrillated chains, or relatively thick chains might form in the system. Thus the increase in light intensity becomes smaller than that in particle concentration of 5 wt%. At the particle concentration less than 5 wt%, slim chains may form in the system due to less particles being available for fibrillation. In any case the electric field will induce the formation of fibrillated chains and the thickness of chain increases with the increase of the electric field, leading to a more bright display.

An electro-optical switch was proposed on the basis of the negative ER effect [62]. As one may know already, for the negative ER fluid the particles do not form a fibrillated chain structure under an electric field. Instead, the particles tend to move either to the anode or cathode and pack on the electrode surface. If the polarity of the electrode is changed, then the particles can move away from the electrode or back dependent on the polarity of the electrode. If the electrode is transparent and a beam of light passes through the electrode gap, the transmitted light can be controlled by whether the electrode surface is covered by the negative particle or not. Such an electro-optical switch device is schematically illustrated in Figure 21.

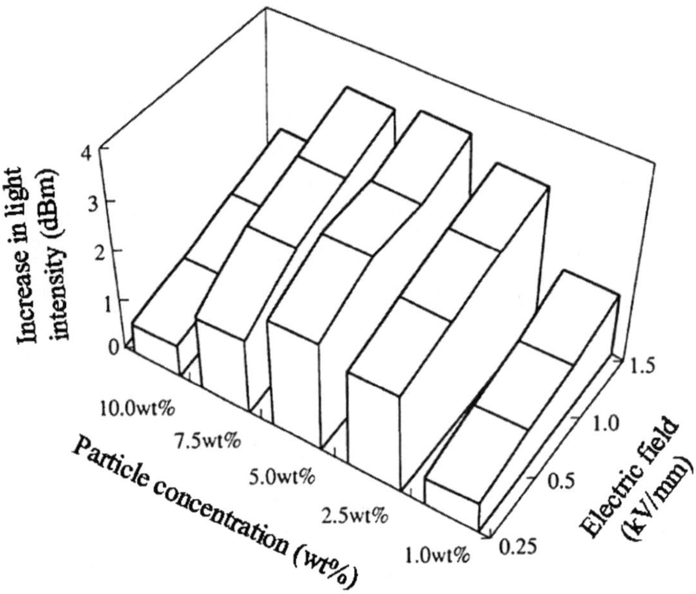

Figure 20 Increment of the light intensity of an electro-optical display using the ER fluid against the particle weight fraction and the electric field strength. The ER fluid is an inorganic/organic composite particle with TiO_2 core coated with copolymer such as polyacrylic acid ester dispersed in the 100 cSt silicone oil. Reproduced with permission from ref. K. Akashi, H. Anzai, K. Edamura, and Y. Otsubo, European Patent 0 697 615, 1996.

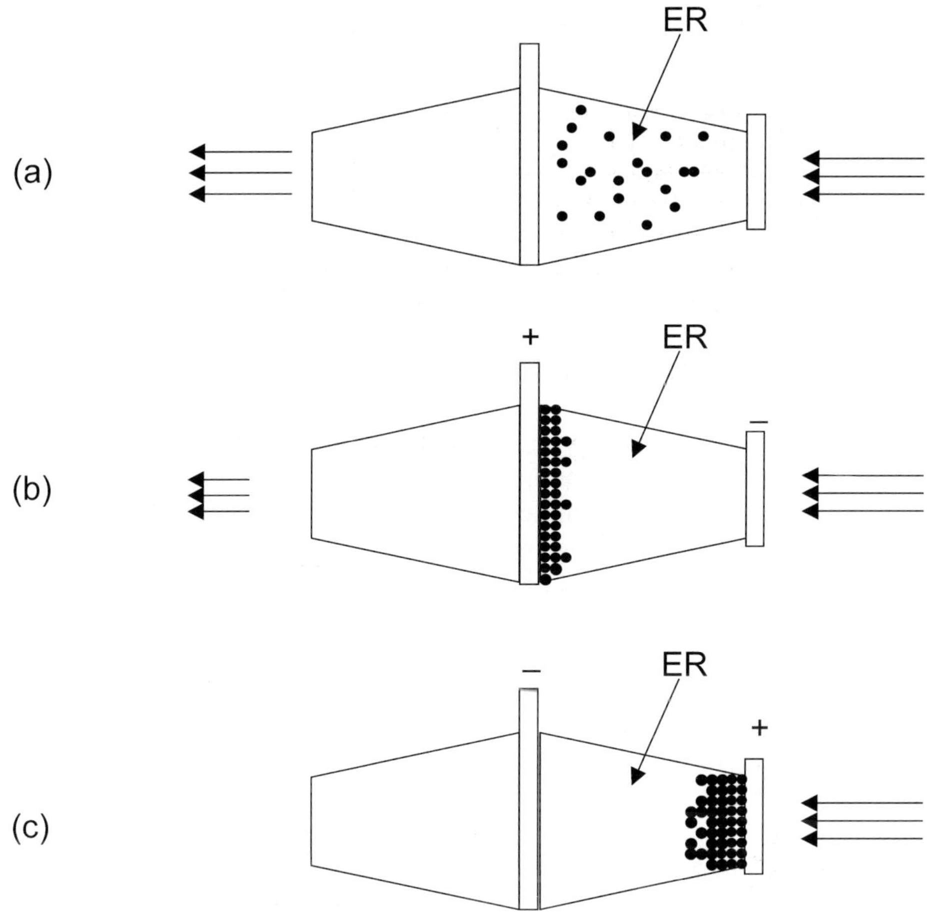

Figure 21 Schematic illustration of an electro-optical switch on the basis of the negative ER effect. (a) Without an electric field, the ER particle is randomly distributed in the medium and the incident light can be scattered out the system; (b) Under an electric field, the negative ER particle moves to the anode, blocking the light. The light may still get through the system with decreased intensity, depending on the particle concentration and the applied electric field; (c) After the polarity of the electrode is changed, the negative ER particle may accumulate on the surface of the small electrode, fully blocking the light. Redrawn on the basis of F.G. Schmidt, WO 02/05025 A1, 2002

7. ER SENSORS

DC current passing through the ER fluid was found to sinusoidally oscillate with the applied mechanical field [63,64]. This phenomenon was theoretically addressed [65] on the basis of the fibrillated chain structure framework and the quasi-one-dimensional variable range hopping conductive model. An ER sensor for real-time monitoring the mechanical field based on this phenomenon was thus proposed. It can be primarily used as a detector of seismography and a monitor for bridge vibration conditions

8. ER APPLICATION FOR DRUG DELIVERY

The ER effect was also proposed for controlling the drug delivery process [66]. The numerical simulation indicates that partial ordering of the particle may reduce the diffusion pathlength (tortuosity) of the ER system [67]. Based on this result, the drug molecules may have less distance to travel before being released, resulting in higher release rates. Since the particle ordering degree can be controlled by the electric field, the release rate thus can also be controlled by an electric field. Figure 22 shows the experimental setup for a drug delivery test. The drug, benzocaine, is dissolved in olive oil, which is mixed with nano-sized starch particle to form the ER fluid. This ER fluid is stored in the left chamber, which is separated by the nylon membrane with the right chamber containing 1N HCl. To prevent the electric current from flowing through the system, an insulating layer is placed on the inner side of both electrodes. Since benzocaine is soluble in 1N HCl, the amount of benzocaine released from the left chamber to the right can be easily determined by a UV spectrophotometer. Without the starch particle, there is no appreciable release rate change under an electric field. In the presence of starch particle, the release rate increases by 52.8 % at 25 °C and 290 V after 180 min. At 30 °C the release rate increases by 80% after 20 min. With further increase of temperature, the release rate increment decreases in comparison with that at 30 °C. The release rate after 20 min. is plotted against the reciprocal of temperature in Figure 23 at zero and 290 V.

The electric field facilitates the release rate in the whole temperature range from 25 to 45 °C. The increment becomes less significant at high temperatures, which may result from the weakening ER effect or defective chain structure at high temperatures. Typically thermal energy may intensify the particle Brownian motion that may compete with the particle fibrillation process. It appears that the fibrillated chains provide a bridge for easy movement of drug ions.

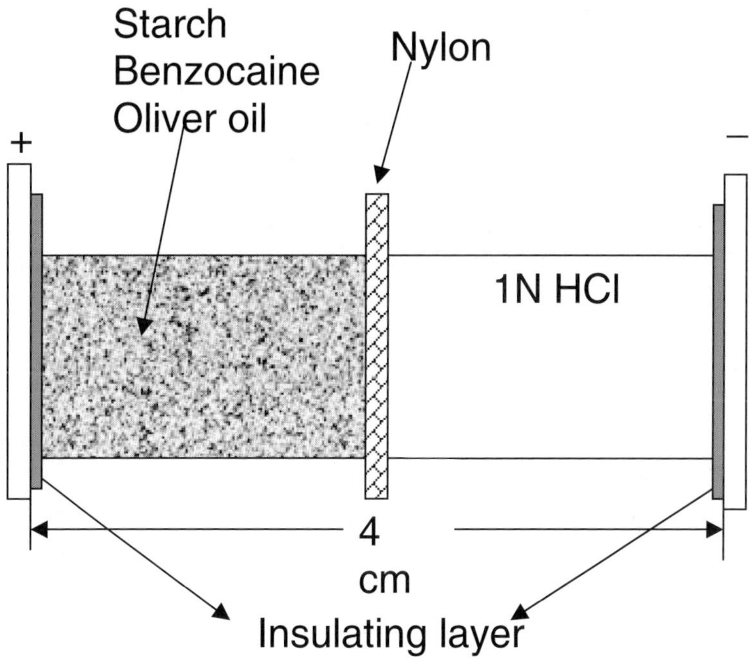

Figure 22 The setup for ER drug delivery test. The starch particle of mean size 198 nm dispersed in olive oil constitutes the ER fluid.

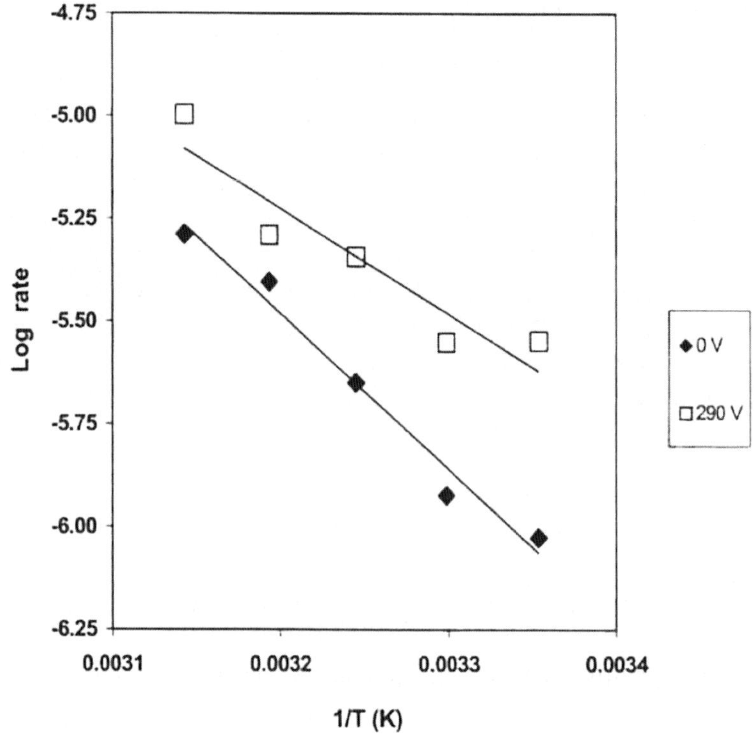

Figure 23 the drug release rate vs. the reciprocal of temperature at zero and 290 V electric field. Reproduced with permission from A.K. Kamal, S.K. Nutalapati, T. Jochsberger, F.M. Plakogiannis, and R.A. Bellantone, PDA J Pharm Sci Technol. 55(2001)248

9 SUMMARY AND OUTLOOK

ER fluids are smart materials of vast potential applications in many fields. A lot of materials ranging from inorganic to liquid crystal and to semiconductive polymeric materials can be used for fabricating the ER fluids, forming the heterogeneous and homogeneous fluids in view of the miscibility of dispersed/dispersing phases, or hydrous and anhydrous fluids in view of the water content. The main problem for the heterogeneous ER fluids is particle sedimentation, and the homogeneous ER fluids are promising in this regard, though they sometimes do not show a strong enough transmitted shear stress and have a relatively high viscosity at zero field. Anhydrous ER fluids could work in a wide temperature range, and are thought to be superior to the hydrous ER fluids. Three major ER effects are discussed in this chapter, including the positive ER effect, the negative ER effect and photo-ER effect. Applications of utilizing the positive ER effect in material fabrication, tactile display and sensors are introduced, exemplifying how the ER fluids will be used in practice. Applications of using the negative and photo-ER effects are few; however, they will emerge soon in the foreseeable future.

The positive ER effect is usually used for transmitting electric-mechanical signals, while the negative ER effect may be mainly used for material processing. For example, processing a high loading ceramic or polymeric composite system would be extremely difficult due to the high viscosity. In such a case, the negative ER effect can be used to reduce the viscosity, allowing the processing to become possible and easier.

Developing high performance ER fluids having a strong ER effect and without the particle sedimentation problem will be continuously focused on. Many approaches such as particle surface modification (coating and grafting) and particle blending (doping) will be developed and used for improving ER performances. Developing high performance ER materials is still key to the ER technology. A breakthrough in fabricating the durable and stable ER fluids will boost the ER technology and speed up the commercialization of the ER devices. The ER sensors, the ER damping devices, the ER inks and the ER slurries for polishing may become the front runners, because they do not require the ER fluids to work in extreme conditions, and they have vast industrial needs.

REFERENCES

[1] D.L. Hartsock, R.F. Novak, and G.J. Chaundy, J. Rheol., 35(1991)1305
[2] H. Block, J.P. Kelley, J. Phys. D.: Appl. Phys., 21(1988)1661
[3] T.C. Jordan, M.T. Shaw, IEEE Trans. Electr. Insul., 24(1989)849
[4] R. EL-Kassouf, European Patent 1 070 871, 2001
[5] W. Niaura, P. Pakdel, European Patent 1 065 405, 2001
[6] P. Andreas,W. Eckhardt, R. Horst, US Patent 5 988 336, 1999
[7] W. Eckhardt, B. Klaus, S. Horst, R. Horst, European Patent EP0898093, 1999
[8] B. Chadwick, M. Nowak, US Patent 5 970 581, 1999
[9] Y. L. Xu, W. L. Qu, J. M. Ko, Earthquake Eng. Struct. Dyn., 29(2000)557
[10] C. Norbert, R. Axel, German Patent DE 19 830 559, 2000
[11] K. Bohon, S. Krause, J. Polym. Sci., Part B: Polym. Phys., 36 (1998)1091
[12] J. Starkovich, E. Shtarkman, L. Rosales, European Patent 0 872 665, 1998
[13] K. Yoshida, M.Kikuchi, J.H.Park, and S.Yokota, Sensors Actuators, 95(2002)227
[14] S.B.Choi, J.H. Bang, M.S.Cho, and Y.S.Lee, Proc. Inst. Mech. Engrs. Part D, 216(2002)897
[15] F.E. Filisko, Encyclopedia of Smart Materials, Edited by: M. Schwartz, John Wiley & Sons, 2002, p376
[16] N.M. Wereley, J. Lindler, N. Rosenfeld, and Y. Choi, Smart. Mater. Struct. 13 (2004) 743
[17] S.B. Choi, Y.T. Choi, and D.W. Park, Trans. AMSE: J. Dynm. Syst. Meas. Contr., 122(2000)114;
[18] S.B. Choi, Encyclopedia of Smart Materials, Edited by: M. Schwartz, John Wiley & Sons, 2002, p1085
[19] J.M. Vance, and D.Ying, Transc. ASME: J. Eng.Gas Turbines and Power, 122(2000)337
[20] H. P. Gavin, Smart Materials and Structures, 7(1998)664
[21] H. P.Gavin, R. D. Hanson, and F.E. Filisko, ASME J. Applied Mechanics, 63(1996) 669
[22] H. P. Gavin, R.D. Hanson, and N.H. McClamroch, ASME J. Appl. Mechanics, 63(1996)676
[23] T. Hong, X. Zhao, and S. Liu, Smart Mater. Struct. 12(2003)347
[24] V. Noresson, N.G. Ohlson, and M. Nilsson, Materials and Design 23 (2002)361
[25] W. Wen, X. Huang, S. Yang, K. Lu, and P. Shen, Nature Materials, 2(2003)727
[26] T. Kawakami, R. Aizawa, et al., J. Mod. Phys. B 13(1999)1721
[27] Y.L. Zhang, K.Q. Lu, et al., Appl. Phys. Lett. 80 (2002) 888
[28] W. Wen, P. Shen, K.L. Chang, and C.K. Nam, EP 1528278 A1(2005)
[29] W.A. Bullough, R. Firoozian, A. Hosseini-Sianaki, A.R. Johnson, J. Makin, and S.Xiao, US 5524743 (1996)
[30] J.L.Sproston, N.G.Stevens, and I.M.Page, US 4782927, 1988
[31] J. Hollerbach, I. Hunter, and J. Ballantyne, The Robotics Review 2, O. Khatib, J. Craig and Losano-Perez Eds., MIT Press, Cambridge, MA, pp. 299-342, 1992
[32] X. Chu, H. Qiu, L. Li, and Z. Gui, Smart Mater. Struct. 13 (2004) N51–N56
[33] R. Hanaoka, S. Takata, H. Fujita,T. Fukami, K. Sakurai, K. Isobe, and T. Hino, Int.J. Mod. Phys. B, 16(2002)2433

[34] R. Hanaoka, S. Tanaka, Y. Nakazawa, T. Fukami, and K. Sakurai, Electr. Eng. Jpn, 142(2003)1
[35] T. Hao, Ph.D. Thesis, Institute of Chemistry, Chinese Academy of Sciences, Beijing, 1995
[36] W. Kordonsky, A. Matsepuro, V. Makatun, Z. Novicova, S. Demachuk, US Patent 5 810 126, 1998
[37] C.Park, and R.E. Robertson, J. Mat. Sci., 33(1998)3541
[38] Y.An, B. Liu, and M.T. Shaw, Int.. J. Mod. Phys. B, 16(2002)2440
[39] Jia-Yi Yeh , Lien-Wen Chen,, Ching-Cheng Wang, Composite Structures, 64 (2004) 47
[40] Y. An, and M. T. Shaw, Smart Mater. Struct. 12 (2003)157
[41] R. W. Gundlach, E. G. Rawson, US Patent 6 048 050, 2000
[42] S. Sohn, US Patent 5 510 817, 1996
[43] X. Wen, US Patent 6 102 513, 2000
[44] K. Nakatsuka, T. Atarashi, European Patent 0 913 432, 1999
[45] Y. Suda, H. Kuno, Eur. Patent 0 824 227, 1998
[46] D. Carlson, D. Weiss, E. Bares, World Patent WO9 312 192, 1993
[47] J. Rogers, Sciences, 291(2001)1502
[48] R. Tao, Proc. Int. Conf. Electrorheological Fluids, M. Nakano, and K. Koyama, Eds., World Scientific, Singapore 1998, p. 811
[49] T. Gong, D. Wu, and D. M. Marr, Langmuir 19(2003)5967
[50] M. Trau, D.A.Saville, and I.A. Aksay, Science 272 (1996)706
[51] M. Bohmer, Langmuir 12 (1996)5747
[52] R. C. Hayward, D. A. Saville, and I. A. Aksay, Nature, 404(2000)56
[53] Y. Akagami, S. Nishimura, Y. Ogasawana, T. Fujita, B. Jeyadevan, K. Nuri, K. Itoh, Proc. Int. Conf. Electrorheological Fluids, M. Nakano and K. Koyama, Eds., World Scientific, Singapore 1998, p. 803
[54] T. Kuriyagawa, M. Saeki, K. Syoji, Precision Eng., 5282(2002)1
[55] W.B. Kim, S.J. Lee, Y.J. Kim, E.S. Lee, Int. J. Machine Tools Manufacture, 43 (2003) 81
[56] N. Umehara, K. Kato, S. Mizuguchi, and S. Nakamura, Int. J. Jpn. Soc. Precision Eng. 60 (11) (1994)1606
[57] W. I. Kordonski and S. D. Jacobs, Int. J. Mod. Phys. B, 10(1996)2937
[58] J. Fricke, H. Baeching, J. Microcomput. 16(1993)259
[59] P. Taylor, D. Poller, A. Sianaki, and C. Varley, Displays, 18(1998)135
[60] D. Klein, D. Rensink, H. Freimuth, G. J. Monkman, S. Egersdorfer, H. Bose, and M. Baumann, J. Phys. D: Appl. Phys. 37 (2004)794
[61] K. Akashi, H. Anzai, K. Edamura, and Y. Otsubo, European Patent 0 697 615, 1996
[62] F.G. Schmidt, WO 02/05025 A1, 2002
[63] T. Hao, Y. Xu, J. Colloid Interface Sci. 185(1997)324
[64] T. Hao, Y. Xu, J. Colloid Interface Sci. 181(1996)581
[65] T. Hao, J. Phys. Chem. 102(1998)1
[66] A.K. Kamal, S.K. Nutalapati, T. Jochsberger, F.M. Plakogiannis, and R.A. Bellantone, PDA J. Pharm. Sci. Technol. 55(2001)248
[67] S.K.Nutalapati, and R.A.Bellantone, AAPS annual meeting, New Orleans, LA, November, 1999

INDEX